T0343155

CARBOHYDRATE CHEMISTRY, BIOLOGY AND MEDICAL APPLICATIONS

CARBOHYDRATE CHEMISTRY, BIOLOGY AND MEDICAL APPLICATIONS

Edited by

Hari G. Garg

Harvard Medical School, Pulmonary and Critical Care Unit
Massachusetts General Hospital, Boston, MA, USA

Mary K. Cowman

Department of Chemical and Biological Sciences,
Polytechnic University, Brooklyn, NY, USA

and

Charles A. Hales

Harvard Medical School, Pulmonary and Critical Care Unit
Massachusetts General Hospital, Boston, MA, USA

ELSEVIER

Amsterdam – Boston – Heidelberg – London – New York – Oxford
Paris – San Diego – San Francisco – Singapore – Sydney – Tokyo

Elsevier
Linacre House, Jordan Hill, Oxford OX2 8DP, UK
Radarweg 29, PO Box 211, 1000 AE Amsterdam, The Netherlands

First edition 2008

British Library Cataloguing in Publication Data
A catalogue record for this book is available from the British Library

Library of Congress Cataloging-in-Publication Data
A catalog record for this book is available from the Library of Congress

ISBN: 978-0-08-054816-6

For information on all Elsevier publications
visit our website at books.elsevier.com

Printed and bound by CPI Group (UK) Ltd, Croydon, CR0 4YY

Transferred to Digital Print 2011

Dedication

To my wife, Mithlesh, and our children, Ashima and Arvin, for their loyal support and generous love.

Hari G. Garg

To my husband, Gene, and our son Rudd, for their loyal support and generous love.

Mary K. Cowman

To my wife, Mary Ann, and our sons, Sam, Chris, and John, for their loyal support and generous love.

Charles A. Hales

Contributors

Tomoya O. Akama, Glycobiology Program, Cancer Research Center, Burnham Institute for Medical Research, La Jolla, California, USA

Luis Z. Avila, Genzyme Corporation, Drug and Biomaterial R&D, Waltham, Massachusetts, USA

Endre A. Balazs, Matrix Biology Institute, Edgewater, New Jersey, USA

Philip A. Band, Departments of Pharmacology and Orthopedic Surgery, New York University School of Medicine, New York, USA

Günther Boehm, Numico Research, Friedrichsdorf, Germany, and Sophia Children's Hospital, Erasmus University, Rotterdam, The Netherlands

Andrew Burd, Division of Plastic and Reconstructive Surgery, Department of Surgery, Chinese University of Hong Kong, Prince of Wales Hospital, Hong Kong, SAR

Gregory T. Carroll, Department of Chemistry, Columbia University, New York, USA

Catherine E. Costello, Center for Biomedical Mass Spectrometry, and Departments of Biochemistry and Biophysics, Boston University School of Medicine, Boston, Massachusetts, USA

Hironobu Eguchi, Departments of Chemistry and Biochemistry, Ohio State University, Columbus, Ohio, USA

Michiko N. Fukuda, Glycobiology Program, Cancer Research Center, Burnham Institute for Medical Research, La Jolla, California, USA

Johann Garssen, Numico Research, Wageningen, and Utrecht Institute for Pharmacological Sciences, Utrecht University, The Netherlands

Diego A. Gianolio, Genzyme Corporation, Drug and Biomaterial R&D, Waltham, Massachusetts, USA

Zhongwu Guo, Department of Chemistry, Wayne State University, Detroit, Michigan, USA

Tim Horlacher, Laboratory for Organic Chemistry, Swiss Federal Institute of Technology (ETH), Zürich, Switzerland

Derek Horton, Department of Chemistry, American University, Washington DC, USA

Lin Huang, Division of Plastic and Reconstructive Surgery, Department of Surgery, Chinese University of Hong Kong, Prince of Wales Hospital, Hong Kong, SAR

George Karakiulakis, Department of Pharmacology, School of Medicine, Aristotle University, Thessaloniki, Greece

Jan Knol, Numico Research, Wageningen, The Netherlands

Paul A. Konowicz, Genzyme Corporation, Drug and Biomaterial R&D, Waltham, Massachusetts, USA

Robert J. Linhardt, Departments of Chemistry and Chemical Biology, Chemical and Biological Engineering, and Biology, Rensselaer Polytechnic Institute, Troy, New York, USA

Mara S. Ludwig, Meakins-Christie Laboratories, McGill University, Montreal, Canada

Robert J. Miller, Genzyme Corporation, Drug and Biomaterial R&D, Waltham, Massachusetts, USA

Saravanababu Murugesan, Department of Chemical and Biological Engineering, Rensselaer Polytechnic Institute, Troy, New York, USA

Eleni Papakonstantinou, Department Pharmacology, School of Medicine, Aristotle University, Thessaloniki, Greece

Jose L. de Paz, Laboratory for Organic Chemistry, Swiss Federal Institute of Technology (ETH), Zürich, Switzerland

Michael Philbrook, Genzyme Corporation, Drug and Biomaterial R&D, Framingham, Massachusetts, USA

Michael Roth, Department of Internal Medicine, Pulmonary Cell Research, University Hospital, Basel, Swizerland, and Woolcock Institute for Medical Research, Molecular Medicine, Camperdown, Australia

Peter J. Roughley, Genetics Unit, Shriners Hospital for Children and Department of Surgery, McGill University, Montreal, Canada

Michael R. Santos, Genzyme Corporation, Drug and Biomaterial R&D, Framingham, Massachusetts, USA

Paul G. Scott, Department of Biochemistry, University of Alberta, Edmonton, Alberta, Canada

Peter H. Seeberger, Laboratory for Organic Chemistry, Swiss Federal Institute of Technology (ETH), Zürich, Switzerland

Jing Song, Departments of Chemistry and Biochemistry, Ohio State University, Columbus, Ohio, USA

Bernd Stahl, Numico Research, Friedrichsdorf, Germany

Doris M. Su, Departments of Chemistry and Biochemistry, Ohio State University, Columbus, Ohio, USA

Kazuhiro Sugihara, Department of Obstetrics and Gynecology, Hamamatsu University, School of Medicine, Himamatsu, Japan

Craig Turchi, ADA Technologies, Inc., Littleton, Colorado, USA

Denong Wang, Stanford Tumor Glycome Laboratory, Stanford University School of Medicine, Stanford, California, USA

Peng G. Wang, Departments of Chemistry and Biochemistry, Ohio State University, Columbus, Ohio, USA

Robert L. Woodward, Departments of Chemistry and Biochemistry, Ohio State University, Columbus, Ohio, USA

Chengfeng Xia, Departments of Chemistry and Biochemistry, Ohio State University, Columbus, Ohio, USA

Bo Xie, Center for Biomedical Mass Spectrometry, Boston University School of Medicine, and Department of Chemistry, Boston University, Boston, Massachusetts, USA

Jin Xie, Department of Chemistry and Chemical Biology, Rensselaer Polytechnic Institute, Troy, New York, USA

Wen Yi, Departments of Chemistry and Biochemistry, Ohio State University, Columbus, Ohio, USA

Masahiko Yoneda, Biochemistry and Molecular Biology Laboratory, Aichi Prefectural College of Nursing and Health, Nagoya, Japan

Masahiro Zako, Department of Opthalmology, Aichi Medical University, Aichi, Japan

Xichun Zhou, ADA Technologies, Inc., Littleton, Colorado, USA

Contents

Preface

Carbohydrates are one of four major classes of biologically important molecules, the others being nucleic acids, proteins, and lipids. Historically, carbohydrate studies have dealt mainly with the structure and role of monosaccharides and small oligosaccharides (such as glucose, fructose, sucrose, and lactose) in cellular metabolism and fermentation processes, and the roles of polysaccharides such as cellulose in cell wall organization, and starch and glycogen as storage forms of glucose. The last fifty years have seen an explosion of knowledge about complex carbohydrate structures and their biological functions. First came structural studies that elucidated the role of cellulose in plant cell walls and chitin in invertebrates, as fibrillar assemblies providing mechanical strength. The glycosaminoglycans, their proteoglycans, and other network-forming polysaccharides such as alginates were found to give support to the polysaccharide or protein fibrillar components of the extracellular matrices of both plant and animal tissues. Thus mechanical and physical properties (tensile strength, viscoelasticity, osmotic pressure) were understood to depend on polysaccharides. More recently, other important roles have been elucidated for oligosaccharides, polysaccharides, glycoproteins, glycolipids and other complex glycoconjugates. These roles depend on specific receptor-ligand interactions, and control not only the organization of the pericellular and extracellular matrices, but also a myriad of cellular activities such as changes in gene expression through signaling pathways. Pathogen-host interactions, response to growth factors, inflammation, wound healing, a number of genetically linked disorders, and many other physiological processes depend on specific carbohydrate structures and interactions. This book is intended to assist physicians, glycobiologists, pharmaceutical scientists, and graduate students to gain an understanding the complex nature and functions of carbohydrates. The 17 chapters in this volume, written by renowned scientists and physicians working in the carbohydrate field, provide a broad overview of the chemistry, biology, and medical applications of carbohydrates.

The first four chapters deal with different aspects of carbohydrate chemistry. In Chapter 1, Derek Horton discusses the development of carbohydrate chemistry and biology from antiquity to the present and beyond and summarizes the structures and methods for structural analysis of complex carbohydrates. In Chapter 2, Bo Xie and Catherine Costello review the state-of-art application of mass spectrometry to the structural analysis of complex carbohydrates. In Chapter 3, Zhongwu Guo briefly summarizes the types of glycosylation methods and synthetic strategies that are commonly used in the chemical synthesis of glycoconjugates. Enzymes have simplified the synthesis of some oligosaccharides, and in Chapter 4,

Doris Su *et al.* review the enzymatic synthesis of oligosaccharides and their conversion to glycolipids.

Chapters 5–10 focus on the GAGs and heavily glycosylated proteins, that is, PGs. Malfunction either in the synthesis or in the breakdown of these macromolecules is associated with numerous human diseases. In Chapter 5, Mara Ludwig describes the contribution of different PGs to lung biology. Chapter 6, which includes the PGs of intervertebral disk, is by Peter Roughley. In Chapter 7, Paul Scott focuses on skin and its small leucine-rich repeat PGs. In Chapter 8, Masahiro Zako and Masahiko Yoneda discuss the role of GAGs in ocular pathogenic conditions. In Chapter 9, Michael Roth *et al.* summarize the biological functions of GAGs. In Chapter 10, Jin Xie *et al.* describe the physiological, pathophysiological, and therapeutic roles of heparin and heparan sulfate.

Carbohydrates are critical for biological activity. Chapters 11–17 address the therapeutic and diagnostic medical applications of carbohydrates. In Chapter 11, Andrew Burd and Lin Huang describe the use of carbohydrate polymers in wound dressings. In Chapter 12, Günther Boehm *et al.* review oligosaccharides in human milk and infant formulas. In Chapter 13, Michiko Fukuda *et al.* describe the role of cell surface carbohydrates in development and disease. In Chapters 14 and 15, Endre Balazs and Philip Band, and Luis Avila *et al.*, respectively, detail the medical use of hyaluronan-based therapeutic products and drug delivery and medical applications of chemically modified hyaluronan. Chapters 16 and 17 by Xichun Zhou *et al.*, and Tim Horlacher *et al.*, respectively, describe carbohydrate microarrays, which are being used to address specific challenges in carbohydrate research to provide novel medical diagnostic approaches and to unravel the functions of carbohydrates in health and disease.

Hari G. Garg
Mary K. Cowman
Charles A. Hales

Carbohydrate Chemistry, Biology and Medical Applications
Hari G. Garg, Mary K. Cowman and Charles A. Hales

DOI: 10.1016/B978-0-08-054816-6.00001-X

Chapter 1

The Development of Carbohydrate Chemistry and Biology

DEREK HORTON

Department of Chemistry, American University, Washington, DC 20016, USA

I. Early History

A major proportion of the organic matter on Earth is plant tissue ("biomass") and is composed of carbohydrates, principally cellulose. This is the structural support polymer of land plants and the material used since ancient times in the form of cotton and linen textiles, and later as paper. Chitin is a polymer related to cellulose that has skeletal function in arthropods and fungi. Other polymeric carbohydrates constitute the structural support framework for marine plants and the cell walls of microorganisms. The sweet carbohydrate of sugar cane, now termed sucrose, has been a dietary item for at least 10 millennia.

Ranking alongside cellulose in abundance is starch, a biopolymer that is the food-reserve carbohydrate of photosynthetic plants, and the closely related glycogen, the storage carbohydrate in the animal kingdom. Starch occurs as microscopic granules in plant storage tissue (cereal grains, tubers), and the process for its isolation was clearly described by Cato the Elder (1): steeping the grains in cold water to swell them, straining off the husks, and allowing the milky suspension to settle to afford, after drying, a white powder. This procedure is essentially the same as the modern corn wet-milling process, and the resultant starch powder is essentially pure carbohydrate whose molecular formula can be expressed as $C_6(H_2O)_5$, hence the term carbohydrate. It was used as early as 4000 BCE in Egypt as an adhesive with the cellulosic fiber of the papyrus plant to make writing material, and early in the first millennium CE for sizing paper and to stiffen cloth (2).

The photosynthetic apparatus in the green plant utilizes solar energy to effect the reduction of atmospheric carbon dioxide in a complex sequence of reactions (3,4) whose net result is the formation of glucose, a simple sugar (monosaccharide) having the molecular formula $C_6H_{12}O_6$. Subsequent *in vivo* conversions afford carbohydrate polymers (polysaccharides), notably starch and cellulose, along with many related polymers formed from other monosaccharide sugars. In 1811, it was demonstrated (5) that acid hydrolysis of starch led to near-quantitative conversion into glucose, and later (6,7) it was shown that hydrolysis of cellulose likewise afforded glucose.

In the early nineteenth century, individual sugars were often named after their source, for example grape sugar (Traubenzucker) for glucose and cane sugar (Rohrzucker) for saccharose (the name sucrose was coined much later). The latter sugar was subsequently obtained from the juice of the sugar beet. During that century, a number of other simple sugars (monosaccharides) were isolated from acid hydrolyzates of other polysaccharide sources and were given names with the ending "-ose"; these include mannose, xylose, arabinose, and fucose. A nitrogen-containing sugar obtained by hydrolysis of chitin was termed glucosamine, and hydrolysis of milk sugar (lactose) led to galactose.

The name glucose was coined by Dumas in 1838 for the sugar obtained from honey, grapes, starch, and cellulose; 20 years later its molecular formula of $C_6H_{12}O_6$ was established, and subsequently it was shown to be a linear six-carbon aldehyde having hydroxyl groups on each of the chain carbon atoms. Polarimetric studies led Kekulé in 1866 to propose the name "dextrose" because glucose is dextrorotatory, and the levorotatory "fruit sugar" (Fruchtzucker, fructose) obtained by hydrolysis of cane sugar (sucrose) was for some time named "levulose." The French word "cellule" for cell and the "-ose" suffix led to the term cellulose, long before its structure was known. The term "carbohydrate" (French "hydrate de carbone," German "Kohlenhydrate") was applied originally to monosaccharides in recognition of the fact that their empirical composition can be expressed as $C_n(H_2O)_m$. However, the term is now used generically in a wider sense.

II. The Contribution of Emil Fischer

During the last two decades of the nineteenth century, Emil Fischer (8) began his fundamental studies on carbohydrates, showing that phenylhydrazine reacts with glucose, mannose, and fructose to give the same crystalline phenylosazone, and he utilized the reaction introduced by Kiliani, the addition of hydrogen cyanide to a sugar, to give two isomeric acids. In his 1890 address to the German Chemical Society (9), he showed that "Traubenzucker" is a 2,3,4,5,6-pentahydroxyhexanal and that "Fruchtzucker" is a 1,3,4,5,6-pentahydroxy-2-hexanone.

The concept of stereochemistry, developed since 1874 by Van't Hoff and Le Bel, had a great impact on carbohydrate chemistry because it could easily explain isomerism. From extensive chemical manipulations conducted with great skill, coupled with brilliant reasoning, Fischer correlated the families of sugars, and within 10 years, he was able to assign the relative configurations of most known sugars and to synthesize many additional examples (8). He introduced the "lock and key" hypothesis to interpret the action of an enzyme on a glycosidic substrate, and his monumental achievements inspired a whole generation of researchers in the field of natural products (10).

To name the various compounds, Fischer and others laid the foundations of a terminology still in use, based on the terms triose, tetrose, pentose, and hexose. He endorsed Armstrong's proposal to classify sugars into aldoses and ketoses, and proposed the name fructose for levulose because he found that the sign of optical rotation was not a suitable criterion for grouping sugars into families.

The major transformations of the aldoses utilized by Fischer are illustrated in Scheme 1, wherein aldoses are converted by mild oxidation (bromine water) into aldonic acids, by reduction (originally sodium amalgam in ethanol, nowadays sodium borohydride) into alditols, by reaction with phenylhydrazine to form phenylosazones, and by vigorous oxidation (nitric acid) into aldaric acids. Reaction of an aldose with hydrogen cyanide creates a new asymmetric center and affords a pair of isomeric nitriles (termed epimers), convertible by a sequence of transformations (Fischer–Kiliani chain-ascent reaction) into a pair of higher aldoses. A reaction sequence (Ruff degradation) that essentially effects the reverse process converts an aldose into the corresponding lower aldose having one fewer carbon atom in the chain.

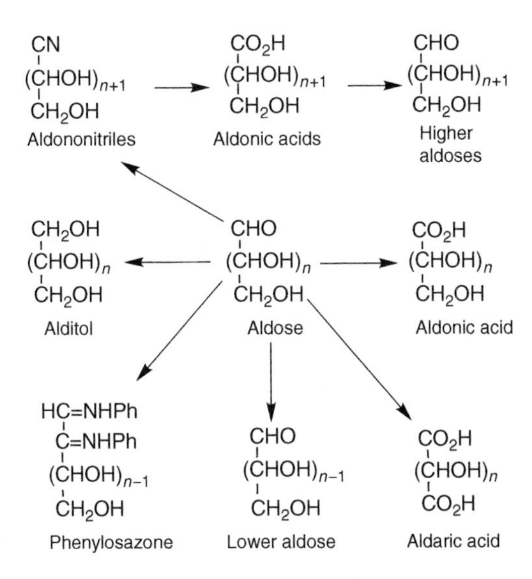

Scheme 1 Major transformations of aldoses.

Emil Fischer introduced the classical projection formulas for sugars, with a standard orientation (carbon chain vertical, carbonyl group at the top, Scheme 2); as he used models with flexible bonds between the atoms, he could easily "stretch" his sugar models into a position suitable for projection.

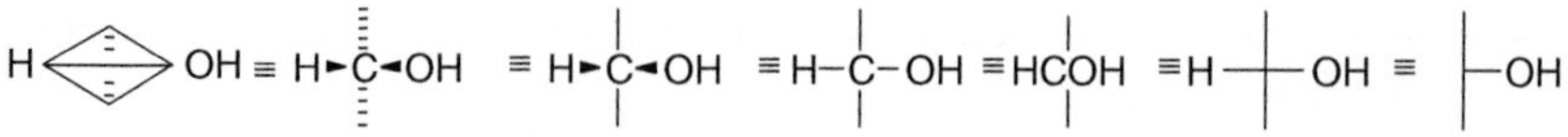

H–C–OH ≡ H–C–OH ≡ H–C–OH ≡ H–C–OH ≡ HCOH ≡ H–|–OH ≡ |–OH

Scheme 2 The Fischer projection formula for a tetrahedrally substituted carbon atom.

He assigned to the dextrorotatory glucose (via the derived glucaric acid) the projection having the OH group at C-5 pointing to the right (Scheme 3), well knowing that there was a 50% chance that this was wrong. Much later, Bijvoet (11) proved Fischer's arbitrary assumption to be correct in the absolute sense, by using a special X-ray technique.

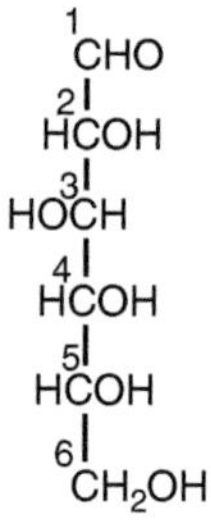

1 CHO
2 HCOH
3 HOCH
4 HCOH
5 HCOH
6 CH_2OH

Scheme 3 D-Glucose, Fischer projection.

Rosanoff in 1906 (12) selected the enantiomeric glyceraldehydes as the point of reference; any sugar derivable by chain lengthening from what is now known as D-glyceraldehyde belongs to the D series, a convention still in use (Scheme 4).

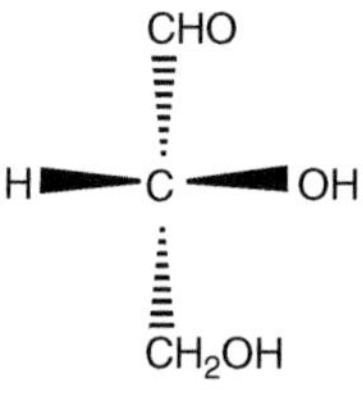

CHO
H–C–OH
CH_2OH

Scheme 4 D-Glyceraldehyde.

For a sugar having n stereocenters, there are 2^n isomers, so that there are $2^4=16$ aldohexoses, 8 in the D series and 8 in the L series. The names of the D-aldoses and their stereochemical relationships are illustrated in Scheme 5. These names form the basis of the group-configurational designators: *allo*, *altro*, *gluco*, *manno*, *gulo*, *ido*, *galacto*, and *talo* for four stereocenters; *ribo*, *arabino*, *xylo*, and *lyxo* for three stereocenters; and *erythro* and *threo* for two stereocenters.

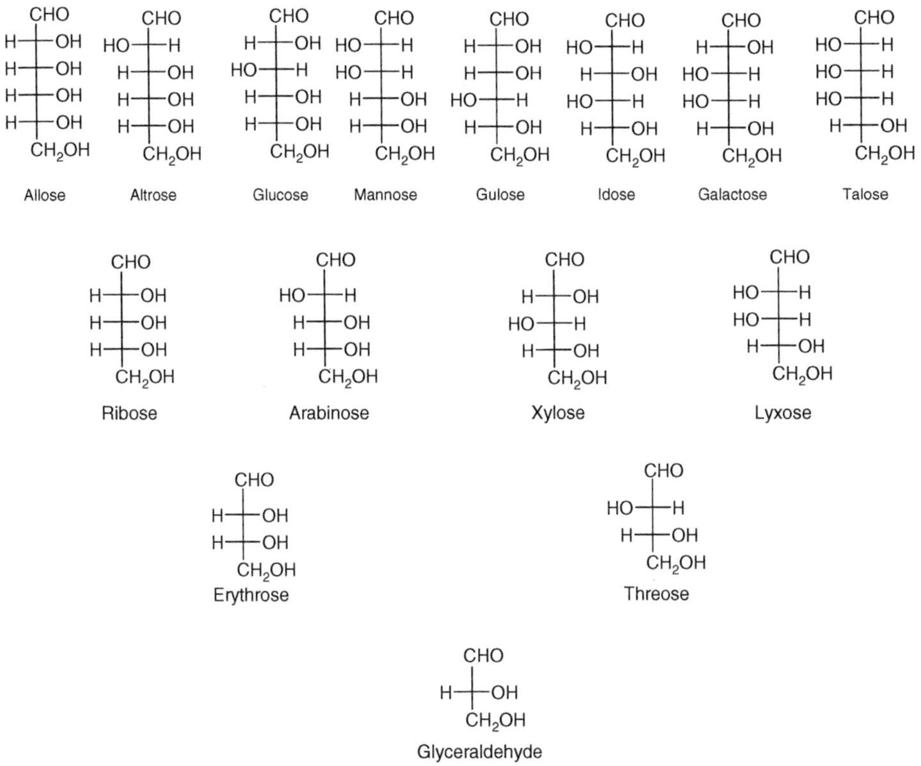

Allose Altrose Glucose Mannose Gulose Idose Galactose Talose

Ribose Arabinose Xylose Lyxose

Erythrose Threose

Glyceraldehyde

Scheme 5 The D family of aldoses.

The Cahn–Ingold–Prelog *R*,*S* system of stereodesignators (13), introduced much later for naming other natural products, is not used for sugars as it would lead to unwieldy names.

III. Cyclic Forms

Toward the end of the nineteenth century, it was realized that the free sugars exist as cyclic hemiacetals or hemiketals. Individual sugars react with methanol under acid catalysis to give stable products termed methyl glycosides, which are mixed full acetals. These have a new asymmetric center at the original carbonyl atom and were thus isolated as pairs of isomers that were designated as the α and β forms, later termed anomers. In the D series, the α anomer was defined as the one having the more-positive specific rotation, and in the L series, the α anomer has the more-negative specific rotation. The phenomenon of mutarotation, discovered in 1846 by Dubrunfaut, wherein the specific rotation of a free sugar in water solution changes with time, was now interpreted as being due to equilibration of the anomeric hemiacetals formed through reaction of one of the chain hydroxyl groups with the carbonyl group. Emil Fischer assumed the cyclic form to be a five-membered ring, which Tollens designated by the symbol <1,4>, while the six-membered ring received the symbol <1,5>. The α anomers of D-glucose in these two ring forms are here depicted in the Fischer projection (Scheme 6).

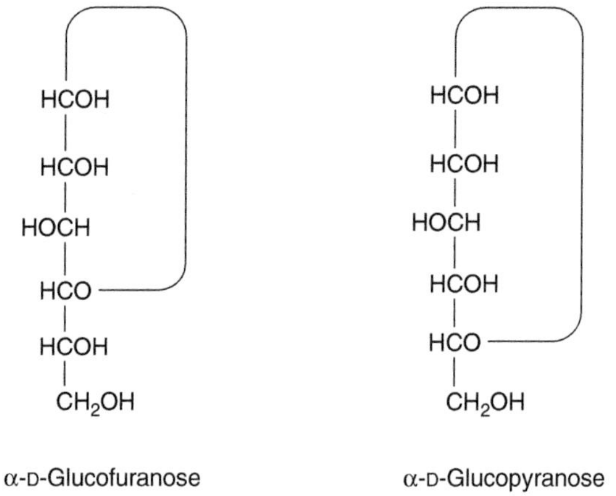

α-D-Glucofuranose α-D-Glucopyranose

Scheme 6 Fischer projection of α-D-glucofuranose and α-D-glucopyranose.

In the 1920s, Haworth and his school (14) proposed the terms "furanose" and "pyranose" for the two forms, and introduced what became known as the "Haworth depiction" for writing structural formulas, a convention that was soon widely followed (Scheme 7).

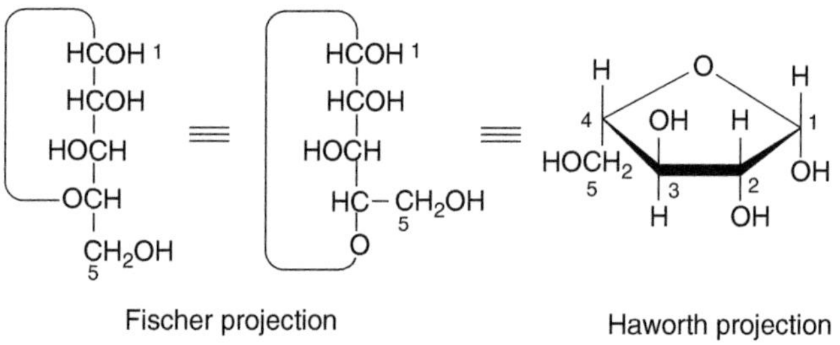

Fischer projection Haworth projection

Scheme 7 Relationship of the Fischer and Haworth projection formulas illustrated for β-L-arabinofuranose.

The Haworth projection formula presents a nominally planar ring with planes of substituents above and below the ring, whereas the Haworth conformational formula depicts the actual shape of the molecule with tetrahedral bond angles (Scheme 8). The Mills formula also depicts the ring arbitrarily as a plane, with substituents respectively above and below the plane. All three types of representation remain widely in use.

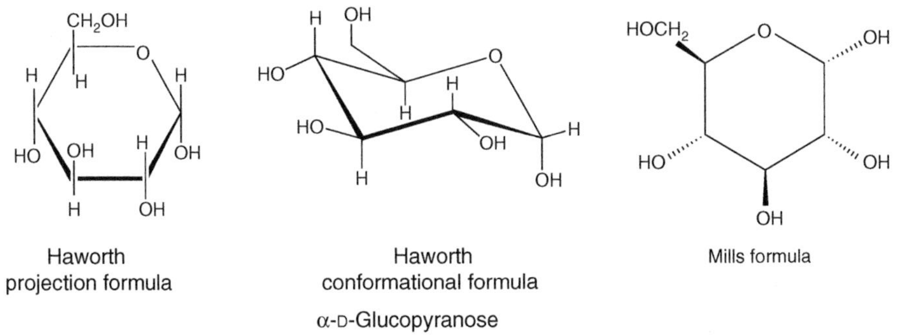

Haworth projection formula

Haworth conformational formula

Mills formula

α-D-Glucopyranose

Scheme 8 Cyclic representations of α-D-glucopyranose.

The question of ring size in the common glycosides and the free sugars remained controversial in the early part of the twentieth century, with polarimetric studies leading Hudson (15) to support Fischer's suggestion of a five-membered ring, but Haworth provided decisive evidence based on methylation studies (16) that Fischer's "α-methylglucosid" has a six-membered ring, and today it would be named methyl α-D-glucopyranoside. Strong evidence for the six-membered ring form in the free aldoses was also provided by Isbell (17) from studies on their oxidation to 1,5-lactones by bromine water. Subsequently, Jackson and Hudson applied Malaprade's glycol-cleavage reaction by periodate (18) to several related glycosides, and Hudson accepted (19) that Haworth's pyranoside formulation had been correct. In extending the periodate method, Jackson and Hudson (20) also were able to show that the glucose monomeric units in starch and cellulose likewise had the pyranose ring structure.

These two classic methods for structure determination in the carbohydrate field, methylation linkage-analysis and periodate oxidation, have remained key tools up to the present day in applications with numerous polysaccharides and glycoconjugates. The methylation reaction to introduce acid-stable methyl ether groups, originally performed with methyl iodide–silver oxide (21) or dimethyl sulfate–sodium hydroxide (22), is now more effectively conducted on a microscale by use of powerful deprotonating agents and dipolar aprotic solvents prior to acid hydrolysis of the permethylated carbohydrate (23,24). Analysis of the partially methylated products as their alditol acetates by gas–liquid chromatography–mass spectrometry is a rapid and routine technique (25). Likewise, the fragmentation of complex carbohydrates by periodate glycol cleavage (26) is a standard tool for structural analysis, especially when combined with reduction of the "dialdehydes" initially produced to the alditols followed by mild acid hydrolysis to cleave open-chain acetal groups [Smith degradation (27)].

IV. Nomenclature of Carbohydrates

Up to the 1940s, nomenclature proposals were made by individuals; in some cases, they were followed by the scientific community and in some cases not. Official bodies like the International Union of Chemistry, though developing and expanding the Geneva nomenclature for organic compounds, made little progress with carbohydrate nomenclature. The International Union of Pure and Applied Chemistry (IUPAC) Commission on Nomenclature of Biological Chemistry put forward a classification scheme for carbohydrates, but the new terms proposed have not survived. However, in 1939, the American Chemical Society (ACS) formed a committee to look into this matter, since rapid progress in the field had led to various misnomers arising from the lack of guidelines (28). Within this committee, the foundations of modern systematic nomenclature for carbohydrates and derivatives were laid: the numbering of the sugar chain, the use of D and L and α and β, and the designation of stereochemistry by italicized prefixes (multiple prefixes for longer chains). Some preliminary communications appeared, and the final report, prepared by M.L. Wolfrom, was approved by the ACS Council and published in 1948.

Not all problems were solved, however, and different usages were encountered on the two sides of the Atlantic. A joint British–American committee was therefore set up, and in 1952, it published "Rules for Carbohydrate Nomenclature." This work was continued, and a revised version was published in 1963 with endorsement by the ACS and by the Chemical Society in Britain. The publication of this report led the IUPAC Commission on Nomenclature of Organic Chemistry, jointly with the IUPAC–IUB Commission on Biochemical Nomenclature, to issue the "Tentative Rules for Carbohydrate Nomenclature, Part I, 1969," published in 1971/72 in several journals. A major revision of this 1971 document, published in 1996 (29), details current usage, and the Web version (30) of that document includes revisions and corrections.

For graphical representation of the structures of complex oligosaccharides and polysaccharides and for encoding in electronic databases, the individual sugar residues are assigned three-letter abbreviations, for example Glc (glucose), Man, Gal, Rib, Xyl, with appropriate modifiers, for instance Man*p* (mannopyranose), Fru*f* (fructofuranose), GlcA (glucuronic acid), GalNAc (*N*-acetylgalactosamine). Inter-residue linkages are designated by position locants and connecting arrows, for example β-D-Gal*p*-(1→4)-α-D-Glc*p* (lactose). For polysaccharides, the generic suffix "-an" is used rather than the "poly" prefix employed with synthetic polymers, thus mannan, xylan, carrageenan, xanthan, dextran, and the general term glycan. A few nonsystematic traditional names remain: for instance, cellulose, starch, inulin, glycogen, heparin, pectin. The borderline with polysaccharides cannot be drawn strictly; however, the term "oligosaccharide" is commonly used to refer to a defined structure as opposed to a polymer of unspecified length or a homologous mixture.

V. General Elucidation of Carbohydrate Structure

The generic term "carbohydrate" includes monosaccharides, oligosaccharides, and polysaccharides as well as substances derived from monosaccharides by reduction of the carbonyl group (alditols), by oxidation of one or more terminal groups to carboxylic acids (aldonic and uronic acids), or by replacement of one or more hydroxyl group(s) by a hydrogen atom (deoxy sugars), an amino group (amino sugars), a thiol group, or other functional group. It also includes numerous derivatives of these compounds formed by acylation, etherification, acetalation, and other reactions. The term "sugar" is frequently applied to monosaccharides and lower oligosaccharides. There are probably close to a million known compounds that conform to these definitions.

The chemistry and biochemistry of carbohydrates in general is detailed in a four-volume series (31), and a comprehensive treatment of polysaccharides can be found in the three volumes edited by Aspinall (32). A multiauthored treatise on polysaccharides published in 2005 (33) covers a wide range of aspects of the structural diversity, biological relevance, and technological applications of polysaccharides. A volume in the series *Comprehensive Natural Product Chemistry* edited by Pinto (34) provides extensive detail on the biosynthesis and biological role of carbohydrates.

Numerous natural products are glycosides, constituted of a sugar component (the glycon, frequently D-glucose, but many other sugars may be encountered) coupled through the anomeric position to an alcohol or phenol (the aglycon), a component often of considerable complexity. Examples include amygdalin (a cyanogenic glycoside), the anthocyanin plant pigments, the digitalis glycosides, and numerous antibiotics. Many of these have wide applications in medicine.

The chemical synthesis of glycosides of simple alcohols may be accomplished by acid-catalyzed reaction of a monosaccharide with the alcohol to yield anomeric mixtures of alkyl glycosides. This generally proceeds by a kinetic phase giving mainly the furanosides, followed by a thermodynamic phase where the pyranosides predominate (35,36). However, the controlled coupling of a sugar to a variety of aglycons has presented a major challenge to chemists. The "acetohalogen" synthesis introduced by Koenigs and Knorr (37) employed a peracetylated glycosyl halide reacting with an alcohol or phenol to give the peracetylated glycoside, readily de-esterified to give the product glycoside (Scheme 9).

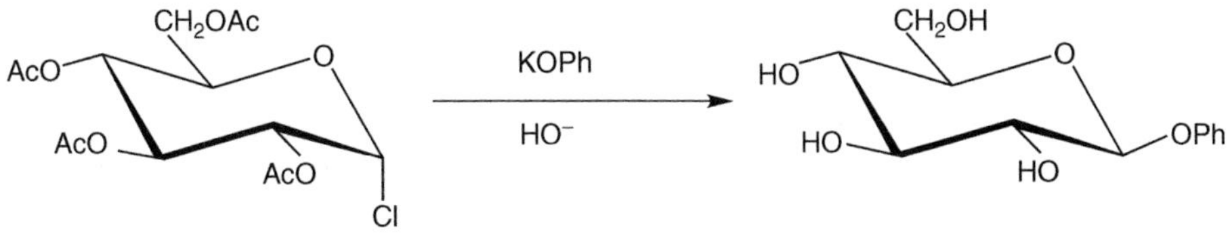

Scheme 9 Koenigs–Knorr synthesis.

This reaction affords the glycoside as the anomeric form having the aglycon and the substituent at the C-2 position in trans disposition [the *trans* rule (38)]. However, effective control of the anomeric specificity and the realization of high yields with a wide range of acceptor alcohols have presented major obstacles to synthetic chemists, later resolved by the development of numerous methodologies involving a variety of leaving groups at the anomeric position, different protecting groups on the sugar, and various catalysts and promoters of the reaction. These are treated in detail in Chapter 3. The synthesis of specific oligosaccharides remained difficult up to the middle of the twentieth century; synthesis of even the most abundant pure organic compound, namely sucrose, was not realized until 1953 (39), and it was not until the last decades of the twentieth century that the synthesis of moderately complex oligosaccharides became routine.

VI. Monosaccharide Structure

Although the early studies established the relationships and properties of the classic families of aldoses and ketoses, and their simple derivatives, some unusual monosaccharide structures were later encountered that were of importance in the biomedical area. The antiscorbutic vitamin isolated from paprika in 1928 by Szent-Gyorgi (40) and initially termed "hexuronic acid" was subsequently shown to be a six-carbon furanose lactone containing an enediol structure (Scheme 10) (41), and its synthesis was achieved independently by Reichstein and coworkers (42) and by Haworth and coworkers (43).

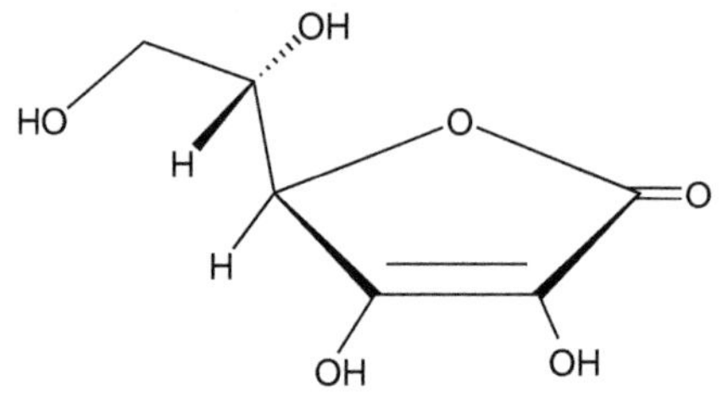

Scheme 10 L-Ascorbic acid, vitamin C, L-*threo*-hex-2-enono-1,4-lactone.

Many ingenious syntheses of L-ascorbic acid have been devised (44) and an adaptation of the Reichstein synthesis (45) from D-glucose constitutes the procedure used on the industrial scale.

Among the higher-carbon monosaccharides, the most important examples are the nine-carbon amino sugar *N*-acetylneuraminic acid (5-acetamido-3,5-dideoxy-D-*glycero*-α-D-*galacto*-non-2-ulopyranosonic acid, Neu5Ac; Scheme 11), the parent member of the family of sialic acids, and the eight-carbon sugar 3-deoxy-D-*manno*-oct-2-ulosonic acid, Kdo. The former is widely distributed as a constituent of glycoconjugates in the animal kingdom as well as in microorganisms, whereas the latter is encountered in Gram-negative bacteria.

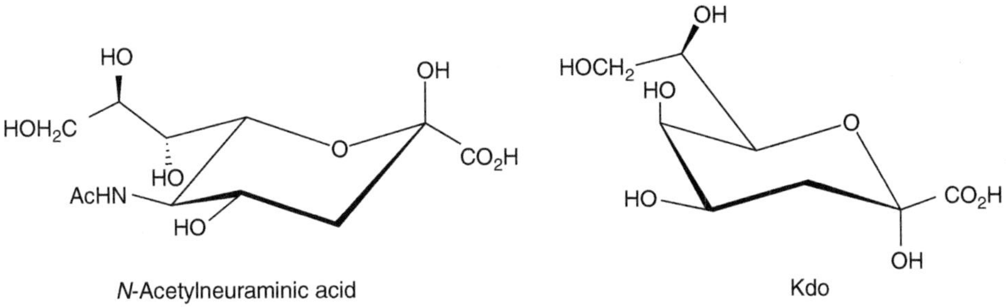

N-Acetylneuraminic acid   Kdo

Scheme 11 *N*-Acetylneuraminic acid (Neu5Ac) and 3-deoxy-D-*manno*-oct-2-ulosonic acid (Kdo).

N-Acetylneuraminic acid (46) was isolated by Gottschalk (47) from glycoproteins by the action of an enzyme (sialidase) from influenza virus, and was recognized as part of a group of related nine-carbon sugars (48). Encountered as a component of certain oligosaccharides of human milk (see Chapter 12) in the extensive researches of Richard Kuhn (49), the broad significance of this unusual sugar in biological processes (50) as a key feature of cell-surface carbohydrates continues to unfold, especially in its role as a component of the tetrasaccharide determinants (sialyl-Lewis^x and sialyl-Lewis^a) involved in cell adhesion and the inflammatory response (51). The eight-carbon sugar Kdo is a component of the "core region" of the complex cell-surface lipopolysaccharides of Gram-negative bacteria, serving to link the endotoxin lipid A to the remainder of the carbohydrate component (52,53).

VII. Oligosaccharide Structure

Sucrose (β-D-fructofuranosyl α-D-glucopyranoside), a nonreducing dextrorotatory disaccharide, is the most abundant pure (99.95%) organic compound (Scheme 12),

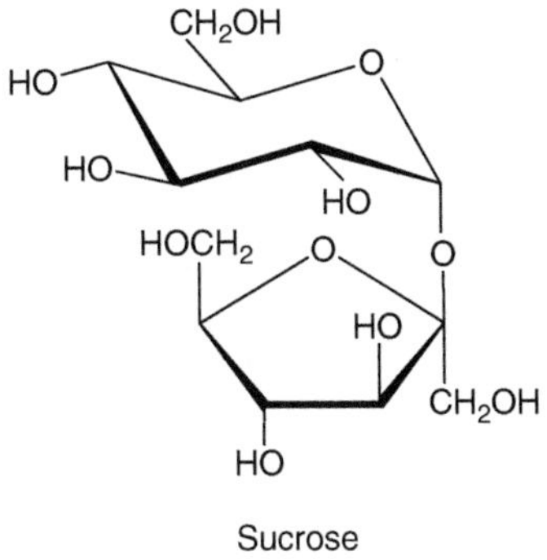

Sucrose

Scheme 12 Sucrose: β-D-fructofuranosyl α-D-glucopyranoside.

with an annual world production of almost 10^9 tonnes, and is used primarily as a nutritive sweetening agent. It is rapidly hydrolyzed into its component monosaccharides by the enzyme invertase or by dilute acid [the Arrhenius equation (54) relating temperature and reaction rate was developed from this reaction], but its decomposition by alkali is much slower and takes place by a different mechanism (55).

The low cost of sucrose makes it an attractive synthetic precursor for a wide range of applications (56), and a chlorinated derivative (1,6-dichloro-6-deoxy-β-D-fructofuranosyl 6-chloro-4,6-dideoxy-α-D-galactopyranoside, sucralose) is widely used as a noncaloric sweetener (Splenda®). Various other oligosaccharides occur naturally in the free form (57), but far more have been isolated as fragmentation products from larger biomolecules; their chemical or enzymatic synthesis is a very active current area of research, as detailed in Chapters 3 and 4.

Certain products of microbial fermentation having wide utility as antibiotics may be formally regarded as oligosaccharides, albeit with extensively modified monosaccharide components; the first example to be introduced (streptomycin) is briefly noted later in this chapter and more detail is presented elsewhere in the references cited.

VIII. Polysaccharide Structure

Application of the methylation linkage-analysis procedure to cellulose established (58) that cellulose is a linear polymer having the glucose units (1→4)-linked, and partial hydrolysis gave a series of oligosaccharides (cello-oligosaccharides) uniformly β-linked. The technique of partial hydrolysis of a polysaccharide to afford small oligosaccharides whose structure can be determined (fragmentation analysis) remains another general method for structure determination of polysaccharides, alongside methylation and periodate oxidation procedures. Periodate studies applied to chitin (59) disclosed that it has the same basic architecture (60) as cellulose, except that the hydroxyl group at C-2 in cellulose is replaced by an acetamido group in chitin. Chitin finds numerous applications in the biochemical and medical fields (61); see Chapter 11.

Although cellulose and chitin have a primary structure that is a linear chain, interchain hydrogen-bonding leads to a secondary level of structural organization giving water-insoluble fibers that form the basis of textile technology (cotton) and the paper industry (62).

Early structural studies (63) on starch indicated an α-(1→4)-linked glucan with some α-(1→6) branches, but it was not realized until considerably later (64) that starch is composed of two polymers, a fully linear component termed amylose and a branched component in which α-(1→6) branches lead to a tree-like structure (amylopectin). The chemistry and technology of starch is detailed in two major treatises (65,66). The animal storage polysaccharide glycogen has a structure similar to that of amylopectin, but is more highly branched (67).

The techniques of methylation linkage-analysis, periodate oxidation and the Smith degradation, and fragmentation analysis have been applied (32) to the structural elucidation of a wide range of polysaccharides isolated from plants, many finding practical applications (68) as gelling agents (agar, carrageenan, and other algal polysaccharides). Certain microorganisms convert simple sugars into polysaccharides, and many such microbial polysaccharides have useful properties (dextran, xanthan).

The polysaccharides from animal tissues are usually more complex than those from plant sources, and aside from glycogen they generally contain amino sugar components, typically in alternating sequence with uronic acid residues (glycosaminoglycuronoglycans). These are discussed in detail elsewhere in this book. Major examples include hyaluronic acid and the sulfated polysaccharides chondroitin, dermatan, keratan, and heparin (69). Hyaluronic acid plays an important role in connective tissue in controlling interstitial permeability and serving as a lubricant in joints, and finds use in ophthalmic surgery (see Chapter 14). Isolated in 1934 (70) from vitreous humor, it is composed of glucuronic acid and *N*-acetylglucosamine residues (71) in a linear polymer of very high molecular weight. A combination of methylation studies (72,73), fragmentation analysis by acid and enzymatic methods, and synthesis of fragments showed it to have the disaccharide repeating sequence:

→ 4)-β-D-Glc*p*A-(1 → 3)-β-D-Glc*p*NAc-(1 →

From cartilage, there were isolated two related glycosaminoglycuronans, chondroitin 4-sulfate (74) and chondroitin 6-sulfate (75), *O*-sulfated at positions 4 or 6 of the amino sugar component, which in this case is *N*-acetylgalactosamine. The repeating sequence of chondroitin 6-sulfate may thus be depicted as:

→ 4)-β-D-Glc*p*A-(1 → 3)-β-D-Gal*p*NAc6S-(1 →

In the tissues, these two polymers exist covalently linked to protein as proteoglycans (69), as does the related dermatan sulfate (older names β-heparin, chondroitin sulfate B), whose sequence is the following:

→ 4)-α-L-IdoA-(1 → 3)-β-D-Gal*p*NAc6S-(1 →

A product extracted from dog liver having a marked anticoagulant effect (76), preventing the formation of thrombin from prothrombin, was termed heparin (77). Elucidation of its detailed structure was a major challenge to chemists, and its fine structure and full range of biological activities were not unfolded until many years later. It was shown to be a highly sulfated polysaccharide (78) containing glucosamine (78,79) and uronic acid (80) residues in equimolar proportion, and in alternating sequence with (1→4) linkages throughout (81,82) and the α configuration at the anomeric position of glucosamine. In addition to

O-sulfate groups on both components, the uronic acid component was shown to be not only D-glucuronic acid (80) but also L-iduronic acid (83,84), linked α-L or β-D to the amino sugar (85).

Subsequent studies revealed much greater complexity in the heparin structure and a multitude of physiological functions (86,87), as discussed in Chapters 5 and 10. The native structure in the tissues is a high molecular weight proteoglycan consisting of a protein chain having an alternating sequence of glycine and L-serine residues to which are attached the sulfated glycosaminoglycan chains through a tetrasaccharide "linkage region" GlcA-Gal-Gal-Xyl to the hydroxyl groups of the serine residues (88). This proteoglycan structure is disrupted in the process for the isolation of heparin from tissue sources (commonly pig intestinal mucosa); this involves proteolysis and/or aqueous alkali treatment, and yields clinical heparin as a highly anionic polyelectrolyte. Studies on its fine structure have revealed microheterogeneity in the distribution of sulfate groups, the placement of the two uronic acid components, and other features, and the anticoagulant activity has been shown to be localized in a specific pentasaccharide region of the chain. In very recent years, advances in synthetic methodology for complex oligosaccharides have permitted practical synthesis of this subunit of heparin for use as an anticoagulant [fondaparinux, Arixtra® (89)].

IX. Carbohydrate Antibiotics

A revolution in therapies to combat infectious pathogens took place in 1937 with the discovery by Fleming of penicillin, a microbial metabolite, and the first of a class of β-lactam antibiotics. The field of carbohydrate-structured antibiotics of major clinical utility opened subsequently with Waxman's discovery (90) of streptomycin in the fermentation broth of an *Actinomyces* species. Streptomycin was the first example of the class of aminocyclitol (also termed aminoglycoside) antibiotics, and it was found particularly useful in the treatment of tuberculosis, although a high incidence of ototoxicity is a limitation in its modern clinical use. Elucidation of its chemical structure provided a portent of the rich diversity of unusual sugar structures in microbial metabolites, as later revealed in numerous other antibiotics in this class and such other groups as the macrolides (for example erythromycin), nucleoside antibiotics such as puromycin, glycosylated aromatic structures such as doxorubicin (adriamycin), and many others. Structure elucidation of the monosaccharide components of these antibiotics divulged a profusion of novel structural types: deoxy sugars, amino sugars, branched-chain sugars, higher-carbon sugars, sugars of unusual stereochemistry, to name just a few (91), and provided the stimulus for a great deal of innovative work in developing procedures for their chemical synthesis.

Streptomycin is a pseudotrisaccharide having an amino sugar, *N*-methylglucosamine as the unusual L enantiomer, linked by way of a branched-chain sugar component streptose, in turn connected to a guanidinocyclohexanepolyol termed streptidine, as shown in Scheme 13:

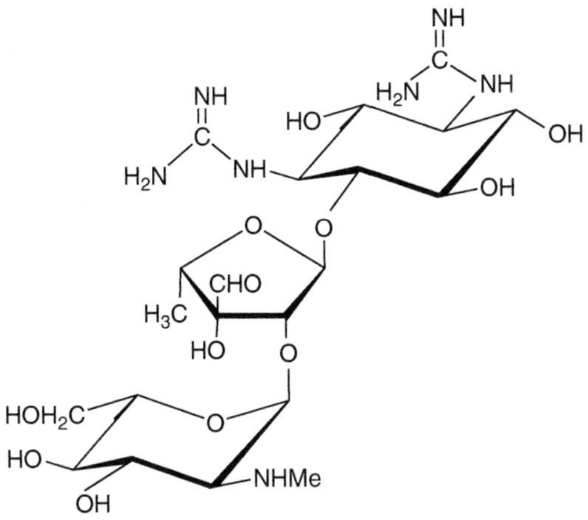

Scheme 13 Streptomycin.

The structure was largely elucidated by Wolfrom and his group (92). This pioneering work presaged the subsequent discovery of numerous other aminocyclitol antibiotics produced by microorganisms, notably the neomycins, kanamycins, and gentamicins. The discovery, structure determination, and understanding of the mode of action of these agents owe much to the work of Sumio Umezawa and Hamao Umezawa (93,94), who also showed that synthetic modification of parent antibiotic structures could overcome the problem caused by development of drug-resistant strains of bacteria. Later work has disclosed the role of aminocyclitol–nucleic acid recognition, and in particular, ribosomal binding in the mechanism of bactericidal action (95).

Access to carbohydrate antibiotics by chemical assembly from their component sugars has presented exercises in synthetic virtuosity, but in general their production through microbial fermentation remains the most economically viable approach.

X. Structural Methodology

A. General

The progress of research in the carbohydrate field has been greatly enhanced by the introduction of new reagents for functional-group transformations, by greater understanding of reaction mechanisms in the light of detailed conformational properties of these polyfunctional molecules, and above all in the development of powerful physical tools for structural elucidation, most notably nuclear magnetic resonance (NMR) spectroscopy, X-ray crystallography, and mass spectrometry, to complement the older tools of polarimetry (96) and infrared spectroscopy.

Among the chemical reagents of value in the carbohydrate field, the introduction (97,98) of sodium borohydride as a water-stable reducing agent for conversion of aldehydes, ketones, and hemiacetals into alcohols is particularly noteworthy, especially when used in conjunction with the periodate glycol-cleavage reaction already mentioned. It also features in the alkaline-reductive cleavage reaction applied with *O*-linked glycoproteins to detach the carbohydrate chains from the protein component (99).

Haworth in his 1929 book (14) mentioned that consideration of the conformational properties of sugars would lead to a better understanding of their reactivity, but at that time there were no experimental tools to develop this concept further. In 1937, Isbell interpreted (100) the difference in rates of bromine oxidation of α- and β-D-glucopyranose to D-glucono-1,5-lactone as a conformational effect resulting from the orientation (now termed axial and equatorial) of the hydroxyl group at C-1 in the chair form (now termed the $^4C_1$ chair), thereby pioneering the use of conformational analysis as a tool in the understanding of organic chemical reactions. He also interpreted the role of neighboring groups in directing the stereochemical outcome of substitution reactions. Both of these concepts were developed by Isbell long before they became recognized by the broader organic chemical community (101), and form part of an extensive body of mechanistic and synthetic studies on sugars conducted by Isbell and a small group of coworkers during more than six decades (102).

B. X-Ray Crystallography

Some of the earliest X-ray crystallographic studies on carbohydrates were conducted on oriented fibers of cellulose, and they demonstrated four hexopyranose rings in a monoclinic unit cell (103). A key crystallographic study on a small sugar molecule was that performed by Cox and Jeffrey (104) on α-D-glucosamine hydrobromide, undertaken to settle the controversy as to the orientation of the amino group at C-2. Not only was this point resolved, but the three-dimensional architecture of the nonhydrogen atoms in the unit cell clearly demonstrated the familiar $^4C_1$ chair conformation of the pyranose ring and the axial orientation of the hydroxyl group at C-1, long before the advent of NMR spectroscopy. The computational burden of manually resolving the diffraction data was enormous in early work, but the introduction of digital computers has greatly facilitated the task to the point that the three-dimensional structure of most sugar derivatives for which a small single crystal is available can be obtained rapidly and routinely, greatly reducing the need for traditional degradative methods for structure determination.

Since that pioneering 1939 crystallographic study on glucosamine, Jeffrey and Sundaralingam have compiled critical reviews of all published crystal structures dealing with sugars, nucleosides, and nucleotides (105) up through 1980, correcting where necessary the authors' original interpretations and rendering the structures in familiar conformational depictions. It should be noted that most crystal-structure determinations do not differentiate between enantiomers, and require reference to a known center of chirality in the molecule. In addition to presenting the generalized chair conformations of the pyranose ring system, precise angular deviations from the ideal chair are represented as Cremer–Pople (106) puckering parameters. The large volume of subsequent crystal structure data on sugars is accessible in the Cambridge Crystallographic Data Bank (107), and the

Glycoscience database (108) provides a comprehensive resource on a wide range of structural data on carbohydrates of biological interest.

Crystallographic studies on oriented fibers of polysaccharides provide unit-cell dimensions, but precise interpretations of overall polymer conformation continue to invite debate, and often require correlation with data on small-molecule components of the polymer and on computerized modeling. From the early studies on cellulose fibers (103,109), there has developed extensive and often controversial literature on this most abundant organic molecule (61), as well as on starch, the second-most abundant glucan. As early as 1943, the concept of a helical conformation in a biomolecule was advanced by Rundle and French (110) from crystallographic studies on the amylose component of starch, long before helical conformations were accepted as important structures in proteins and nucleic acids. Short bibliographic summaries on all crystallographic studies on polysaccharides up to 1979 have been compiled by Sundararajan and Marchessault (111), and subsequent work is detailed in a current monograph (112) and a Web site (113).

C. Mass Spectrometry

The use of mass spectrometry in the structural analysis of carbohydrates, first reported in 1958 (114), was developed in detail by Kochetkov and Chizhov (115). They showed that, under electron impact, the acetylated and methyl ether derivatives of monosaccharides provided a wealth of structural information through analysis of typical fragmentation pathways of the initial molecular ion. This has proved of enormous utility in the structural elucidation of polysaccharides and complex oligosaccharides: sequential permethylation, hydrolysis, reduction to the alditol, and acetylation, affords mixtures of peracetylated, partially methylated alditol acetates that can be separated and analyzed by use of a gas chromatograph coupled directly to a mass spectrometer (25). The mass spectra of stereoisomers are normally identical, while the gas chromatographic retention times readily permit differentiation of stereoisomers.

The early studies on mass spectrometry of sugars by electron impact were limited to derivatives of adequate volatility; these included trimethylsilyl ethers, various acetal derivatives, along with acylated and methylated derivatives. However, with the advent of "soft" ionization techniques (116) such as chemical ionization, field-desorption, fast-atom bombardment, electrospray ionization, and especially matrix-assisted, laser-desorption time-of-flight mass spectrometry, it became possible to examine unsubstituted sugars along with complex oligosaccharides, carbohydrate antibiotics, nucleosides, and related large molecules. High-resolution mass spectrometry giving exact mass data allows differentiation between isomers of the same unit mass, and provides molecular weight data on such large molecules as the glycan components of glycoproteins (25).

D. NMR Spectrometry

The advent of NMR spectroscopy revolutionized research in the carbohydrate field by providing a rapid method for structural identification and for study of the dynamic behavior of sugars in solution. The seminal 1958 paper by Lemieux et al. (117), introducing proton NMR in the study of acetylated aldoses, demonstrated that the proton at the anomeric position gives its signal at a distinctive field position as a doublet through coupling with its vicinal neighbor at C-2. The magnitude of this coupling reflects the dihedral angle between H-1 and H-2, permitting assignment of favored conformation for pyranose sugars. Much of the early proton NMR studies have been discussed by Hall (118,119) and their implications in understanding the conformational behavior of sugars have been reviewed (120). Major advances in instrumental methodology, especially the introduction of superconducting solenoids and pulsed Fourier-transform techniques, have overcome earlier problems of inadequate resolution and the necessity for relatively large samples. A wide array of pulse-sequence procedures makes the structural attribution of simple sugar derivatives a routine procedure and permits detailed investigations on the conformations of oligosaccharides (121,122). Structural studies on the carbohydrate components of glycoproteins (123), of profound significance as biological recognition markers, have particularly benefited from the application of high-resolution proton NMR.

The ^{12}C nucleus does not have a magnetic moment, and the low (1%) abundance of the magnetic ^{13}C nucleus presented an early handicap in applications of C-13 NMR with carbohydrates, but modern instrumentation has made studies on the carbon framework of sugars and polysaccharides a routine complement to proton NMR investigations. Bock and Pedersen (124) have provided extensive tabulation of ^{13}C chemical shifts for the whole series of free sugars and their principal derivatives in solution.

The technique of ^{13}C cross-polarization magic-angle spinning spectroscopy extends NMR spectroscopy to applications with carbohydrates in the solid state, notably with cellulose (125) as well as with monosaccharides (126).

XI. Carbohydrate Biochemistry

This vast topic can only be covered in the briefest outline within the scope of the current chapter, but the short book by Lehmann (127) presents an excellent modern overview of the biological aspects of carbohydrates, dealing with their metabolism and biosynthesis, their role in biological recognition, their functions in cell walls and cell membranes, and as energy sources.

Credit again goes to the great pioneer Emil Fischer for the earliest studies on the reactions of carbohydrates with enzymes (8). He conducted extensive experiments with enzymes isolated from plant, animal, and microbial sources on a range of glycosides, oligosaccharides, and polysaccharides, and formulated the celebrated "lock and key" concept for the mode whereby the enzyme recognizes its substrate, stating: *Um ein Bild zu gebrauchen, will ich sagen, daß Enzym und Glucosid wie Schloß und Schlüssel zueinander passen müssen, um eine chemische Wirkung aufeninander ausüben zu können* (128). Subsequent work by Hudson (129) on the hydrolysis of sucrose to glucose and fructose by the enzyme invertase, and by Pigman (130) on the action of glycoside-cleaving enzymes (glycosidases) laid the basis of much of our understanding of enzyme kinetics.

Classic work on the pathways of catabolism of monosaccharides, largely through use of radiolabeled (^{14}C and ^{3}H) intermediates as tracers, led to the elucidation of the glycolysis pathway for the breakdown of monosaccharides (127), as well as details of the anabolic pathway to sugars in photosynthesis (3,127). It became recognized that, although the action of many enzymes is reversible, anabolic processes generally proceed by separate pathways from those involved in catabolism.

The glycoside-cleaving enzymes effectively transfer a glycosyl group to a water molecule (glycosidases) or a different acceptor molecule (glycosyltransferases). Early work on these enzymes has been reviewed in detail (131,132) and the mechanism of their action (133) constitutes a very active current area of research (134,135), with "enzyme engineering" targeting the development of modified enzymes for practical applications (136). Much further detail is presented in Chapter 4.

Although the catabolism of oligo- and poly-saccharides takes place by the action of glycosidases, the mechanism of their anabolism was not decisively revealed until the work of Leloir and his associates in the 1950s (137). They demonstrated that the process involves glycosyltransferase enzymes that transfer glycosyl groups from nucleoside 5′-(glycosyl diphosphates) ("sugar nucleotides") to an acceptor (138). The ability of uridine 5′-(α-D-glucopyranosyl diphosphate) ("UDP-D-glucose"; Scheme 14) to serve as a glucosyl donor presaged the discovery of a wide range of related sugar nucleotides incorporating uridine, guanosine, cytidine, adenosine, and thymidine as the nucleoside component, and various monosaccharides as the glycosyl component. The discovery led to rapid progress in elucidating the biosynthetic mechanism of numerous glycosides, oligosaccharides, and polysaccharides in animal, microbial, and plant cells. Furthermore, it was established that such nucleotide sugars as UDP-D-glucose can undergo transformation of the sugar component via oxidation–reduction and deoxygenation reactions to afford a wide range of nucleotide sugars bearing structurally modified monosaccharides that can take part in transferase reactions.

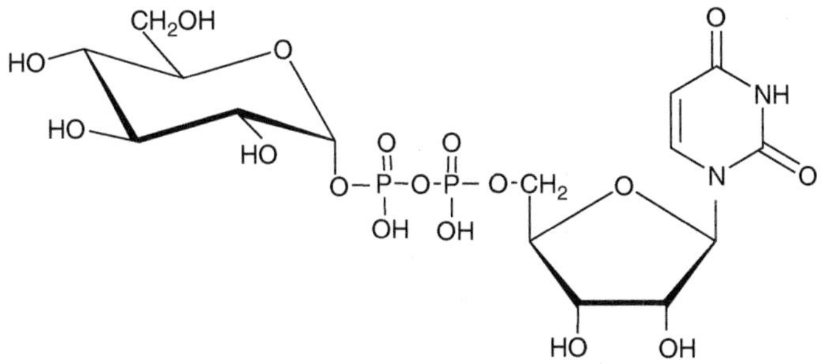

Scheme 14 UDP-D-glucose.

The groundbreaking work of Leloir et al. (139) laid the basis for much of our understanding of the biosynthetic pathways via nucleotide sugars in the formation of, for example, sucrose from UDP-D-glucose and fructose 6-phosphate (140). Extensive studies on the biosynthesis of glycogen, starch, cellulose, along with numerous glycoproteins, proteoglycans, glycolipids, steroidal glycosides, the ABO blood-group determinants, teichoic acids, and other glycoconjugates underscore the broad involvement of nucleotide sugars in the elaboration of complex carbohydrates.

It was subsequently shown that lipid-linked intermediates also play a key role in the biosynthesis of complex carbohydrates. In bacterial systems, undecaprenol (a C_{55} polyprenol), as its phosphate ester, serves as a membrane-soluble carrier to which is transferred a sugar derivative in the biosynthesis of cell-envelope components (141). In animal glycoproteins, as exemplified by the work of Behrens and Leloir (142), UDP-D-glucose transfers the sugar to dolichol phosphate, and the D-glucosyl group is subsequently transferred to a protein acceptor. Dolichols are polyprenols composed of 13–22 isoprene units (143) and are involved in eukaryotes in the glycosylation of lipids and proteins in the endoplasmic reticulum membrane.

XII. Carbohydrate Antigens and Vaccines

The highly specific biological functions of carbohydrates were recognized by Heidelberger (144) from his studies on the capsular antigens of different strains of *Streptococcus pneumoniae.* These were not proteins, but turned out to be polysaccharides. The antigen of the type 3 strain proved to be a polysaccharide having a disaccharide repeating unit →4)-β-D-Glc*p*-(1→3)-β-D-Glc*p*A-(1→ (145). The factors responsible for the ABO human blood groups were shown by Morgan and Watkins (146) to be oligosaccharide components of various glycoconjugates built from a disaccharide α-L-Fuc*p*-(1→2)-β-D-Gal*p*-(1→ (H factor) to which an α-D-Gal*p*NAc group is linked (1→3) to the Gal*p* component (A factor) or an α-D-Gal*p* group is similarly linked (B factor). These studies opened up a large field of study of the interactions of carbohydrates and proteins, of major importance in pharmacology and medicine as it relates to antigen–antibody interactions, cell–cell recognition phenomena, and the fixation of bacteria, viruses, and toxins on cell surfaces.

As early as 1918, Landsteiner and Lampl (147) observed that nonimmunogenic small molecules (haptens) could be rendered immunogenic by covalent coupling to proteins. The pneumococcal capsular polysaccharides by themselves are relatively weak stimulators of the immune system in producing antibodies, but when conjugated to a protein carrier they can serve as effective vaccines for targeting virulent strains of pathogenic bacteria. The validity of this concept was demonstrated in 1931 by Goebel and Avery (148), who coupled the type 3 pneumococcal polysaccharide to horse-serum globulin to produce a conjugate that induced specific antibody to the polysaccharide antigen in experimental animals.

The success of this approach is amply demonstrated in the development of effective vaccines for clinical use, with the notable example of antibacterial prophylaxis against childhood meningitis using a conjugate vaccine prepared by coupling the capsular material of the Gram-negative bacterium *Haemophilus influenzae* type b (Hib) with tetanus toxoid (149). Since these early discoveries, many other conjugate vaccines of clinical utility have been developed (150,151) that provide immunoprophylaxis against virulent strains of pathogens, both Gram positive and Gram negative. They offer a complementary therapeutic approach to the broad-spectrum antibiotics, which target a wide range of Gram-positive organisms, in providing narrowly focused protection against individual strains of pathogens. Multivalent vaccines, incorporating the capsular polysaccharide haptens of several serotypes of a given pathogen in the same conjugate, have been developed against *S. pneumoniae* (152), and others in clinical use are targeted against *Neisseria meningitides*, group B *Streptococcus, Salmonella typhi*, and *Staphylococcus aureus* (151). The field remains under active development with efforts to optimize the specific carbohydrate hapten component, the nature of the protein employed, and the mode of covalent linking of the two components (153).

XIII. Conclusion

The foregoing survey of the enormous field of carbohydrate chemistry and biology is of necessity brief and selective, but an attempt has been made to cite the seminal work of major pioneers, along with key review articles of later developments, together with leads to the most current activity in the field. Large, multiauthor compendia published in recent years having much detailed information include the 1999 volume in the series *Comprehensive Natural Product Chemistry* (34) and the 2005 edition of *Polysaccharides: Structural Diversity and Functional Versatility* (33). The synthesis of complex oligosaccharides of biological importance is a particularly active area of current research; the 2008 volume *Glycoscience: Chemistry and Chemical Biology* (154) brings together wide-ranging results on chemical and enzymatic approaches to synthesis, and the 4-volume set Comprehensive Glycoscience (112) published in 2007 is particular focused on biological aspects.

The *Dictionary of Carbohydrates* (155) has a listing of some 24,000 individual carbohydrates, giving names, structures, sources, data, and literature references for each, together with a complementary electronic database. Another valuable reference work that includes a wealth of information on carbohydrate derivatives important in biochemistry and biology is the *Oxford Dictionary of Biochemistry and Molecular Biology* (156).

Among the smaller single-author texts on carbohydrates may be noted the Ferrier–Collins book *Monosaccharide Chemistry* (157) and Stick's *Carbohydrates: The Sweet Molecules of Life* (158), which includes a brief treatment of biological aspects. Lehmann's book *Carbohydrates: Structure and Biology* (127) has the biological and biochemical aspects of carbohydrates as its central focus, and is particularly useful for gaining a broad overview of the subject.

References

1. Marcus Porcius Cato (Cato the Elder). De Agricultura. *ca* 184 BCE, http://soilandhealth.org/01aglibrary/010121cato/catofarmtext.htm, chapter 87.
2. Robyt JF. Essentials of Carbohydrate Chemistry. Springer, New York, 1997; 42–44.
3. Bassham J, Benson A, Calvin M. The path of carbon in photosynthesis. J Biol Chem 1950; 185:781–787.
4. Bassham JA. Mapping the carbon reduction cycle: a personal retrospective. Photosynthesis Res 2003; 76:25–52.
5. Kirchoff GSC. Acad Imperale St Petersbourg, Mem 1811; 4:27.

6. Willstätter R, Zechmeister L. Hydrolysis of cellulose I. Ber 1913; 46:2401–2412.
7. Monier-Williams GW. Hydrolysis of cotton cellulose. J Chem Soc 1921; 119:803–805.
8. Fischer E. Untersuchungen über Kohlenhydrate und Fermente, 1884–1908. Berlin: Julius Springer, 1909.
9. Fischer E. Synthesen in der Zuckergruppe I. Ber 1890; 23:2114–2143.
10. Lichtenthaler FW. Emil Fischer, his personality, his achievements, and his scientific progeny. Eur J Org Chem 2002; 4095–4122.
11. Trommel J, Bijvoet JM. Crystal structure and absolute configuration of the hydrochloride and hydrobromide of D(–)-isoleucine. Acta Crystallogr 1954; 7:703–709.
12. Rosanoff MA. On Fischer's classification of stereo-isomers. J Am Chem Soc 1906; 28:114–121.
13. Cahn RS, Ingold CK, Prelog V. Specification of asymmetric configuration in organic chemistry. Experientia 1956; 12:81–94.
14. Haworth WN. The Constitution of Sugars. London: Edward Arnold & Co, 1929.
15. Hudson CS. The significance of certain numerical relationships in the sugar group. J Am Chem Soc 1909; 31:66–86.
16. Haworth WN, Hirst EL, Miller EJ. Structure of the normal and γ-forms of tetramethylglucose. Oxidation of tetramethyl-δ- and γ-gluconolactone. J Chem Soc 1927; 2436–2443.
17. Isbell HS, Hudson CS. The course of the oxidation of the aldose sugars by bromine water. Bur Stand J Res 1932; 8:327–338.
18. Malaprade L. Action of polyalcohols on periodic acid. Analytical applications. Bull Soc Chim Fr 1928; 43:683–696.
19. Jackson EL, Hudson CS. Studies on the cleavage of the carbon chain of glycosides by oxidation. A new method for determining ring structures and alpha and beta configurations of glycosides. J Am Chem Soc 1937; 59:994–1003.
20. Jackson EL, Hudson CS. The structure of the products of the periodic acid oxidation of starch and cellulose. J Am Chem Soc 1938; 60:989–991.
21. Purdie T, Irvine JC. The alkylation of sugars. J Chem Soc 1903; 83:1021–1037.
22. Haworth WN. A new method of preparing alkylated sugars. J Chem Soc 1915; 107:8–16.
23. Hakomori S. Rapid permethylation of glycolipids and polysaccharides, catalyzed by methylsulfinyl carbanion in dimethyl sulfoxide. J Biochem (Tokyo) 1964; 55:205–208.
24. Ciucanu I, Kerek F. A simple and rapid method for the permethylation of carbohydrates. Carbohydr Res 1984; 131:209–217.
25. Rodrigues JA, Taylor AM, Sumpton DP, Reynolds JC, Pickford R, Thomas-Oates J. Mass spectrometry of carbohydrates: newer aspects. Adv Carbohydr Chem Biochem 2007; 61:61–147.
26. Perlin A. Glycol-cleavage oxidation. Adv Carbohydr Chem Biochem 2006; 60:183–250.
27. Goldstein IJ, Hay GW, Lewis BA, Smith F. Controlled degradation of polysaccharides by periodate oxidation, reduction, and hydrolysis. Methods Carbohydr Chem 1965; 5:361–370.
28. Horton D. Development of carbohydrate nomenclature. In: Loening K ed. The Terminology of Biotechnology: A Multidisciplinary Problem, Berlin: Springer-Verlag, 1990; 41–49.

29. IUPAC-IUBMB Nomenclature of carbohydrates. Adv Carbohydr Chem Biochem 1997; 52:199743–177.
30. Nomenclature of carbohydrates, http://www.chem.qmul.ac.uk/iupac/2carb/.
31. Pigman W, Horton D, eds. The Carbohydrates, Chemistry and Biochemistry, 2nd edn. Academic Press, New York, 1972; Vol. IA, IIA, IIB, 1980; Vol. IB.
32. Aspinall GO, ed. The Polysaccharides. Academic Press, New York, 1982; Vol. 1. 1983; Vol. 2, 1985; Vol. 3.
33. Dumitriu S, ed. Polysaccharides: Structural Diversity and Functional Versatility, 2nd edn., Marcel Dekker, New York, 2005.
34. Pinto M, ed. Comprehensive Natural Products Chemistry: Carbohydrates and Their Derivatives, Including Tannins, Cellulose, and Related Lignins: Pergamon Elsevier, Oxford, 1999; Vol. 3.
35. Levene PA, Raymond AL, Dillon RT. Glucoside formation in the commoner monoses. J Biol Chem 1932; 95:699–713.
36. Bishop CT, Cooper FP. Glycosidation of sugars. I. Formation of methyl D-xylosides. Can J Chem 1962; 40:224–232.
37. Koenigs W, Knorr E. Some derivatives of grape sugars and galactose. Ber 1901; 34:957–981.
38. Tipson RS. Action of silver salts of organic acids on bromoacetyl sugars. A new form of tetraacetyl-*l*-rhamnose. J Biol Chem 1939; 130:55–59.
39. Lemieux RU, Huber G. A chemical synthesis of sucrose. J Am Chem Soc 1953; 75:4118.
40. Szent-Gyorgyi A. Observations on the function of peroxidase systems and the chemistry of the adrenal cortex: description of a new carbohydrate derivative. Biochem J 1928; 22:1387–1409.
41. Herbert RW, Hirst EL, Percival EGV, Reynolds RJW, Smith F. Constitution of ascorbic acid. J Chem Soc 1933; 1270–1290.
42. Reichstein T, Grüssner A, Oppenauer A. Synthesis of d-ascorbic acid (d-form of vitamin C). Helv Chim Acta 1933; 16:561–565.
43. Ault RG, Baird DK, Carrington HC, Haworth WN, Herbert RW, Hirst EL, Percival EGV, Smith F, Stacey M. Synthesis of d- and l-ascorbic acid and of analogous substances. J Chem Soc 1933; 1419–1423.
44. Crawford T, Crawford SA. Synthesis of L-ascorbic acid. Adv Carbohydr Chem Biochem 1980; 37:79–155.
45. Reichstein T, Grüssner A. A high-yield synthesis of l-ascorbic acid (vitamin C). Helv Chim Acta 1934; 17:311–328.
46. Klenk E. Neuraminic acid, the cleavage product of a new brain lipoid. Z Physiol Chem 1941; 268:50–58.
47. Gottschalk A. *N*-substituted isoglucosamine released from mucoproteins by the influenza virus enzyme. Nature 1951; 167:845–847.
48. Blix FG, Gottschalk A, Klenk E. Proposed nomenclature in the field of neuraminic and sialic acids. Nature 1957; 179:1088.
49. Baer HH, Richard Kuhn. Adv Carbohydr Chem Biochem 1969; 24:1–12.
50. Schauer R. Chemistry, metabolism, and biological functions of sialic acids. Adv Carbohydr Chem Biochem 1982; 40:131–234.
51. Unger FM. The chemistry of oligosaccharide ligands of selectins. Adv Carbohydr Chem Biochem 2001; 57:207–435.
52. Mahmat U, Seydel U, Grimmecke D, Holst O, Rietschel ET. Lipopolysaccharides. In Pinto M, ed. Comprehensive Natural Products Chemistry:

Carbohydrates and their Derivatives, Including Tannins, Cellulose, and Related Lignins. Pergamon Elsevier, Oxford, 1999; Vol. 3, pp 179–239.
53. Morrison DC and Ryan JL, eds. Bacterial Endotoxic Lipopolysaccharides,Vol 1: Molecular Biochemistry and Cellular Biology: CRC Press, Boca Raton, FL, 1992; Vol. 1:449.
54. Arrhenius S. Z Phys Chem (Leipzig) 1889; 4:226–248.
55. Clarke MA, Edye LA, Eggleston G. Sucrose decomposition in aqueous solution. Adv Carbohydr Chem Biochem 1997; 52:441–470.
56. Queneau Y, Jarosz S, Lewandowski B, Fitremann J. Sucrose chemistry and applications of sucrochemicals. Adv Carbohydr Chem Biochem 2007; 61: 218–292.
57. Staněk J, Černý M, Pacák J. The Oligosaccharides. Czechoslovak Academy of Sciences, Prague, 1965.
58. Haworth WN, Machamer H. Polysaccharides X. Molecular structure of cellulose. J Chem Soc 1932; 2270–2277.
59. Jeanloz R, Forchielli E. Hyaluronic acid and related compounds III. Determination of the structure of chitin by periodate oxidation. Helv Chim Acta 1950; 33:1690–1697.
60. Meyer KH, Pankow GW. The constitution and structure of chitin. Helv Chim Acta 1935; 18:589–598.
61. Muzzarelli RAA. In Aspinall GO, ed. The Polysaccharides. Academic Press, New York, 1985; Vol. 3:417–450.
62. Ott E, Spurlin HM, Grafflin M. Cellulose and Cellulose Derivatives, 2nd edn. 1954; Parts I, II, and III, Interscience, New York, 1954.
63. Haworth WN, Heath RL, Peat S. Constitution of the starch synthesized by the agency of potato phosphorylase. J Chem Soc 1942; 55–58.
64. Schoch T. Fractionation of starch by selective precipitation with butanol. J Am Chem Soc 1942; 64:2957–2961.
65. Kerr RW, ed. Chemistry and Industry of Starch, 2nd edn. Academic Press, New York, 1950.
66. Radley JA, ed. Starch and Its Derivatives, 4th edn. Chapman & Hall, London, 1968.
67. Geddes R. Glycogen: a structural viewpoint. In Aspinall GO, ed. The Polysaccharides. Academic Press, New York, 1985; Vol. 3:283–336.
68. Whistler RL, ed. Industrial Gums: Polysaccharides and Their Derivatives, 2nd edn. Academic Press, New York, 1973.
69. Jeanloz RW. Mucopolysaccharides of higher animals. In: Pigman W, Horton D, eds. The Carbohydrates, Chemistry and Biochemistry, 2nd edn. Academic Press, New York, 1970; Vol. IIB: 589–625.
70. Meyer K, Palmer JW. The polysaccharide of the vitreous humor. J Biol Chem 1934; 107:629–634.
71. Meyer K, Palmer JW, Smyth EM. Glycoproteins. II. The polysaccharides of vitreous humor and of umbilical cord. J Biol Chem 1936; 114:689–703.
72. Weissman B, Meyer K. The structure of hyalobiuronic acid and of hyaluronic acid from umbilical cord. J Am Chem Soc 1954; 76:1753–1757.
73. Jeanloz RW. Forchielli. Hyaluronic acid and related substances. IV. Periodate oxidation. J Biol Chem 1951; 190:537–546.
74. Winter W. Studies on the quantitative composition of cartilage tissue. Biochem Z 1932; 246:10–28.
75. Meyer K, Rapport MM. The mucopolysaccharides of the ground substance of connective tissue. Science 1951; 113:596–599.

76. McLean J. The thromboplastic action of cephalin. Am J Physiol 1916; 41:250–257.
77. Howell WH. The purification of heparin and its presence in blood. Am J Physiol 1925; 71:553–562.
78. Jorpes JE. The chemistry of heparin. Biochem J 1935; 29:1817–1830.
79. Wolfrom ML, Weisblat DI, Karabinos JV, McNeely WH, McLean J. Chemical studies on crystalline barium acid heparinate. J Am Chem Soc 1943; 65:2077–2085.
80. Wolfrom ML, Rice FAH. Uronic acid component of heparin. J Am Chem Soc 1946; 68:532.
81. Wolfrom ML, Vercellotti JR, Horton D. Two disaccharides from carboxyl-reduced heparin. Linkage sequence in heparin. J Org Chem 1964; 29:540–547.
82. Wolfrom ML, Tomomatsu H, Szarek WA. Configuration of the glycosidic linkage of 2-amino-2-deoxy-D-glucopyranose to D-glucuronic acid in heparin. J Org Chem 1966; 31:1173–1178.
83. Perlin AS, Mazurek M, Jaques LB, Kavanagh LW. A proton magnetic resonance spectral study of heparin. L-Iduronic acid residues in commercial heparins. Carbohydr Res 1968; 7:369–379.
84. Wolfrom ML, Honda S, Wang PY. Isolation of L-iduronic acid from the crystalline barium acid salt of heparin. Carbohydr Res 1969; 10:259–265.
85. Perlin AS, Casu B, Sanderson GR, Johnson LF. 220 MHz spectra of heparin, chondroitins, and other mucopolysaccharides. Can J Chem 1970; 48:2260–2268.
86. Casu B. Structure and biological activity of heparin. Adv Carbohydr Chem Biochem 1985; 43:51–134.
87. Casu B, Lindahl U. Structure and biological interactions of heparin and heparan sulfate. Adv Carbohydr Chem Biochem 2001; 57:159–206.
88. Rama Krishna N, Agrawal PK. Molecular structure of the carbohydrate–protein linkage region fragments from connective tissue proteoglycans. Adv Carbohydr Chem Biochem 2000; 56:201–234.
89. Petitou M, van Boeckel CAA. A synthetic antithrombin III binding pentasaccharide is now a drug! What comes next? Angew Chem Int Ed 2004; 43:3118–3133.
90. Schatz A, Bugie E, Waksman SA. Streptomycin, a substance exhibiting antibiotic activity against gram-positive and gram-negative bacteria. Proc Soc Exptl Biol Med 1944; 55:66–69.
91. Hanessian S, Haskell TH. Antibiotics containing sugars. In: Pigman W, Horton D, eds. The Carbohydrates, Chemistry and Biochemistry, 2nd edn. Academic Press, 1970; Vol. IIA:139–211.
92. Lemieux RU, Wolfrom ML. The chemistry of streptomycin. Adv Carbohydr Chem 1948; 3:337–384.
93. Umezawa S. Aminoglycoside antibiotics. Adv Carbohydr Chem Biochem 1974; 30:111–182.
94. Umezawa H. Biochemical mechanism of resistance to aminoglycoside antibiotics. Adv Carbohydr Chem Biochem 1974; 30:183–225.
95. Wills B, Arya DP. An expanding view of aminoglycoside–nucleic acid recognition. Adv Carbohydr Chem Biochem 2006; 60:251–302.
96. Bates FJ and Associates. Polarimetry, saccharimetry, and the sugars. Circular C44 of the National Bureau of Standards, US Government Printing Office. 1942.
97. Wolfrom ML, Anno K. Sodium borohydride as a reducing agent in the sugar series. J Am Chem Soc 1952; 74:5583–5584.

98. Wolfrom ML, Thompson A. Reduction with sodium borohydride. Methods Carbohydr Chem 1963; 2:65–68.
99. Carlson D. Structures and immunochemical properties of oligosaccharides isolated from pig submaxillary mucins. J Biol Chem 1968; 243:616–626.
100. Isbell HS. Configuration of the pyranoses in relation to their properties and nomenclature. J Res Natl Bur Stand 1937; 18:505–534.
101. El Khadem HS, Horace S. Isbell. Adv Carbohydr Chem Biochem 1995; 51:1–13.
102. El Khadem HS, Frush HL. The collected papers of H.S. Isbell. Carbohydrate Division, American Chemical Society, 1988.
103. Meyer KH, Misch L. The constitution of the crystalline part of cellulose. VI. The positions of the atoms in the new space model of cellulose. Helv Chim Acta 1937; 20:232–244.
104. Cox EG, Jeffrey GA. Crystal structure of glucosamine hydrobromide. Nature 1939; 143:894–895.
105. Jeffrey GA, Sundaralingam M. Bibliography of crystal structures of carbohydrates, nucleosides, and nucleotides for 1979 and 1980: addenda and errata for 1970–1978; and index for 1935–1980. Adv Carbohydr Chem Biochem 1985; 43:203–421.
106. Cremer D, Pople JA. General definition of ring puckering coordinates. J Am Chem Soc 1975; 97:1354–1358.
107. Cambridge Crystallographic Data Bank: http://www.ccdc.cam.ac.uk/
108. Glycoscience database: http://www.glycosciences.de/index.php
109. Marchessault RH, Sarko A. X-Ray structure of polysaccharides. Adv Carbohyr Chem 1967; 22:421–482.
110. Rundle RE, French D. Configuration of starch and the starch–iodine complex. II. Optical properties of crystalline starch fractions. J Am Chem Soc 1943; 65:558–561.
111. Sundararajan PR, Marchessault RH. Bibliography of crystal structures of polysaccharides. Adv Carbohydr Chem Biochem 1976; 33:387–404; 1978; 35:377–385, 1979; 36:315–332, 1982; 40:381–399.
112. Perez S. Oligosaccharide and polysaccharide conformations by diffraction methods, In: Kamerling JP, Boons GJ, Lee YC, Suzuki A, Taniguchi N, Voragen AGJ, Comprehensive Glycoscience, Elsevier, Amsterdam, 2007, Vol. 2.
113. GLYCO3D. A site for glycosciences. http://www.cermav.cnrs.fr/glyco3d/.
114. Finan PA, Reed RI, Snedden W. Application of the mass spectrometer to carbohydrate chemistry. Chem Ind London 1958; 1172.
115. Kochetkov NK, Chizhov OS. Mass spectrometry of carbohydrate derivatives. Adv Carbohydr Chem 1966; 21:39–93.
116. Dell A. F.A.B.-Mass spectrometry of carbohydrates. Adv Carbohydr Chem Biochem 1987; 45:19–72.
117. Lemieux RU, Kullnig RK, Bernstein HJ, Schneider WG. Configuration effects on the proton magnetic resonance spectra of six-membered ring compounds. J Am Chem Soc 1958; 80:6098–6105.
118. Hall LD. Solutions to the hidden-resonance problem in proton nuclear magnetic resonance spectroscopy. Adv Carbohydr Chem Biochem 1974; 29:11–40.
119. Hall LD. High-resolution nuclear magnetic resonance spectroscopy. In: Pigman W, Horton D, eds. The Carbohydrates, Chemistry and Biochemistry. Academic Press, New York, 1980; Vol. IB:1299–1326.
120. Durette PL, Horton D. Conformational analysis of sugars and their derivatives. Adv Carbohydr Chem Biochem 1971; 26:49–125.

121. Coxon B, Sari N, Mulard LA, Kovac P, Pozsgay V, Glaudemans CPJ. Investigation by NMR spectroscopy and molecular modeling of the conformations of some modified disaccharide antigens for *Shigella dysenteriae* type 1. J Carbohydr Chem 1997; 16:927–946.
122. Pozsgay V, Nese S, Coxon B. Measurement of interglycosidic $^{3}J_{CH}$ coupling constants of selectively ^{13}C labeled oligosaccharides by 2D *J*-resolved ^{1}H NMR spectroscopy. Carbohydr Res 1998; 308:229–238.
123. Vliegenthart JFG, Dorland L, van Halbeek H. High-resolution ^{1}H nuclear magnetic resonance spectroscopy as a tool in the structural analysis of carbohydrates related to glycoproteins. Adv Carbohydr Chem Biochem 1983; 41:209–374.
124. Bock K, Pedersen C. Carbon-13 nuclear magnetic resonance of monosaccharides. Adv Carbohydr Chem Biochem 1983; 41:27–66.
125. Atalla RH, VanderHart DL. The role of solid-state carbon-13 NMR spectroscopy in studies of native celluloses. Solid State NMR 1999; 15:1–19.
126. Chen Y-Y, Luo S-Y, Hung S-C, Sunney I, Tzou D-LM. ^{13}C solid-state NMR chemical shift anisotropy analysis of the anomeric carbon in carbohydrates. Carbohydr Res 2005; 340:723–729.
127. Lehmann J. Carbohydrates: Structure and Biology, 2nd edn. Thieme, Stuttgart, 1998;xiv–274.
128. Fischer E. Einfluß der Konfiguration auf die Wirkung der Enzyme. I. Ber 1894; 27:2985–2993.
129. Hudson CS. The inversion of cane sugar by invertase. III. J Am Chem Soc 1909; 31:655–664.
130. Pigman WW. Action of almond emulsion on the phenyl glycosides of synthetic sugars and on β-thiophenyl-*d*-glucoside. J Res Natl Bur Stand 1941; 26:197–204.
131. Nisizawa K, Hashimoto Y. Glycoside hydrolases and glycosyl transferases. In Pigman W, Horton D, eds. The Carbohydrates, Chemistry and Biochemistry, 2nd edn. Academic Press, 1972; Vol. 2A:241–300.
132. Matheson NK, McCleary B. Enzymes metabolizing polysaccharides and their application to the analysis of structure and function of glycans. In Aspinall GO, ed. The Polysaccharides. Academic Press, New York, 1985; Vol. 3, 107–207.
133. Hehre EJ. A fresh understanding of the stereochemical behavior of glycosylases: structural distinction of "inverting" (2-MCO-type) versus "retaining" (1-MCO-type) enzymes. Adv Carbohydr Chem Biochem 1999; 55:265–310.
134. Liarison LL, Withers SG. Mechanistic analogies amongst carbohydrate modifying enzymes. Chem Comm 2004; 2243–2248.
135. Davies GJ, Gloster TM, Henrissat B. Recent structural insights into the expanding world of carbohydrate-active enzymes. Curr Opinion Struct Biol 2005; 15:637–645.
136. Hancock SM, Vaughan MD, Withers SG. Engineering of glycosidases and glycosyltransferases. Curr Opinion Struct Biol 2006; 10:509–515.
137. Caputto R, Leloir LF, Cardini E, Paladini AC. Isolation of the coenzyme of the galactose phosphate–glucose phosphate transformation. J Biol Chem 1950; 184:333–350.
138. Nikaido H, Hassid WZ. Biosynthesis of saccharides. Adv Carbohydr Chem Biochem 1971; 26:351–483.
139. Leloir LF, Cardini CE, Cabib E. Utilization of free energy for the biosynthesis of saccharides. In: Florkin M, Mason HS, eds. Comprehensive Biochemistry. Academic Press, New York, 1960; Vol. 2:97–138.

140. Frydman RB, Hassid WZ. Biosynthesis of sucrose with sugar cane leaf preparations. Nature 1963; 199:382–383.
141. Higashi Y, Strominger JL, Sweeley CC. Biosynthesis of the peptidoglycan of bacterial cell walls. XXI. Isolation of free C_{55}-isoprenoid alcohol and of lipid intermediates in peptidoglycan synthesis from *Staphylococcus aureus*. J Biol Chem 1970; 245–3697.
142. Behrens NH, Leloir LF. Dolichol monophosphate glucose: intermediate in glucose transfer in liver. Proc Nat Acad Sci USA 1970; 66:153–159.
143. Schwarz RT, Datema R. The lipid pathway of protein glycosylation and its inhibitors: the biological significance of protein-bound carbohydrates. Adv Carbohydr Chem Biochem 1982; 40:287–379.
144. Heidelberger M, Dilapi MM, Siegel M, Walter AW. Presence of antibodies in human subjects injected with pneumococcal polysaccharides. J Immunol 1950; 65:535–541.
145. Reeves RE, Goebel WF. Chemoimmunological studies on the soluble specific substance of pneumococcus. V. The structure of the type III polysaccharide. J Biol Chem 1941; 139:511–519.
146. Watkins W. Chemical structure, biosynthesis, and genetic regulation of carbohydrate antigens: retrospect and prospect. Pure Appl Chem 1991; 63:561–568.
147. Landsteiner K, Lampl H, Antigens. XII. The relationship between serological specificity and chemical structure (Preparation of antigens with specific groups of known chemical structure). Biochem Zeitschr 1918; 86:343–394.
148. Goebel WF, Avery OT. Chemo-immunological studies on conjugated carbohydrate-proteins: IV. The synthesis of the *p*-aminobenzyl ether of the soluble specific substance of type III pneumococcus and its coupling with protein. J Exp Med 1931; 54:431–436.
149. Schneerson R, Barrera O, Sutton A, Robbins JB. Preparation, characterization, and immunogenicity of *Haemophilus influenzae* type b polysaccharide–protein conjugates. J Exp Med 1980; 152:361–376.
150. Jennings HJ. Capsular polysaccharides as human vaccines. Adv Carbohydr Chem Biochem 1983; 43:155–208.
151. Pozsgay V. Oligosaccharide–protein conjugates. Adv Carbohydr Chem Biochem 2000; 56:153–199.
152. Jones C. Carbohydrates in Europe. 1998; 21:17–23.
153. Levine MM, Woodrow GC, Kaper JB, Cohon GS, eds. New Generation Vaccines. New York: Marcel Dekker, 1997.
154. Fraser-Reid BO, Tatsuta K, Thiem J, Côté GL, Flitsch S, Ito Y, Kondo H, Nishimura S, Yu B, eds. Glycoscience, Chemistry and Chemical Biology, 2nd edn. Springer, Berlin, 2008; 3000 pp.
155. Collins PM, ed. Dictionary of Carbohydrates, 2nd edn. Chapman & Hall/CRC, Boca Raton, FL, 2006; xxi+ 1282 pp.
156. Cammack R, Attwood TK, Campbell PN, Parish JH, Smith AD, Stirling JL, Vella F, eds. Oxford Dictionary of Biochemistry and Molecular Biology, 2nd edn. Oxford University Press, 2006; xv+ 720 pp.
157. Collins PM, Ferrier R. Monosaccharides: Their Chemistry and Their Roles in Natural Products. Wiley, Chichester, UK, 1995; xix+ 574 pp.
158. Stick RV. Carbohydrates: The Sweet Molecules of Life. 2001; Academic Press, New York, 2001; ix + 256.

Carbohydrate Chemistry, Biology and Medical Applications
Hari G. Garg, Mary K. Cowman and Charles A. Hales

DOI: 10.1016/B978-0-08-054816-6.00002-1

Chapter 2

Carbohydrate Structure Determination by Mass Spectrometry

BO XIE*,† AND CATHERINE E. COSTELLO*,‡

**Center for Biomedical Mass Spectrometry, Boston University School of Medicine, Boston, MA 02118, USA*
†Department of Chemistry, Boston University, Boston, MA 02215, USA
‡Departments of Biochemistry and Biophysics, Boston University School of Medicine, Boston, MA 02118, USA

I. Glycoproteins and Other Glycoconjugates

Glycoconjugates are biomolecules that contain one or more sugar residues linked to a noncarbohydrate moiety such as protein or lipid. They are ubiquitous components of extracellular matrices and cellular surfaces where their sugar moieties are implicated in a wide range of cell–cell and cell–matrix recognition events (1–3). Glycosylation is directed and regulated by specific enzymes (1). The glycosylation process is one of the four principal cotranslational and posttranslational modification (PTM) steps in the synthesis of membrane and secreted proteins, and the majority of proteins synthesized in the rough endoplasmic reticulum (RER) undergo glycosylation. Elaboration of the final structures takes place in the ER and Golgi. Glycolipids undergo similar processing steps in the Golgi. Glycosylation of proteins or lipids can, in many cases, significantly alter the chemical and physical properties, including solubility, shape, viscosity, and resistance to digestion by enzymes (e.g., proteases, ceramidases). Glycosylation also influences biological properties, such as site localization and turnover, and the addition of sugars greatly expands the ability of biomolecules to be recognized specifically by other molecules (2,4,5).

A. Glycans and Monosaccharides

The glycan chains of glycoconjugates are composed of monosaccharides. Because the number of oligosaccharide structures associated with a specific glycoconjugate may be highly variable, glycoproteins and glycolipids usually exist as complex mixtures of glycosylated variants (glycoforms). This heterogeneity significantly complicates the structural analysis of glycoconjugates. For example, proteins having the same sequence but different glycosylation patterns show different bands or spots on SDS-PAGE gels and additional efforts are thus necessary to identify the protein expression patterns.

Mannose, *N*-acetylglucosamine (GlcNAc), galactose, fucose, and sialic acid are the most common monosaccharides in the glycans of vertebrate glycoproteins. Figure 1 shows the structures and the symbols generally used in this field. These monosaccharides connect with one another to form the glycan chains.

■	○	△	●	◆	□
N-Acetyl-glucosamine (GlcNAc)	**Mannose (Man)**	**Fucose (Fuc)**	**Galactose (Gal)**	**Sialic Acid (NANA)**	**N-Acetyl-galactosamine (GalNAc)**

Figure 1 The most common monosaccharides in the glycans of glycoproteins with corresponding structures and symbols. Most species other than humans may have *N*-glycolyl as well as *N*-acetyl substitution on sialic acid. For sialic acid, R=$CHOHCHOHCH_2OH$.

B. Major Glycan Classes in Glycoproteins and Glycoconjugates

The majority of glycans from glycoproteins fall into two groups, depending on the linkage between the glycan and the protein: *N*-linked glycans and *O*-linked glycans, as shown in Fig. 2. In glycolipids, the glycans are generally *O*-linked to carbon.

1. Structural Features of N-linked Glycans

As shown in Fig. 2, *N*-linked glycans are attached to the amide nitrogens (*N*) of asparagine side chains, and the glycosylated residues are almost invariably found in the sequences Asn-Xxx-Ser (NXS) or Asn-Xxx-Thr (NXT), where Xxx (X) can be any amino acid except proline. In animal cells, the monosaccharide linked to an asparagine residue is inevitably GlcNAc, and the diverse *N*-linked glycans have a common core structure $Man_3GlcNAc_2$ attached to Asn but differ in the terminal elaborations that extend from this trimannosyl core, as shown in Fig. 3.

N-glycans in vertebrates exist as high-mannose, hybrid, or complex types (Fig. 3). The high-mannose type is defined as glycans with only mannose residues as terminal elaboration, based on the core structure. Hybrid glycans contain both

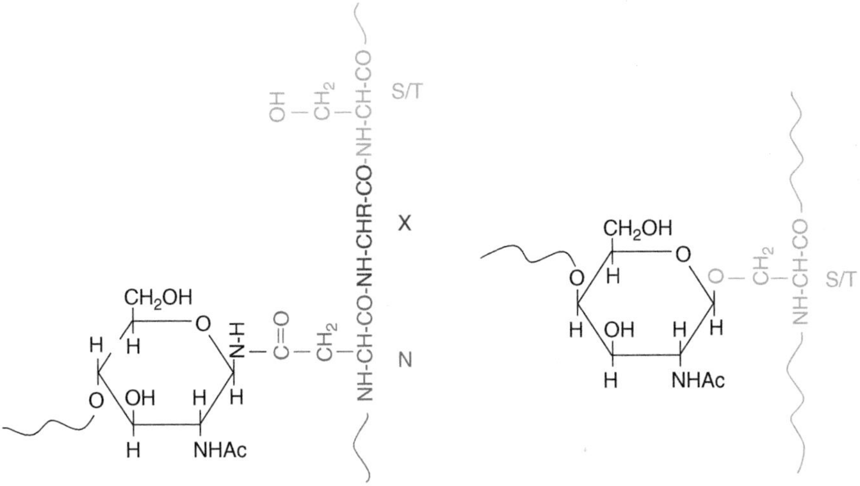

Figure 2 *N*-linked (left) and *O*-linked (right) glycosylation of glycoproteins. Adapted from figures available on www.ionsource.com/Card/carbo/nolink.htm.

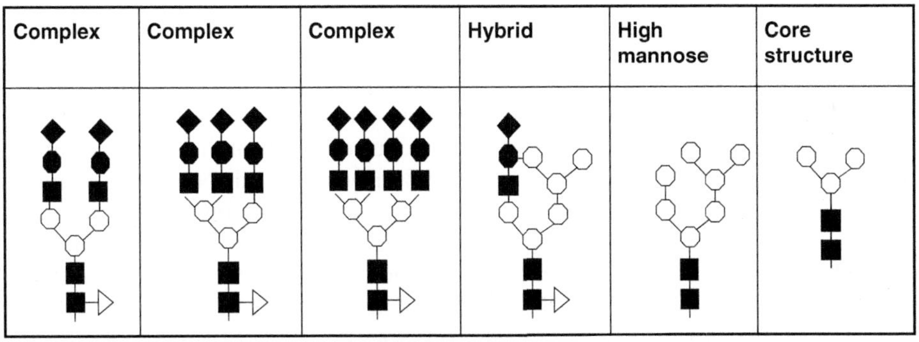

Complex	Complex	Complex	Hybrid	High mannose	Core structure

Figure 3 The diversity of common *N*-linked glycan structures and core structure subunits.

unsubstituted and substituted mannose residues. In complex glycans, a GlcNAc residue is added to the trimannosyl core structure as the first residue in each of the antennae.

2. Biosynthetic Pathways of N-linked Glycans

The biosynthetic pathway of *N*-linked glycans has been defined by the extensive research efforts of multiple groups, and summaries of the present understanding of the process are now readily available. The pathway for biosynthesis of *N*-linked glycans includes three major steps.

a. The synthesis of the dolichol oligosaccharide precursor. The dolichol oligosaccharide precursor consists of dolichol lipid Dol (Fig. 4) carrying a pyrophosphate linkage to an oligosaccharide chain that consists of 14 monosaccharide residues. The assembly process for this precursor starts at the cytosolic side of the RER; the two GlcNAc residues are linked to the Dol-P lipid by GlcNAc-1-phosphotransferase and then by a GlcNAc-transferase; five mannose residues are added successively by GDP-Man as donor. This $Man_5GlcNAc_2$-P-P-Dol is transferred to the luminal side of the RER, where additional processing occurs. Then another four mannoses are added by Dol-P-Man as donor; finally, three glucose residues are added by Dol-P-Glc as donor. After assembly of the dolichol lipid-linked oligosaccharide precursor is completed, this precursor is ready for transfer to an asparagine residue as the protein translation occurs.

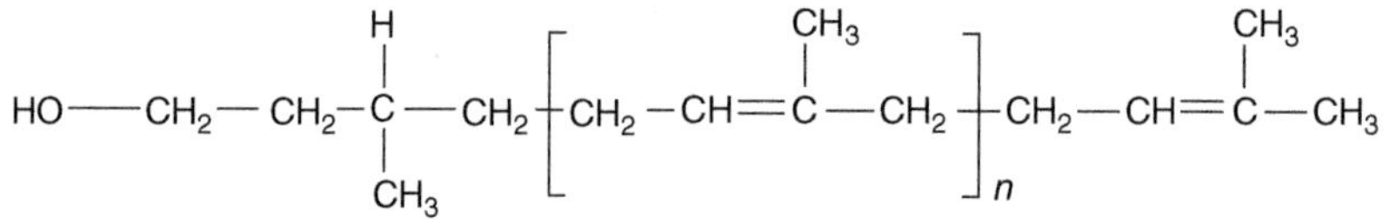

$$HO-CH_2-CH_2-\underset{CH_3}{\overset{H}{C}}-CH_2-\left[CH_2-CH=\overset{CH_3}{C}-CH_2\right]_n-CH_2-CH=\overset{CH_3}{C}-CH_3$$

Figure 4 The structure of Dolichol.

b. Transportation of the precursor and following trimming. Oligosaccharyltransferase (OST), a multisubunit protein complex in the ER membrane of eukaryotes, carries out this transportation. In the process, the OST complex binds to the lipid-linked oligosaccharide and transfers it to a nascently translated protein by releasing cleaved Dol-P-P. Although the consensus sequence (NXS or NXT, where X can be any amino acid except proline) is required for protein *N*-glycosylation, not all potential glycosylation sites are necessarily occupied. Next, the terminal Glc residues are sequentially removed by two membrane-bound glucosidases, α-1,2-glucosidase I and α-1,3-glucosidase II. After all the Glc residues are removed, one mannose on the Man (α-1,6) branch is cleaved by α-1,2-mannosidase (ER). The glycoprotein with the linked $Man_9GlcNAc_2$ (high-mannose type) side chain is then transported from the RER to the *cis*-cisternae of the Golgi.

c. Final trimming of glycan. Diversification of *N*-glycans occurs in this step and depends on the action of glycosyltransferases, as the glycoprotein transits in the sequential Golgi compartments. Extracellular *N*-glycans in vertebrates exist as high-mannose, hybrid, or complex glycans according to the results of the different trimming pathways.

For example, for the synthesis of a complex glycan, *N*-acetylglucosaminyltransferases I, II, and IV are responsible for adding GlcNAc residues to the terminal Man residues of different branches. Then galactosyltransferase enables Gal residues to link to the terminal GlcNAc residues, followed by the sialidase-dependent addition of sialic acid residues as termini of the antennae. Core fucosylation is very common in *N*-glycans, except for the high-mannose type. In many vertebrates, *N*-glycans may have a core fucose (α-1,6 linkage) attached to the reducing end GlcNAc residue by the action of a fucosyltransferase. The structures of complex glycans are determined by the order of action of the transferases. A bisecting GlcNAc residue may be introduced by *N*-acetylglucosaminyltransferase III, prior to the action of mannosidase II.

The complex mechanism of glycan synthesis is a factor that contributes to the difficulty of studying glycan structures, but recent advances in mass spectrometry (MS) technologies have made it possible to determine a growing number of glycan structures experimentally.

3. O-Linked Glycans

The *O*-linked glycans of glycoproteins are usually attached to a serine or threonine residue through the side chain oxygen atom by starting with a GalNAc residue, but there is no simple consensus amino acid sequence for *O*-linked glycosylation and there are eight or more oligosaccharide core structures (6). The biosynthesis of *O*-linked glycans is also dependent on the presence of essential glycosyltransferases, transporters, and monosaccharide residues. In glycolipids, assembly occurs according to well-defined pathways (see www.lipidmaps.org) and the glycan is *O*-linked; for glycosphingolipids, the attachment site is the 1-hydroxyl of the long-chain base. Structures such as lipid A may contain multiple and varied acyl groups esterified to the carbohydrate hydroxyl position; the specific locations and details of the acyl groups strongly affect the activity of the lipid A molecule and so must be carefully defined. Other glycoconjugates, such as saponins, contain diverse glycans that may be linked to several positions of the aglycon. For such classes of glycoconjugates, multistage tandem MS enables full determinations of each of the structures that usually occur as complex mixtures of closely related species. Not all of these categories of important and interesting molecules can be discussed here; the reader is encouraged to seek out classic and recent publications.

C. Diverse Biological Roles of Glycans

Glycans play key roles in protein folding, biological lifetime, recognition of binding pattern, cancer metastasis, signaling, and the immune system (7). For example, surface carbohydrates provide the interface between the cell and its environment, and serve to define self versus nonself. Many pathogens recognize particular cell surface carbohydrates, and structural studies have recently led to progress in this field (4). Glycans can serve as intermediates in generating energy, as signaling molecules, or as structural components. For example, the structural roles of glycans become particularly important in constructing complex multicellular organs and organisms, through steps which require interactions of cells with one another and surrounding matrix (8).

D. Species-Dependent Glycosylation

Protein glycosylation depends on the overall protein conformation, the effect of local conformation, and the glycosylation-processing enzymes for the particular cell type (2). It is known that the glycan profile is protein-specific, site-specific, and tissue- or cell-specific. Besides, glycosylation is species-specific and also varies with growth, development, and disease.

II. Instrumentation for MS Applied to Carbohydrate Structural Determinations

MS is an analytical technique used to find the composition of a chemical or biological sample by generating a spectrum representing the mass-to-charge (m/z) ratios of sample components. The technique has many applications that include (a) identifying unknown compounds by the masses of the compound molecules or their fragments, (b) determining the isotopic compositions of elements in a compound, (c) determining the structure of a compound by observing its fragmentation, (d) quantifying the amount of a compound in a sample, (e) studying the fundamentals of gas phase ion chemistry, (f) determining other physical, chemical, or even biological properties of compounds.

Throughout its various applications, all MS is based on the measurement of the m/z ratios of charged particles in a vacuum. In order to measure the m/z ratio of samples, all mass spectrometers contain three main components: ionization source, mass analyzer, and ion detector.

The ionization source is the mechanical device that allows ionization to occur. Matrix-assisted laser desorption/ionization (MALDI) and electrospray ionization (ESI) are now the most common ionization sources for biomolecular MS, because both of them are soft ionization techniques; that is, the sample ionization process generates few or no fragments even for large biomolecules, such as proteins and oligosaccharides, so that the intact molecular ions can be easily observed.

The mass analyzer is used to separate sample ions. Commonly used analyzers include time-of-flight (TOF), quadrupole (Q), quadrupole ion trap (QIT), and Fourier-transform ion cyclotron resonance (FT-ICR) (9–14). These analyzers provide a wide mass range, high accuracy, and resolution for biomolecular analysis.

In MS, resolution is the ability of a mass spectrometer to distinguish ions with slightly different m/z ratios. The resolution can be measured from a single peak based on the following equation:

$$R = \frac{M}{\Delta M} \tag{1}$$

where M corresponds to m/z, ΔM is the peak width at half maximum of peak height. The narrower the peak is, the higher the resolution. Resolution calculated in this way is called full width at half maximum (FWHM). For example, a peak with m/z 800 and full width at half maximum 0.5 Da has a resolution of 1600.

The detectors are used to generate a signal current from the incident ions to record the m/z ratios and abundances of the ions. A variety of methods are used to detect ions depending on the type of mass spectrometer, such as channel electron multipliers for quadrupoles and ion traps, and microchannel plate for TOF mass spectrometers, inductive detector for Fourier-transform mass spectrometry (FTMS).

A. "Soft" Ionization Processes: ESI and MALDI

1. Electrospray Ionization

John Fenn, who shared the Nobel Prize in Chemistry in 2002, developed the methodology for ESI MS of large and fragile polar biomolecules (15,16). Note that a simultaneous development occurred in Russia (17), although it was not well known at the time. ESI is an atmospheric pressure ionization technique now widely used for analysis of peptides, proteins, carbohydrates, and so on. It produces gaseous ionized molecules directly from a liquid solution (Fig. 5). During ESI, the sample solution continuously flows through a charged metal capillary, to which a potential difference (usually 3–5 kV) is applied to the tip and the surrounding atmospheric region. This potential difference can disperse the solution into a fine spray of charged droplets. At the same time, a flow of dry inert gas is introduced to the droplets at atmospheric pressure, resulting in the evaporation of solvent from each droplet, leaving the charged macromolecules as gas phase ions. Samples are usually prepared in a solution that is much more volatile than the analyte. The derived gaseous ions are subsequently introduced into the vacuum of the mass analyzer for

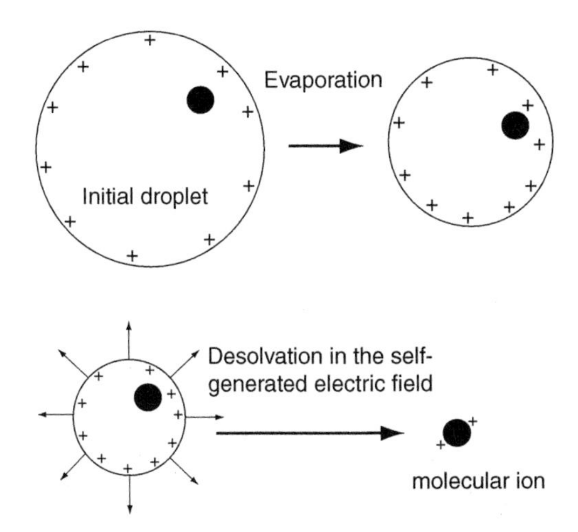

Figure 5 Illustration of ion formation in an electrospray ion source (16). When the size of the charged droplet decreases due to evaporation of the solvent, the charge density on its surface increases. Then strong mutual coulombic repulsions cause ions to be ejected from the droplet until the droplet is small enough that the surface charge density field can desorb ions from the droplet into the ambient gas. The molecular ions desorb with noncovalently associated solvent. After the ions form at atmospheric pressure, they are electrostatically directed into the mass analyzer.

separation and detection. As ESI is a continuous-flow operation, it can be easily interfaced with high-performance liquid chromatography (HPLC) or capillary electrophoresis.

ESI generates multiply charged ions from larger molecules, and this feature makes it possible to observe very large molecules (extending to the megadalton range), even with a mass analyzer having a limited mass range, because the mass spectrometer measures *m/z*. However, ESI has little tolerance for salts or detergents. These contaminants can form adducts with the analytes and/or compete with them during ionization, thus reducing sensitivity. Therefore, sample purification is very important before the sample is subjected to ESI. In addition, the presence of multiply charged ions can make the spectrum complex to interpret, especially in mixture analysis.

2. Matrix-Assisted Laser Desorption/Ionization

MALDI was introduced by Hillenkamp and Karas (18), and has become a widespread technique for peptides, proteins, and most other biomolecules. Tanaka introduced a procedure that used a dispersion of fine metal particles to desorb proteins, but this method did not prove to be practical (19). For the MALDI ionization method, analyte is mixed with matrix, and then deposited onto a metal target; the droplet is allowed to dry. Then the plate is introduced into a vacuum ion source, and an intense, high energy laser beam is used to irradiate the sample spot to induce ionization of intact analyte via efficient energy transfer through matrix-assisted laser-induced desorption, as illustrated in Fig. 6 (20,21). The high ion yields of MALDI provide subpicomole sensitivity for the measurement of biological compounds, so it is ideal for low abundance sample analysis. The mechanisms underlying the MALDI mechanism(s) are still under investigation,

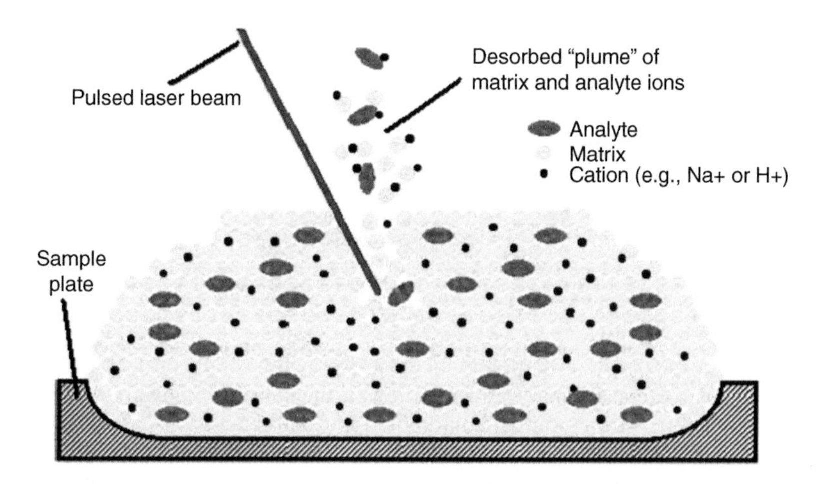

Figure 6 Schematic diagram of ion formation from a MALDI source. Adapted from figures available at www.udel.edu/chem/bahnson/chem645/websites/Lloyd/MALDI.jpg.

but it is generally accepted that the laser energy is first absorbed by the matrix molecules and then transferred to the analytes; this results in their desorption/vaporization from the condensed phase to form gas phase ions.

In contrast to ESI, usually one or a few protons (or cations, anions) can be attached to (or removed from) the analyte molecules in MALDI, to form singly or low-charge state multiply charged ions, so the mass spectra are more straightforward to interpret. MALDI is thus suitable for the analysis of complex mixtures. The usual mass range for MALDI-MS is up to 300,000 Da, but it can be used for even larger analytes. Compared with ESI, it can tolerate relatively high concentrations of contaminants (salts, lipids, detergents, or polymers), even when present in millimolar concentration.

a. Matrices and their applications. The matrix plays a very critical role in the efficient ionization of analytes by MALDI-MS because it absorbs the laser radiation and facilitates the desorption/ionization process (22). Usually the matrix is a nonvolatile solid material which should have the following properties: sufficient light absorption at the laser wavelength used (easy to ionize), good solubility in the commonly used solvents (easy to prepare), adequate stability during laser irradiation, and, most often, an acid functional group (H^+ donor). Commonly used matrices are 2,5-dihydroxybenzoic acid (DHB), α-cyano-4-hydroxycinnamic acid (CHCA), and sinapinic acid (SA). CHCA is the matrix most often chosen for peptide or small protein analysis (molecular mass less than 15,000 Da). It generates very intense signals from proteins and peptides; compared with other matrices, it is called a "hot" matrix because it deposits excess energy on the analyte. DHB is a "mild" matrix compared to CHCA, therefore it is suitable for the studies of glycans and proteins or peptides that bear labile PTMs, such as glycosylation or phosphorylation, because the modifying groups are not cleaved from the protein/peptide backbone during ionization.

B. Mass Analyzers

1. Time-of-Flight

The TOF mass analyzer determines the mass of biomolecular samples according to the time required for the ions to travel from the ion source to the detector. It is the most simple and robust mass analyzer, and has practically unlimited mass range. It has a pulsed mode of operation. Because MALDI generates ions in short, nanosecond pulses, TOF analysis works very well with MALDI. TOF analyzers have also been applied to ESI and gas chromatography electron ionization mass spectrometry (GC-MS).

In the TOF mass analyzer, the accelerating potential applies the same amount of energy to the entire ion packet generated after laser irradiation, and then the energy is transformed into kinetic energy. The mass, charge, and kinetic energy of the ions each affect the time of arrival at the detector. An additional constant value contributes to the total flight time; this constant depends on the laser pulse width and desorption time. Ions with different m/z ratios have different velocities, and thus reach the detector at different times; the ions with low *m/z* have the highest velocity and thus the shortest flight time.

TOF analyzers include both linear and reflectron ion optics as shown in Fig. 7, with the TOF reflectron being widely used now. An electrostatic mirror (reflectron) is placed at the linear detector end. This ion mirror consists of a series of parallel plates, to which successively higher voltages are applied to reverse the flight direction of the ions. Ions with slightly higher velocities penetrate deeper into the ion mirror, while ions with slightly lower velocities penetrate less, and thus their flight times are compensated and they ultimately reach the reflectron detector at the same time and generate a spectrum that has higher resolution. Most advanced MALDI-MS instruments can be operated in both linear and reflectron modes.

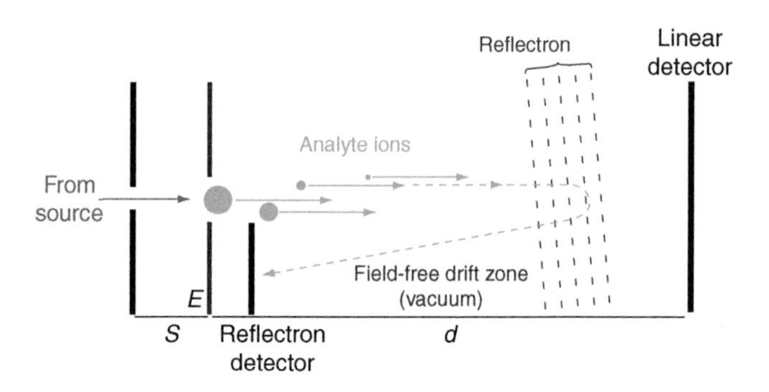

Figure 7 The scheme of TOF mass analyzer in linear and reflectron detection mode. Adapted from figures available at www.chm.bris.ac.uk/ms/theory/tof-massspec.html.

2. Quadrupoles and Quadrupole Ion Traps

The quadrupole is the most common type of mass analyzer today. As shown in Fig. 8, the linear quadrupole is composed of four parallel metal rods. Each opposing rod pair is connected together electrically, and a radio frequency (*rf*) voltage is applied between one pair of rods and the other. A direct current voltage is then superimposed on the *rf* voltage. Analyte ions travel down the quadrupole mass filter in between the rods. Only ions of a certain *m/z* will reach the detector within a given ratio of voltages: other ions have unstable trajectories and will collide with the rods. This allows selection of a particular ion or scanning by varying the voltages. The quadrupolar fields can also be generated in a cubic or rectangular geometry, in which case the mass analyzer is called a QIT.

Quadrupole mass analyzers are usually coupled with an ESI source for the following reasons: tolerance for high pressure (5×10^{-5} Torr), which makes them work well with atmospheric pressure ESI sources; mass range extending to *m/z* 4000, suitable for observing the usual charge distribution in ESI (*m/z* 1000–3500); and a relatively accessible price.

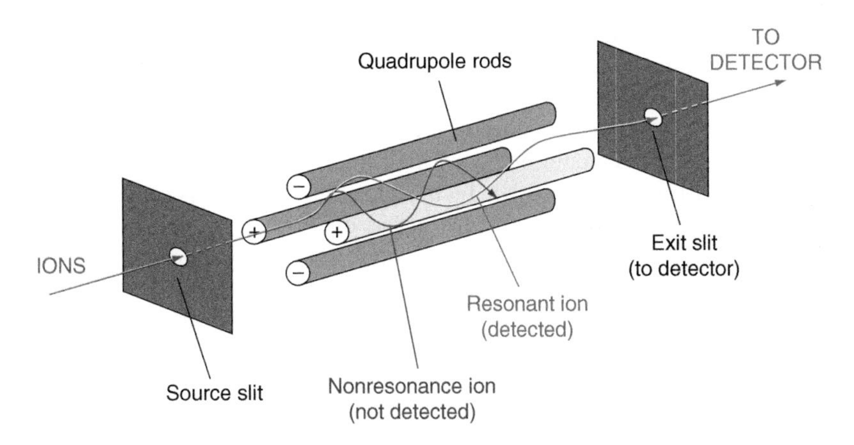

Figure 8 Scheme of a quadrupole mass analyzer. Adapted from figures available at www.bris.ac.uk/nerclsmsf/techniques/gcms.html.

3. Fourier-Transform Mass Spectrometry (FT-ICR, FT-Orbitrap)

The FT-ICR mass analyzer determines the *m/z* values of ions based on monitoring of the cyclotron frequency of the ions in a uniform magnetic field. After the charged ions pass from the ionization source into the magnetic field, all ions in the analyzer cell are excited into a higher cyclotron orbit by a pulsed radio frequency (*rf*) (Fig. 9). The image currents (the flow of electrons in the external

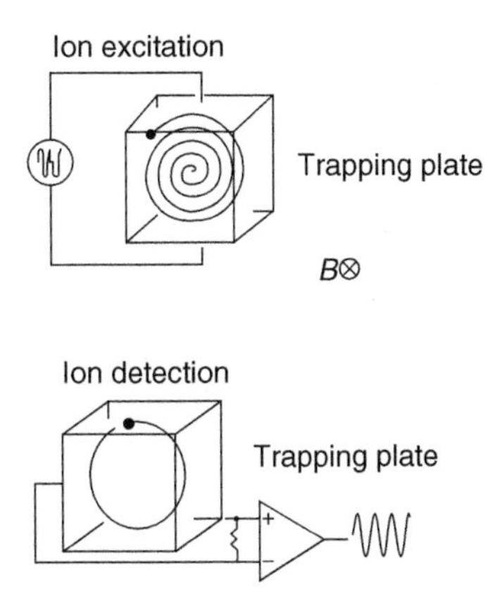

Figure 9 Illustration of ion excitation to a cyclotron orbit and subsequent image current detection in an FT-ICR MS analyzer cell (23). In the upper drawing, B represents the magnetic field directed into the plane and a pulsed radio frequency (*rf*) is applied to the excitation plates. Ions that are in resonance with the excitation frequency gain kinetic energy and spiral outwards from the center of the cell into a larger cyclotron orbit. In the lower drawing, ions continue to undergo cyclotron motion with a large radius orbit. An image signal is produced on the detection plates that are connected to the amplifier.

circuit) generated by all of the charged ions are measured and Fourier transformed to obtain the frequencies of all the ions, which correspond to their *m/z* values.

Compared with other mass analyzers, in FT-ICR MS, *m/z* values for ionized biomolecules can be resolved and detected with high accuracy (<1–5 ppm) because the masses are calculated from high accuracy frequency measurements, but not in space or time as with other techniques. FT-ICR MS is often used to help determine the composition of molecules based on the highly accurate mass measurements. In addition, FT-ICR MS has high sensitivity and high resolution. A very high vacuum is required to allow ion transient lifetimes sufficiently long to achieve very high resolution. FT-ICR MS is particularly valuable for analysis of complex mixtures because its high resolution (narrow peak width) allows signals having very similar *m/z* to be detected as distinct ions.

A second type of Fourier-transform mass analyzer that uses the principle of orbital trapping in electrostatic fields (24) is now commercially available (25). The Orbitrap™ is configured together with a QIT MS and offers high performance in terms of resolution (typically 50–60 K) and mass accuracy (low ppm), together with more convenience of operation and lower cost than the FT-ICR MS systems. Although this mass analyzer has some limitations in terms of available range of fragmentation modes and ultimate performance when compared to the FT-ICR MS, it has significant advantages for many types of laboratories and is rapidly coming into wide use.

C. Key MS Systems

1. *MALDI- and ESI-TOF MS*

As mentioned above, the TOF mass analyzer works well with the MALDI ion source because both of them operate in the pulsed mode. A great advantage of MALDI-TOF MS is that the soft ionization process causes little or no fragmentation of analytes, allowing the intact biomolecules to be identified; this property is especially important for analysis of mixtures. In addition, MALDI-TOF MS analysis is sensitive, being able to detect molecules at the low femtomole level and even below. A spectrum that is the sum of many laser shots can be generated within seconds from each of the hundreds of spots where a mixture of sample and matrix have been dried on the MALDI target. Hence, the majority of high throughput proteomics, genomics, and glycomics analyses are carried out by MALDI-TOF MS. MALDI-TOF/TOF MS instruments also provide fragment ion information that can provide sequence information for biopolymers.

Recently, simple yet high-performance ESI-TOF MS systems that have mass accuracy below 10 ppm have become commercially available. Although they do not have MS/MS capability, these are very useful for accurate mass determination of single analytes and mixture profiling.

2. *ESI- and MALDI-Q-o-TOF MS and MS/MS*

ESI-Q-o-TOF (or ESI-Qq-TOF) MS, where Q refers to a mass-resolving quadrupole, q refers to an *rf*-only quadrupole or hexapole collision cell, o indicates the orthogonal orientation of the components, and TOF refers to a TOF mass analyzer (26), is presently one of the most powerful and robust instruments. It combines the high resolution and mass accuracy of the TOF reflectron with the widely used ESI ionization source for operation in MS and MS/MS modes (Fig. 10). MALDI ion sources are also available for some of these instruments.

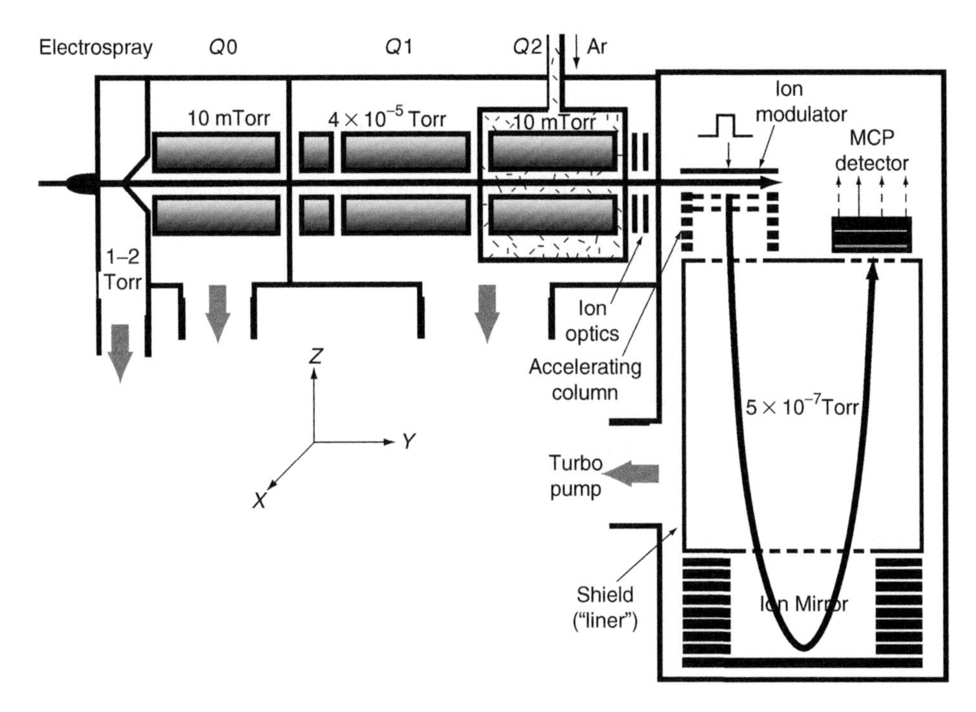

Figure 10 Schematic diagram of the ESI tandem Q-o-TOF MS (26).

In this type of mass spectrometer, as shown in Fig. 5, ions are generated with multiple charges during ESI at atmospheric pressure. While in the single MS mode, all three quadrupoles serves as ion guides, and the TOF is used to record the mass spectra. In order to perform tandem mass analysis, the three quadrupoles Q0, Q1, and Q2 have separate functions: Q0 provides collisional damping because it is used for collisional cooling and focusing of the ions entering the mass analyzer. Q1 is used to scan across a preset *m/z* range or, when acting as a mass filter, to select an ion of interest, then the ion is accelerated to an energy between 20 and 200 eV before it enters Q2. Q2 is a collision cell and transmits the ions while also introducing a collision gas (argon or nitrogen) into the flight path of the selected ion to cause collision-induced dissociation (CID). The resulting fragment (product) ions as well as any surviving parent (precursor) ions are collisionally cooled and focused. After cooling and focusing, the ions are reaccelerated and are focused by the ion optics into a parallel beam. Following their original direction, the ions in this beam continuously enter the ion modulator in the TOF. Then the ions are pushed by a pulsed electric field in a direction orthogonal to their original trajectory into the accelerating column. Finally, the ions enter the field-free drift region, where the TOF provides *m/z* analysis for intact precursors and fragment ions generated in the collision cell Q2.

3. ESI- and MALDI-FT MS/MS

After the development of ESI, the coupling of ESI and FTMS became an important analytical technique. As noted above, ESI produces multiply-charged ions. This not only expands the mass range accessible for analysis of the biomolecules but also takes the advantage of that fact that the multiply-charged ions are easily fragmented in MS/MS experiments because the electrical repulsion from the multiple charge sites lowers the barrier for collisionally activated decomposition. Figure 11 is the diagram of a custom made ESI-qQq-FT-ICR instrument (27) used for the studies illustrated here, where qQq refers to three quadrupoles with different functions.

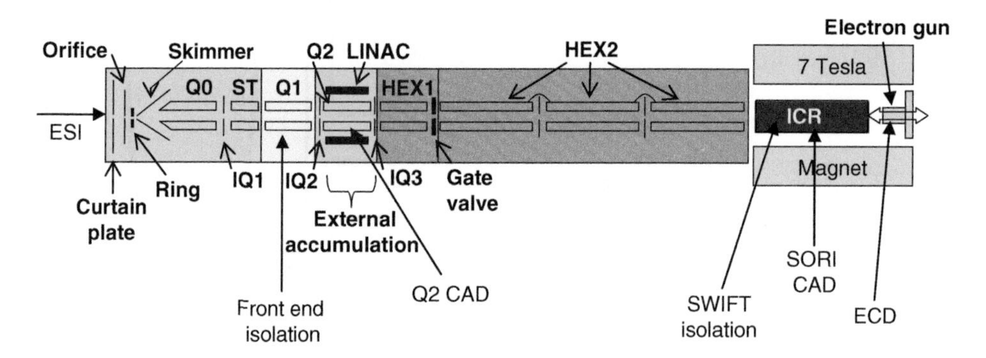

Figure 11 Schematic representation of the home-built ESI-qQq-FTMS with a 7-Tesla actively shielded magnet, used to generate some of the data shown in Fig. 17 (27). Q0, *rf* only quadrupole; Q1, resolving quadrupole; Q2, collision cell; HEX1 and HEX2, *rf*-only transfer hexapoles; ICR, ion cyclotron resonance cell.

This novel custom ESI-qQq-FT-ICR instrument allows the performance of several types of MS/MS experiments including Q2 collisionally activated dissociation (Q2 CAD), electron capture dissociation (ECD), and infrared multiphoton dissociation (IRMPD).

For MALDI-FT-ICR MS experiment, vibrational cooling has been introduced in order to assure that the analyte ions will survive long enough to reach the cell without cleavage of labile linkages, such as the easily broken glycosidic bond of sialic acid residues. A pulse of gas is injected in the region immediately surrounding the sample spot, just prior to the laser pulse. Interaction with the gas allows excess vibrational energy to be dissipated from the desorbed ions, and metastable decomposition is thus prevented (28,29). This approach can also be used for glycan analysis on orthogonal TOF instruments (30). As demonstrated in these papers, vibrational cooling is particularly useful for the direct analysis of glycoconjugates separated by thin layer chromatography.

D. Dissociation Methods

1. Collision-Induced Dissociation

In CID, also referred to as CAD, ions separated by the first stage of mass analysis are subjected to decomposition by interaction with a gas (usually helium, argon, or nitrogen) and the products are separated by a second stage of mass analysis. In some instruments, this process may be repeated multiple times, in what are called MS^n experiments, n being the number of stages of precursor ion generation/selection and product ion separation. In the qQq-FT MS instrument, ions of interest can be isolated during transmission through Q1, then the selected precursor ions are fragmented in collision cell Q2 by controlling the collision energy (typically 15–35 eV) as in a Qq-TOF MS. Then the product ions are accumulated in Q3 before being injected into the FT-ICR analyzer cell. For a peptide, Q2 CAD generates primarily b and y fragment ions that result from cleavages of the amide bonds; for an oligosaccharide, b and y ions generated by cleavages at the glycosidic linkages usually have the highest abundance.

Alternatively, the dissociation process may take place within the FT-ICR cell. For highest efficiency, the ions of interest are subjected to sustained off-resonance irradiation immediately prior to the gas collisions (SORI-CAD).

2. Electron Capture Dissociation

ECD is a method used to fragment ions in FT MS/MS by allowing them to capture low-energy electrons emitted by a heated dispenser cathode (31). Ions generated in the ESI ion source are injected into the FTMS analyzer cell, where they stay close to the center axis of the cell and oscillate rapidly back and forth between the two trapping plates. A low-energy (1–10 eV) electron beam with a narrow energy bandwidth (<1 eV) is directed into the ICR cell and irradiates the trapped ions.

The detailed mechanisms of ECD are not thoroughly understood. It has been proposed that a positive ion captures a low-energy electron and forms a reactive hydrogen atom radical. The intermediate produced from the ion–electron recombination reaction has high excess internal energy. Fragmentation occurs rapidly, usually before the energy is able to redistribute through the molecule. For peptides, ECD produces extensive c- and z-type fragmentation ions that result from the cleavage at the N–Cα bond.

ECD has a big advantage in that PTMs such as phosphorylation and glycosylation are retained in the fragments, while these PTMs are usually lost in CAD. Because these PTMs generally form weak covalent bonds with the amino acid side chains, the bonds are labile. At higher ECD energies ("hot" ECD), fragmentation of the carbohydrate can be induced.

A functionally similar dissociation method, electron transfer dissociation (ETD) was more recently reported (32), specifically for use on QIT instruments, although it is also beginning to be adapted to other mass analyzers. ETD is accomplished by electron transfer to the analyte from a negatively charged species that is produced in a chemical ionization source and directed into the region where the analyte ions are trapped. For peptides and proteins, it produces spectra that strongly resemble ECD spectra. Although no reports of ETD on glycans have yet been published, it seems likely that this method will also be applicable for carbohydrate analyses.

3. *Photodissociation*

An additional option for generation of fragments from glycan precursors is dissociation with photons, usually generated from a CO_2 laser at 10.6 μm (33). Irradiation of the trapped ions results in multiphoton absorption and dissociation of all species, both primary and secondary, yielding spectra that are rich in fragment ions (34).

III. Glycan Structural Determination by MS

A. Introduction

Compared with the sequencing of proteins and nucleic acids, analyzing the covalent structures of glycans attached to glycoproteins (glycan sequencing) is a more challenging task because of the branched nature in the structures and the extensive heterogeneity of the glycans. Methods available for determining the covalent structures of oligosaccharides include the use of enzymes, lectins, NMR, and MS (7,35–37).

MS is an important tool for the structural analysis of carbohydrates and offers precise results, analytical versatility, and very high sensitivity. Whereas MS analysis options for proteins and peptides are well defined relative to those for carbohydrates, tandem MS product–ion patterns are more complex and the pathways are still being elucidated.

B. Release of N-Linked Glycans from Glycoproteins

To analyze the pool of glycans, the glycans are usually released from the polypeptide either by enzyme digestion or by chemical degradation followed by MS.

1. *Enzymatic Digestion Methods*

The two enzymes usually used to digest and release the *N*-linked glycans from glycoproteins are peptide-*N*4-(acetyl-β-glucosaminyl)-asparagine amidase (PNGase F) (Fig. 12) and endoglycosidase H (endo H) (Fig. 13). PNGase F cleaves between the innermost GlcNAc and the asparagine residues to release high-mannose, hybrid, and complex oligosaccharides from *N*-linked glycoproteins (38,39). It leaves the

Figure 12 PNGase F hydrolyzes nearly all types of *N*-glycan chains from glycopeptides/proteins. x=H or sugar(s).

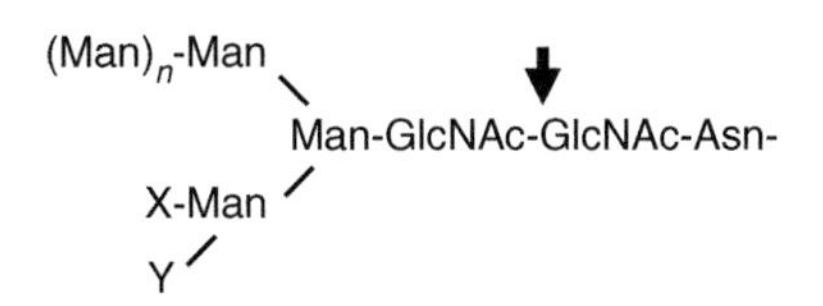

Figure 13 Endo H cleaves only high-mannose structures (n=2–150, x=(Man)1–2, y=H) and hybrid structures (e.g., n=2, x and/or y=NeuAc-Gal-GlcNAc).

oligosaccharide intact but deamidates the asparagine (*N*) residue to aspartic acid (D). However, PNGase F will not remove oligosaccharides containing α(1,3)-linked core fucose commonly found on plant glycoproteins; PNGase A can be used for this purpose.

Endo H is commercially available as a recombinant glycosidase that cleaves within the chitobiose core of high-mannose and some hybrid oligosaccharides from N-linked glycoproteins as shown in Fig. 13 (38).

PNGase F and Endo H have different specificities for *N*-linked glycans, hence they can be used to provide information about the type of glycans present on glycoproteins. For detailed characterization of glycans and the protein moieties of glycoproteins by MS, PNGase F is the most widely used glycosidase; it releases most glycans to yield the intact glycan and the unglycosylated protein (with Asn converted to Asp at the former glycosylation site).

C. Profiling of Expressed Glycans

Profiling of expressed glycans released from glycoproteins includes the analysis of glycan composition and sequence, the linkages between monosaccharides, and the distribution of glycoforms.

1. *Composition Analysis*

Exoglycosidases cleave specific terminal monosaccharides and are used to sequentially degrade the terminal residues from glycans. Use of exoglycosidases of known specificity can achieve structural analysis of individual glycans (40,41). However, the information about the correlation between monosaccharides and their specific attachment points will be lost. Exoglycosidases are only practical for determinations of oligosaccharides with known structures because these approaches rely on the known specificities of the enzymes. Thus, exoglycosidases are limited to analysis of structures that resemble those already characterized by other approaches.

MS provides measurement of the molecular weight for glycans. The saccharide composition for glycans released from glycoproteins (or glycolipids) can be determined based on the accurate mass, the structures of known *N*- or *O*-linked glycans, and the core structure, as well as the masses of the monosaccharides. In addition, tandem MS can confirm the proposed composition and provide further details of the structure.

The use of MALDI as the ionization method for determination of molecular weights in oligosaccharide analysis has advantages over ESI, particularly for applications that involve the profiling of mixtures. Quantification of derivatized glycan mixtures (especially for permethylated glycans) with MALDI has been demonstrated to be reliable and reproducible, and the variations for those analyses are very small compared with those obtained for the same oligosaccharide mixtures after derivatization with a chromophore and chromatographic quantitation (42). MALDI-TOF MS is thus very useful for profiling glycan distributions, for example, on samples obtained from biological sources or from batches of recombinant proteins.

In a recent investigation of the receptor for FimH *E. coli* that are responsible for urinary tract infections in humans and other mammals, we determined the glycosylation patterns of Uroplakin Ia (UPIa) and Ib, two glycoproteins that are major components of the inner surface of the bladder (43). Each has a single potential *N*-linked glycosylation site.

Figure 14 presents the MALDI-TOF MS profiles obtained for the oligosaccharides of murine UPIa and Ib, after release by PNGase F and permethylation. The results indicated that the glycans of UPIa have high-mannose structures, whereas those of UPIb are primarily complex. Determination of the molecular weight distributions does not, however, provide conclusive information on the isomer content of the glycans. The specific structures were assigned on the basis of tandem MS, as discussed briefly below and more fully in Ref. (43). The preference of the bacteria for UPIa could be directly related to the presence of the high-mannose oligosaccharides.

2. Sequence and Linkage Analysis

Tandem MS methods such as CID of glycans result in the observation of ions that correspond to cleavage within the oligosaccharide portion of the molecule. The nomenclature for oligosaccharide fragmentation described by Domon and Costello and now used throughout the MS field is shown in Fig. 15 (44). Fragment ions that contain a nonreducing terminus are labeled with upper case letters from the beginning of the alphabet (A, B, C), and those that contain the reducing end of the oligosaccharide or the aglycon are labeled with X, Y, Z; subscripts indicate the order of the cleavage sites. The A and X ions are produced by cleavage across a glycosidic ring (cross-ring cleavages), and are labeled by assigning each ring bond a number, counting clockwise and beginning with 0 for the O–C_1 bond. Cleaved bonds are indicated by a superscript preceding the letter designation. Ions produced from cleavage of successive residues are labeled: A_m, B_m, and C_m, with $m=1$ for the nonreducing end and X_n, Y_n, and Z_n, with $n=1$ for the reducing end residue. B, C, Y, and Z fragments are ions produced via glycosidic cleavages and provide composition and sequence information of glycans. (Note that the Y_0 and Z_0 fragments result from cleavage of the glycosidic bonds adjacent to the aglycon and do not contain a sugar residue.) A and X ions, which arise from cross-ring cleavage, supply information not only about composition and sequence but also about the linkage position.

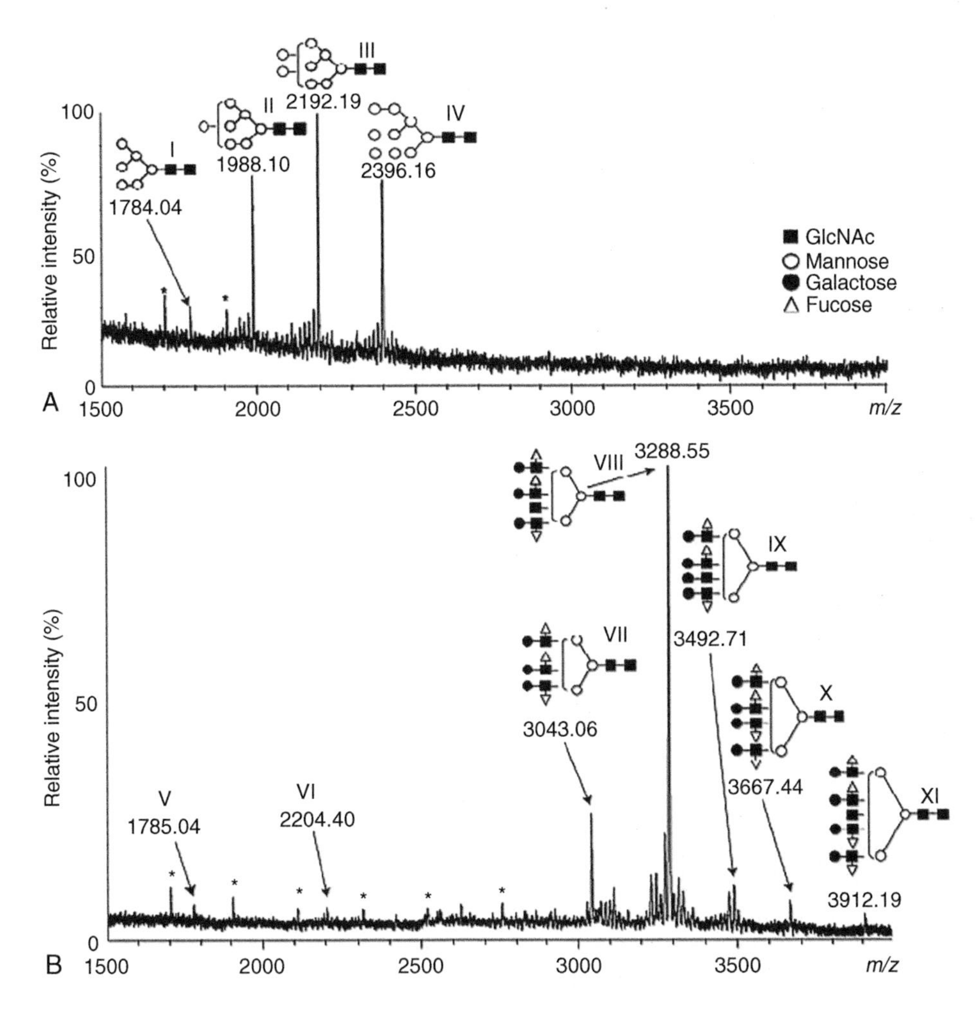

Figure 14 MALDI-TOF MS of glycans permethylated after PNGase F release from murine (A) UPIa and (B) UPIb. Structures were assigned on the basis of tandem mass spectrometry (43).

The confident assignment of glycan structures depends on the determination of the sequence and branching structure. For example, for the released and permethylated UP Ia glycans whose MALDI-TOF MS profile is shown in Fig. 14A, the assignments of the observed $[M+Na]^+$ peaks to individual high-mannose structures could be made from subsequent ESI MS/MS analyses. For the component with $[M+2Na]^{2+}$ *m/z* 1209.6 in the ESI mass spectrum (corresponding to $[M+Na]^+$ *m/z* 2419.16 in the MALDI-TOF mass spectrum shown in Fig. 14A),

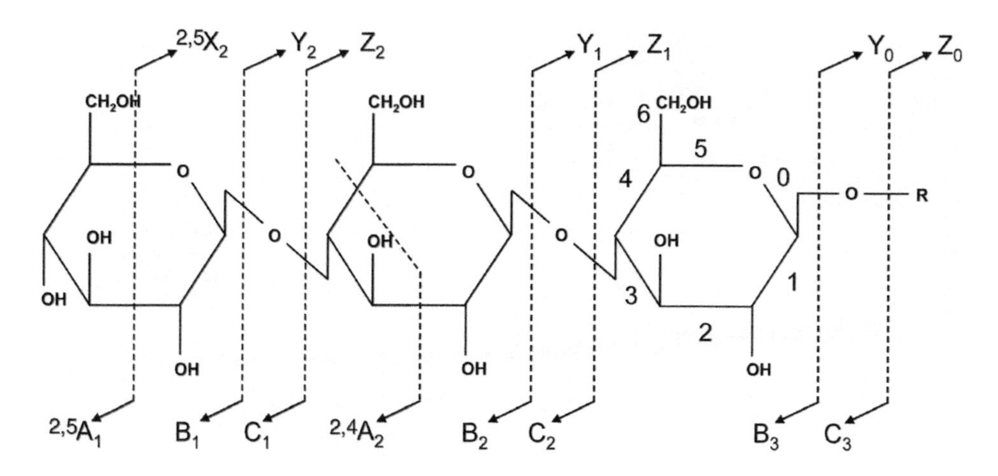

Figure 15 Nomenclature for glycan product ions generated by tandem MS. Reproduced from Domon Costello (44) with their permission.

the CID spectrum and the assigned structure are shown in Fig. 16. As indicated on the fragmentation scheme, A-type cross-ring cleavages provided the information necessary to ascertain the linkage and branching (43).

In the course of studying less well-known structures, it is valuable to be able to determine very accurately the masses of the components, so that unambiguous assignments can be made. During a recent investigation of the multimerization of P0 proteins, glycosylated single-pass transmembrane proteins associated with peripheral neuropathies which are major components of the myelin of peripheral nerves, it was noted that the glycoform distribution of the dimeric P0 from *Xenopus* differed from that of other tetrameric P0s whose glycan structures had already been reported in the literature.

We employed tandem MS on an FTMS to obtain data that would enable confident assignment of all the glycans (45). Figure 17 shows the CAD spectrum obtained for one of the hybrid oligosaccharides released from *Xenopus* P0 and permethylated. As indicated in the scheme, the MS/MS spectrum definitively indicates the location of the arm containing the complex sugars, as well as their sequence. The inset illustrates the high resolution at which the spectrum was recorded.

3. Derivatization Methods

Two procedures are most commonly used for the conversion of native glycans to hydrophobic derivatives, to improve their signal strengths regardless of the ionization technique used: peracetylation (46) and permethylation (47,48). Both methods can be achieved with high yields (49). Hydrolysis of large glycans and conversion of the released monosaccharides to volatile derivatives through acetylation and methylation enables their analysis by GC-MS. The combination of retention index and electron ionization fragmentation pattern is unique for each saccharide. GC-MS is thus used to define monosaccharide composition and perform linkage analysis.

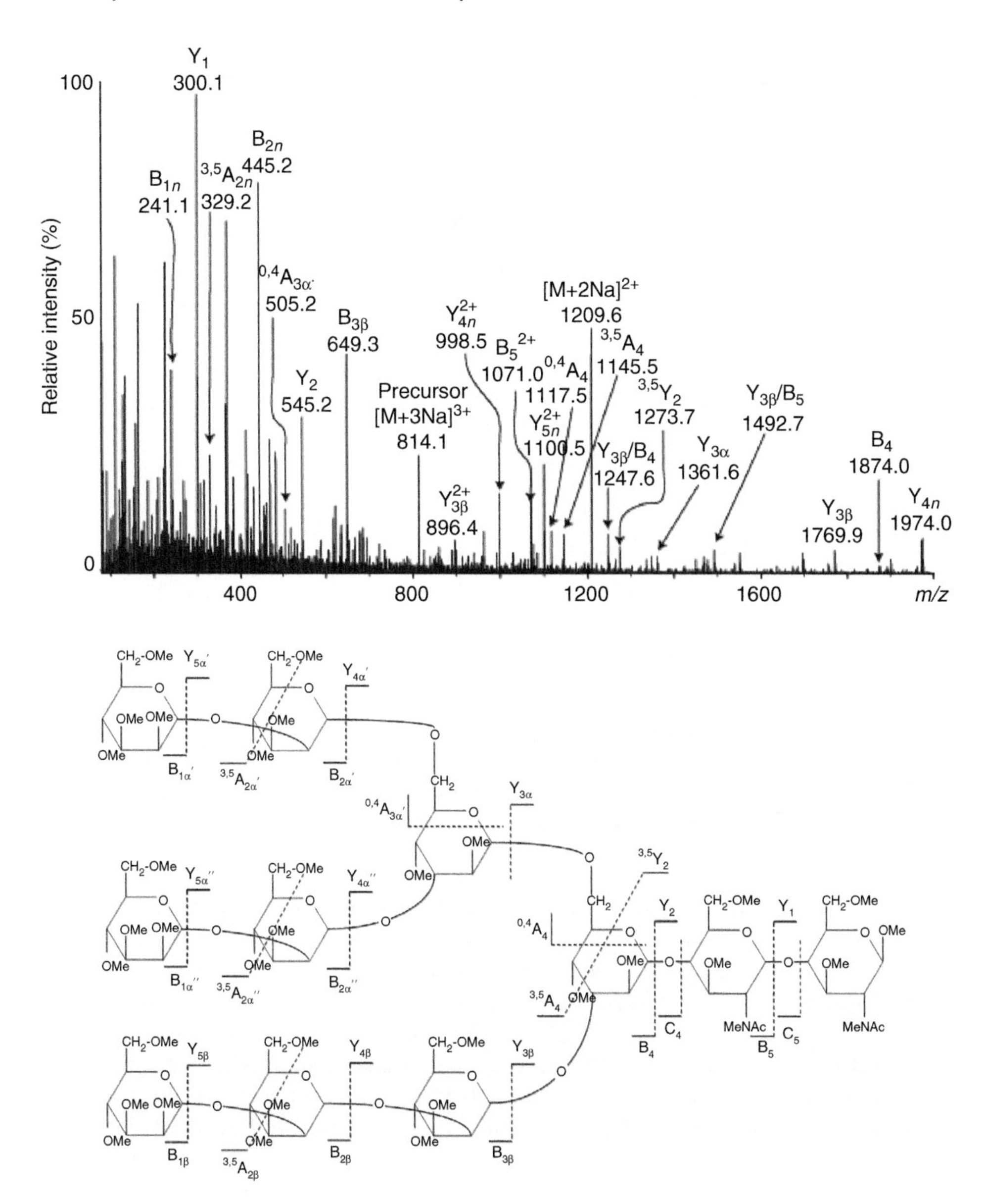

Figure 16 ESI-Q-o-TOF mass spectrum of $[M+2Na]^{2+}$ m/z 1209.6 from the mixture of the glycans from murine Uroplakin Ia, obtained after release by PGNase F and permethylation (43). Scheme indicates assignments of the product ions.

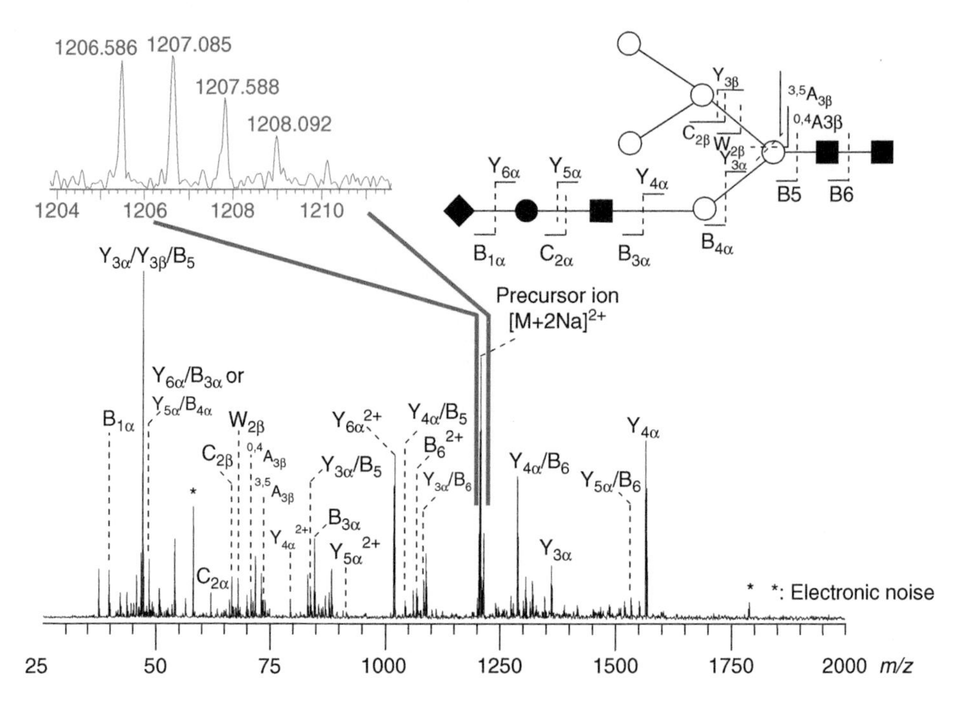

Figure 17 Nanospray tandem mass spectrum, obtained with CAD in Q1 of a qQq FTMS, for a hybrid glycan from *Xenopus* P0, after release by PNGase F and permethylation (45). Inset shows expansion of isotopic cluster in the molecular ion region of the precursor, $[M+2Na]^{2+}$ m/z 1206.586. Reproduced with permission from (45).

For the profiling and structural determination of intact glycans, permethylation has become the preferred method because it produces a smaller mass increase compared with peracetylation, and the products have greater volatility (50–52). Permethylated glycans are most often ionized and detected as $[M+Na]^+$ ions in the positive mode. Tandem MS (CID) fragmentation produces reducing and nonreducing terminal product ions with approximately equal abundances, with preferential fragmentation to the reducing side of HexNAc residues (53). In addition, tandem MS analysis of permethylated *N*-linked oligosaccharides is facilitated by permethylation of labile groups such as sialic acid (NeuAc) and fucose (Fuc) residues.

An additional type of derivative that is often used for glycan profiling and analysis is the modification of the reducing end with a chromophore, usually achieved by the formation of a Schiff base with an aromatic amine, and subsequent reduction with sodium borohydride to stabilize the initially formed product by its conversion to a saturated amine. This procedure was initially developed for HPLC analysis with ultraviolet detection, but it is also appropriate for MS analysis because the substituted amino group represents a site for charge localization that simplifies the fragmentation pattern and also provides the opportunity to introduce a stable isotope label for quantitative purposes (54).

Formation of the aminobenzamide derivative facilitated the structural determination of a monosialylated biantennary glycan found to be present on the serum transferrin of a patient with a new type of congenital disorder of glycosylation, CDG IIh. In this case, the patient sample and appropriate standards were analyzed by negative ion nanospray tandem MS on a quadrupole orthogonal TOF instrument, and the single sialic acid residue could be clearly located at the terminus of the 6-arm (55).

4. MS Analysis of Native and Released Glycopeptides

For glycoproteins with multiple *N*-glycosylation sites, proteases can be used to digest the glycoprotein into glycopeptides for further chromatography and MS analysis, in order to obtain site-specific information about the glycan profile and peptides. The choice of enzyme will be sample-dependent. For glycoproteins observed to produce tryptic peptides unfavorable for analysis (e.g., too small, too large, or exhibiting incomplete cleavage), a complementary protease such as endoproteinase Asp-N or a nonspecific protease such as proteinase K can be used.

Glycopeptides have high mass, hydrophilicity, and limited surface activities in liquid matrices (56). The development of ESI has resulted in a significantly better glycopeptide signal response and has been widely applied, and is particularly successful when used for LC/MS^n analysis. For example, this approach has been used to provide detailed insight into site-specific distribution of *N*-linked glycoforms on immunoglobulin A (57).

Tandem MS, such as CID of *N*-linked glycopeptides, is dominated by the formation of ions that correspond to fragmentation within the glycan (58–60). Shown in Fig. 18 is the CID profile obtained from an *N*-linked glycopeptide with a high-mannose glycan ($Man_9GlcNAc_2$) (61). Although the glycan is the same as that for which data is shown in Fig. 16, the fragmentation pattern obtained from the native glycopeptide is not nearly so rich as that of the released and permethylated glycan. Nevertheless, substantial information about the glycan structure is available, as well as partial sequence information on the attached peptide. Because HexNAc residues are not present in the antennae, fragmentation of the glycan portion of the molecule required a collision energy high enough to cause some peptide-backbone fragmentation, as shown in the panel A.

MALDI-TOF MS is also very useful for analysis of glycoproteins, particularly when an infrared (2.94 μm) laser is employed (62). For instance, for α_1-acid glycoprotein with heterogeneous glycan profiles, the MALDI-TOF mass spectrum shows a broader peak than would be observed for a typical unmodified protein in the same mass range; this broadness reflects the glycoform heterogeneity.

5. Profiling and Quantitative Analysis of Glycans

As mass-spectrometry-based glycan analysis has become well-established, it has begun to be used for the profiling of glycans and glycoconjugates. A recent interlaboratory comparison showed that the MALDI-TOF MS and/or ESI MS/MS methods employed for profiling of *N*-linked glycans, from a common sample analyzed at 20 different sites, produced results that are quite comparable (63).

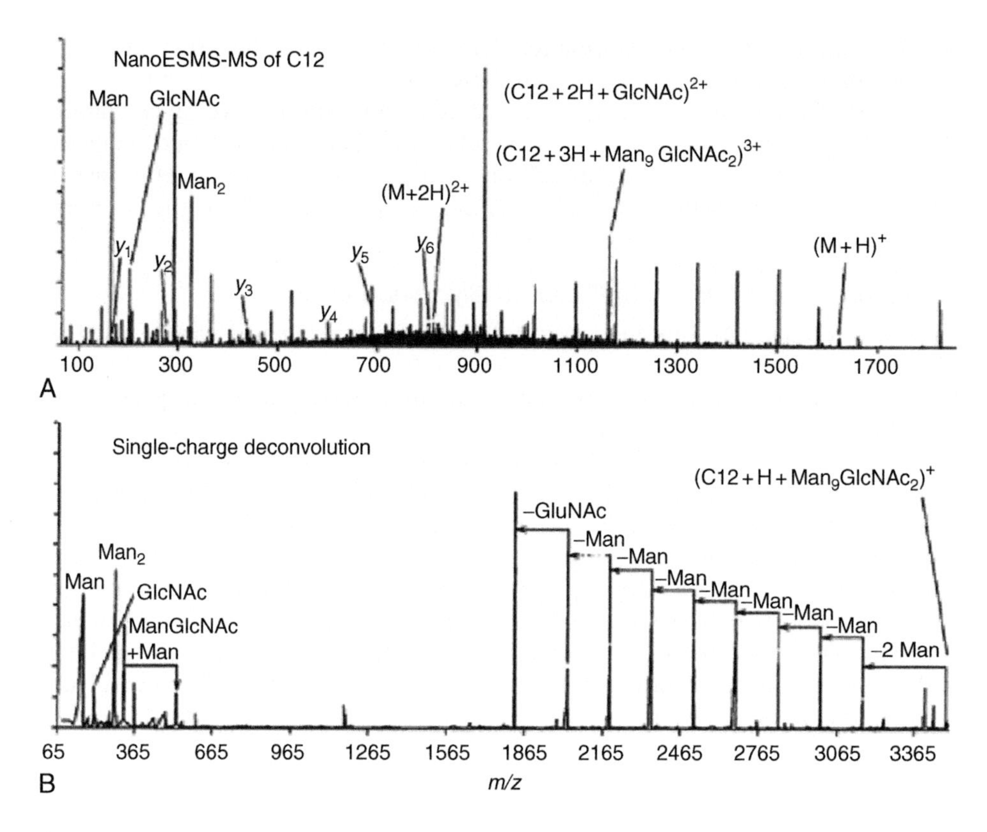

Figure 18 Nano-ESI MS/MS analysis of a triply protonated $Man_9GlcNAc_2$ *N*-linked glycopeptide. (A) The mass spectrum with peptide y_n ions indicated and (B) the singly charged deconvoluted mass spectrum of panel A where the sugar-loss ion series is indicated (61).

Analysis of glycans, even glycosaminoglycans that are extensively modified with labile groups such as sulfate, can yield highly accurate results, when careful attention is paid to each procedural step (64). Interestingly, the high-resolution MS profiling of glycoconjugates in urine can be used as a means to diagnose disease (65).

IV. Summary

Since the first glycoprotein primary structure was determined successfully by MS about 30 years ago, structural glycobiology has undergone rapid expansion because of the application of modern ultrahigh-sensitivity MS methods capable of low-femtomole MS/MS analysis, which provide detailed sequence and site-of-attachment data for glycoproteins. The methods and strategies being developed are compatible with the ubiquitous natural microheterogeneity, allowing characterization of even very complex minor components.

An important current goal of MS in carbohydrate analysis is the development and application of sensitive multiple MS approaches to the structural analysis of glycoproteins, glycolipids, and free oligosaccharides related with diseases, using only the limited sample amounts available from biological sources. The detailed glycan profiles and identification of attachment sites for glycoproteins and other glycoconjugates achieved with ultrahigh-sensitivity MS strategies can be used to investigate glycan structural biology with respect to biological function and disease states.

References

1. Drickamer K, Taylor ME. Evolving views of protein glycosylation. Trends Biochem Sci 1998; 23:321–324.
2. Dwek RA. Glycobiology: "towards understanding the function of sugars." Biochem Soc Trans 1995; 23:1–25.
3. Dwek RA. Glycobiology: more functions for oligosaccharides. Science 1995; 269:1234–1235.
4. Dell A, Morris HR. Glycoprotein structure determination by mass spectrometry. Science 2001; 291:2351–2356.
5. Fukuda M. The biology of glycoconjugates. Glycoconj J 1993; 10:127–130.
6. Lis H, Sharon N. Protein glycosylation—structural and functional-aspects. Eur J Biochem 1993; 218:1–27.
7. Rudd PM, Dwek RA. Rapid, sensitive sequencing of oligosaccharides from glycoproteins. Curr Opin Biotechnol 1997; 8:488–497.
8. Varki A, Cummings R, Esko J, Freeze H, Hart G, Marth J, eds. Essentials of Glycobiology. Cold Spring Harbor: Cold Spring Harbor Laboratory Press, 1999.
9. Jonscher KR, Yates JR. The quadrupole ion trap mass spectrometer—a small solution to a big challenge. Anal Biochem 1997; 244:1–15.
10. Dass C. Principles and Practice of Biological Mass Spectrometry. New York: John Wiley & Sons, Inc., 2001.
11. Cotter RJ, Russell DH. Time-of-flight mass spectrometry: instrumentation and applications in biological research. J Am Soc Mass Spectrom 1998; 9:1104–1105.
12. Sauter AD, Betowski LD, Ballard JM. Comparison of priority pollutant response factors for triple and single quadrupole mass spectrometers. Anal Chem 1983; 55:116–119.
13. Marshall AG, Hendrickson CL, Jackson GS. Fourier transform ion cyclotron resonance mass spectrometry: a primer. Mass Spectrom Rev 1998; 17:1–35.
14. March RE, Hughes RJ, Todd JFJ. Quadrupole Storage Mass Spectrometry. New York: Wiley, 1989.
15. Fenn JB. Electrospray wings for molecular elephants (Nobel lecture). Angew Chem Int Ed Engl 2003; 42:3871–3894.
16. Fenn JB, Mann M, Meng CK, Wong SF, Whitehouse CM. Electrospray ionization for mass spectrometry of large biomolecules. Science 1989; 246:64–71.
17. Aleksandrov ML, Gall LM, Krasnov MV, Nikolaev VI, Shkurov VA. Metod mass-spectrometricheskogo analiza trudnoletuchix termicheski nestabilnyx vezcestv osnovannyi na extrakcii ionov iz rastvora pri atmosfernom davlenii

(Method of mass spectrometric analysis of non-volatile thermally unstable compounds based on ion extraction from solution under atmospheric pressure). Zhur Anal Khimii 1985; 40:160–172.
18. Karas M, Hillenkamp F. Laser desorption ionization of proteins with molecular masses exceeding 10000 Daltons. Anal Chem 1988; 60:2299–2301.
19. Tanaka K, Waki H, Ido Y, Akita S, Yoshida Y, Yoshida T, Matsuo T. Protein and polymer analyses up to *m/z* 100,000 by laser ionization time-of-flight mass spectrometry. Rapid Commun Mass Spectrom 1988; 2:151–153.
20. Knochenmuss R, Dubois F, Dale MJ, Zenobi R. The matrix suppression effect and ionization mechanisms in matrix-assisted laser desorption/ionization. Rapid Commun Mass Spectrom 1996; 10:871–877.
21. Zenobi R, Knochenmuss R. Ion formation in MALDI mass spectrometry. Mass Spectrom Rev 1998; 17:337–366.
22. Fitzgerald MC, Parr GR, Smith LM. Basic matrices for the matrix-assisted laser-desorption ionization mass-spectrometry of proteins and oligonucleotides. Anal Chem 1993; 65:3204–3211.
23. Amster IJ. Fourier transform mass spectrometry. J Mass Spectrom 1996; 31:1325–1337.
24. Makarov A. Electrostatic axially harmonic orbital trapping: a high-performance technique of mass analysis. Anal Chem 2000; 72:1156–1162.
25. Makarov A, Denisov E, Kholomeev A, Balschun W, Lange O, Strupat K, Horning S. Performance evaluation of a hybrid linear ion trap/orbitrap mass spectrometer. Anal Chem 2006; 78:2113–2120.
26. Chernushevich IV, Loboda AV, Thomson BA. An introduction to quadrupole-time-of-flight mass spectrometry. J Mass Spectrom 2001; 36:849–865.
27. Jebanathirajah JA, Pittman JL, Thomson BA, Budnik BA, Kaur P, Rape M, Kirschner M, Costello CE, O'Connor PB. Characterization of a new qQq-FTICR mass spectrometer for post-translational modification analysis and top-down tandem mass spectrometry of whole proteins. J Am Soc Mass Spectrom 2005; 16:1985–1999.
28. O'Connor PB, Mirgorodskaya E, Costello CE. High pressure matrix-assisted laser desorption/ionization Fourier transform mass spectrometry for minimization of ganglioside fragmentation. J Am Soc Mass Spectrom 2002; 13:402–407.
29. Ivleva VB, Elkin YN, Budnik BA, Moyer SC, O'Connor PB, Costello CE. Coupling thin layer chromatography with vibrational cooling matrix-assisted laser desorption/ionization Fourier transform mass spectrometry for the analysis of ganglioside mixtures. Anal Chem 2004; 76:6484–6491.
30. Ivleva VB, Sapp LM, O'Connor PB, Costello CE. Ganglioside analysis by TLC-MALDI-oTOF MS. J Am Soc Mass Spectrom 2005; 16:1552–1560.
31. Zubarev RA, Horn DM, Fridriksson EK, Kelleher NL, Kruger NA, Lewis MA, Carpenter BK, McLafferty FW. Electron capture dissociation for structural characterization of multiply charged protein cations. Anal Chem 2000; 72:563–573.
32. Syka JEP, Coon JJ, Schroeder MJ, Shabanowitz J, Hunt DF. Peptide and protein sequence analysis by electron transfer dissociation mass spectrometry. Proc Natl Acad Sci USA 2004; 101:9528–9533.

33. Little DP, Speir JP, Senko MW, O'Connor PB, McLafferty FW. Infrared multiphoton dissociation of large multiply charged ions for biomolecule sequencing. Anal Chem 1994; 66:2809–2815.
34. Hakansson K, Cooper HJ, Emmett MR, Costello CE, Marshall AG, Nilsson CL. Electron capture dissociation and infrared multiphoton dissociation MS/MS of an N-glycosylated tryptic peptide to yield complementary sequence information. Anal Chem 2001; 73:4530–4536.
35. Morris HR. Biomolecular structure determination by mass spectrometry. Nature 1980; 286:447–452.
36. Dwek RA, Edge CJ, Harvey DJ, Wormald MR, Parekh RB. Analysis of glycoprotein-associated oligosaccharides. Annu Rev Biochem 1993; 62:65–100.
37. Zaia J. Mass spectrometry of oligosaccharides. Mass Spectrom Rev 2004; 23:161–227.
38. Maley F, Trimble RB, Tarentino AL, Plummer TH. Characterization of glycoproteins and their sssociated oligosaccharides through the use of endoglycosidases. Anal Biochem 1989; 180:195–204.
39. Trimble RB, Tarentino AL. Identification of distinct endoglycosidase (Endo) activities in Flavobacterium-Meningosepticum—Endo-F1, Endo-F2, and Endo-F3—Endo-F1 and Endo-H hydrolyze only high mannose and hybrid glycans. J Biol Chem 1991; 266:1646–1651.
40. Abelson JN, Simon MI, Lennarz WJ, Hart GW. Guide to Techniques in Glycobiology. New York: Academic Press, 1994.
41. Rudd PM, Guile GR, Kuster B, Harvey DJ, Opdenakker G, Dwek RA. Oligosaccharide sequencing technology. Nature 1997; 388:205–207.
42. Viseux N, Hronowski X, Delaney J, Domon B. Qualitative and quantitative analysis of the glycosylation pattern of recombinant proteins. Anal Chem 2001; 73:4755–4762.
43. Xie B, Zhou G, Chan S-Y, Shapiro E, Kong X-P, Wu X-R, Sun TT, Costello CE. Distinct glycan structures of Uroplakins Ia and Ib: structural basis for the selective binding of FimH adhesin to Uroplakin Ia. J Biol Chem 2006; 281:14644–14653.
44. Domon B, Costello CE. A systematic nomenclature for carbohydrate fragmentations in FAB-MS/MS spectra of glycoconjugates. Glycoconjugate J 1988; 5:397–409.
45. Xie B, Luo X, Zhao C, Priest CM, Chan S-Y, O'Connor PB, Kirschner DA, Costello CE. Molecular characterization of myelin protein zero in Xenopus laevis peripheral nerve: equilibrium between non-covalently associated dimer and monomer. Int J Mass Spectrom 2007; 268:304–315.
46. Bourne EJ, Stacey M, Tatlow JC, Tedder JM. Studies on trifluoroacetic acid. Part I. Trifluoroacetic anhydride as a promotor of ester formation between hydroxy-compounds and carboxylic acids. J Chem Soc 1949; 2976–2979.
47. Ciucanu I, Kerek F. A simple and rapid method for the permethylation of carbohydrates. Carbohydr Res 1984; 131:209–217.
48. Ciucanu I, Costello CE. Elimination of oxidative degradation during per-*O*-methylation of carbohydrates. J Am Chem Soc 2003; 125:16213–16219.
49. Dell A. Preparation and desorption mass spectrometry of permethyl and peracetyl derivatives of oligosaccharides. Methods Enzymol 1990; 193:647–660.

50. McNeil M, Darvill AG, Aman P, Franzen LE, Albersheim P. Structural analysis of complex carbohydrates using high-performance liquid chromatography, gas chromatography, and mass spectrometry. Methods Enzymol 1982; 83:3–45.
51. Dell A, Ballou CE. Fast atom bombardment mass-spectrometry of a 6-*O*-methylglucose polysaccharide. Biomed Mass Spectrom 1983; 10:50–56.
52. Dell A, Oates JE, Morris HR, Egge H. Structure determination of carbohydrates and glycosphingolipids by fast atom bombardment mass-spectrometry. Int J Mass Spectrom Ion Processes 1983; 46:415–418.
53. Egge H, Dell A, Vonnicolai H. Fucose containing oligosaccharides from human-milk. 1. Separation and identification of new constituents. Arch Biochem Biophys 1983; 224:235–253.
54. Bowman MJ, Zaia J. Tags for the stable isotopic labeling of carbohydrates and quantitative analysis by mass spectrometry. Anal Chem 2007; 79:5777–5784.
55. Kranz C, Ng BG, Sun L, Sharma V, Eklund EA, Miura Y, Ungar D, Lupashin V, Winkel DR, Cipollo JF, Costello CE, Loh E, Hong W, Freeze HH. COG8 deficiency causes new congenital disorder of glycosylation Type IIh. Hum Mol Genet 2007; 16:731–741.
56. Carr SA, Barr JR, Roberts GD, Anumula KR, Taylor PB. Identification of attachment sites and structural classes of asparagine-linked carbohydrates in glycoproteins. Methods Enzymol 1990; 193:501–518.
57. Kragten EA, Bergwerff AA, van Oostrum J, Muller DR, Richter WJ. Site-specific analysis of the *N*-glycans on murine polymeric immunoglobulin A using liquid chromatography/electrospray mass spectrometry. J Mass Spectrom 1995; 30:1679–1688.
58. Mansoori BA, Dyer EW, Lock CM, Bateman K, Boyd RK, Thomson BA. Analytical performance of a high-pressure radio frequency-only quadrupole collision cell with an axial field applied by using conical rods. J Am Soc Mass Spectrom 1998; 9:775–788.
59. Hirayama K, Yuji R, Yamada N, Kato K, Arata Y, Shimada I. Complete and rapid peptide and glycopeptide mapping of mouse monoclonal antibody by LC/MS/MS using ion trap mass spectrometry. Anal Chem 1998; 70:2718–2725.
60. Zhu X, Borchers C, Bienstock RJ, Tomer KB. Mass spectrometric characterization of the glycosylation pattern of HIV-gp120 expressed in CHO cells. Biochemistry 2000; 39:11194–11204.
61. Nemeth JF, Hochensang GP, Marnett LJ, Caprioli RM. Characterization of the glycosylation sites in cyclooxygenase-2 using mass spectrometry. Biochemistry 2001; 40:3109–3116.
62. Overberg A, Karas M, Bahr U, Kaufmann R, Hillenkamp F. Matrix-assisted infrared-laser (2.94 m) desorption/ionization mass spectrometry of large biomolecules. Rapid Commun Mass Spectrom 1990; 4:293–296.
63. Wada Y, Azadi P, Costello CE, Dell A, Geyer H, Geyer R, Kakehi K, Karlsson NG, Kato K, Kawasaki N, Khoo K-H, Kim S, Kondo A, Lattova E, Mechref Y, Nakamura K, Narimatsu H, Novotny MV, Packer NH, Perreault H, Peter-Katalinić J, Pohlentz G, Reinhold VN, Rudd PM, Suzuki A, Taniguchi N. Mass spectrometry of glycoprotein glycans: HUPO HGPI

(Human Proteome Organisation Human Disease Glycomics/ Proteome Initiative) multi-institutional study. Glycobiology 2007; 17:411–422.
64. Hitchcock A, Costello CE, Zaia J. Glycoform quantification of chondroitin/dermatan sulfate using a liquid chromatography-tandem mass spectrometry platform. Biochemistry 2006; 45:2350–2361.
65. Vakhrushev SY, Mormann M, Peter-Katalinic J. Identification of glycoconjugates in the urine of a patient with congenital disorder of glycosylation by high-resolution mass spectrometry. Proteomics 2006; 6:983–992.

Carbohydrate Chemistry, Biology and Medical Applications
Hari G. Garg, Mary K. Cowman and Charles A. Hales

DOI: 10.1016/B978-0-08-054816-6.00003-3

Chapter 3

Chemical Synthesis of Complex Carbohydrates

ZHONGWU GUO

Department of Chemistry, Wayne State University, Detroit, MI 48202, USA

I. Introduction

Carbohydrates are arguably the most structurally complex and diverse molecules in nature, not only because carbohydrate monomers can couple to each other to form complex oligomers and/or polymers but more notably because each carbohydrate monomer can accommodate multiple linkages, branches, and modifications and each glycosidic linkage can have two different anomeric configurations. As a result, the potential number of structures of even small oligosaccharides can be astronomical. A calculation reported by Laine (1) has revealed that a hexasaccharide can have at least 1.5×10^{15} potential linear and branched isomers. On top of this, carbohydrates are frequently modified by various functionalities in nature that further increases the number of potential isomers of an oligosaccharide by magnitudes.

The exceptional complexity and diversity of carbohydrate structures render them the perfect carrier of biological signals because they can use the shortest sequence to store the most information. Naturally, carbohydrates play an important role in many biological and pathological processes (2). On the other hand, the structural complexity and diversity of carbohydrates make their chemical synthesis much more difficult than the chemical synthesis of other biopolymers such as peptides and nucleic acids.

The chemical synthesis of carbohydrates is subject to a number of difficulties. First, to realize the regioselective reactions at a particular position of a sugar unit, the hydroxyl group at this position has to be distinguished from all other hydroxyl groups in the structure, while all these hydroxyl groups have similar properties. Additionally, when a sugar unit has one or more branches attached to it, several types of protecting groups must be utilized to discriminate different positions, and in this situation, the design of a proper protection tactic can be a challenge. Another problem is related to the glycosylation reactions used to couple sugars. Compared to the reactions used in peptide and nucleic acid syntheses, glycosylation reactions are definitely much more difficult and, consequently, afford rather poor yields. More importantly, glycosylation reactions have a unique issue, namely that each glycosylation can produce two stereoisomers, the α- and β-anomers, an issue that is not present in peptide or nucleic acid synthesis. How to achieve the desired glycosidic linkage in a stereoselective manner remains one of the most important challenges in carbohydrate chemistry. On the other hand, because many biologically significant carbohydrates exist as glycopeptides, glycolipids, and other glycoconjugates in nature, most synthetic oligosaccharides are finally linked to a peptide and/or lipid chain. In this case, the protection tactics, the reactions used to assemble carbohydrates, and the sequence of molecular assembly must be carefully designed to ensure compatibility with carbohydrate, peptide, and lipid chemistry. Moreover, there is no synthetic method or protocol that is universally applicable to all carbohydrates, thus each oligosaccharide synthesis must be taken as a unique challenge.

Nevertheless, carbohydrate synthesis has achieved great progress in the last decades. For example, many regioselective reactions and protecting groups or tactics have been developed for discrimination of different hydroxyl groups (3). It is essentially possible to differentiate any hydroxyl group from all others in a sugar unit, though the procedure may need many protection and deprotection transformations and thus can be tedious and time-consuming. There are also many glycosylation reactions and methods established for stereoselective formation of the desired glycosides. The development of new glycosylation methods that can be achieved under different conditions makes it possible to establish many innovative strategies for complex oligosaccharide synthesis, such as solid-phase synthetic (4) and one-pot synthesis built on armed-disarmed glycosylation (5), orthogonal glycosylation (6,7), selective activation of glycosyl donors, preactivation of donors (8–10). As a result, it is currently possible to achieve extremely complex oligosaccharides by carefully planned schemes and strategies (11,12).

It is impossible in a short review like this to cover all aspects of development in carbohydrate synthesis. Instead, this review will briefly summarize the types of glycosylation methods and synthetic strategies that are commonly used in carbohydrate synthesis. It will also illustrate the applications of some of these strategies in the chemical synthesis of a class of complex and biologically important glycoconjugates, namely, glycosylphosphatidylinositols (GPIs).

II. Glycosylation Methods in Carbohydrate Synthesis

A. Some General Considerations Concerning the Stereochemistry of Glycosylation Reactions

Since the first glycosylation method was reported by Koenigs and Knorr (13) a hundred years ago, many different glycosylation methods have been developed

(3,14,15). As each glycosylation reaction may produce two anomers, many studies are directing at the development of strategies to control the stereochemistry of glycosylation reactions. However, the latter topic is not the focus of this chapter, so only general principles of the stereochemistry of glycosylation reactions are discussed.

Under certain conditions a glycosylation reaction can be highly stereoselective to give one glycosyl anomer. For example, it is relatively easy to achieve 1,2-*trans*-glycosylation for both *gluco* and *manno* sugars. As depicted in Scheme 1, the presence of an acyl group, such as an acetyl or a benzoyl group, at the 2-*O*-position of a glycosyl donor can ensure stereoselective 1,2-*trans*-glycosylation. It is believed that, through the participation of the neighboring acyl group, these glycosylation reactions can form a dioxolanylium cation that can only be attacked by a nucleophile from the opposite side of the dioxolanylium ring to afford 1,2-*trans*-glycosides. Thus, the protecting group at 2-*O*-position of the glycosyl donor often has a critical influence on the rate and stereochemistry of its reactions.

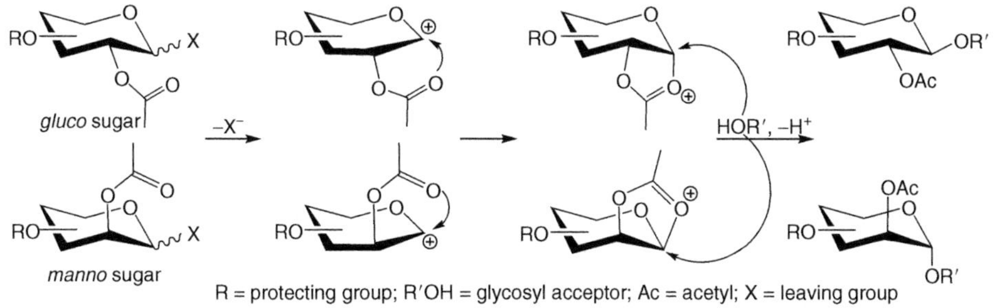

R = protecting group; R′OH = glycosyl acceptor; Ac = acetyl; X = leaving group

Scheme 1 Stereoselective 1,2-*trans*-glycosylation of sugars assisted by the participating neighboring group.

In contrast, stereoselective 1,2-*cis*-glycosylation is much more difficult to achieve. To obtain the 1,2-*cis*-glycoside from a glycosylation reaction, the 2-*O*-position of the glycosyl donor must not be protected by a participating functional group. Under this condition, the stereoselectivity of a glycosylation reaction is usually dominated by anomeric effects to favor the formation of α-glycosides (1,2-*cis*) with limited stereoselectivity. For *manno* sugars, however, both anomeric and steric effects will favor the α-anomer (1,2-*trans*), thus the reactions usually give the α-glycosides (1,2-*trans*) with high stereoselectivity. The chemical synthesis of β-*manno*-glycosides (1,2-*cis*) is then a significant challenge. Progresses concerning stereoselective synthesis of 1,2-*cis* glycosides have been reviewed recently (16–20).

B. Glycosylation Methods Using Glycosyl Halides as Donors

The Koenigs–Knorr glycosylation method using glycosyl bromides or chlorides as donors dates back to 1901 (13), and it is still in wide use. The glycosylation reaction was originally achieved with silver carbonate as a promoter, but other heavy metal salts, especially silver triflate (21), have been introduced as more effective promoters. The stereoselectivity of this reaction is largely determined by the properties of the protecting group at the 2-*O*-position of the glycosyl donor. However, Lemieux et al. (22) observed that glycosyl bromides could anomerize in the presence of a bromide and that the β-glycosyl bromide was more reactive than the α-glycosyl bromide. Therefore, in the absence of promoters, the β-glycosyl bromide, but not the α-bromide, could slowly react with an alcohol by S_N2 displacement to give the α-glycosides in high stereoselectivity. This provides a convenient α-glycosylation method.

Later on, Mukaiyama et al. (23) introduced glycosyl fluorides as more versatile donors. Glycosyl fluorides can be easily activated by transition metal complexes, such as silver perchlorate-hafnocene dichloride (24), under mild conditions to offer smooth and effective glycosylations. Thus, this method has been widely adopted to synthesize complex oligosaccharides. Meanwhile, as its reaction condition is orthogonal to many other glycosylation methods, it has also been widely used in the development of new one-pot carbohydrate synthetic strategies. Recently, Gervay-Hague et al. (25) have demonstrated that glycosyl iodides can undergo S_N2 displacement under almost neutral conditions to afford glycosides in good yields and stereoselectivity in the synthesis of a number of complex oligosaccharides (26,27).

C. Glycosylation Methods Using Glycosyl Imidates or Esters as Glycosyl Donors

Schmidt glycosylation (28), which uses glycosyl trichloroacetimidates as donors, is one of the most broadly adopted glycosylation methods. It is still among the best glycosylation methods for complex carbohydrate synthesis. The activators for this reaction are various Lewis acids such as trifluoroboron (BF_3) and trimethylsilyl triflate (TMSOTf), and usually, the reactions are fast even at low temperatures. This method had a significant impact on the progress of carbohydrate research (29–31).

Glycosyl acetates are easily available, but as glycosyl donors they are mainly used in the glycosylation of simple acceptors due to their relative low reactivity. However, glycosyl phosphates and phosphites have proven to be more versatile donors. Through systematic studies, Seeberger and coworkers (32,33) have revealed that both α- and β-glycosyl phosphates could be activated by TMSOTf to afford β-glycosides in high yields and stereoselectivity. This method has been applied to the synthesis of β-*manno*-glycosides (34) and oligosaccharides (35) and solid-phase carbohydrate synthesis (36). Glycosyl phosphites are highly reactive, so they can be activated under mild conditions (37) to afford good stereoselectivity (38).

D. Glycosylation Methods Using Thioglycosides and Related Derivatives as Glycosyl Donors

Various forms of thioglycosides are probably the most extensively studied glycosyl donors. As shown in Scheme 2, the functional group R′ can vary to have a significant influence on the reactivity of the resultant thioglycosides as donors.

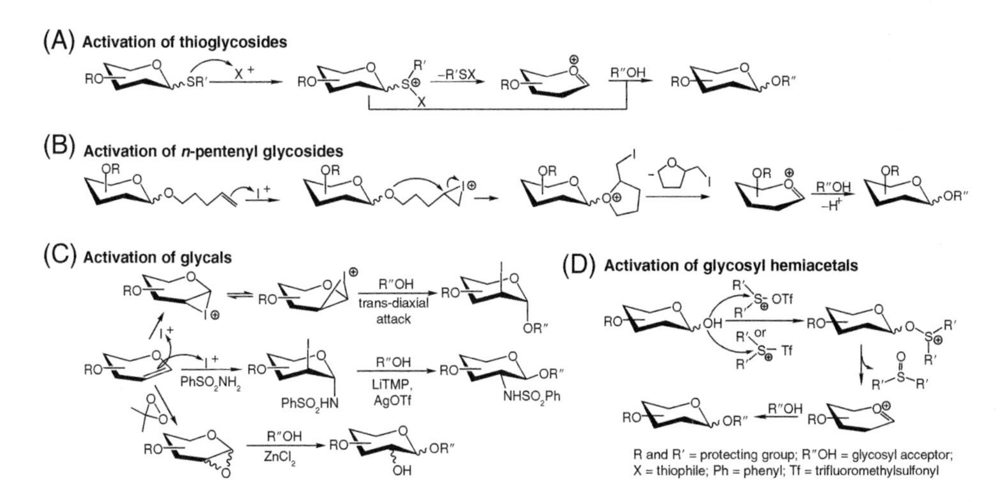

Scheme 2 Glycosylation method using thioglycosides, *n*-pentenyl glycosides, glycals, and glycosyl hemiacetals as glycosyl donors.

Therefore, through the modification of R′, one can easily tune the reactivity of thioglycosides (39,40) to meet specific demands from a glycosylation or a synthetic design. Moreover, like glycosyl donors, thioglycosides are exceptionally stable, but they can be easily activated under special conditions. These properties make thioglycosides very attractive in the synthesis of complex oligosaccharides and in the design of convergent synthetic strategies.

For the activation of thioglycosides, many thiophilic promoters have been developed (41), among which the most commonly used are methyl triflate (MeOTf), dimethyl-methylthiosulfonium triflate (DMTST) (42), *N*-iodosuccinimide/triflic acid (NIS/TfOH) (43), iodonium dicollidine perchlorate (IDCP), and sulfenyl triflate or halides (44). Recently, Crich et al (45) have developed another useful promoter, 1-benzenesulfinyl piperidine and applied it to stereoselective formation of β-mannosides, which is difficult to achieve. Methyl, ethyl, phenyl, and *p*-methylphenyl thioglycosides are most commonly used glycosyl donors. Recently, *p*-methylphenyl thioglycosides have received significant attention in the development of innovative synthetic strategies (8,46). Heterocyclic aryl thioglycosides have also been extensively explored. The 2-pyridyl thioglycosides introduced by Hanessian and coworkers (47) can be activated by "remote activation" with methyl iodide as the promoter. Demchenko and coworkers (48,49) have recently introduced benzoxazolyl and thiazolinyl thioglycosides that can be selectively activated by MeOTf or silver triflate (AgOTf) in the presence of alkyl thioglycosides for stereoselective 1,2-*cis*-glycosylation and one-pot oligosaccharide synthesis (49).

After oxidation, thioglycosides were converted into glycosyl sulfoxides that can be activated by triflic anhydride (Tf_2O) at low temperature, hence, offering another powerful glycosylation method by Kahne et al. (50). This method has been employed to synthesize complex oligosaccharides both in solution (51) and by solid-phase strategy (52). Its application has been further expanded by Crich et al. (53) in the synthesis of β-mannosides. The seleno analogues of thioglycosides are more reactive than thioglycosides, so they can be selectively activated under specific conditions (54).

E. Glycosylation Methods Using *n*-Pentenyl Glycosides as Glycosyl Donors

This is a glycosylation method (Scheme 2) developed by Fraser-Reid and coworkers (55) and has been utilized to synthesize a number of complex oligosaccharides (56,57). For these reactions, the commonly used promoters are NIS and *N*-bromosuccinimide in combination with a Lewis acid, such as TfOH. It has been demonstrated that these glycosylation reactions are sensitive to the protecting groups in the glycosyl donors, based on which the armed–disarmed (5) glycosylation concept was proposed and explored in the design of new carbohydrate synthetic strategies.

F. Glycosylation Methods Using Glycals as Glycosyl Donors

Systematic studies from several groups, especially Danishefsky and coworkers (58,59), have shown that glycals are versatile glycosyl donors. As shown in Scheme 2, there are several activation methods for glycals. When only halogenium cations are used as the activators, both α- and β-halogenium intermediates are formed. When glycosyl acceptors attack these intermediates, *trans*-diaxial addition will prevail to give the α-glycosylation product stereoselectively. Once the desired oligosaccharides are obtained, the halogen atoms can be removed by a reduction reaction to offer 2-deoxyglycosides (60). Alternatively, glycals can be activated in the presence of a sulfonamide, leading to an iodosulfonamide glycoside intermediate (61). When this intermediate is treated with an alcoholic acceptor in the presence of lithium tetramethylpiperidide and AgOTf, a 2-deoxy-2-amino-β-glycoside product is formed. It was proven useful for synthesis of 2-amino-2-deoxy glycosides (62). The third extensively explored protocol involves the epoxidation of glycals by 3,3-dimethyldioxirane to give 1,2-anhydrosugars as key intermediates. The reaction between the 1,2-dehydrosugars and acceptors can be promoted by anhydrous zinc chloride to give the glycosylation products. This method has been applied to complex oligosaccharides (59) and solid-phase synthesis (63). Recently, Gin and coworkers (64,65) reported another approach to activate glycals using a sulfonium cation as the activator. These reactions involve reactive species such as 1,2-dehydrosugar and oxazoline as the reaction intermediates.

G. Dehydrative Glycosylation Using Glycosyl Hemiacetals as Glycosyl Donors

A dehydrative glycosylation method developed by Gin (66) is mechanistically very interesting. As outlined in Scheme 2, Gin's glycosylation method uses a sulfonium cation for hemiacetal activation, and the sulfonium cation has been created by two similar but different approaches, namely, by reaction of sulfoxide with Tf_2O or by reaction of sulfide with Tf_2O (67–69). The reaction between the donor and the sulfonium cation gives rise to an oxosulfonium cation, which itself is a potential glycosylating species. The oxosulfonium cation can also convert into a glycosyl cation that reacts with the alcoholic acceptor to afford the desired glycosides.

In summary, many glycosylation methods have developed for chemical synthesis of oligosaccharides, and each may be particularly suitable for certain synthetic designs. However, there is not a general rule that can be used to predict which glycosylation method is the best for which specific substrate. In fact, the stereochemical outcome of a glycosylation method can significantly depend on conditions such as the reaction temperature, the protecting groups used, the reactivity of involved glycosyl donors and acceptors, the reagents used to promote the reaction, and the reaction solvent. Therefore, it is always worthy to examine different glycosylation methods to find out the best one for a specific synthesis, especially in the preparation of large oligosaccharides and oligosaccharides with difficult glycosidic linkages.

III. Strategies for Oligosaccharide Assembly

For carbohydrate synthesis, though traditional linear assembly method is still used to construct short and simple oligosaccharides, it can be problematic for complex or branched oligosaccharides. Moreover, the yield and stereoselectivity of current glycosylation reactions are far from satisfactory that leads to the drastic decrease of synthetic efficiency with the chain elongation and drastic increase of side products. To overcome these problems, much research effort is currently focused on the development of synthetic strategies that may improve the efficiency of the final stage of oligosaccharide assembly (9,10).

A. Linear Assembly of Carbohydrates

Linear assembly is the oldest strategy, but it is still in wide use, especially in the synthesis of linear and short oligosaccharides. As outlined in Scheme 3, there are two possible ways for chain elongation. One (A) starts from the reducing end with the carbohydrate chain elongated along the nonreducing end direction. The glycosyl acceptor needs to contain a free hydroxyl group at the position of glycosylation, while the glycosyl donor needs to have an easily removable temporary protection at the position of subsequent glycosylation. Each cycle of chain elongation involves a step of glycosylation followed by a step of selective deprotection of the glycosylation product. The other (B) starts from the nonreducing end, and the carbohydrate chain is elongated along its reducing end direction. In this case, the glycosyl donor is fully protected except its reducing end which should be in the reactive form. The glycosyl acceptor needs to have a free hydroxyl group at the position of glycosylation, while its anomeric center needs to be protected by an easily removable functional group to facilitate subsequent deprotection. Therefore, each cycle of chain elongation in the latter strategy will involve a step of glycosylation, a step of selective deprotection of the reducing end of the

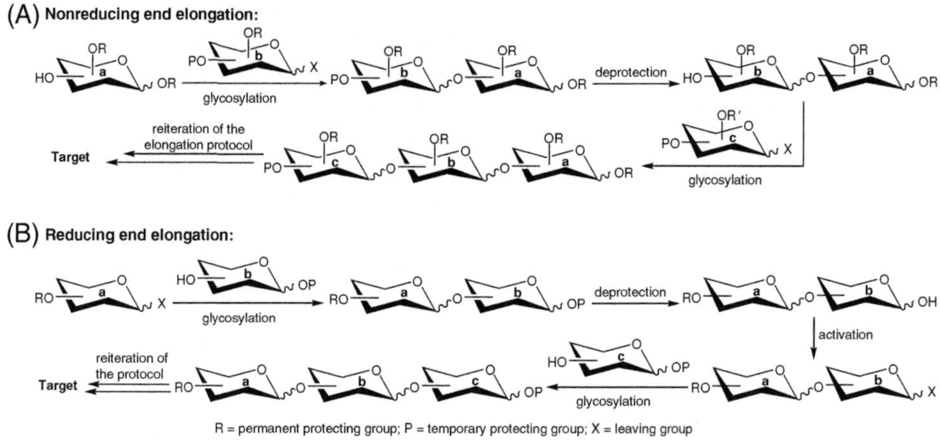

(A) Nonreducing end elongation:

(B) Reducing end elongation:

R = permanent protecting group; P = temporary protecting group; X = leaving group

Scheme 3 Linear assembly of oligosaccharides.

glycosylation product, and a step of anomeric activation. In both cases, the chain elongation protocol can be reiterated to obtain the desired oligosaccharides. It is also possible to combine these two ways of assembly in one synthesis if necessary.

Traditionally, the strategy of chain elongation along the nonreducing end (strategy A) is more popular because it has some potential advantages over the strategy B. First, the glycosyl donors used in the past were not particularly stable; their decomposition would significantly affect the synthetic efficiency and/or product purification for strategy B. Moreover, the harsh reaction conditions employed to activate sugar anomeric centers may sometimes affect the glycosidic linkages of the oligosaccharide intermediates involved in strategy B. Second, for strategy A, the glycosyl donors are simple and relatively easily available, and they can be used in large excess to push the glycosylation reaction to the desired direction. Third, in case that the glycosylation is unsatisfactory, the precious oligosaccharide acceptor of strategy A is readily recoverable, which is not easy with the precious oligosaccharyl donor of strategy B. However, with the development of new and effective glycosylation methods, oligosaccharide assembly based on carbohydrate chain reducing end elongation has become more practical than before. Moreover, linear assembly also seems to be the natural choice for some new synthetic techniques based on iterative glycosylations, such as solid-phase synthesis, one-pot synthesis.

B. Convergent Carbohydrate Assembly with Short Oligosaccharides as Building Blocks

A serious problem concerning the linear assembly strategy is the extended synthetic route, which is extremely damaging for oligosaccharide synthesis. As a result, convergent carbohydrate assembly has become more and more popular, especially for complex oligosaccharides (11,12). The basic design is to use short oligosaccharides as building blocks to assemble the final synthetic target as outlined in Scheme 4. For example, in the synthesis of a tetrasaccharide **a**-**b**-**c**-**d**, first, two disaccharide blocks **a**-**b** and **c**-**d** are constructed. In these building blocks, the anomeric center of sugar **b** and the future glycosylation site of sugar **c** are blocked by temporary protecting groups **Y** and **P**, respectively. Once **a**-**b** is obtained, its reducing end is deprotected, and then the resultant disaccharide is converted into a glycosyl donor. On the other hand, after the temporary protecting group in **c**-**d** is removed, which will expose a free hydroxyl group, the resultant disaccharide will be employed as the glycosyl acceptor. Finally, the two disaccharide blocks are coupled together by the designated glycosylation reaction to give the desired tetrasaccharide. Although this strategy is applicable to an oligosaccharide as simple as a tetrasaccharide, it is more useful for more complicated structures, especially those that contain branches and those that are composed of repeating oligosaccharide units.

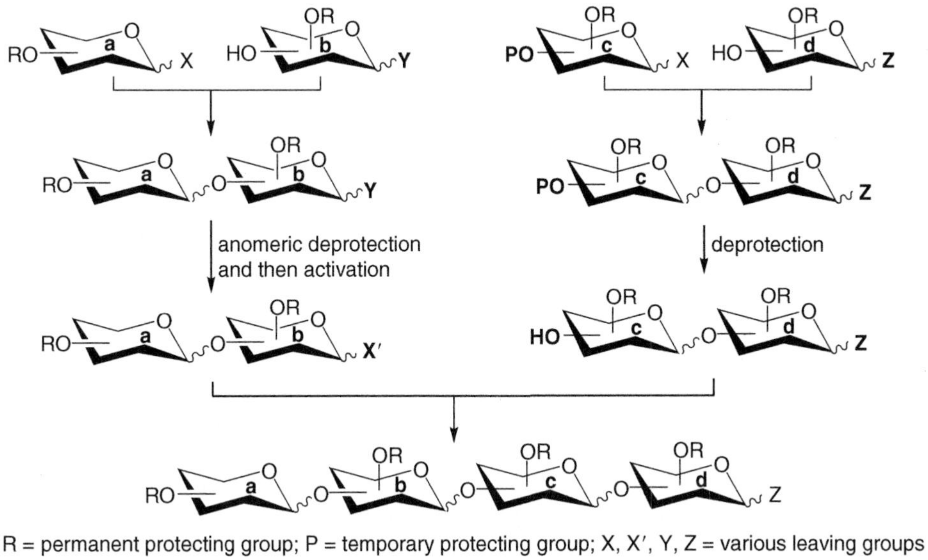

R = permanent protecting group; P = temporary protecting group; X, X′, Y, Z = various leaving groups

Scheme 4 Convergent assembly of oligosaccharides.

C. Oligosaccharide Assembly Based on Discriminative Activation of Glycosyl Donors

There are several basic concepts based on which different strategies have been designed. One is to take advantage of different reaction conditions required to activate glycosyl donors that bear different leaving groups at their anomeric centers. As a result, the involved anomeric centers can be activated sequentially. A direct application of this concept is to have the anomeric centers of substrates linked to different functionalities, while each can be directly activated under a specific condition. A good example for this case is Ogawa and Ito's orthogonal glycosylation strategy (6). As outlined in Scheme 5, it was based on the fact that the reaction conditions used to activate thioglycosides (NIS/AgOTf) and glycosyl fluorides (silver perchlorate/hafnocene dichloride) do not interfere with each other. Based on this concept, numerous strategies have been explored, as reviewed by Demchenko (9).

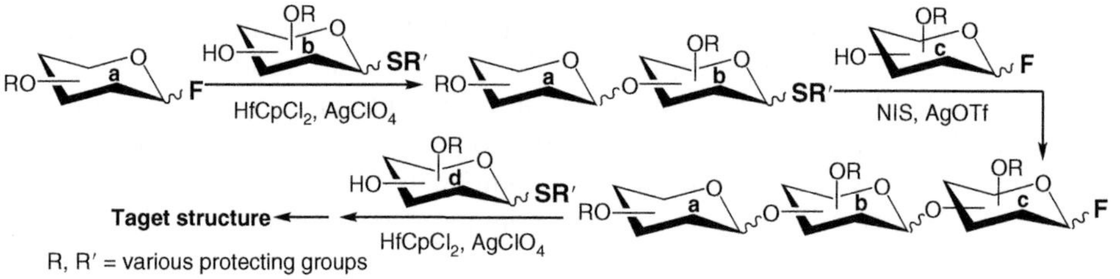

Scheme 5 Synthesis of oligosaccharides based on orthogonal glycosylations.

Another concept is "armed–disarmed" glycosylation proposed by Fraser-Reid et al. (5). Glycosyl donors with an electron donating group, such as a benzyl group, at the 2-*O*-position are "armed" (more reactive), and donors with an electron withdrawing group, such as an acetyl or benzoyl group, at the 2-*O*-position are "disarmed" (less reactive). Armed donors can be selectively activated in the presence of disarmed donors to achieve specific glycosylations (Scheme 6). The rationale for this difference is that an electron withdrawing group destabilizes the cationic glycosylation intermediate and makes those glycosyl donors disarmed. However, once the selective glycosylation by the armed donor is finished, the resultant oligosaccharide, which is disarmed, can be activated by a stronger promoter to realize the next step of glycosylation and further elongation of the carbohydrate chain. Substituents at other than 2-*O*-positions (70) or torsional strains (71) in a sugar can also have a decisive influence on its reactivity. Moreover, the armed–disarmed effect has also been observed with other glycosyl donors such as thioglycosides. Thus, this concept has exhibited a broad impact on the progress of carbohydrate chemistry (72).

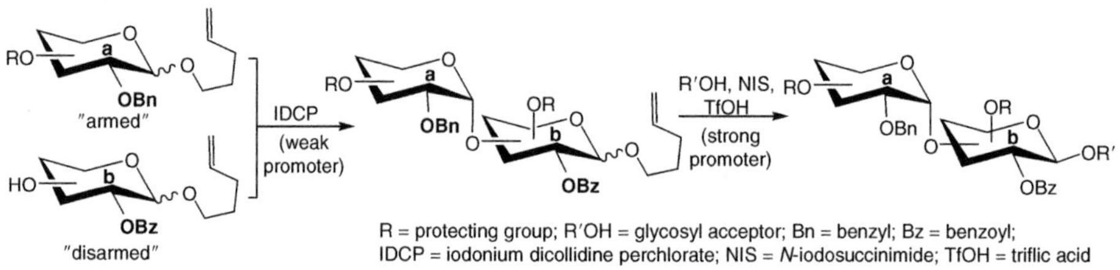

Scheme 6 Oligosaccharide synthesis based on "armed–disarmed" glycosylation.

The third concept is "programmed glycosylation" based on tuning the reactivity of various glycosyl donors (73). This concept is rather similar to the "armed–disarmed" concept in that both make use of the distinct reactivity of differently protected glycosyl donors. However, the programmed glycosylation strategy needs more well-defined reactivity information concerning the substrates. After comparing the relative reactivity values (RRVs) of a large number of different sugars, Wong and coworkers (46) concluded that a sugar bearing different functional groups, as well as different sugars bearing same or different functional groups, may have sufficiently different reaction rates to assure their selective activation and glycosylation. For example, as glycosyl donors, peracetylated toluyl thiogalactoside (RRV=14.3) is about 6 times more reactive than peracetylated toluyl thioglucoside (RRV=2.7), whereas perbenzylated toluyl thioglucoside (RRV=2656) is about 1000 times more reactive than peracetylated thioglucoside. With the help of this information, optimized building blocks can be designed for the synthesis of a complex oligosaccharide by a series of selective glycosylations. Thus, these glycosylations can be achieved in one pot by adding the glycosyl donors sequentially. The progress in this and related areas has been reviewed in ref. (73).

Recently, a new and potentially powerful strategy for one-pot carbohydrate synthesis, which is closely related to but conceptually different from the above strategies, has been developed based on preactivation of glycosyl donors, which has been explored by Ye, Huang and coworkers (8,74). Its protocol is to first convert a relatively stable glycosyl donor, for example a thioglycoside, to an extremely reactive species, for example a glycosyl triflate, at low temperature and then introduce the glycosyl acceptor for glycosylation. The anomeric center of the acceptor can bear the same functionality as that of the donor. So, the activation and glycosylation protocol can be reiterated for one-pot synthesis. This strategy has been used to prepare several rather complex and difficult oligosaccharides with excellent overall yields (8,74).

D. Solid-Phase and Polymer-Supported Solution-Phase Oligosaccharide Synthesis

Solution- and solid-phase polymer supported carbohydrate synthesis (63,75–78), including automated solid-phase synthesis (79), has also achieved great progress. This progress largely relies on the significant advancement in glycosylation methods in the last three decades. For polymer-supported syntheses, the strategy of linear assembly seems to be the natural choice for chain elongation. Both reducing end elongation and nonreducing end elongation strategies have been explored, with the latter being more popular due to the potential advantages such as the possibility of using a large excess of donors to improve the speed and efficiency of glycosylation reactions.

In solid-phase carbohydrate synthesis, it is impossible to purify the synthetic intermediates, whereas glycosylation reactions, especially the two-phase reactions, usually give relatively poor efficiency and stereoselectivity. As a result, a significant number and quantity of side products will be formed together with the desired product on the polymer. After carbohydrate chains are cut off from the polymer, a complex mixture is obtained, and the final product purification becomes difficult. To deal with this problem, several innovative strategies have been explored recently, which include Fukase's capture–release strategy (80,81), Seeberger's cap-scavenger strategy (82), and our group's cap-capture-release strategy (83). These strategies not only help the purification of final products, but some strategies (82,83) may also reduce the number and amount of side products formed during the solid-phase synthesis. For polymer-supported solution-phase synthesis, an interesting development was its combination with orthogonal glycosylation reported by Ito et al. (7). Ito et al. have also developed some innovative methods to facilitate the monitoring of the reaction progress (84) and product isolation (85) in these syntheses.

E. Chemoenzymatic Synthesis of Carbohydrates

Chemoenzymatic carbohydrate synthesis has been emerging as an extremely powerful tool to obtain complex oligosaccharides. Chemoenzymatic synthesis has many potential advantages, of which the most significant ones are probably that it offers perfect stereoselectivity and renders the tedious work of synthesizing various selectively protected monosaccharide units unnecessary. It also makes the most difficult glycosylation reactions seem easy (86). Chemoenzymatic synthesis of oligosaccharides is out of the scope of this chapter, but it was the topic of many excellent reviews (73,87–89).

In summary, many strategies have been developed for carbohydrate synthesis. These strategies have significantly improved the overall synthetic efficiency and resulted in more effective use of synthetic building blocks in oligosaccharide synthesis. Among all the strategies available, it is difficult to tell which one is the best. The synthetic target determines the strategy. Usually, linear assembly is particularly suitable for simple structures and for strategies including one-pot synthesis, polymer-supported synthesis, and so on. Convergent assembly may be necessary for long and highly branched oligosaccharides and for oligosaccharides that are composed of repeating units. Sometimes, the combination of different strategies will give the best results. The application of various strategies is illustrated with the synthesis of GPIs.

IV. Chemical Synthesis of GPIs

GPIs are a class of glycolipids ubiquitously expressed by all types of eukaryotic cells (90,91). One of the most important functions of GPIs is to anchor proteins, glycoproteins, and polysaccharides onto cell membranes. Thus, GPIs and GPI-anchored structures play a pivotal role in many biological events, such as cell recognition, binding, and transmembrane signal transduction (91–93).

All protein/glycoprotein-anchoring GPIs share a common core: Man$\alpha(1\rightarrow4)$GlcNH$_2\alpha(1\rightarrow6)$-*myo*-Inositol-1-PO$_4$-lipid (94). The structural diversity of GPIs is created mainly by the variation in their lipid structures as well as the additional sugar chains and other functionalities linked to the core GPI glycan (95). Proteins and glycoproteins are invariably attached to GPIs through a phosphoethanolamine functionality that bridges the peptide C-terminus and the nonreducing end

6-*O*-position of the GPI glycan core. The complex structure of GPIs is a fantastic challenge for synthetic chemists (96), as GPI synthesis requires the integration of methods developed in different areas in organic chemistry, including carbohydrate, inositol, lipid, and phosphate chemistry.

After the first complete GPI structure was revealed in 1988 (97,98), it immediately attracted the attention of carbohydrate chemists. The first total synthesis of a GPI, that is the GPI anchor of *Trypanosoma brucei*, was achieved by Ogawa and coworkers (99–101). Since then, a number of GPIs have been synthesized by groups such as Fraser-Reid and coworkers (102–105), Guo and coworkers (106,107), Ley and coworkers (71,108), Martin-Lomas and coworkers (109,110), Nikolaev and coworkers (111), Schmidt and coworkers (112–115), and Seeberger and coworkers (116–118). Already there are reviews concerning the chemical synthesis of GPIs (96). This chapter is not intended to cover all synthetic studies in the field; instead, it is mainly focused on the strategies that have been adopted to assemble GPIs. These strategies include linear, convergent, and solid-phase assembly of the carbohydrate chains.

A. GPI Synthesis by Linear Assembly of the Carbohydrate Chain

The synthesis of a nonlipidated structure of *Plasmodium falciparum* GPI (Scheme 7) reported by Seeberger and coworkers (117) is a representative example. After an optically pure derivative **1** of inositol was prepared, it was coupled to a glucosamine derivative **2** to afford the α-linked pseudodisaccharide **3**. Like most cases, an azido sugar **2** was utilized as a latent glucosamine. As the azido group is a nonparticipating protecting group, α-glycosylation is assured and this glycosylation was achieved by the thioglycoside method. Thereafter, four α-mannose residues were introduced sequentially using **5, 8, 11**, and **14** as glycosyl donors. These glycosylations were realized by the Schmidt method, and the protocol for carbohydrate chain elongation includes a deprotection step after each glycosylation. Eventually, the GPI glycan core, namely, **15** was obtained, in which the inositol 1,2-*O*-positions were protected by a ketal and the 6-*O*-position of the third mannose residue was protected by a silyl group. Following selective deprotections of these positions, a cyclic phosphate functionality and a phosphoethanolamine group were installed. Finally, global deprotection afforded the desired GPI analogue **16**, which was then linked to a carrier protein to formulate a useful antimalarial vaccine. As shown, the carbohydrate chain was elongated by strictly linear assembly along the oligosaccharide nonreducing end.

B. GPI Synthesis by Convergent Assembly of the Carbohydrate Chain

The synthesis of *T. brucei* GPI anchor reported by Ley and coworkers (71) is a representative example (Scheme 8). After the properly protected inositol residue and various monosaccharide units were obtained, they were first assembled to give several simple oligosaccharide blocks including a pseudodisaccharide **26**, a disaccharide **24**, and a trisaccharide **21**. The coupling of the disaccharide and the trisaccharide afforded a pentasaccharide **25**, which was then coupled to the pseudodisaccharide. In this way, a complex glycosyl inositol **27** was rapidly assembled in a highly

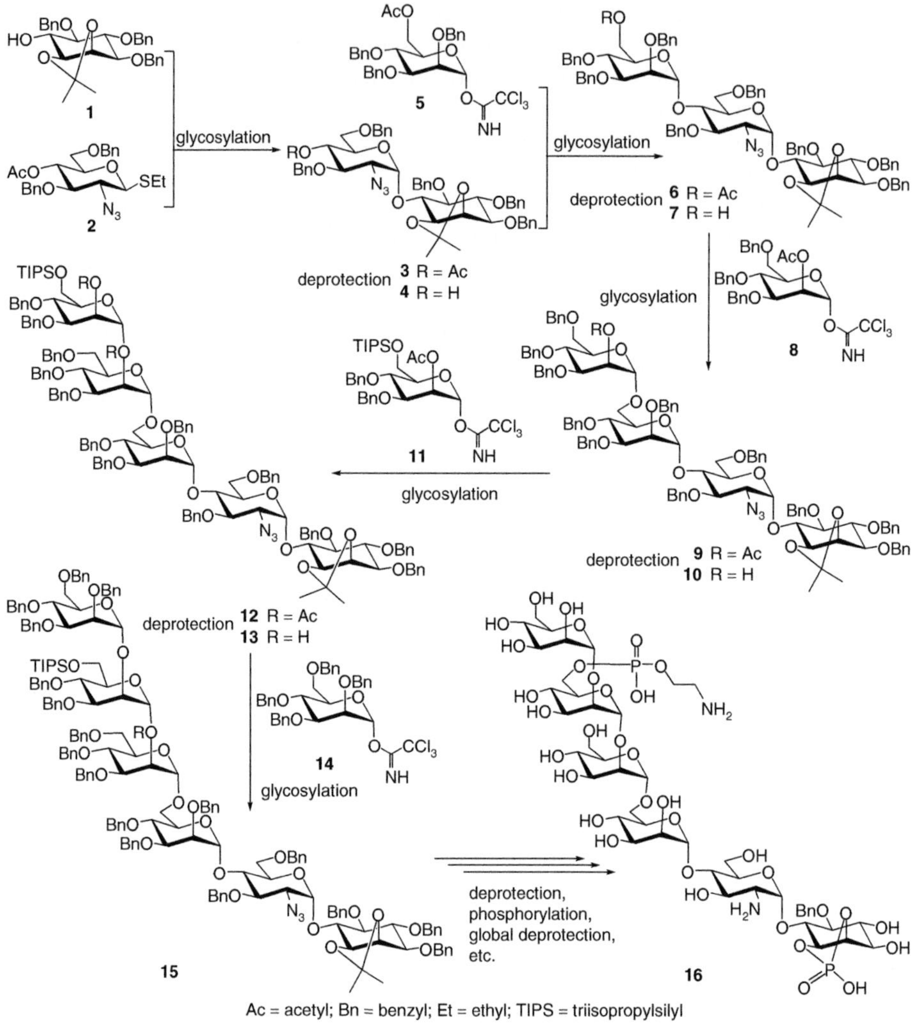

Ac = acetyl; Bn = benzyl; Et = ethyl; TIPS = triisopropylsilyl

Scheme 7 Linear assembly of the core of *Plasmodium falciparum* GPI anchor.

convergent manner. At the final stage, TBS and All as protecting groups of the non-reducing end mannose and the inositol residue were selectively removed for regiospecific introduction of the phosphoethanolamine and phospholipid moieties, respectively, which was followed by global deprotection to afford the final synthetic target **28**.

This highly efficient synthesis was made possible owing to the successful use of glycosylation strategies based on selective glycosyl donor activation involving "armed–disarmed" and semiorthogonal concepts. For example, the presence of a cyclic 1,2-diketal protection in **18** and **23** structures causes torsional strains, which made them disarmed; therefore, the armed glycosyl donors **17** and **22** could be

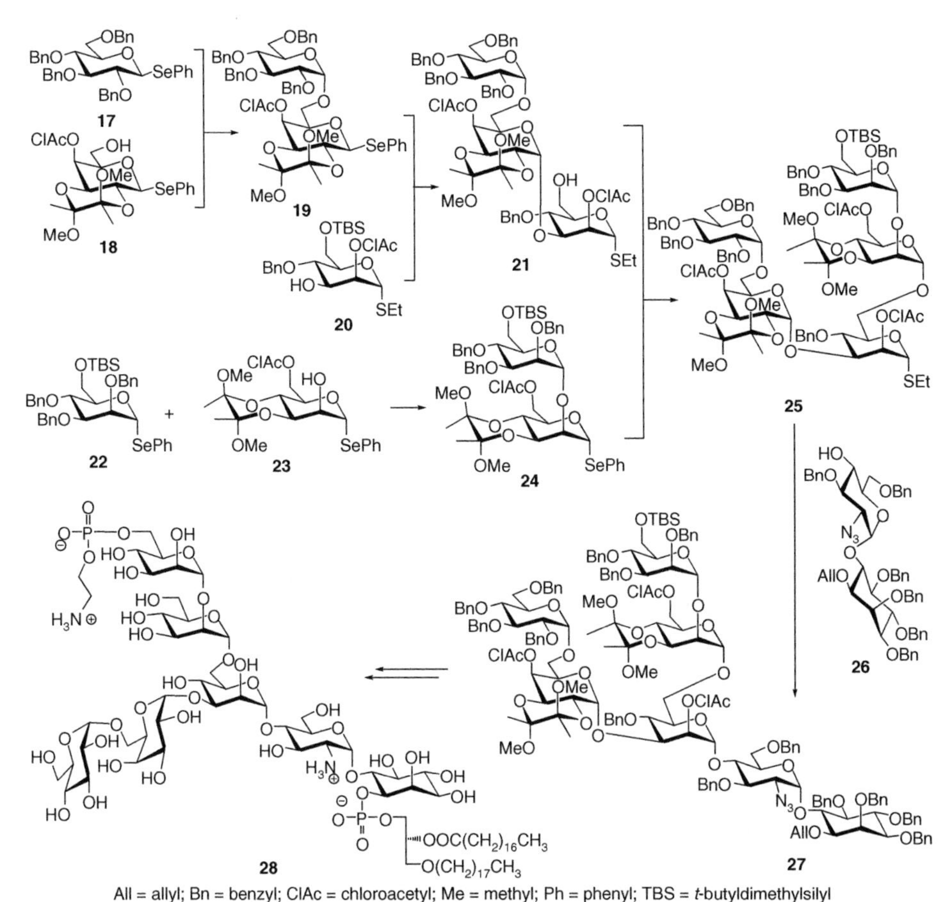

Scheme 8 Convergent assembly of the GPI anchor of *Trypanosoma brucei*.

selectively activated in the presence of **18** and **23**, even though **17, 18, 22**, and **23** all had the same phenylseleno group at their anomeric centers. On the other hand, because selenoglycosides are more reactive than thioglycosides, it was possible to activate glycosyl donors **19** and **24** in the presence of **20** and **21** for selective glycosylation. Finally, thioglycoside **25** was activated as a donor by a stronger promoter to achieve the last glycosylation.

C. GPI Synthesis by Solid-Phase Assembly of the Carbohydrate Chain

There are several reports in this area, but here we only present the solid-phase synthesis of a GPI glycan (Scheme 9) reported by Lopez-Prados and Martin-Lomas (110). In this synthesis, the carbohydrate chain was elongated along the nonreducing end, and glycosylation reactions were realized by Schmidt method. Notably, the difficult glycosidic linkage between glucosamine and inositol was accomplished

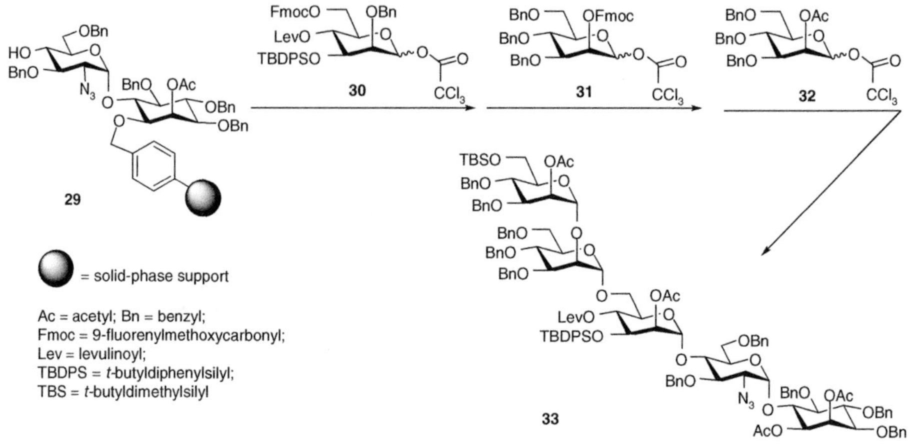

= solid-phase support

Ac = acetyl; Bn = benzyl;
Fmoc = 9-fluorenylmethoxycarbonyl;
Lev = levulinoyl;
TBDPS = *t*-butyldiphenylsilyl;
TBS = *t*-butyldimethylsilyl

Scheme 9 Solid-phase assembly of the glycan core of GPI anchors.

in solution first, and then, the resultant pseudodisaccharide was attached to a polyethylene glycol-grafted polystyrene resin to get **29**. The carbohydrate chain was assembled starting from **29** with glycosyl trichloroimmidates **30, 31**, and **32** as donors. Finally, the desired glycan **33** was released from the resin and isolated in a 20% overall yield.

D. Convergent Synthesis of a GPI with 2-*O*-Acylated Inositol

In the synthesis of sperm CD52 GPI anchor that contains a 2-*O*-acylated inositol, we met a special problem. The presence of a large acyl group at the inositol 2-*O*-position adds more steric hindrance to the already rather crowded structure of GPIs around the inositol residue. Therefore, the traditional synthetic design, namely, to introduce the phospholipid moiety to the inositol 1-*O*-position at the final stage failed, and a new synthetic design was developed to solve the problem (106). This is a good manifestation that presently there is not a universally applicable synthetic design for all carbohydrates and that the synthesis of each oligosaccharide must be treated as an individual challenge with the synthesis planned accordingly, even when there are precedent syntheses of similar structures.

Our solution to the above problem was to introduce the phospholipid moiety at an early stage, that is, before the elongation of the carbohydrate chain (106) (Scheme 10). Therefore, after the pseudodisaccharide **36** was obtained, its inositol 1-*O*-position was deprotected and phospholipidated immediately, which offered **39** in a good yield. Thereafter, the strategy of convergent assembly was employed to elongate the carbohydrate chain. As shown, the trimannose fragment **46** was constructed by a process based on semiorthogonal glycosylations, and **46** was directly used to glycosylate **39** to obtain the GPI core **47**. It was proven that the phospholipid moiety was stable to subsequent transformations. Finally, a phosphoethanolamine group was introduced to the carbohydrate nonreducing end that

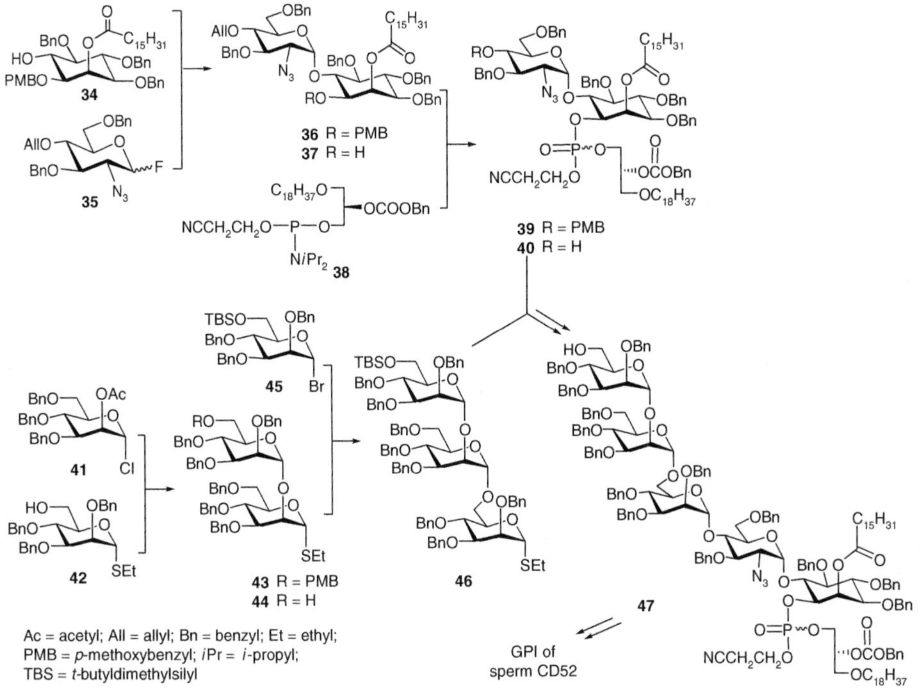

34
35
36 R = PMB
37 R = H
38
39 R = PMB
40 R = H
41
42
43 R = PMB
44 R = H
45
46
47
GPI of sperm CD52
Ac = acetyl; All = allyl; Bn = benzyl; Et = ethyl;
PMB = p-methoxybenzyl; iPr = i-propyl;
TBS = t-butyldimethylsilyl

Scheme 10 Convergent assembly of the GPI anchor of sperm CD52 antigen.

was followed by global deprotection to eventually produce the GPI anchor of sperm CD52.

V. Conclusion

During the last three decades, carbohydrate synthesis has witnessed tremendous progress, manifested by the enormous number of protecting groups and tactics, glycosylation methodologies, and synthetic strategies developed in the area, as well as the complex structures that have been achieved. However, presently there is not a magical strategy that is generally applicable to various oligosaccharides, unlike in oligopeptide and oligonucleotide synthesis. The problems are severalfold. First, the availability of monosaccharide building blocks is still difficult. Although it is currently possible to prepare all kinds of building blocks in the laboratory, it is extremely time-consuming. In fact, for most oligosaccharide syntheses, the largest part of time is spent in the preparation of monosaccharide building blocks. This situation will not change very soon, as it is determined by the structural property of carbohydrates. For example, each amino acid or nucleoside may need only several protecting forms to meet all kinds of synthetic demands, but for a monosaccharide, this number must be one to two orders of magnitudes higher. On top of this, the cost of synthesizing a regioselectively protected monosaccharide is extremely high, which at least presently prohibits the commercialization of complex monosaccharide derivatives. Second, the results of many glycosylation reactions are still difficult to predict. Compared to the chemistry involved in peptide and nucleotide synthesis, glycosylation reactions are much more sensitive to reaction conditions, such as moisture, temperature, the structure of substrates including both donors and acceptors, and solvent. Presently, there is no method that can give consistently high coupling efficiency and stereoselectivity for a wide range of substrates. Additionally, carbohydrate synthesis is a specialized discipline that requires a lot of special training and experience for one to deal with related problems. It is difficult for laboratories that are not specialized in organic synthesis to perform carbohydrate synthesis efficiently. On the other hand, it has become more and more evident that carbohydrates play a critical role in many biological processes and that the demand for carbohydrates from various research areas has been ever growing. Consequently, studies to search for new glycosylation methods and synthetic strategies to improve the overall efficiency of carbohydrate synthesis and studies to establish robust and general synthetic protocols to simplify carbohydrate synthesis will remain hot research topics in years to come.

Acknowledgments

I thank my devoted students and research associates who have made great contributions to the research program of our laboratory. We appreciate the US National Science Foundation (CHE-0407144) and National Institutes of Health (R01-CA95142) for their financial support of our research projects.

References

1. Laine RA. A calculation of all possible oligosaccharide isomers both branched and linear yields 1.05×10^{12} structures for a reducing hexasaccharide: the isomer barrier to development of single-method saccharide sequencing or synthesis systems. Glycobiology 1994; 4:759–767.
2. Varki A. Biological roles of oligosaccharides: all of the theories are correct. Glycobiology 1993; 3:97–130.
3. Hanessian S ed. In: Preparative Carbohydrate Chemistry, New York: Marcel Dekker, Inc., 1997.
4. Seeberger PH, Haase W-C. Solid-phase oligosaccharide synthesis and combinatorial carbohydrate libraries. Chem Rev 2000; 100:4349–4394.
5. Mootoo DR, Konradsson P, Udodong U, Fraser-Reid B. Armed and disarmed n-pentenyl glycosides in saccharide couplings leading to oligosaccharides. J Am Chem Soc 1988; 110:5583–5584.
6. Kanie O, Ito Y, Ogawa T. Orthogonal glycosylation strategy in oligosaccharide synthesis. J Am Chem Soc 1994; 116:12073–12074.
7. Ito Y, Kanie O, Ogawa T. Orthogonal glycosylation strategy for rapid assembly of oligosaccharides on a polymer support. Angew Chem Int Ed 1996; 35:2510–2512.

8. Huang X, Huang L, Wang H, Ye X. Iterative one-pot synthesis of oligosaccharides. Angew Chem Int Ed 2004; 43:5221–5224.
9. Demchenko AV. Strategic approach to the chemical oligosaccharide synthesis. Lett Org Chem 2005; 2:580–589.
10. Boons G-J. Strategies in oligosaccharide synthesis. Tetrahedron 1996; 54:1095–1121.
11. Matsuzaki Y, Ito Y, Nakahara Y, Ogawa T. Synthetic studies on cell-surface glycans. 93. Synthesis of branched poly-N-acetyllactosamine type pentaantennary pentacosasaccharide: glycan part of a glycosyl ceramide from rabbit erythrocyte membrane. Tetrahedron Lett 1993; 34:1061–1064.
12. Fraser-Reid B, Lu J, Jayaprakash KN, Lopez JC. Synthesis of a 28-mer oligosaccharide core of mycobacterial lipoarabinomannan (LAM) requires only two n-pentenyl ortho-ester progenitors. Tetrahedron: Asymmetry 2006; 17: 2449–2463.
13. Koenigs W, Knorr E. Uber einige derivate des traubenzuckers und der galactose. Ber 1901; 34:957–981.
14. Khan SH, O'Neil RA, eds. In: Modern Methods in Carbohydrate Synthesis, New York: Harwood Academic Publishers, 1996.
15. Ernst B, Hart GW, Sinay P, eds. In: Carbohydrates in Chemistry and Biology, Weinheim: Wiley, 2000.
16. Barresi F, Hindsgaul O. Synthesis of β-D-mannose containing oligosaccharides. In: Khan SH, O'Neil RA, eds. Modern Methods in Carbohydrate Synthesis. New York: Harwood Academic Publishers, 1996; 251–276.
17. Crich D. Chemistry of glycosyl triflates: synthesis of β-mannopyranosides. J Carbohydr Chem 2002; 21:667–690.
18. El Ashry ESH, Rashed N, Ibrahim ESI. Strategies of synthetic methodologies for constructing β-mannosidic linkage. Curr Org Syn 2005; 2:175–213.
19. Ito Y, Ohnishi Y. Stereoselective synthesis of β-manno-glycosides. Glycoscience 2001; 2:1589–1619.
20. Demchenko AV. Stereoselective chemical 1,2-cis O-glycosylation: from "Sugar Ray" to modern techniques of the 21st century. Synlett 2003; 1225–1240.
21. Kronzer FJ, Schuerch C. Methanolysis of some derivatives of 2,3,4-tri-O-benzyl-a-D-glucopyranosyl bromide in the presence and absence of silver salts. Carbohydr Res 1973; 27:379–390.
22. Lemieux RU, Hendriks KB, Stick RV, James K. Halide ion catalyzed glycosidation reactions. Syntheses of α-linked disaccharides. J Am Chem Soc 1975; 97:4056–4062.
23. Mukaiyama T, Murai Y, Shoda S. An efficient method for glucosylation of hydroxy compounds using glucopyranosyl fluoride. Chem Lett 1981; 431–432.
24. Suzuki K, Maeta H, Matsumoto T. An improved procedure for metallocene-promoted glycosidation. Enhanced reactivity by employing 1:2-ratio of Cp_2HfCl_2-$AgClO_4$. Tetrahedron Lett 1989; 30:4853–4856.
25. Hadd MJ, Gervay J. Glycosyl iodides are highly efficient donors under neutral conditions. Carbohydr Res 1999; 320:61–69.
26. Lam SN, Gervay-Hague J. Solution-phase hexasaccharide synthesis using glucosyl iodides. Org Lett 2002; 4:2039–2042.
27. Lam SN, Gervay-Hague J. Efficient synthesis of Man2, Man3, and Man5 oligosaccharides, using mannosyl iodide donors. J Org Chem 2005; 70:8772–8779.

28. Schmidt RR. New methods of glycoside and oligosaccharide syntheses: are there alternatives to the Koenigs-Knorr method? Angew Chem, Int Ed Engl 1986; 25:212–215.
29. Schmidt RR. Recent developments in the synthesis of glycoconjugates. Pure Appl Chem 1989; 61:1257.
30. Schmidt RR, Kinzy W. Anomeric-oxygen activation for glycoside synthesis: the trichloroacetimidate method. Adv Carbohydr Chem Biochem 1994; 50:21–56.
31. Schmidt RR, Jung K. Trichloroacetimidates. In: Ernst B, Hart GW, Sinay P, eds. Carbohydrates in Chemistry and Biology. Weinheim: Wiley, 2000; 5–59.
32. Plante OJ, Andrade RB, Seeberger PH. Synthesis and use of glycosyl phosphates as glycosyl donors. Org Lett 1999; 1:211–214.
33. Love KR, Seeberger PH. Synthesis and use of glycosyl phosphates as glycosyl donors. Org Syn 2005; 81:225–234.
34. Plante OJ, Palmacci ER, Seeberger PH. Formation of β-glucosamine and β-mannose linkages using glycosyl phosphates. Org Lett 2000; 2:3841–3843.
35. Plante OJ, Palmacci ER, Andrade RB, Seeberger PH. Oligosaccharide synthesis with glycosyl phosphate and dithiophosphate triesters as glycosylating agents. J Am Chem Soc 2001; 123:9545–9554.
36. Palmacci ER, Plante OJ, Seeberger PH. Oligosaccharide synthesis in solution and on solid support with glycosyl phosphates. Euro J Org Chem 2002; 2002:595–606.
37. Tanaka H, Sakamoto H, Sano A, Nakamura S, Nakajima M, Hashimoto S. An extremely mild and stereocontrolled construction of 1,2-cis-α-glycosidic linkages via benzyl-protected glycopyranosyl diethyl phosphites. Chem Commun 1999; 1259–1260.
38. Zhang Z, Wong C-H. Glycosylation methods: use of phosphites. In: Ernst B, Hart GW, Sinay P, eds. Carbohydrates in Chemistry and Biology. Weinheim: Wiley, 2000; 117–134.
39. Roy R, Andersson FO, Letellier M. "Active" and "latent" thioglycosyl donors in oligosaccharide synthesis. Application to the synthesis of a-sialosides. Tetrahedron Lett 1992; 33:6053–6056.
40. Huang L, Wang Z, Huang X. One-pot oligosaccharide synthesis: reactivity tuning by post-synthetic modification of aglycon. Chem Commun 2004; 1960–1961.
41. Garegg PJ. Thioglycosides as glycosyl donors in oligosaccharide synthesis. Adv Carbohydr Chem Biochem 1997; 52:179–205.
42. Fugedi P, Garegg PJ. A novel promoter for the efficient construction of 1,2-trans linkages in glycoside synthesis, using thioglycosides as glycosyl donors. Carbohydr Res 1986; 149:C9–C12.
43. Konradsson P, Mootoo DR, McDevitt RE, Fraser-Reid B. Iodonium ion generated *in situ* from *N*-iodosuccinimide and trifluoromethanesulfonic acid promotes direct linkage of disarmed pent-4-enyl glycosides. J Chem Soc Chem Commun 1990; 270–272.
44. Dasgupta F, Garegg PJ. Use of sulfenyl halides in carbohydrate reactions. Part I. Alkyl sulfenyl triflate as activator in the thioglycoside-mediated formation of β-glycosidic linkages during oligosaccharide synthesis. Carbohydr Res 1988; 177:C13–C17.

45. Crich D, Banerjee A, Li W, Yao Q. Improved synthesis of 1-benzenesulfinyl piperidine and analogs for the activation of thioglycosides in conjunction with trifluoromethanesulfonic anhydride. J Carbohydr Chem 2005; 24:415–424.
46. Zhang Z, Ollmann IR, Ye X-S, Wischnat R, Baasov T, Wong C-H. Programmable one-pot oligosaccharide synthesis. J Am Chem Soc 1999; 121:734–753.
47. Lou B, Huynh HK, Hanessian S. Oligosaccharide synthesis by remote activation: O-protected glycosyl 2-thiopyridylcarbonate donors. In: Hanessian S ed., Preparative Carbohydrate Chemistry. New York: Marcel Dekker, Inc., 1997; 431–448.
48. Demchenko AV, Malysheva NN, De Meo C. S-Benzoxazolyl (SBox) glycosides as novel, versatile glycosyl donors for stereoselective 1,2-cis glycosylation. Org Lett 2003; 455–458.
49. Pornsuriyasak P, Demchenko AV. S-thiazolinyl (STaz) glycosides as versatile building blocks for convergent selective, chemoselective, and orthogonal oligosaccharide synthesis. Chem Eur J 2006; 12:6630–6646.
50. Kahne D, Walker S, Cheng Y, van Engen D. Glycosylation of unreactive substrates. J Am Chem Soc 1989; 111:6881–6882.
51. Kim S-H, Augeri D, Yang D, Kahne D. Concise synthesis of the calicheamicin oligosaccharide using sulfoxide glycosylation method. J Am Chem Soc 1994; 116:1766–1775.
52. Yan L, Taylor CM, Goodnow R, Kahne D. Glycosylation on the merrifield resin using anomeric sulfoxides. J Am Chem Soc 1994; 116:6953–6954.
53. Crich D, Sun S. Direct synthesis of β-mannopuranosides by the sulfoxide method. J Org Chem 1997; 62:1198–1199.
54. Mehta S, Pinto M. Phenyl selenoglycosides as versatile glycosylation agents in oligosaccharide synthesis and the chemical synthesis of disaccharides containing sulfur and selenium. In: Hanessian S ed., Preparative Carbohydrate Chemistry. New York: Marcel Dekker, Inc., 1997; 106–129.
55. Mootoo DR, Date V, Fraser-Reid B. n-Pentenyl glycosides permit the chemospecific liberation of the anomeric center. J Am Chem Soc 1988; 110:2662–2663.
56. Fraser-Reid B, Merritt JR, Handlon AL, Andrews CW. The chemistry of N-pentenyl glycosides: synthetic, theoretical, and mechanistic ramifications. Pure Appl Chem 1993; 65:779–786.
57. Fraser-Reid B, Anilkumar G, Gilbert MR, Joshi S, Kraehmer R. Glycosylation methods: use of n-pentenyl glycosides. In: Ernst B, Hart GW, Sinay P, eds. Carbohydrates in Chemistry and Biology. Weinheim: Wiley, 2000; 135–154.
58. Seeberger PH, Bilodeau MT, Danishefsky SJ. Synthesis of biologically important oligosaccharides and other glycoconjugates by the glycal assembly method. Aldrichimica Acta 1997; 30:75–92.
59. Danishefsky SJ, Bilodeau MT. Glycals in organic synthesis: the evolution of comprehensive strategies for the assembly of oligosaccharides and glycoconjugates of biological consequence. Angew Chem Int Ed 1996; 35:1380–1419.
60. Toshima K, Tatsuta K. Recent progress in O-glycosylation methods and its application to natural products synthesis. Chem Rev 1993; 93:1503–1531.

61. Griffith DA, Danishefsky SJ. Sulfonamidoglycosylation of glycals. A route to oligosaccharides with 2-aminohexose subunits. J Am Chem Soc 1990; 112:5811–5819.
62. Banoub J, Boullanger P, Lafont D. Synthesis of oligosaccharides of 2-amino-2-deoxy sugars. Chem Rev 1992; 92:1167–1195.
63. Seeberger PH, Danishefsky SJ. Solid-phase synthesis of oligosaccharides and glycoconjugates by the glycal assembly method: a five year retrospective. Acc Chem Res 1998; 31:685–695.
64. Valeria DB, Kim Y, Gin DY. Direct oxidative glycosylations with glycal donors. J Am Chem Soc 1998; 120:13515–13516.
65. Valeria DB, Liu J, Huffman LGJ, Gin DY. Acetamidoglycosylation with glycal donors: a one-pot glycosidic coupling with direct installation of the natural C(2)-N-acetylamino functionality. Angew Chem Int Ed 2000; 39:204–207.
66. Gin DY. Dehydrative glycosylation with 1-hydroxy donors. J Carbohydr Chem 2002; 21:645–665.
67. Garcia BA, Poole JL, Gin DY. Direct glycosylations with 1-hydroxy glycosyl donors using trifluoromethanesulfonic anhydride and diphenyl sulfoxide. J Am Chem Soc 1997; 119:7597–7598.
68. Garcia BA, Gin DY. Dehydrative glycosylation with activated diphenyl sulfonium reagents. Scope, mode of C(1)-hemiacetal activation, and detection of reactive glycosyl Intermediates. J Am Chem Soc 2000; 122:4269–4279.
69. Nguyen HM, Chen Y, Duron SG, Gin DY. Sulfide-mediated dehydrative glycosylation. J Am Chem Soc 2001; 123:8766–8772.
70. Fraser-Reid B, López JC, Gómez AM, Uriel C. Reciprocal donor acceptor selectivity (RDAS) and Paulsens concept of match in saccharide coupling. Eur J Org Chem 2004; 1387–1395.
71. Baeschlin DK, Chaperon AR, Green LG, Hahn MG, Ince SJ, Ley SV. 1,2-Diacetals in synthesis: total synthesis of a glycosylphosphatidylinositol anchor of Trypanosoma brucei. Chem Eur J 2000; 6:172–186.
72. Yu B, Yang Z, Cao H. One-pot glycosylation (OPG) for the chemical synthesis of oligosaccharides. Curr Org Chem 2005; 9:179–194.
73. Koeller KM, Wong C-H. Synthesis of complex carbohydrates and glycoconjugates: enzyme-based and programmable one-pot strategies. Chem Rev 2000; 100:4465–4494.
74. Wang Y, Huang X, Zhang L, Ye X. A four-component one-pot synthesis of alpha-gal pentasaccharide. Org Lett 2004; 6:4415–4417.
75. Osborn HMI, Khan TH. Recent developments in polymer supported synthesis of oligosaccharides and glycopeptides. Tetrahedron 1999; 55:1807–1850.
76. Seeberger PH. Solid phase oligosaccharide synthesis. J Carbohydr Chem 2002; 21:613–643.
77. Krepinsky JJ. Advances in polymer-supported solution synthesis of oligosaccharides. In: Khan SH, O'Neil RA, eds. Modern Methods in Carbohydrate Synthesis. New York: Harwood Academic Publishers, 1996; 194–224.
78. Ito Y, Manabe S. Solid-phase oligosaccharide synthesis and related technologies. Curr Opin Chem Biol 1998; 2:701–708.
79. Plante OJ, Seeberger PH. Recent advances in automated solid-phase carbohydrate synthesis: from screening to vaccines. Curr Opin Drug Dis Dev 2003; 6:521–525.

80. Egusa K, Kusumoto S, Fukase K. A new catch-and-release purification method using a 4-azido-3-chlorobenzyl group. Synlett 2001; 777–780.
81. Egusa K, Kusumoto S, Fukase K. Solid-phase synthesis of a phytoalexin elicitor pentasaccharide using a 4-azido-3-chlorobenzyl group as the key for temporary protection and catch-and-release purification. Eur J Org Chem 2003; 3435–3445.
82. Palmacci ER, Hewitt MC, Seeberger PH. "Cap-Tag"–novel methods for the rapid purification of oligosaccharides prepared by automated solid-phase synthesis. Angew Chem Int Ed 2001; 40:4433–4437.
83. Wu J, Guo Z. Cap and capture-release techniques applied to solid-phase oligosaccharide synthesis. J Org Chem 2006; 71:7067–7070.
84. Ito Y, Ogawa T. Intramolecular aglycon delivery on polymer support: gatekeeper monitored glycosylation. J Am Chem Soc 1997; 119:5562–5566.
85. Matsuya Y, Itoh T, Nemoto H. Total synthesis of A-nor B-aromatic OSW-1 aglycon: a highly effective approach to optically active trans-4,5-benzhydrindane. Eur J Org Chem 2003; 2221–2224.
86. Yu H, Chokhawala H, Karpel R, Yu H, Wu B, Zhang J, Zhang Y, Jia Q, Chen X. A multifunctional Pasteurella multocida sialyltransferase: a powerful tool for the synthesis of sialoside libraries. J Am Chem Soc 2005; 127:17618–17619.
87. Gijsen HJM, Qiao L, Fitz W, Wong C-H. Recent advances in the chemoenzymatic synthesis of carbohydrates and carbohydrate mimetics. Chem Rev 1996; 96:443–473.
88. Wymer N, Toone EJ. Enzyme-catalyzed synthesis of carbohydrates. Curr Opin Chem Biol 2000; 4:110–119.
89. Song J, Zhang H, Li L, Bi Z, Chen M, Wang W, Yao Q, Guo H, Tian M, Li H, Yi W, Wang PG. Enzymatic biosynthesis of oligosaccharides and glycoconjugates. Curr Org Syn 2006; 3:159–168.
90. Ferguson MAJ. The structure, biosynthesis and functions of glycosylphosphatidylinositol anchors, and the contributions of trypanosome research. J Cell Sci 1999; 112:2799–2809.
91. Hwa K. Glycosyl phosphatidylinositol-linked glycoconjugates: structure, biosynthesis and function. Adv Exp Med Biology 2001; 491:207–214.
92. Menon AK, Baumann NA, Rancour DM. Functions of glycosyl phosphatidylinositols. Carbohydr Chem Biol 2000; 4:757–769.
93. Gaulton GN, Pratt JC. Glycosylated phosphatidylinositol molecules as second messengers. Semin Immunol 1994; 6:97–104.
94. Ferguson MAJ, Williams AF. Cell-surface anchoring of proteins via glycosylphosphatidylinositol structure. Ann Rev Biochem 1988; 57:285–320.
95. Treumann A, Lifely MR, Schneider P, Ferguson MAJ. Primary structure of CD52. J Biol Chem 1995; 270:6088–6099.
96. Guo Z, Bishop L. Chemical synthesis of GPIs and GPI-anchored glycopeptides. Euro J Org Chem 2004; 3585–3596.
97. Ferguson MAJ, Homans SW, Dwek RA, Rademacher TW. Glycosyl-phosphatidylinositol moiety that anchors Trypanosoma brucei variant surface glycoprotein to the membrane. Science 1988; 239:753–759.
98. Homans SW, Ferguson MAJ, Dwek RA, Rademacher TW, Anand R, Williams AF. Complete structure of the glycosyl phosphatidylinositol membrane anchor of rat brain Thy-1 glycoprotein. Nature 1988; 333:269–272.

99. Murakata C, Ogawa T. A total synthesis of GPI anchor of Trypanosoma brucei. Tetrahedron Lett 1991; 32:671–674.
100. Murakata C, Ogawa T. Stereoselective total synthesis of the glycosylphosphatidylinositol (GPI) anchor of Trypanosoma brucei. Carbohydr Res 1992; 235:95–114.
101. Murakata C, Ogawa T. Stereoselective synthesis of glycobiosyl phosphatidylinositol, a part structure of the glycosylphosphatidylinositol (GPI) anchor of Trypanosoma brucei. Carbohydr Res 1992; 234:75–91.
102. Campbell AS, Fraser-Reid B. First synthesis of a fully phosphorylated GPI membrane anchor: rat brain Thy-1. J Am Chem Soc 1995; 117:10387–10388.
103. Udodong UE, Madsen R, Roberts C, Fraser-Reid B. A ready, convergent synthesis of the heptasaccharide GPI membrane anchor of rat brain Thy-1. J Am Chem Soc 1993; 115:7886–7887.
104. Lu J, Jayaprakash KN, Fraser-Reid B. First synthesis of a malarial prototype: a fully lipidated and phosphorylated GPI membrane anchor. Tetrahedron Lett 2004; 45:879–882.
105. Lu J, Jayaprakash KN, Schlueter U, Fraser-Reid B. Synthesis of a malaria candidate glycosylphosphatidylinositol (GPI) structure: a strategy for fully inositol acylated and phosphorylated GPIs. J Am Chem Soc 2004; 126:7540–7547.
106. Xue J, Guo Z. Convergent synthesis of a GPI containing an acylated inositol. J Am Chem Soc 2003; 125:16334–16339.
107. Xue J, Shao N, Guo Z. First total synthesis of a GPI-anchored peptide. J Org Chem 2003; 68:4020–4029.
108. Baeschlin DK, Chaperon AR, Charbonneau V, Green LG, Ley SV, Lucking U, Walther E. Rapid assembly of oligosaccharides: total synthesis of a glycosylphosphatidylinositol anchor of Trypanosoma brucei. Angew Chem Int Ed 1998; 37:3423–3428.
109. Lopez-Prados J, Martin-Lomas M. Inositolphosphoglycan mediators: an effective synthesis of the conserved linear GPI anchor structure. J Carbohydr Chem 2006; 24:393–414.
110. Reichardt N-C, Martin-Lomas M. A practical solid-phase synthesis of glycosylphosphatidylinositol precursors. Angew Chem Int Ed 2003; 42:4674–4677.
111. Yashunsky DV, Borodkin VS, Ferguson MAJ, Nikolaev AV. The chemical synthesis of bioactive glycosylphosphatidylinositols from Trypanosoma cruzi containing an unsaturated fatty acid in the lipid. Angew Chem Int Ed 2006; 45(3):468–474.
112. Pekari K, Schmidt RR. A variable concept for the preparation of branched glycosyl phosphatidyl inositol anchors. J Org Chem 2003; 68:1295–1308.
113. Mayer TG, Kratzer B, Schmidt RR. Synthesis of a GPI anchor of yeast (Saccharomyces cerevisiae). Angew Chem, Int Ed 1994; 33:2177–2181.
114. Mayer TG, Schmidt RR. Glycosyl phosphatidylinositol (GPI) anchor synthesis based on versatile building blocks: total synthesis of a GPI anchor of yeast. Eur J Org Chem 1999; 1153–1165.
115. Pekari K, Tailler D, Weingart R, Schmidt RR. Synthesis of the fully phosphorylated GPI anchor pseudohexasaccharide of Toxoplasma gondii. J Org Chem 2001; 66:7432–7442.

116. Hewitt MC, Snyder DA, Seeberger PH. Rapid synthesis of a glycosylphosphatidylinositol-based malaria vaccine using automated solid-phase oligosaccharide synthesis. J Am Chem Soc 2002; 124:13434–13436.
117. Schofield L, Hewitt MC, Evans K, Siomos M-A, Seeberger PH. Synthetic GPI as a candidate antitoxic vaccine in a model of malaria. Nature 2002; 418:785–789.
118. Kwon Y, Soucy RL, Snyder DA, Seeberger PH. Assembly of a series of malarial glycosylphosphatidylinositol anchor oligosaccharides. Chem Euro J 2005; 11:2493–2504.

Carbohydrate Chemistry, Biology and Medical Applications
Hari G. Garg, Mary K. Cowman and Charles A. Hales

DOI: 10.1016/B978-0-08-054816-6.00004-5

Chapter 4

Enzymatic Synthesis of Oligosaccharides and Conversion to Glycolipids

DORIS M. SU, HIRONOBU EGUCHI, CHENGFENG XIA, JING SONG, WEN YI, ROBERT L. WOODWARD AND PENG G. WANG

Departments of Chemistry and Biochemistry, The Ohio State University, Columbus, OH 43210, USA

I. Introduction

Glycobiology has drawn a tremendous amount of attention and interest because oligosaccharides play essential roles in physiological and pathological processes such as molecular recognition, signal transduction, cell communication, tumor metastasis, and immune responses (1–5). There has also been a growing interest in the application of carbohydrates as pharmaceutical drugs in the field of invasive bacterial diseases, inflammatory diseases, and viral infections (6–10). However, research on the application of carbohydrates in modern medicine is limited due

Abbreviations: α1,3GalT, α-1,3-galactosyltransferase; Agm1, phosphoacetylglucosamine mutase; ATP, adenosine triphosphate; CK, creatine kinase; CMK, CMP kinase; CTP, cytidine triphosphate; Gal-1-P, galactose-1-phosphate; GalK, galactokinase; GalPUT, galactose-1-phosphate uridylyltransferase; GalU, glucose-1-phosphate uridylyltransferase; GFS, GDP-fucose synthetase; GMP, GDP-mannose pyrophosphorylase; LgtA, β-1,3-*N*-acetylglucosaminyltransferase; LgtC, α-1,4-galactosyltransferase; LgtD, β-1,3-*N*-acetylgalactosaminyltransferase; NDP, nucleoside diphosphate; NTP, nucleoside triphosphate; PEP, phosphoenolpyruvate; PMI, phosphomannose isomerase; PMM, phosphomannose mutase; PPi, pyrophosphate; PPA, inorganic pyrophosphatase; PTA, phosphate acetyltransferase (EC 2.3.1.8); PykF, pyruvate kinase; UDP, uridine diphosphate; UGD, UDP-glucose 6-dehydrogenase; UTP, uridine triphosphate; WbgU, UDP-GlcNAc 4-epimerase.

to the lack of availability of bioactive carbohydrates (11–13). Most of these glycoconjugates are rather difficult to obtain in large quantities, partially due to considerable limitations in oligosaccharide synthesis. Unlike DNA or peptide syntheses, which are commonly performed on solid phase with commercial instruments, comparable methods for oligosaccharides are still in their infancy (14,15).

Chemical oligosaccharide synthetic procedures often require tedious protection and deprotection steps. Enzymatic synthesis, on the other hand, offers tremendous benefits because of their high regio- and stereoselectivities, stability, organic solvent compatibility, and low cost (16–24). Glycosyltransferase-catalyzed synthesis of carbohydrates has been recognized as one of the most practical approaches for its high efficiency and specificity under very mild conditions. The two main obstacles in the enzymatic synthesis of carbohydrates are the inadequate availability of glycosyltransferases and the high cost of sugar nucleotides, which serve as intermediates.

II. Oligosaccharide Biosynthesis-Associated Enzymes

Enzymes have been used extensively to simplify the synthesis of complex oligosaccharides and glycoconjugates. Glycosyltransferases and glycosidases, for example, are valuable catalysts for the formation of specific glycosidic linkages (18,25). Other enzymes such as aldolases and sulfotransferases can also be exploited for the synthesis of distinct structures that are critical to oligosaccharide functions. Among the numerous enzymes associated with carbohydrate processing in cells, the enzymes used in enzymatic synthesis belong to three categories: glycosidases, glycosynthases, and glycosyltransferases (Table 1).

A. Glycosidases

Glycosidases are responsible for glycan-processing reactions that take place during glycoprotein synthesis. The physiological function of these enzymes is the cleavage of glycosidic linkages. However, under controlled conditions, glycosidases can be used to synthesize glycosidic bonds, in which a carbohydrate hydroxyl moiety acts as a more efficient nucleophile than water itself. Nevertheless, traditional glycosidase-catalyzed transglycosylations still suffer from low yields and unpredictable regioselectivities.

B. Glycosynthases

Glycosynthases, a class of glycosidase mutants, have been developed to enhance enzymatic activity toward the synthesis of oligosaccharides through the mutation of a single catalytic carboxylate nucleophile to a neutral amino acid residue (Ala or Ser). The resulting enzymes have no hydrolytic activity, but instead increased activity toward the synthesis of oligosaccharides when using glycosyl fluorides as activated donors (26–29).

C. Glycosyltransferases

Glycosyltransferases are enzymes that can transfer a sugar moiety to a defined acceptor to construct a specific glycosidic linkage. This "one enzyme–one linkage"

Table 1 Enzymatic Formations of Glycosidic Bonds

Donor + Acceptor $\xrightarrow{\text{Enzymes}}$ Product

Enzyme	Glycosyl donor	Advantages	Disadvantages
Glycosidase	Sugar-NP[a]	• Easy to perform • Low cost	• Low yield • Low regioselectivity
Glycosynthase	Sugar-F[b]	• High yield	• Low availability • Unpredictable regioselectivity
Leloir glycosyltransferase[c]	Sugar nucleotide	• High yield • High regio- and stereoselectivity • Essential for important sugar sequences	• High cost
Non-Leloir glycosyltransferase[d]	Sugar phosphate or glycoside	• High yield • High regio- and stereoselectivity	• Not essential for important sugar sequences

[a]Sugar-NP: nitrophenyl glycoside.
[b]Sugar-F: glycosyl fluoride.
[c]Leloir glycosyltransferase: glycosyltransferases of the Leloir biosynthetic pathway.
[d]Non-Leloir glycosyltransferase: glycosyltransferases of non-Leloir biosynthetic pathways.

concept makes glycosyltransferases extremely useful and important in the construction of glycosidic linkages in carbohydrates (30–32). These enzymes can be further divided into two groups, the transferases of the Leloir pathway and those of non-Leloir pathways. The Leloir pathway enzymes require sugar nucleotides as glycosylation donors, while non-Leloir glycosyltransferases typically utilize glycosylphosphates or glycosides. The Leloir transferases are responsible for the synthesis of most glycoconjugates in cells, especially in mammalian systems (33–35). Leloir transferases transfer a given carbohydrate from the corresponding sugar nucleotide donor substrate to a specific hydroxyl group of the acceptor sugar (Scheme 1).

Glycosyltransferases exhibit very strict stereospecificity and regiospecificity. Moreover, they can transfer with either retention or inversion of configuration at

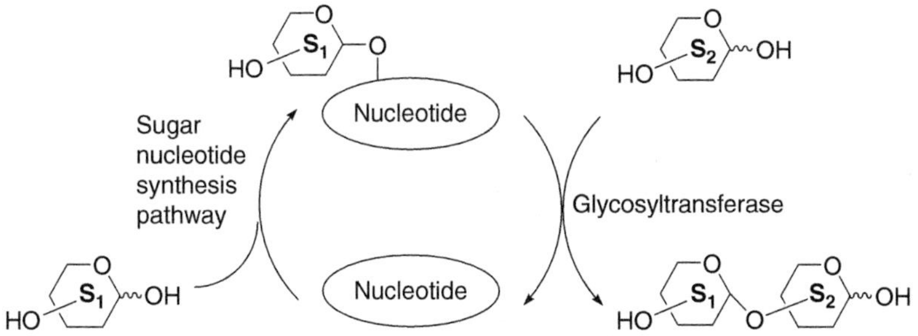

Scheme 1 Enzymatic glycosylation with regeneration of sugar nucleotide.

the anomeric carbon of the sugar residue. Although they are generally specific to substrates, minor modifications on donor and acceptor structures can be tolerated (36,37). Despite these merits, the preparative application of glycosyltransferases is limited by their inadequate availability. The amount of glycosyltransferases that can be isolated from natural resources is often limited by the low concentrations of these enzymes in most tissues. Furthermore, the purification procedures are quite complicated due to their relative instability because glycosyltransferases are often membrane bound or membrane associated. For these reasons, many efforts have been geared toward the genetic engineering of glycosyltransferases and consideration of their recombinant sources (38–42).

Recent advances in the area of enzymatic oligosaccharide synthesis are emerging from the identification and cloning of a large number of bacterial glycosyltransferases with many different donor, acceptor, and linkage specificities (43–47). Bacterial glycosyltransferases, compared to mammalian glycosyltransferases, are more easily cloned and expressed in large quantities in *Escherichia coli* (44). Furthermore, some bacterial transferases have been found to produce mammalian-like oligosaccharide structures that make these enzymes quite promising in the synthesis of biologically important oligosaccharides (43,48,49). In addition, bacterial glycosyltransferases seem to have relatively broader acceptor substrate specificities, thereby offering tremendous advantages over mammalian enzymes in the chemoenzymatic synthesis of oligosaccharides and their analogues for the development of antiadhesion therapies for infectious diseases (50,51).

III. Sugar Nucleotide Regeneration Systems

Although all the common sugar nucleotides are commercially available, these materials are prohibitively expensive. Because sugar nucleotides only serve as intermediates in enzymatic glycosylation, the most efficient synthetic approach is to regenerate them *in situ*. In addition to the benefit of cost reduction, the low concentration of sugar nucleotides regenerated *in situ* can reduce its inhibitory effect on the glycosyltransferase and increase the synthetic efficiency. The idea of *in situ* regeneration of sugar nucleotides was first demonstrated in 1982 by Whitesides and Wong with their work on uridine 5′-diphosphogalactose (UDP-Gal) regeneration (52). Since then, this revolutionary concept has been adopted in other regeneration systems and further developed in glycoconjugate syntheses (53–63).

A. Basic Principle

The eight common monosaccharide building blocks for important oligosaccharide sequences are glucose (Glc), galactose (Gal), glucuronic acid (GlcA), *N*-acetylglucosamine (GlcNAc), *N*-acetylgalactosamine (GalNAc), mannose (Man), fucose (Fuc), and sialic acid *N*-acetylneuraminic acid, Neu5Ac. In order to construct the sugar chains, the monosaccharides must be activated by attachment to nucleotides. The role of sugar nucleotides in glycoconjugate synthesis was first discovered in 1950 by Nobel laureate L. F. Leloir. Leloir demonstrated that a nucleoside triphosphate (NTP) such as uridine triphosphate (UTP) reacts with a glycosyl-1-phosphate to form a high-energy donor that participates in glycoconjugate synthesis. Half a century later, the biosynthetic pathways of the eight common sugar nucleotides (UDP-Glc, UDP-GlcNAc, UDP-GlcA, UDP-Gal, UDP-GalNAc, GDP-Man, GDP-Fuc, and CMP-NeuNAc synthetase (CMP-NeuAc)) have become well established. The general biosynthetic pathways of sugar nucleotides provide the basic guidelines on how to construct sugar nucleotides *in vitro* (64,65) (Scheme 2).

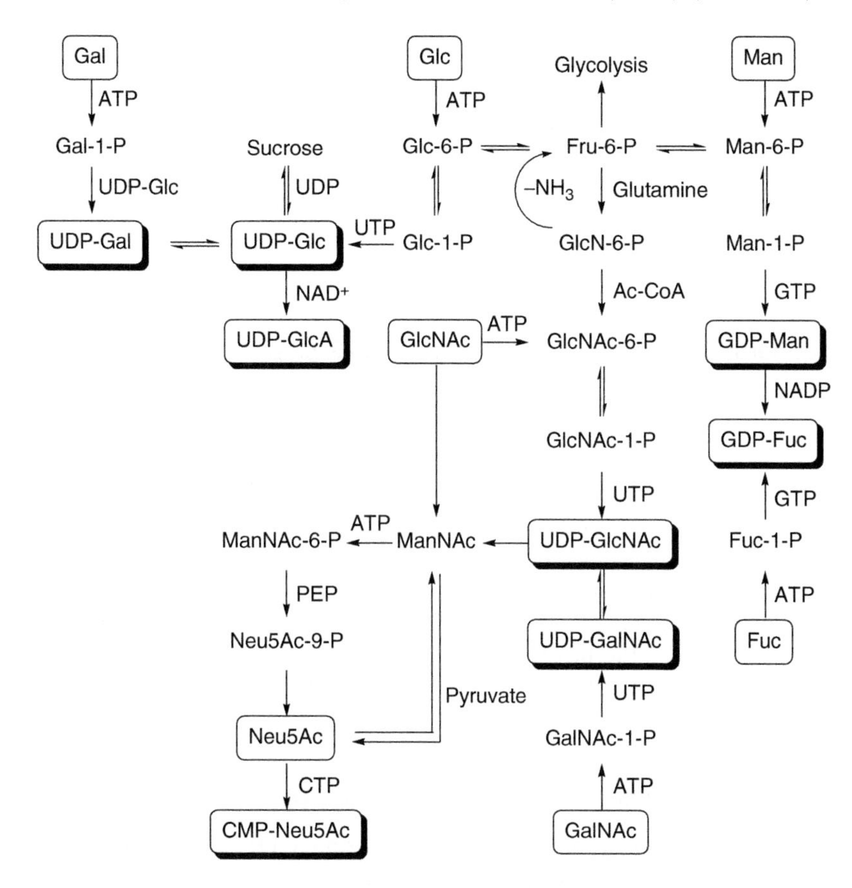

Scheme 2 The integrated biosynthetic pathway of sugar nucleotides.

As shown in Scheme 2, sugar nucleotides are formed by one or more of the three reaction pathways:

1. Sugar + ATP → Sugar-P + NTP → NDP-Sugar + PPi
2. $\text{NDP-Sugar}_A \leftrightarrow \text{NDP-Sugar}_B$
3. $\text{NDP-Sugar}_A + \text{Sugar}_B\text{-1-P} \leftrightarrow \text{NDP-Sugar}_B + \text{Sugar}_A\text{-1-P}$

As in reaction pathway 1, the steps to generating the corresponding sugar nucleotides for Glc, GlcNAc, and Man are (1) phosphorylation of sugar to sugar-6-P by a kinase; (2) epimerization of sugar-6-P to sugar-1-P by a mutase; and (3) conversion of sugar-1-P to nucleoside diphospho-sugar (NDP-Sugar) by a pyrophosphorylase. It should be noted that UDP-GlcA is universally generated by the oxidation of UDP-Glc. There are no C6 sugar kinases reported for Gal, GalNAc, and Fuc. However, the 6-OH group of Gal and GalNAc is quite sterically hindered for enzymatic phosphorylation because of the existence of the axial 4-OH group. Thus, Gal, GalNAc, and Fuc are phosphorylated at the anomeric position, converting them directly to sugar-1-phosphates.

In reaction pathways 2 and 3, there are several enzymes that interconvert different sugar nucleotides, such as UDP-glucose 4-epimerase (GalE, EC 5.1.3.2), UDP-GlcNAc 4-epimerase (WbgU, EC 5.1.3.7), and galactose-1-phosphate uridylyltransferase (GalPUT, EC 2.7.7.12). These enzymes are highly useful in sugar nucleotide regeneration systems because they can be employed to convert one sugar nucleotide into another or provide both sugar nucleotides at the same time.

The biosynthesis of CMP-Neu5Ac is distinct from the rest of the common sugar nucleotides because CMP-Neu5Ac is synthesized directly from Neu5Ac and cytidine triphosphate (CTP) without a sugar-1-phosphate intermediate.

B. Regeneration Systems of UDP-Gal, UDP-Glc, and UDP-GlcA

UDP-Glc, the central sugar nucleotide in cells, can be prepared from UTP and Glc-1-P in the presence of glucose-1-phosphate uridylyltransferase (GalU, EC 2.7.7.9) (66,67) or from sucrose and UDP with sucrose synthase (SusA, EC 2.4.1.13) (Scheme 3) (59,68–70). UDP-Gal can be prepared from UDP-Glc by C4 epimerization using GalE (71), from galactose-1-phosphate (Gal-1-P) and UTP, or from UDP-Glc and Gal-1-P using GalPUT (54,72,73). Besides the pioneering work of Whitesides, Wong, and their collaborators, Wang *et al.* have also applied five recombinant enzymes to regenerate UDP-Gal (Scheme 4) (74,75). Finally, UDP-GlcA can be readily prepared by the nicotinamide adenine dinucleotide-dependent oxidation of the C6 hydroxyl group of UDP-Glc catalyzed by UDP-glucose 6-dehydrogenase (UGD, EC 1.1.1.22) (76) (Scheme 5).

C. Regeneration Systems of UDP-GlcNAc and UDP-GalNAc

Following the biosynthetic pathway of synthesizing UDP-GlcNAc in eukaryotes (Scheme 6), GlcNAc is first phosphorylated by *N*-acetylglucosamine kinase (GlcNAcK, EC 2.7.1.59) (77). Phosphoacetylglucosamine mutase (Agm1, EC 5.4.2.3) (78) then converts GlcNAc-6-P to GlcNAc-1-P, which is subsequently

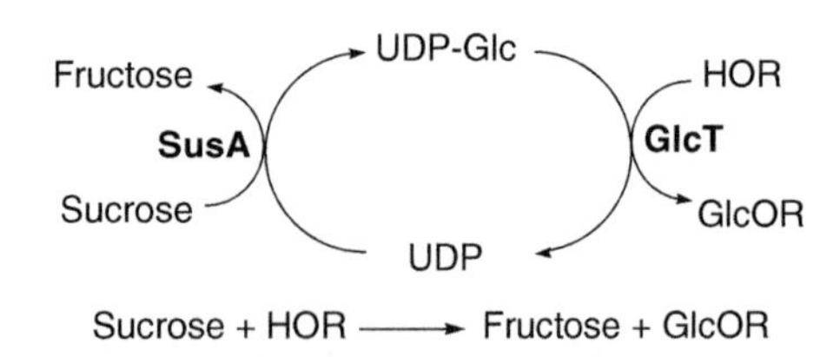

Scheme 3 Regeneration cycle of UDP-Glc by SusA.

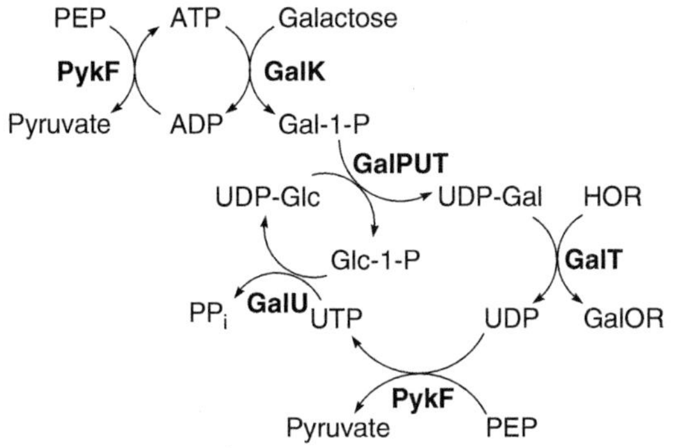

PEP, ATP, Galactose, **PykF**, **GalK**, Pyruvate, ADP, Gal-1-P, **GalPUT**, UDP-Glc, UDP-Gal, HOR, Glc-1-P, **GalT**, PP_i, **GalU**, UTP, UDP, GalOR, **PykF**, Pyruvate, PEP

Galactose + HOR + 2PEP ⟶ GalOR + 2Pyruvate + PP_i

Scheme 4 Regeneration cycle of UDP-Gal.

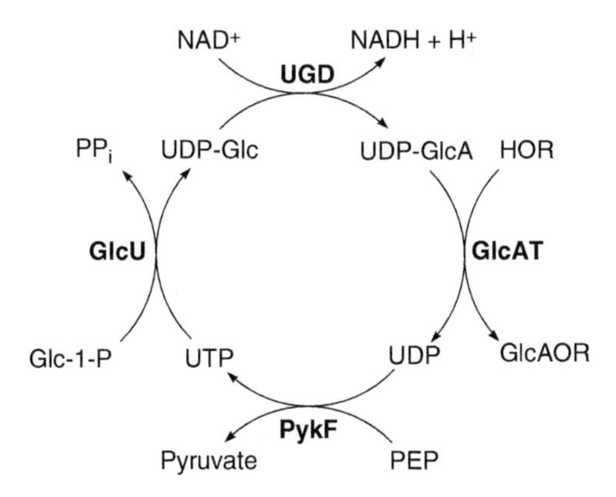

Glc-1-P + HOR + PEP + NAD^+ ⟶ GlcAOR + Pyruvate + PP_i + NADH + H^+

Scheme 5 Regeneration cycle of UDP-GlcA.

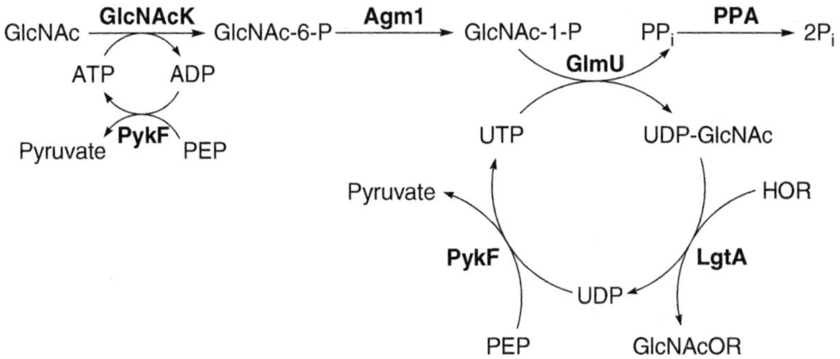

GlcNAc, **GlcNAcK**, GlcNAc-6-P, **Agm1**, GlcNAc-1-P, PP_i, **PPA**, $2P_i$, ATP, ADP, **GlmU**, Pyruvate, **PykF**, PEP, UTP, UDP-GlcNAc, Pyruvate, HOR, **PykF**, **LgtA**, UDP, PEP, GlcNAcOR

GlcNAc + HOR + 2PEP ⟶ GlcNAcOR + 2Pyruvate + $2P_i$

Scheme 6 Regeneration cycle of UDP-GlcNAc from GlcNAc.

uridylated to form UDP-GlcNAc by a truncated UDP-GlcNAc pyrophosphorylase (GlmU, EC 2.7.7.23) (79). The resulting ADP can be reconverted to adenosine triphosphate (ATP) by pyruvate kinase (PykF, EC 2.7.1.40) (80) with the consumption of one equivalent of phosphoenolpyruvate (PEP). The by-product pyrophosphate (PPi) is finally hydrolyzed by inorganic pyrophosphatase (PPA, EC 3.6.1.1) (81) which provides driving force. Wang *et al.* have developed "superbeads" for the synthesis of UDP-GlcNAc by immobilizing these enzymes along the biosynthetic pathway (82).

UDP-GalNAc can be generated by the addition of WbgU to the regeneration cycle of UDP-GlcNAc. Wang *et al.* have in fact cloned and overexpressed a novel WbgU from *Plesiomonas shigelloide* (83). An efficient UDP-GalNAc regeneration system (Scheme 7) was also established and used in synthesis of globotetraose and a series of derivatives with incorporation of the recombinant β-1,3-*N*-acetylgalactosaminyltransferase (LgtD) from the *Haemophilus influenzae* strain Rd (46).

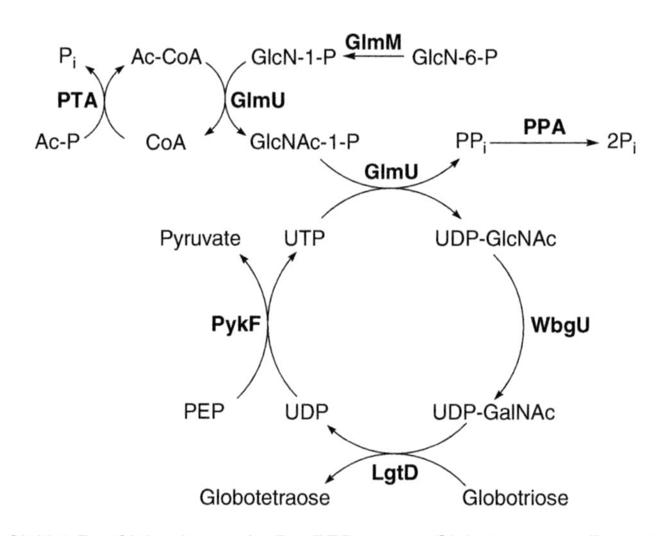

Scheme 7 Synthesis of globotetraose with regeneration of UDP-GalNAc.

D. Regeneration Systems of GDP-Man and GDP-Fuc

Wong *et al.* previously reported a GDP-Man regeneration system starting from Man-1-P (53). However, as Fru-6-P is inexpensive, a new regeneration pathway has been developed by Wang *et al.* starting from Fru-6-P (Scheme 8). As shown in Scheme 8, the cycle comprises of three key enzymes: phosphomannose isomerase (PMI, EC 5.3.1.8) for the conversion of Fru-6-P to Man-6-P (84–87), phosphomannose mutase (PMM, EC 5.4.2.8) for the conversion of Man-6-P to Man-1-P (88), and GDP-mannose pyrophosphorylase (GMP, EC 2.7.7.13) for condensation of Man-1-P with guanosine triphosphate to form GDP-Man (89–94). The recombinant bifunctional PMI/GMP from *Helicobacter pylori* has also been cloned and overexpressed in *E. coli* (95).

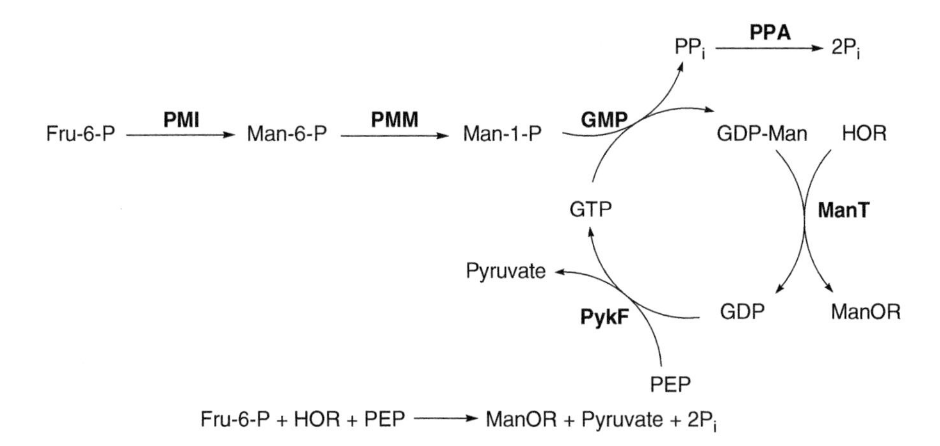

Scheme 8 Regeneration cycle of GDP-Man.

A GDP-Fuc regeneration cycle was constructed by incorporating two more enzymes, GDP-mannose 4,6-dehydratase (EC 4.2.1.47) and GDP-fucose synthetase (GFS, EC 1.1.1.271), into the same biosynthetic pathway of GDP-Man from Fru-6-P (Scheme 9) (96).

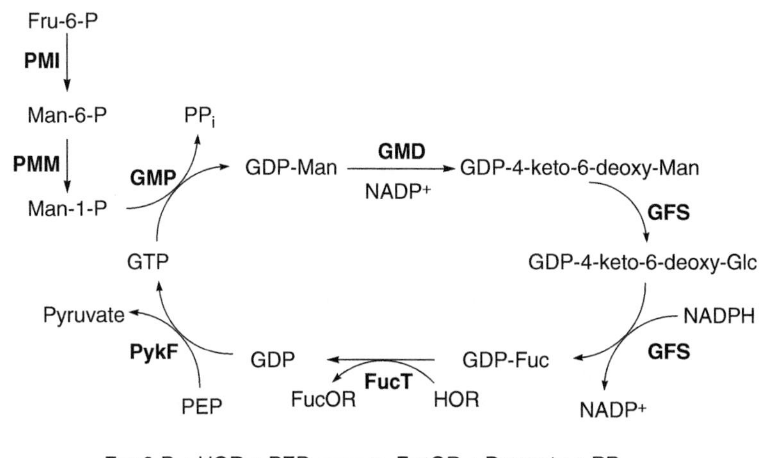

Scheme 9 Regeneration cycle of GDP-Fuc.

E. CMP–Neu5Ac Regeneration

CTP and Neu5Ac are used to synthesize CMP-Neu5Ac, which is a common substrate for the synthesis of sialylated oligosaccharides (97–99). Based on the existing CMP-Neu5Ac regeneration system reported by Wong *et al.* (100,101), the regeneration was extended to a cheaper starting material, ManNAc (Scheme 10). The CMP-Neu5Ac regeneration was constructed with the following five enzymes: NeuAc

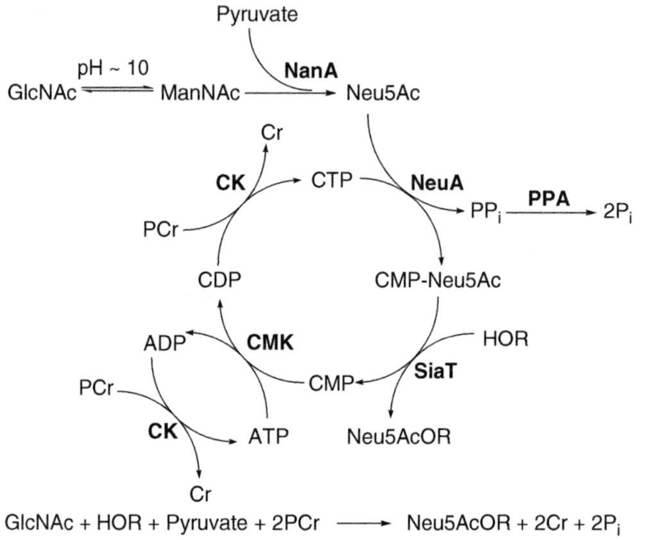

GlcNAc + HOR + Pyruvate + 2PCr ⟶ Neu5AcOR + 2Cr + $2P_i$

Scheme 10 Regeneration cycle of CMP-Neu5Ac.

aldolase (NanA, EC 4.1.3.3) (102,103), CMP-Neu5Ac Synthetase (NeuA, EC 2.7.7.43), CMP kinase (CMK, EC 2.7.4.14), and creatine kinase (CK, EC 2.7.3.2) coupled with an α-2,3-sialyltransferase cloned from *Pasteurella multocida* PM70. As shown in Scheme 10, creatine phosphate was used as an energy source instead of PEP, because creatine phosphate is a cheaper, more efficient, and convenient energy source compared to PEP, especially in large-scale synthesis of glycoconjugates (104). ManNAc can be epimerized chemically from GlcNAc under basic conditions (pH 10), thus GlcNAc can also be used as a starting material in sialyloligosaccharide synthesis.

IV. Enzymatic Oligosaccharide Synthesis Processes

In contrast to the template-driven biosynthesis of nucleic acids and proteins, the biosynthesis of carbohydrates is defined as the cooperation of glycosyltransferase machinery and their cofactors, sugar nucleotides. Chemical methods often require multiple protection–deprotection steps and long synthetic routes. As could be anticipated, the chemical synthesis of oligosaccharides is not an attractive option to industrial and scientific communities (105). Biocatalytic approaches employing enzymes or genetically engineered whole cells are, however, powerful and complementary alternatives to chemical methods (25).

A. Cell-Free Oligosaccharide Synthesis

In the early 1990s, Wong *et al.* established a one-pot multiple-enzyme *in situ* sugar nucleotide regeneration system for carbohydrate synthesis, which was shown to operate very efficiently without problems of product inhibition (55,56,100,101). However, in this one-pot system, purified enzymes were used and the recombinant enzymes remained too expensive. Moreover, the enzymes could not be reused and it was not convenient to separate the product from the system. Alternatively, many research efforts have focused on the immobilization of glycosyltransferases and other enzymes coupled with water-soluble polymers to circumvent these limitations (25,106–111). Immobilized enzyme systems have shown advantages such as ease of product separation, increased stability, and reusability of the catalysts.

1. Glycosyltransferase Immobilization and Water-Soluble Glycopolymers

Recently, two applications of immobilized multienzymes coupled with aqueous polymer supports for synthesis of oligosaccharides have been reported (107,108). Nishimura *et al.* made use of glycosyltransferases expressed as fusion proteins with the maltose-binding protein (MBP). The MBP domain functioned as a specific affinity tag for both purification and immobilization of this engineered biocatalyst onto resin. The acceptor was immobilized on a water-soluble polymer involving an α-chymotrypsin-sensitive linker moiety. Eighty percent of the GlcNAc residue was transferred from UDP-GlcNAc toward the LacCer polymer with immobilized MBP-β1,3GlcNAcT (108). On the other hand, Wong *et al.* developed a homogenous enzymatic synthesis system using thermoresponsive water-soluble polymer support. Several enzymes immobilized on thermoresponsive polyacrylamide polymers were nearly as active as their soluble forms and could be recovered for reuse after gentle heating and precipitation. The trisaccharide Lewis^{X} (Galβ,4(Fucα1,3)GlcNAc) was synthesized (60% yield) with no chromatographic purification of intermediates (107).

In these experiments, the versatility of oligosaccharide synthesis based on immobilized glycosyltransferases was demonstrated by the construction of a simple trisaccharide derivative that can be applied to further conjugation studies to produce glycodrugs or glycomaterials with interesting bioactivities (112). It should also be noted that glycosylation of polymer primers by glycosyltransferases conjugated with a macromolecular support afforded products even in the presence of high concentrations of inhibitors (nucleotides) (112). The availability of immobilized glycosyltransferases should greatly accelerate the development of enzyme-based automated glycosynthesizers (14,113,114).

2. Superbeads

Wang *et al.* have developed the cell-free superbead technology that demonstrated *in vitro* transfer of multiple-enzyme sugar nucleotide regeneration systems onto solid beads, which could be reused as common synthetic reagents for production of glycoconjugates (74). The superbeads were prepared by the following steps (Scheme 11A): (1) cloning and overexpression of individual N-terminal His_6-tagged enzymes along the sugar nucleotide biosynthetic pathway and (2) coimmobilization of the enzymes onto nickel-nitrilotriacetate beads. The first generation of superbeads for UDP-Gal regeneration along the biosynthetic pathway (Scheme 11B) required the following enzymes: galactokinase (GalK, EC

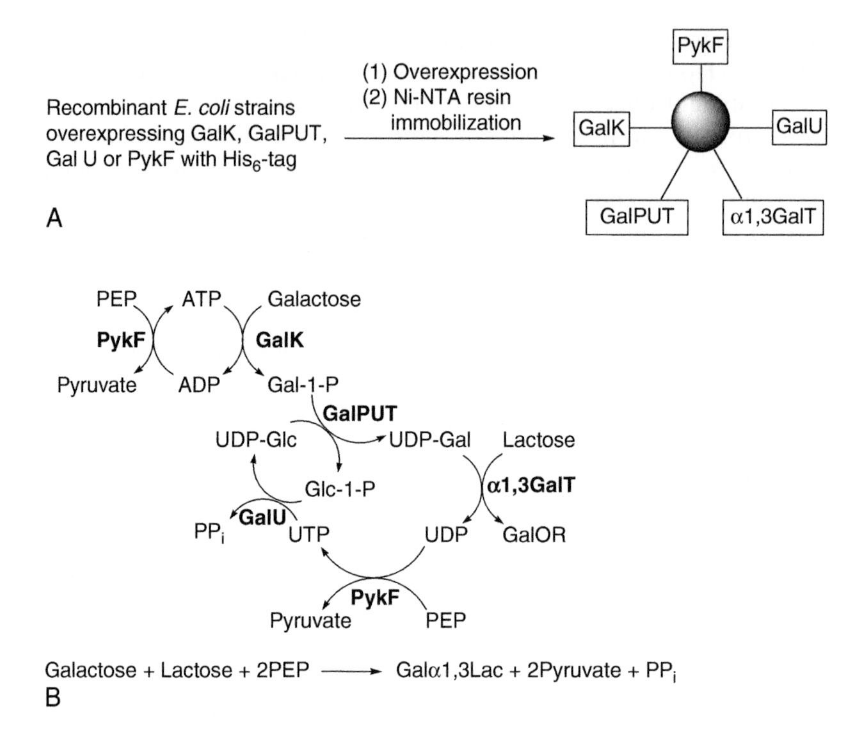

Scheme 11 Production of superbeads and biosynthetic pathway of globotriose with regeneration of UDP-Gal.

2.7.1.6), GalPUT, GalU, and PykF. The feasibility of the superbead approach was demonstrated by gram scale synthesis of Galα1,3Galβ1,4GlcOBn with a truncated bovine α-1,3-galactosyltransferase (α1,3GalT) expressed in *E. coli.* A yield of 72% was achieved based on acceptor LacOBn with 90% enzyme activity retained after reusing three times (yields were 71%, 69%, and 66%, respectively) within three weeks. The versatility was demonstrated in the synthesis of a variety of oligosaccharides such as Galβ1,4GlcNAc (92% yield) and globotriose (86% yield). Recently, an improved three-enzyme system (GalE, GalU, and α1,3GalT) to synthesize isoglobotriose using commercially inexpensive UTP has been reported (115).

Larger oligosaccharide synthesis will involve more corresponding glycosyltransferases coupled with necessary sugar nucleotide regenerating beads. Thus, these superbeads will likely become a new generation of bioreagents for the production of glycoconjugates and their derivatives.

B. Oligosaccharide Synthesis by Metabolically Engineered Bacteria

A rapidly emerging method for the large-scale synthesis of complex carbohydrates is the use of metabolically engineered microorganisms. When using whole cells to produce oligosaccharides, there is no need to isolate enzymes and biotransformations can be carried out with inexpensive precursors (25). Kyowa Hakko Kogyo Co. Ltd. developed a system for large-scale synthesis of oligosaccharides by coupling multiple metabolically engineered bacteria (63,116–118). In addition, Wang *et al.* have developed the "superbug" technology utilizing single recombinant bacteria carrying one engineered recombinant plasmid. Samain et al. have also investigated *in vivo* synthesis of oligosaccharides in recombinant microorganisms (119,120).

1. Bacterial Coupling

The bacterial coupling technology, developed by Kyowa Hakko Kogyo Co. Ltd. in Japan (57,63,116–118), initiated a new era in large-scale enzymatic synthesis of oligosaccharides. The key to Kyowa Hakko's technology was a *Corynebacterium ammoniagenes* bacterial strain engineered to efficiently convert inexpensive orotic acid to UTP (Scheme 12). When combined with an *E. coli* strain engineered to overexpress UDP-Gal biosynthetic genes including *galK, galT, galU*, and *ppa* (pyrophosphatase), UDP-Gal was accumulated in the reaction solution (72 mM/ 21 h). When these two strains were combined with another recombinant *E. coli* strain, overexpression of the α-1,4-galactosyltransferase *LgtC* gene of *Neisseria gonorrhoeae* produced a high concentration of globotriose (188 g/liter) (118).

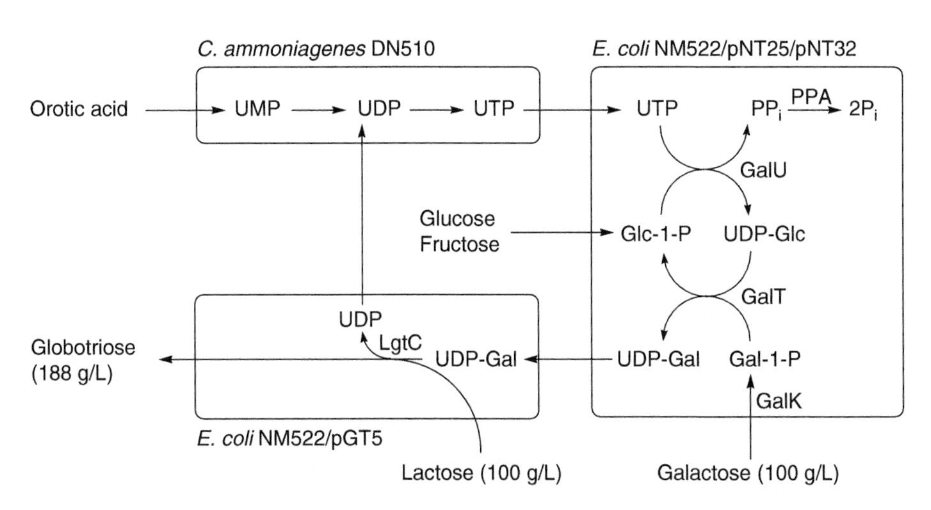

Scheme 12 Large-scale production of globotriose through coupling of engineered bacteria.

Kyowa Hakko also achieved large-scale production of other sugar nucleotides and related oligosaccharides with the utilization of this bacterial coupling concept. For fucosylated carbohydrate synthesis, the combination of genetically engineered *E. coli* overexpressing GDP-fucose biosynthetic genes and *C. ammoniagenes* produced 18.4 g/liter GDP-fucose after 22 h reaction. Total 21 g/liter of Lewis^X were synthesized using this system including α-1,3-fucosyltransferase

(121). Large-scale synthesis of sialylated carbohydrates was also reported (117). However, despite the striking feature of producing sugar nucleotides cost-effectively, the bacterial coupling method still suffers from the need of multiple fermentations of several bacterial strains and transport of substrates between different bacterial strains.

2. *Superbugs*

The superbug technology, developed by Wang *et al.*, made use of engineered bacteria through fermentation to provide all the necessary enzymes along the biosynthetic pathway. A single microbial strain was transformed with a single artificial gene cluster of all the biosynthetic genes. The metabolism of the engineered bacteria can then be exploited to provide the necessary bioenergetics (ATP or PEP) to drive a glycosylation cycle. In this approach, it is unnecessary to purify and immobilize individual enzymes; furthermore, the proteins may be expressed without tags (122).

Around 7.2 g/liter of isoglobotriose (Galα1,3Lac, also known as α-Gal epitope) were synthesized after 36 h using the "superbug" system (75) (Scheme 13). Five enzymes required for the synthesis of isoglobotriose including the sugar nucleotide regeneration pathway were cloned in tandem into single plasmid and transformed into *E. coli*. Similarly, globotriose and its derivatives (Galα1,4LacOBn) were also produced in high yields by using α1,4GalT instead of α1,3GalT (123).

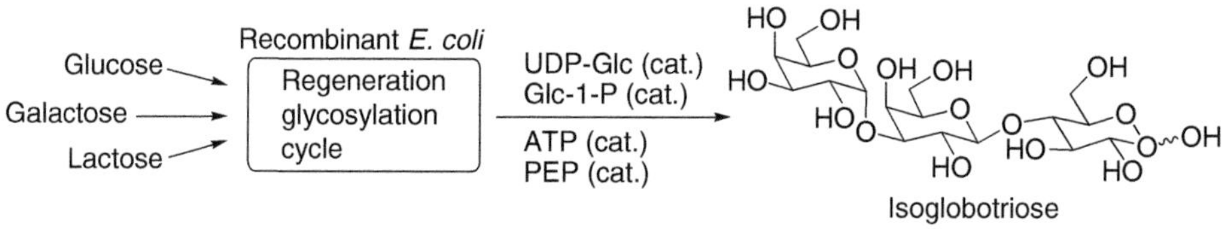

Scheme 13 α-Gal superbug.

An important aspect in the use of this superbug was that only catalytic amounts of ATP or PEP were needed for the whole cell synthesis compared to the need of stoichiometric amounts of ATP or PEP for cell-free *in vivo* synthesis. Obviously, this makes the superbug production of oligosaccharide the most cost-effective method (75,124,125). Another striking feature of the superbug reaction was that both the starting mono- and disaccharides and the trisaccharide products were membrane-permeable. Such superbug technology can be easily adapted and used in a variety of synthetic and biochemical procedures by simply replacing/inserting new glycosyltransferase genes into the plasmid.

3. *The Living Factory Technology*

The living factory technology, developed by Samain *et al.* in France, made use of the bacteria host cell's own ability to produce nucleotide sugars, while the bacteria were simply engineered to incorporate the required glycosyltransferases. In high-cell-density cultures, the oligosaccharide products accumulated intracellularly and have been shown to reach levels on the grams per liter scale (119,120,126–129). For example, high-cell-density cultivation of *lacZ*$^-$ strains that overexpressed the β-1,3-*N*-acetylglucosaminyltransferase *lgtA* gene of *Neisseria meningitides* resulted in the synthesis of 6 g/liter of the trisaccharide GlcNAcβ1,3Galβ1,4Glc (Scheme 14). Furthermore, ganglioside GM2 (GalNAcβ1,4(NeuAcα2,3)Galβ1,4Glc) was produced in recombinant *E. coli* expressing the genes for CMP-NeuAc synthase, α-2,3-sialyltransferase, UDP-GlcNAc C4 epimerase, and LgtD. This system was extended to the synthesis of GM1 (Galβ1,3GalNAcβ1,4(NeuAcα2,3)Galβ1,4Glc) by overexpressing the additional gene, β-1,3-galactosaminyltransferase (126). In this system, 1.25 and 0.89 g/liter of GM2 and GM1 were produced, respectively. Recently, globotetraose (GalNAcβ1,3Galα1,4Galβ1,4Glc) was synthesized in gram-scale quantities by using globotriose generation genes and the *lgtD* gene (130).

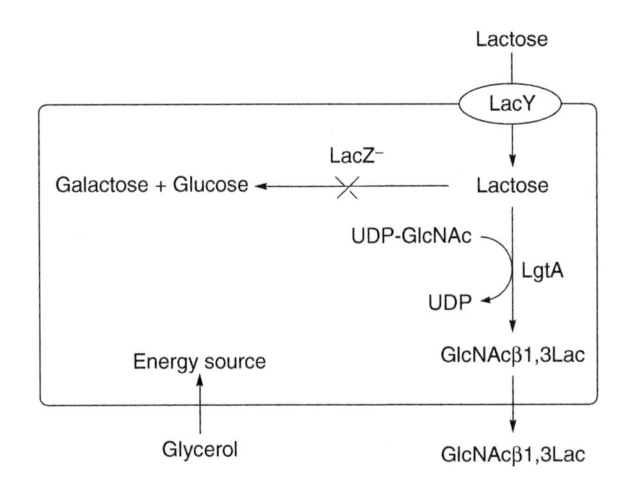

Scheme 14 Production of trisaccharide GlcNAcβ1,3Lac by *E. coli* JM109 expressing the *lgtA* gene that encodes a β-1,3-*N*-acetylglucosaminyltransferase.

This method shows the advantages of having no need for isolation and purification of recombinant glycosyltransferases and the bacteria cells already possess the machinery required for sugar nucleotide synthesis. On the other hand, the efficiency of this system is critically dependent on the intracellular pool of sugar nucleotides.

V. Conversion of Oligosaccharides to Glycolipids

In contrast to the conventional CD4$^+$ and CD8$^+$ T cells of the immune system that recognize specific peptide antigens bound to major histocompatibility complex class II or class I antigen presenting proteins, respectively (131–133), invariant natural killer T cells recognize glycolipid antigens presented by histocompatibility complex class I-like CD1d molecules (134). iNKT cells play a major role as the bridging system between innate and adaptive immunity (135). Upon activation, iNKT cells secrete signaling peptides to regulate a number of disease states *in vivo*, including malignancy, infection as well as autoimmune diseases (136,137). The lysosomal isoglobotrihexosylceramide (iGb3, Galα1,3Galβ1,4Glcβ1,1′Cer) was discovered as an endogenous ligand of iNKT cells (133).

A. Chemoenzymatic Syntheses of Isoglobotrihexosylceramide and Globotrihexosylceramide

The traditional synthetic protocols of glycolipids involved numerous steps of protecting and deprotecting to construct the desired oligosaccharides. However, with enzymatically synthesized oligosaccharides, the conversion of them to glycolipids is quite convenient (115). For example, the enzymatically synthesized isoglobotriose (Galα1,3Galβ1,4Glc) **1** can be used to synthesize the endogenous antigen of invariant natural killer T cells, iGb3 (Scheme 15) (133). The trisaccharide was protected with benzoyl groups by

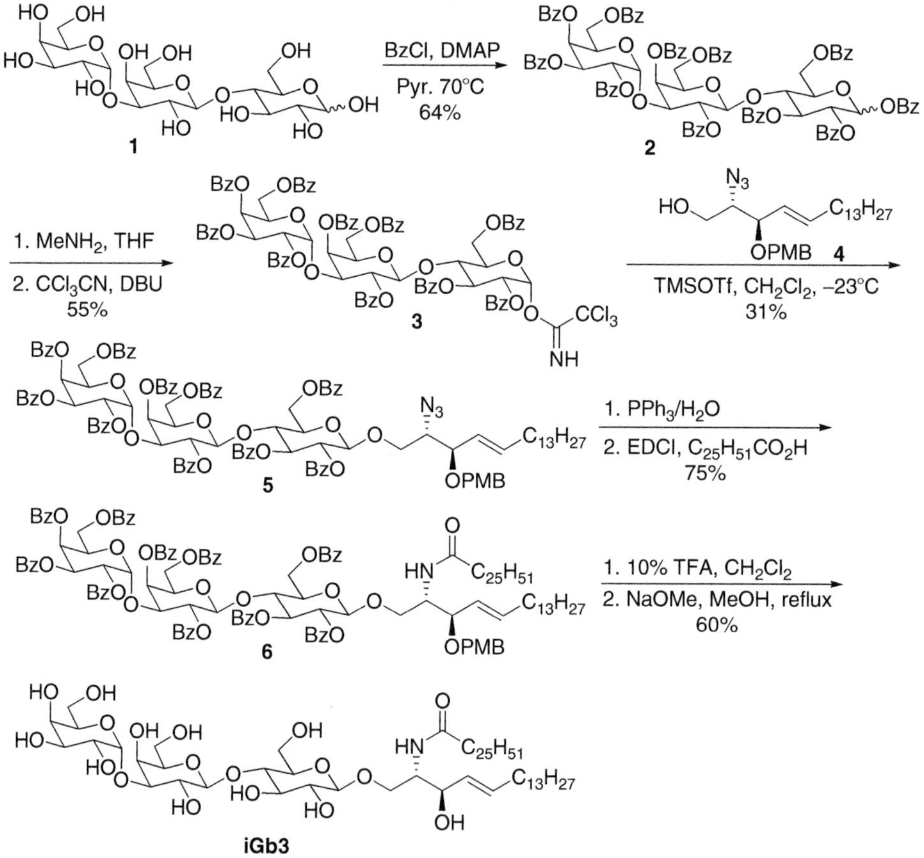

Scheme 15 Conversion of isoglobotriose to glycolipid iGb3.

subjection to benzoyl chloride in the presence of catalytic 4-(dimethylamino)pyridine and the reaction was stirred at 70 °C overnight to afford the perbenzoylated trisaccharide **2** in 64% yield. The ester protecting group of the C2 hydroxyl group is necessary for the formation of a β-configuration linkage between the saccharide and lipid moiety. The 1-OH was then selectively freed by treatment with $MeNH_2$ in tetrahydrofuran solution (138). The product was further converted to trichloroacetimide donor **3** with trichloracetonitrile and 1,8-diazabicyclo[5.4.0]-undec-7-ene using the standard procedure (139). Regarding the lipid acceptor, we used 2-azido-sphingenine **4** (140) instead of the 2-*N*-hexacosanylsphingenine (133) because the amide dramatically decreases the acceptor reactivity and leads to a low yield in glycosylation (141).

The glycosylation of lipid acceptor **4** with donor **3** promoted by trimethylsilyl triflate (TMSOTf) produced the desired β-glycolipid **5** with trace amounts of orthoester by-product. The azido group was reduced with triphenylphosphine and stoichiometric amounts of water in benzene at 70°C (142). The resulting amine was subsequently coupled directly with cerotic acid in the presence of 1-ethyl-3-(3′-dimethylaminopropyl) carbodiimide (EDCI) to provide the protected glycolipid **6** with 75% yield over two steps (143). Finally, the *p*-methoxybenzyl group was removed by trifluoroacetic in dichloromethane and the benzoyl esters were saponified with NaOMe, ultimately providing iGb3 in 60% yield.

Another glycolipid, globotrihexosylceramide (Gb3), was also prepared from globotriose using similar methodology (Scheme 16) (115). In order to avoid the formation of orthoesters in glycosylation, pivalate, a bulkier protecting group, was employed as the ester protecting group instead of benzoate.

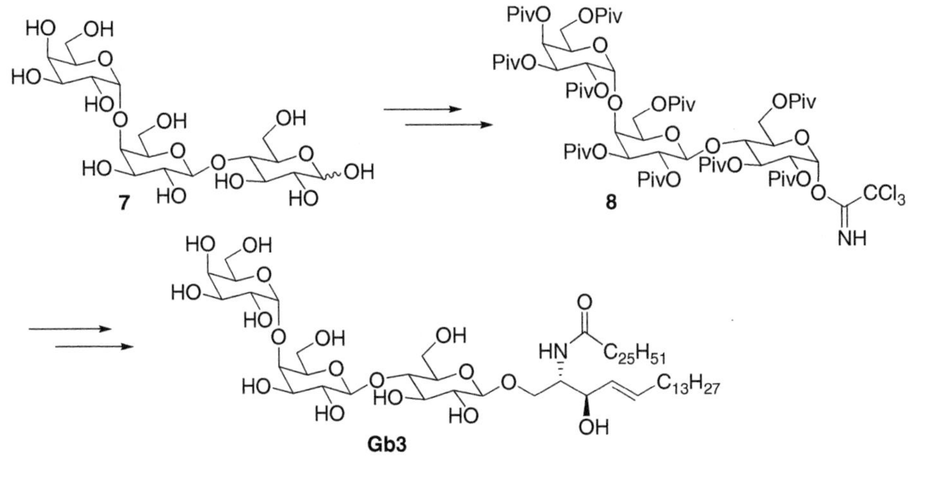

Scheme 16 Conversion of globotriose to glycolipid Gb3.

VI. Conclusion

Significant progress in the study of the enzymatic biosynthesis of complex carbohydrates has been made with developments in protein purification, molecular genetics, and new methods of enzymological analysis. Bioinformatics provide a large amount of putative candidates for carbohydrate active enzymes. The combined enzymatic and genetic approach has overcome formidable obstacles ubiquitous to the study of glycans, and has begun to yield new information that definitively addresses basic enzymological issues relevant to these complex carbohydrates. In fact, complex carbohydrates representing the glycan chain of gangliosides such as GT2, GM1, and GD1a were synthesized using glycosyltransferases (144). Although traditional cell-free enzymatic synthesis of oligosaccharides will continue to be used in laboratory scale synthesis, the future direction for large-scale preparation of complex oligosaccharides, and polysaccharides will be whole-cell systems. Further development of these enzymatic methods will allow synthetic biochemists to create important molecular tools for biochemical, biophysical, and medical applications.

References

1. Nangia-Makker P, Conklin J, Hogan V, Raz A. Carbohydrate-binding proteins in cancer, and their ligands as therapeutic agents. Trends Mol Med 2002; 8:187–192.
2. Varki A. Biological roles of oligosaccharides: all of the theories are correct. Glycobiology 1993; 3:97–130.
3. Kobata A. Structures and functions of the sugar chains of glycoproteins. Eur J Biochem 1992; 209:483–501.
4. Fukuda M. Possible roles of tumor-associated carbohydrate antigens. Cancer Res 1996; 56:2237–2244.
5. Tsuboi S, Fukuda M. Roles of O-linked oligosaccharides in immune responses. Bioessays 2001; 23:46–53.
6. Jennings H. Further approaches for optimizing polysaccharide-protein conjugate vaccines for prevention of invasive bacterial disease. J Infect Dis 1992; 165(Suppl 1):S156–S159.
7. Zhang Y, Yao Q, Xia C, Jiang X, Wang Peng G. Trapping norovirus by glycosylated hydrogels: a potential oral antiviral drug. ChemMedChem 2006; 1:1361–1366.
8. Koeller KM, Wong C-H. Emerging themes in medicinal glycoscience. Nat Biotechnol 2000; 18:835–841.
9. Sears P, Wong C-H. Carbohydrate mimetics: a new strategy for tackling the problem of carbohydrate-mediated biological recognition. Angew Chem Int Ed 1999; 38:2301–2324.
10. Jacob GS. Glycosylation inhibitors in biology and medicine. Curr Opin Struc Biol 1995; 5:605–611.
11. Wong C-H, Halcomb RL, Ichikawa Y, Kajimoto T. Enzymes in organic synthesis: application to the problems of carbohydrate recognition. Part 1. Angew Chem Int Ed 1995; 34:412–432.

12. Wong C-H, Halcomb RL, Ichikawa Y, Kajimoto T. Enzymes in organic synthesis: application to the problems of carbohydrate recognition. Part 2. Angew Chem Int Ed 1995; 34:521–546.
13. Koeller KM, Wong C-H. Complex carbohydrate synthesis tools for glycobiologists: enzyme-based approach and programmable one-pot strategies. Glycobiology 2000; 10:1157–1169.
14. Sears P, Wong CH. Toward automated synthesis of oligosaccharides and glycoproteins. Science 2001; 291:2344–2350.
15. Plante OJ, Palmacci ER, Seeberger PH. Automated solid-phase synthesis of oligosaccharides. Science 2001; 291:1523–1527.
16. Crout DHG, Critchley P, Muller D, Scigelova M, Singh S, Vic G. Application of glycosidases in the synthesis of complex carbohydrates. Spec Publ R Soc Chem 1999; 246:15–23.
17. Wymer N, Toone EJ. Enzyme-catalyzed synthesis of carbohydrates. Curr Opin Chem Biol 2000; 4:110–119.
18. Crout DHG, Vic G. Glycosidases and glycosyl transferases in glycoside and oligosaccharide synthesis. Curr Opin Chem Biol 1998; 2:98–111.
19. Koeller KM, Wong C-H. Enzymes for chemical synthesis. Nature 2001; 409:232–240.
20. Seto NOL, Compston CA, Evans SV, Bundle DR, Narang SA, Palcic MM. Donor substrate specificity of recombinant human blood group A, B and hybrid A/B glycosyltransferases expressed in Escherichia coli. Eur J Biochem 1999; 259:770–775.
21. Whitesides GM, Wong CH. Enzymes as catalysts in organic synthesis. Aldrichim Acta 1983; 16:27–34.
22. Wong CH, Schuster M, Wang P, Sears P. Enzymic synthesis of N- and O-linked glycopeptides. J Am Chem Soc 1993; 115:5893–5901.
23. van Rantwijk F, Woudenberg-van Oosterom M, Sheldon RA. Glycosidase-catalyzed synthesis of alkyl glycosides. J Mol Catal B: Enzym 1999; 6:511–532.
24. Vocadlo DJ, Withers SG. Glycosidase-catalyzed oligosaccharide synthesis. Carbohydr Chem Biol 2000; 2:723–844.
25. Palcic MM. Biocatalytic synthesis of oligosaccharides. Curr Opin Biotechnol 1999; 10:616–624.
26. Jakeman DL, Withers SG. Glycosynthases: new tools for oligosaccharide synthesis. Trends Glycosci Glyc 2002; 14:13–25.
27. Ly HD, Withers SG. Mutagenesis of glycosidases. Annu Rev Biochem 1999; 68:487–522.
28. Tolborg JF, Petersen L, Jensen KJ, Mayer C, Jakeman DL, Warren RA, Withers SG. Solid-phase oligosaccharide and glycopeptide synthesis using glycosynthases. J Org Chem 2002; 67:4143–4149.
29. Withers SG. Understanding and exploiting glycosidases. Can J Chem 1999; 77:1–11.
30. Roseman S. The synthesis of complex carbohydrates by multiglycosyltransferase systems and their potential function in intercellular adhesion. Chem Phys Lipids 1970; 5:270–297.
31. Watkins WM. Glycosyltransferases. Early history, development and future prospects. Carbohydr Res 1986; 149:1–12.
32. Hehre EJ. Glycosyl transfer: a history of the concept's development and view of its major contributions to biochemistry. Carbohydr Res 2001; 331:347–368.

33. Unligil UM, Rini JM. Glycosyltransferase structure and mechanism. Curr Opin Struc Biol 2000; 10:510–517.
34. Davies GJ, Henrissat B. Structural enzymology of carbohydrate-active enzymes: implications for the post-genomic era. Biochem Soc Trans 2002; 30:291–297.
35. Kaneko M, Nishihara S, Narimatsu H, Saitou N. The evolutionary history of glycosyltransferase genes. Trends Glycosci Glyc 2001; 13:147–155.
36. Palcic MM, Sujino K, Qian X. Enzymatic glycosylations with non-natural donors and acceptors. Carbohydr Chem Biol 2000; 2:685–703.
37. Sujino K, Uchiyama T, Hindsgaul O, Seto NOL, Wakarchuk WW, Palcic MM. Enzymatic synthesis of oligosaccharide analogues: evaluation of UDP-Gal analogues as donors for three retaining α-galactosyltransferases. J Am Chem Soc 2000; 122:1261–1269.
38. Fukuda M, Bierhuizen MF, Nakayama J. Expression cloning of glycosyltransferases. Glycobiology 1996; 6:683–689.
39. Fang J-W, Li J, Chen X, Zhang Y-N, Wang J-Q, Guo Z-M, Brew K, Wang PG. A highly efficient chemo-enzymatic synthesis of α-galactosyl epitopes with a recombinant α(1→3)-galactosyltransferase. J Am Chem Soc 1998; 120:6635–6638.
40. Palacpac NQ, Yoshida S, Sakai H, Kimura Y, Fujiyama K, Yoshida T, Seki T. Stable expression of human β1,4-galactosyltransferase in plant cells modifies *N*-linked glycosylation patterns. Proc Natl Acad Sci USA 1999; 96:4692–4697.
41. Malissard M, Zeng S, Berger EG. The yeast expression system for recombinant glycosyltransferases. Glycoconj J 1999; 16:125–139.
42. Shibatani S, Fujiyama K, Nishiguchi S, Seki T, Maekawa Y. Production and characterization of active soluble human β1,4-galactosyltransferase in Escherichia coli as a useful catalyst in synthesis of the gal β1→4 GlcNAc linkage. J Biosci Bioeng 2001; 91:85–87.
43. Blixt O, van Die I, Norberg T, van den Eijnden DH. High-level expression of the Neisseria meningitides lgtA gene in Escherichia coli and characterization of the encoded *N*-acetylglucosaminyltransferase as a useful catalyst in the synthesis of GlcNAcβ1→3Gal and GalNAcβ1→3Gal linkages. Glycobiology 1999; 9:1061–1071.
44. Johnson KF. Synthesis of oligosaccharides by bacterial enzymes. Glycoconj J 1999; 16:141–146.
45. Shao J, Zhang J, Kowal P, Lu Y, Wang PG. Overexpression and biochemical characterization of β-1,3-*N*-acetylgalactosaminyltransferase LgtD from Haemophilus influenzae strain Rd. Biochem Biophys Res Commun 2002; 295:1–8.
46. Shao J, Zhang J, Kowal P, Wang PG. Donor substrate regeneration for efficient synthesis of globotetraose and isoglobotetraose. Appl Environ Microbiol 2002; 68:5634–5640.
47. Zhang J, Kowal P, Fang J, Andreana P, Wang PG. Efficient chemoenzymatic synthesis of globotriose and its derivatives with a recombinant α-(1→4)-galactosyltransferase. Carbohydr Res 2002; 337:969–976.
48. Izumi M, Shen GJ, Wacowich-Sgarbi S, Nakatani T, Plettenburg O, Wong CH. Microbial glycosyltransferases for carbohydrate synthesis: α-2,3-sialyltransferase from Neisseria gonorrheae. J Am Chem Soc 2001; 123: 10909–10918.
49. DeAngelis PL. Microbial glycosaminoglycan glycosyltransferases. Glycobiology 2002; 12:9R–16R.

50. Sharon N, Ofek I. Safe as mother's milk: carbohydrates as future anti-adhesion drugs for bacterial diseases. Glycoconj J 2000; 17:659–664.
51. Sharon N, Ofek I. Fighting infectious diseases with inhibitors of microbial adhesion to host tissues. Crit Rev Food Sci Nutr 2002; 42:267–272.
52. Wong CH, Haynie SL, Whitesides GM. Enzyme-catalyzed synthesis of N-acetyllactosamine with in situ regeneration of uridine 5′-diphosphate glucose and uridine 5′-diphosphate galactose. J Org Chem 1982; 47: 5416–5418.
53. Wang P, Shen GJ, Wang YF, Ichikawa Y, Wong CH. Enzymes in oligosaccharide synthesis: active-domain overproduction, specificity study, and synthetic use of an α-1,2-mannosyltransferase with regeneration of GDP-Man. J Org Chem 1993; 58:3985–3990.
54. Ichikawa Y, Lin YC, Dumas DP, Shen GJ, Garcia-Junceda E, Williams MA, Bayer R, Ketcham C, Walker LE, Paulson JC, Wong CH. Chemical-enzymic synthesis and conformational analysis of sialyl Lewis X and derivatives. J Am Chem Soc 1992; 114:9283–9298.
55. Ichikawa Y, Wang R, Wong C-H. Regeneration of sugar nucleotide for enzymatic oligosaccharide synthesis. Method Enzymol 1994; 247:107–127.
56. Wong C-H, Wang R, Ichikawa Y. Regeneration of sugar nucleotide for enzymatic oligosaccharide synthesis: use of Gal-1-phosphate uridyltransferase in the regeneration of UDP-galactose, UDP-2-deoxygalactose, and UDP-galactosamine. J Org Chem 1992; 57:4343–4344.
57. Endo T, Koizumi S. Large-scale production of oligosaccharides using engineered bacteria. Curr Opin Struct Biol 2000; 10:536–541.
58. Hokke CH, Zervosen A, Elling L, Joziasse DH, van den Eijnden DH. One-pot enzymatic synthesis of the Galα1→3Galβ1→4GlcNAc sequence with in situ UDP-Gal regeneration. Glycoconj J 1996; 13:687–692.
59. Elling L, Grothus M, Kula MR. Investigation of sucrose synthase from rice for the synthesis of various nucleotide sugars and saccharides. Glycobiology 1993; 3:349–355.
60. Herrmann GF, Wang P, Shen G-J, Wong C-H. Recombinant whole cells as catalysts for enzymic synthesis of oligosaccharides and glycopeptides. Angew Chem 1994; 106:1346–1347.
61. Look GC, Ichikawa Y, Shen GJ, Cheng PW, Wong CH. A combined chemical and enzymatic strategy for the construction of carbohydrate-containing antigen core units. J Org Chem 1993; 58:4326–4330.
62. Gygax D, Spies P, Winkler T, Pfaar U. Enzymic synthesis of β-D-glucuronides with in situ regeneration of uridine 5′-diphosphoglucuronic acid. Tetrahedron 1991; 47:5119–5122.
63. Endo T, Koizumi S, Tabata K, Kaita S, Ozaki A. Large-scale production of N-acetyllactosamine through bacterial coupling. Carbohydr Res 1999; 136: 179–183.
64. Freeze HH. Monosaccharide metabolism. In: Varki A, Cummings R, Esko J, Freeze H, Hart G, Marth J, eds. Essentials of Glycobiology, New York: Coldspring Harbor Laboratory Press, 1999; 69–84.
65. Freeze HH. Human glycosylation disorders and sugar supplement therapy. Biochem Biophl Res Commun 1999; 255:189–193.
66. Ma X, Stockigt J. High yielding one-pot enzyme-catalyzed synthesis of UDP-glucose in gram scales. Carbohydr Res 2001; 333:159–163.

67. Haynie SL, Whitesides GM. Enzyme-catalyzed organic synthesis of sucrose and trehalose with in situ regeneration of UDP-glucose. Appl Biochem Biotechnol 1990; 23:155–170.
68. Winter H, Huber SC. Regulation of sucrose metabolism in higher plants: localization and regulation of activity of key enzymes. Crit Rev Biochem Mol 2000; 35:253–289.
69. Nguyen-Quoc B, Foyer CH. A role for "futile cycles" involving invertase and sucrose synthase in sucrose metabolism of tomato fruit. J Exp Bot 2001; 52:881–889.
70. Fernie AR, Willmitzer L, Trethewey RN. Sucrose to starch: a transition in molecular plant physiology. Trends Plant Sci 2002; 7:35–41.
71. Chen X, Kowal P, Hamad S, Fan H, Wang PG. Cloning, expression and characterization of a UDP-galactose 4-epimerase from Escherichia coli. Biotechnol Lett 1999; 21:1131–1135.
72. Heidlas JE, Lees WJ, Whitesides GM. Practical enzyme-based syntheses of uridine 5′-diphosphogalactose and uridine 5′-diphospho-*N*-acetylgalactosamine on a gram scale. J Org Chem 1992; 57:152–157.
73. Chacko CM, McCrone L, Nadler HL. Uridine diphosphoglucose pyrophosphorylase and uridine diphosphogalactose pyrophosphorylase in human skin fibroblasts derived from normal and galactosemic individuals. Biochim Biophys Acta, Enzymol 1972; 268:113–120.
74. Chen X, Fang J, Zhang J, Liu Z-Y, Andreana P, Kowal P, Wang PG. Sugar nucleotide regeneration beads (superbeads): a versatile tool for practical synthesis of oligosaccharides. J Am Chem Soc 2001; 123:2081–2082.
75. Chen X, Liu Z, Zhang J, Zhang W, Kowal P, Wang PG. Reassembled biosynthetic pathway for large-scale carbohydrate synthesis: α-gal epitope producing "superbug". ChemBioChem 2002; 3:47–53.
76. Stewart DC, Copeland L. Uridine 5′-diphosphate-glucose dehydrogenase from soybean nodules. Plant Physiol 1998; 116:349–355.
77. Yamada-Okabe T, Sakamori Y, Mio T, Yamada-Okabe H. Identification and characterization of the genes for *N*-acetylglucosamine kinase and *N*-acetylglucosamine-phosphate deacetylase in the pathogenic fungus Candida albicans. Eur J Biochem 2001; 268:2498–2505.
78. Hofmann M, Boles E, Zimmermann FK. Characterization of the essential yeast gene encoding *N*-acetylglucosamine-phosphate mutase. Eur J Biochem 1994; 221:741–747.
79. Brown K, Pompeo F, Dixon S, Mengin-Lecreulx D, Cambillau C, Bourne Y. Crystal structure of the bifunctional *N*-acetylglucosamine 1-phosphate uridyltransferase from Escherichia coli: a paradigm for the related pyrophosphorylase superfamily. EMBO J 1999; 18:4096–4107.
80. Ponce E, Flores N, Martinez A, Valle F, Bolivar F. Cloning of the two pyruvate kinase isoenzyme structural genes from Escherichia coli: the relative roles of these enzymes in pyruvate biosynthesis. J Bacteriol 1995; 177: 5719–5722.
81. Lahti R, Pitkaranta T, Valve E, Ilta I, Kukko-Kalske E, Heinonen J. Cloning and characterization of the gene encoding inorganic pyrophosphatase of Escherichia coli K-12. J Bacteriol 1988; 170:5901–5907.
82. Shao J, Zhang JB, Nahálkab J, Wang PG. Biocatalytic synthesis of uridine 5′-diphosphate *N*-acetylglucosamine by multiple enzymes co-immobilized on agarose beads. Chem Commun. 2002; 21:2586–2587.

83. Kowal P, Wang PG. A new UDP-GlcNAc C4 epimerase involved in the biosynthesis of 2-acetamino-2-deoxy-L-altruronic acid in the O-antigen repeating units of Plesiomonas shigelloides O17. Biochemistry 2002; 41: 15410–15414.
84. Davis JA, Freeze HH. Studies of mannose metabolism and effects of long-term mannose ingestion in the mouse. Biochim Biophys Acta, Gen Subj 2001; 1528:116–126.
85. Davis JA, Wu X-H, Wang L, DeRossi C, Westphal V, Wu R, Alton G, Srikrishna G, Freeze HH. Molecular cloning, gene organization, and expression of mouse Mpi encoding phosphomannose isomerase. Glycobiology 2002; 12:435–442.
86. Wills EA, Roberts IS, Del Poeta M, Rivera J, Casadevall A, Cox GM, Perfect JR. Identification and characterization of the Cryptococcus neoformans phosphomannose isomerase-encoding gene, MAN1, and its impact on pathogenicity. Mol Microbiol 2001; 40:610–620.
87. Privalle LS. Phosphomannose isomerase, a novel plant selection system: potential allergenicity assessment. Ann NY Acad Sci 2002; 964:129–138.
88. Regni C, Tipton PA, Beamer LJ. Crystal structure of PMM/PGM: an enzyme in the biosynthetic pathway of P. aeruginosa virulence factors. Structure 2002; 10:269–279.
89. Ning B, Elbein AD. Cloning, expression and characterization of the pig liver GDP-mannose pyrophosphorylase: evidence that GDP-mannose and GDP-Glc pyrophosphorylases are different proteins. Eur J Biochem 2000; 267:6866–6874.
90. Ning B, Elbein AD. Purification and properties of mycobacterial GDP-mannose pyrophosphorylase. Arch Biochem Biophys 1999; 362:339–345.
91. Keller R, Springer F, Renz A, Kossmann J. Antisense inhibition of the GDP-mannose pyrophosphorylase reduces the ascorbate content in transgenic plants leading to developmental changes during senescence. Plant J 1999; 19:131–141.
92. Yoda K, Kawada T, Kaibara C, Fujie A, Abe M, Hashimoto H, Shimizu J, Tomishige N, Noda Y, Yamasaki M. Defect in cell wall integrity of the yeast Saccharomyces cerevisiae caused by a mutation of the GDP-mannose pyrophosphorylase gene VIG9. Biosci Biotechnol Biochem 2000; 64:1937–1941.
93. Agaphonov MO, Packeiser AN, Chechenova MB, Choi E-S, Ter-Avanesyan MD. Mutation of the homologue of GDP-mannose pyrophosphorylase alters cell wall structure, protein glycosylation and secretion in Hansenula polymorpha. Yeast 2001; 18:391–402.
94. Garami A, Mehlert A, Ilg T. Glycosylation defects and virulence phenotypes of Leishmania mexicana phosphomannomutase and dolichol phosphate-mannose synthase gene deletion mutants. Mol Cell Biol 2001; 21:8168–8183.
95. Wu BY, Zhang Y-X, Zheng R, Guo C-W, Wang PG. Bifunctional phosphomannose isomerase/GDP-D-mannose pyrophosphorylase is the point of control for GDP-D-mannose biosynthesis in Helicobacter pylori. FEBS lett 2002; 519:87–92.
96. Wu B-Y, Zhang Y-X, Wang PG. Identification and characterization of GDP-D-mannose 4,6-dehydratase and GDP-L-fucose synthetase in a GDP-L-fucose biosynthetic gene cluster from Helicobacter pylori. Biochem Biophys Res Commun 2001; 285:364–371.

97. Lee S-G, Lee J-O, Yi J-K, Kim B-G. Production of cytidine 5′-monophosphate N-acetylneuraminic acid using recombinant Escherichia coli as a biocatalyst. Biotechnol Bioeng 2002; 80:516–524.
98. Liu JLC, Shen GJ, Ichikawa Y, Rutan JF, Zapata G, Vann WF, Wong CH. Overproduction of CMP-sialic acid synthetase for organic synthesis. J Am Chem Soc 1992; 114:3901–3910.
99. Karwaski M-F, Wakarchuk WW, Gilbert M. High-level expression of recombinant Neisseria CMP-sialic acid synthetase in Escherichia coli. Protein Expression Purif 2002; 25:237–240.
100. Ichikawa Y, Shen GJ, Wong CH. Enzyme-catalyzed synthesis of sialyl oligosaccharide with in situ regeneration of CMP-sialic acid. J Am Chem Soc 1991; 113:4698–4700.
101. Ichikawa Y, Liu JLC, Shen G-J, Wong C-H. A highly efficient multienzyme system for the one step synthesis of a sialyl trisaccharide: in situ generation of sialic acid and N-acetyllactosamine coupled with regeneration of UDP-glucose, UDP-galactose, and CMP-sialic acid. J Am Chem Soc 1991; 113:6300–6302.
102. Rodriguez-Aparicio LB, Ferrero MA, Reglero A. N-acetyl-D-neuraminic acid synthesis in Escherichia coli K1 occurs through condensation of *N*-acetyl-D-mannosamine and pyruvate. Biochem J 1995; 308:501–505.
103. Baumann W, Freidenreich J, Weisshaar G, Brossmer R, Friebolin H. Reversible cleavage of sialic acids with aldolase; 1H-NMR investigations on stereochemistry, kinetics and mechanism. Biol Chem H-S 1989; 370: 141–149.
104. Zhang J, Wu B, Zhang Y, Kowal P, Wang PG. Creatine phosphate-creatine kinase in enzymatic synthesis of glycoconjugates. Org Lett 2003; 5:2583–2586.
105. Koeller KM, Wong CH. Synthesis of complex carbohydrates and glycoconjugates: enzyme-based and programmable one-pot strategies. Chem Rev 2000; 100:4465–4494.
106. Nishiguchi S, Yamada K, Fuji Y, Shibatani S, Toda A, Nishimura S-I. Highly efficient oligosaccharide synthesis on water soluble polymeric primers by recombinant glycosyltransferases immobilised on solid supports. Chem Commun 2001; 19:1944–1945.
107. Huang X, Witte KL, Bergbreiter DE, Wong C-H. Homogenous enzymatic synthesis using a thermo-responsive water-soluble polymer support. Adv Synth Catal 2001; 343:675–681.
108. Toda A, Yamada K, Nishimura S-I. An engineered biocatalyst for the synthesis of glycoconjugates: utilization of β1,3-*N*-acetyl-D-glucosaminyltransferase from Streptococcus agalactiae type Ia expressed in Escherichia coli as a fusion with maltose-binding protein. Adv Synth Catal 2002; 344:61–69.
109. Yamada K, Nishimura S-I. An efficient synthesis of sialoglycoconjugates on a peptidase-sensitive polymer support. Tetrahedron Lett 1995; 36:9493–9496.
110. Yan F, Wakarchuk WW, Gilbert M, Richards JC, Whitfield DM. Polymer-supported and chemoenzymatic synthesis of the Neisseria meningitidis pentasaccharide: a methodological comparison. Carbohydr Res 2000; 328:3–16.
111. Yamada K, Fujita E, Nishimura S-I. High performance polymer supports for enzyme-assisted synthesis of glycoconjugates. Carbohydr Res 1997; 305: 443–461.

112. Zhang J, Wang PG. Novel energy source in synthesis of glycoconjugates. In: Abstracts of Papers, 224th ACS National Meeting, Boston, MA: United States, August 18–22, 2002: CARB-059.
113. Nishimura S-I. Combinatorial syntheses of sugar derivatives. Curr Opin Chem Biol 2001; 5:325–335.
114. Nishimura S-I. Artificial golgi apparatus: automated glycoconjugates synthesis by mimicking biosynthetic pathway. In: Abstracts of Papers, 224th ACS National Meeting, Boston, MA: United States, August 18–22, 2002: CARB-032.
115. Yao Q, Song J, Xia C, Zhang W, Wang PG. Chemoenzymatic syntheses of iGb3 and Gb3. Org Lett 2006; 8:911–914.
116. Endo T, Koizumi S, Tabata K, Kakita S, Ozaki A. Large-scale production of the carbohydrate portion of the sialyl–Tn epitope, α-Neup5Ac-(2→6)-D-GalpNAc, through bacterial coupling. Carbohydr Res 2001; 330:439–443.
117. Endo T, Koizumi S, Tabata K, Ozaki A. Large-scale production of CMP-NeuAc and sialylated oligosaccharides through bacterial coupling. Appl Microbiol Biotechnol 2000; 53:257–261.
118. Koizumi S, Endo T, Tabata K, Ozaki A. Large-scale production of UDP-galactose and globotriose by coupling metabolically engineered bacteria. Nat Biotechnol 1998; 16:847–850.
119. Samain E, Drouillard S, Heyraud A, Driguez H, Geremia RA. Gram-scale synthesis of recombinant chitooligosaccharides in Escherichia coli. Carbohydr Res 1997; 302:35–42.
120. Samain E, Chazalet V, Geremia RA. Production of *O*-acetylated and sulfated chitooligosaccharides by recombinant Escherichia coli strains harboring different combinations of nod genes. J Biotechnol 1999; 72:33–47.
121. Koizumi S, Endo T, Tabata K, Nagano H, Ohnishi J, Ozaki A. Large-scale production of GDP-fucose and Lewis X by bacterial coupling. J Ind Microbiol Biotechnol 2000; 25:213–217.
122. Albermann C, Distler J, Piepersberg W. Preparative synthesis of GDP-β-L-fucose by recombinant enzymes from enterobacterial sources. Glycobiology 2000; 10:875–881.
123. Zhang J, Kowal P, Chen X, Wang PG. Large-scale synthesis of globotriose derivatives through recombinant E. coli. Org Biomol Chem 2003; 1:3048–3053.
124. Shao J, Zhang J, Kowal P, Lu Y, Wang PG. Efficient synthesis of globoside and isogloboside tetrasaccharides by using beta(1→3) *N*-acetylgalactosaminyltransferase/UDP-*N*-acetylglucosamine C4 epimerase fusion protein. Chem Commun 2003; 1422–1423.
125. Wang PG, Chen X, Zhang J, Kowal P, Andreana PR. Transferring a biosynthetic cycle into a productive E. coli strain: large-scale synthesis of galactosides. In: Abstracts of Papers, 222nd ACS National Meeting, Chicago, IL, United States, August 26–30, 2001: MEDI-126.
126. Antoine T, Priem B, Heyraud A, Greffe L, Gilbert M, Wakarchuk WW, Lam JS, Samain E. Large-scale in vivo synthesis of the carbohydrate moieties of gangliosides GM1 and GM2 by metabolically engineered Escherichia coli. ChemBioChem 2003; 4:406–412.
127. Dumon C, Priem B, Martin SL, Heyraud A, Bosso C, Samain E. In vivo fucosylation of lacto-*N*-neotetraose and lacto-*N*-neohexaose by

heterologous expression of Helicobacter pylori α-1,3 fucosyltransferase in engineered Escherichia coli. Glycoconj J 2001; 18:465–474.
128. Priem B, Gilbert M, Wakarchuk WW, Heyraud A, Samain E. A new fermentation process allows large-scale production of human milk oligosaccharides by metabolically engineered bacteria. Glycobiology 2002; 12: 235–240.
129. Bettler E, Samain E, Chazalet V, Bosso C, Heyraud A, Joziasse DH, Wakarchuk WW, Imberty A, Geremia AR. The living factory: in vivo production of *N*-acetyllactosamine containing carbohydrates in E. coli. Glycoconj J 1999; 16:205–212.
130. Antoine T, Bosso C, Heyraud A, Samain E. Large scale in vivo synthesis of globotriose and globotetraose by high cell density culture of metabolically engineered Escherichia coli. Biochimie 2005; 87:197–203.
131. Krummel MF, Davis MM. Dynamics of the immunological synapse: finding, establishing and solidifying a connection. Curr Opin Immunol 2002; 14:66–74.
132. Bromley SK, Burack WR, Johnson KG, Somersalo K, Sims TN, Sumen C, Davis MM, Shaw AS, Allen PM, Dustin ML. The immunological synapse. Ann Rev Immunol 2001; 19:375–396.
133. Zhou D, Mattner J, Cantu C, 3rd, Schrantz N, Yin N, Gao Y, Sagiv Y, Hudspeth K, Wu Y-P, Yamashita T, Teneberg S, Wang D, Proia Richard L, Levery Steven B, Savage Paul B, Teyton L, Bendelac A. Lysosomal glycosphingolipid recognition by NKT cells. Science 2004; 306:1786–1789.
134. Van Kaer L. α-Galactosylceramide therapy for autoimmune diseases: prospects and obstacles. Nat Rev Immunol 2005; 5:31–42.
135. Sharif S, Arreaza GA, Zucker P, Mi Q-S, Delovitch TL. Regulation of autoimmune disease by natural killer T cells. J Mol Med 2002; 80:290–300.
136. Smyth MJ, Crowe NY, Hayakawa Y, Takeda K, Yagita H, Godfrey DI. NKT cells—conductors of tumor immunity? Curr Opin Immunol 2002; 14:165–171.
137. Serizawa I, Koezuka Y, Amao H, Saito TR, Takahashi KW. Functional natural killer T cells in experimental mouse strains, including NK1.1-strains. Exp Anim 2000; 49:171–180.
138. Egusa K, Kusumoto S, Fukase K. Solid-phase synthesis of a phytoalexin elicitor pentasaccharide using a 4-azido-3-chlorobenzyl group as the key for temporary protection and catch-and-release purification. Eur J Org Chem 2003; 17:3435–3445.
139. Ogawa T, Sugimoto M, Kitajima T, Sadozai KK, Nukada T. Synthetic studies on cell surface glycans. 51. Total synthesis of an undecasaccharide. A typical carbohydrate sequence for the complex type of glycan chains of a glycoprotein. Tetrahedron Lett 1986; 27:5739–5742.
140. Xia C, Yao Q, Schuemann J, Rossy E, Chen W, Zhu L, Zhang W, De Libero G, Wang PG. Synthesis and biological evaluation of α-galactosylceramide (KRN7000) and isoglobotrihexosylceramide (iGb3). Bioorg Med Chem Lett 2006; 16:2195–2199.
141. Nicolaou KC, Caulfield TJ, Katoaka H. Total synthesis of globotriaosylceramide (Gb3) and lysoglobotriaosylceramide (lysoGb3). Carbohydr Res 1990; 202:177–191.

142. Elchert B, Li J, Wang J, Hui Y, Rai R, Ptak R, Ward P, Takemoto JY, Bensaci M, Chang C-WT. Application of the synthetic amino sugars for glycodiversification: synthesis and antimicrobial studies of pyranmycin. J Org Chem 2004; 69:1513–1523.
143. Lin S, Yang Z-Q, Kwok BHB, Koldobskiy M, Crews CM, Danishefsky SJ. Total synthesis of TMC-95A and -B via a new reaction leading to Z-enamides. Some preliminary findings as to SAR. J Am Chem Soc 2004; 126:6347–6355.
144. Blixt O, Vasiliu D, Allin K, Jacobsen N, Warnock D, Razi N, Paulson JC, Bernatchez S, Gilbert M, Wakarchuk W. Chemoenzymatic synthesis of 2-azidoethyl-ganglio-oligosaccharides GD3, GT3, GM2, GD2, GT2, GM1, and GD1a. Carbohydr Res 2005; 340:1963–1972.

Carbohydrate Chemistry, Biology and Medical Applications
Hari G. Garg, Mary K. Cowman and Charles A. Hales

DOI: 10.1016/B978-0-08-054816-6.00005-7

Chapter 5

Proteoglycans in the Lung

MARA S. LUDWIG

Meakins-Christie Laboratories, McGill University, Montreal, Quebec H2X2P2, Canada

I. Introduction

Proteoglycans (PGs) are integral components of the extracellular matrix (ECM) and are present throughout the lung. PGs are macromolecules consisting of a protein core and glycosaminoglycan (GAG) side chains. The GAG side chains include chondroitin sulfate (CS), keratin sulfate (KS), heparan sulfate (HS), dermatin sulfate (DS), and hyaluronic acid (HA), a GAG which is not bound to a protein core. Different subclasses of PGs have been described, and include large, aggregating PGs such as versican and aggrecan; small leucine-rich repeat PGs such as decorin, biglycan, lumican, and fibromodulin; basement membrane PGs such as perlecan; and cell surface PGs such as syndecan (1). Members of all these PG families have been identified in the lung. This chapter will describe the contribution of these various PGs to normal lung structure and the role individual PGs play in influencing lung cell biology and normal lung mechanics. The importance of PG in lung development will be briefly considered. Finally, changes in PG in clinically relevant lung diseases, such as asthma and pulmonary fibrosis, will be discussed. How changes in these molecules contribute to the pathophysiology of various lung diseases underscore the critical role that PGs play in the lung.

II. PGs and Normal Lung Structure

PGs are critical components of the lung ECM, and all subclasses of PGs are present in the normal lung. The load-bearing components of the lung scaffold include collagen and elastic fibers, and PGs, which represent the "ground substance" within the fiber network. PGs are divided into subclasses, essentially based on the location of the PG within the tissue and on the gene family to which a given PG belongs. Versican is a large, aggregating PG that, in association with HA, helps form the pericellular matrix of the lung interstitium (2). Aggrecan, another large, aggregating PG, is present in the tracheobronchial cartilage, again in association with HA (3). Several small leucine-rich proteoglycans (SLRP) have been identified in the lung parenchyma, airways, and vasculature, including decorin, biglycan, lumican, and fibromodulin (4–9). Decorin binds to collagen and by influencing collagen fibrillogenesis plays an important role in the assembly of the lung fiber network (10,11). Experiments in decorin-deficient mice show disruption and heterogeneity of collagen fibers (12) (Fig. 1). It is likely that biglycan has a similar function, in as much as biglycan-deficient mice also show abnormalities in collagen fibril morphology (13). (These studies in decorin- and biglycan-deficient animals have not directly examined changes in collagen within the lung itself.) We have published data showing that lumican is the most prevalent SLRP within the lung peripheral tissues (5), and lumican is evident, as well, within the normal airway wall (6). Fibromodulin has been described in normal lung tissue extracts (14).

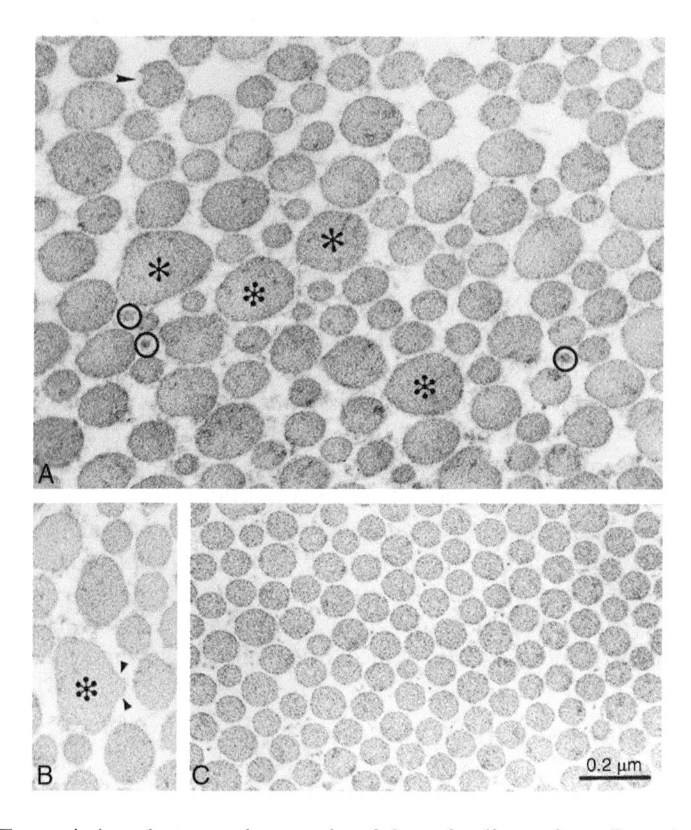

Figure 1 Transmission electron micrographs of dermal collagen from *Dcn−/−* (A, B) and *Dcn+/+* (C) mice. Note the heterogeneity in size and irregularity of fibrils from *Dcn−/−* animals. Reproduced by permission from Ref. 12.

Fibromodulin and lumican bind to type I and II collagens, and influence both the rate of fibrillogenesis and the structure of the resulting fibrils (15). Their site of interaction with collagen, however, is different from that of decorin (16,17). Immunohistochemical study of the pulmonary vasculature shows prominent staining for biglycan, decorin, and lumican (4,5,18). Another subgroup of PGs, basement membrane PGs, are key components of alveolar and pulmonary capillary basal laminae; perlecan is likely the predominant molecule in the lung (19,20). These molecules help define functional compartments and act as filtration barriers. Cell surface PGs, such as syndecan, have been described in lung epithelial cells and function primarily as cell surface receptors for matrix ligands (21). Finally, intracellular and mucus PGs have been identified in the lung.

III. Contribution to Basic Lung Biology

PGs subserve a number of different biological functions. Versican, because of the high ionic charge of its multiple GAG side chains, plays a critical role in determining the water content or turgor of extracellular matrices. Via this mechanism, versican influences tissue viscoelastic behavior, as well as cell migration and proliferation. Decorin and biglycan are molecules that bind to collagen and affect collagen fibrillogenesis and matrix assembly. They can also act to protect the fibrils from cleavage by collagenases through their coating of collagen fibrils (22). These molecules also bind different growth factors, such as transforming growth factor (TGF)-β and fibroblast growth factor (FGF), and by influencing their bioavailability, modulate their ability to influence cell proliferation and matrix deposition (23,24). PGs have also been shown to influence growth factor receptor expression. For example, HS PG increased platelet-derived growth factor receptor expression on human lung fibroblasts (25). PGs may play a role in inflammation not only through cytokine binding but also by acting as ligands for proinflammatory Toll-like receptors (26). PGs may have direct effects on lung cell proliferation and/or apoptosis. HS PGs have been shown to inhibit rat lung pericyte proliferation *in vitro* (27). In our laboratory, we have obtained preliminary data in human airway smooth muscle (ASM) cells, showing that cells grown on decorin matrices show significant growth inhibition (28). Biglycan and decorin can induce cytoskeletal changes in lung fibroblasts, enhancing their ability to migrate (29). Johnson et al. (30) have recently reported that ECM proteins produced by asthmatic ASM cells enhanced ASM proliferation, that is upregulation of ECM proteins may be part of an autocrine loop. While the specific ECM molecules involved were not identified in this study, the observation by these same authors that asthmatic ASM produces more PGs identifies this class of molecules as a potential factor influencing airway structural cell function.

IV. Contribution of PGs to Normal Lung Mechanics

Lung parenchymal tissues display prominent viscoelastic behavior. The lung tissue matrix represents a composite of collagen and elastic fibers, PGs, and GAGs.

Collagen and elastin fibers are essentially elastic in nature (31); there has been discussion in the literature as to whether these fibers may also be responsible for the viscous or resistive properties of the lung tissues (32,33). While networks of fibers may display hysteretic properties (34), it seems more likely that PGs primarily account for the viscous properties of the lung parenchyma. The GAG side chains of PGs are highly hydrophilic, and have the ability to attract ions and fluid into the matrix, thereby affecting tissue turgor and viscoelasticity (1,35). Versican is a large, hydrodynamic molecule with numerous GAG side chains (1). As versican is the predominant PG present in the ECM of the lung parenchyma, it seems likely that this PG plays a key role in determining the turgor of the parenchymal tissues. Lung tissue viscoelasticity has also been attributed to the movement of fibers within the connective tissue matrix. Mijailovich et al. (36,37) have proposed that energy dissipation occurs not at the molecular level within collagen or elastin, but rather at the level of fiber–fiber contact, and by the shearing of GAGs which provide the lubricating film between adjacent fibers. Cavalcante et al. (35) postulated that PGs act to stabilize the collagen–elastin network of connective tissues via their effects on tissue osmolarity. Further, PGs, such as decorin and biglycan, modulate collagen fibrillogenesis, and may affect tissue mechanics through direct effects on collagen fibril formation (1).

We have published experimental data that implicate PGs as key determinants of the viscoelastic behavior of the lung tissues (38). Parenchymal strips were excised from rat lungs, and tissue viscoelastic properties were measured in the organ bath. Exposure of the tissues to the specific degradative enzymes, chondroitinase and heparitinase, which digest GAG side chains, resulted in alterations in tissue viscoelastic behavior (38). We have also obtained data in decorin-deficient mice, characterizing the viscoelastic behavior of the lung both in intact animals and in isolated lung parenchymal strips (39). Decorin-deficient mice showed changes in elastic properties in both *in vivo* and *in vitro* preparations, as compared to wild-type, decorin-replete mice (Fig. 2). In addition, airway resistance was decreased in decorin-deficient mice. These data underscore the contribution of PG to normal lung physiology. As will be described later in this chapter, alterations in PG in different disease processes similarly contribute to abnormal lung mechanics.

V. PGs and Mechanical Strain

Mechanotransduction describes the biological phenomenon wherein cells can alter protein metabolism in response to mechanical stimuli (40). The effect of mechanical strain on cellular systems has been a recent focus of research interest; a number of studies have examined the effects of mechanical forces on ECM remodeling (41). The effects of excessive mechanical strain on ECM production have been studied using experimental systems that either stretch cell monolayers or strain three-dimensional gels of mixed cell cultures. Xu et al. (42,43) have documented that cyclic mechanical strain of mixed fetal lung cell cultures resulted in a modest

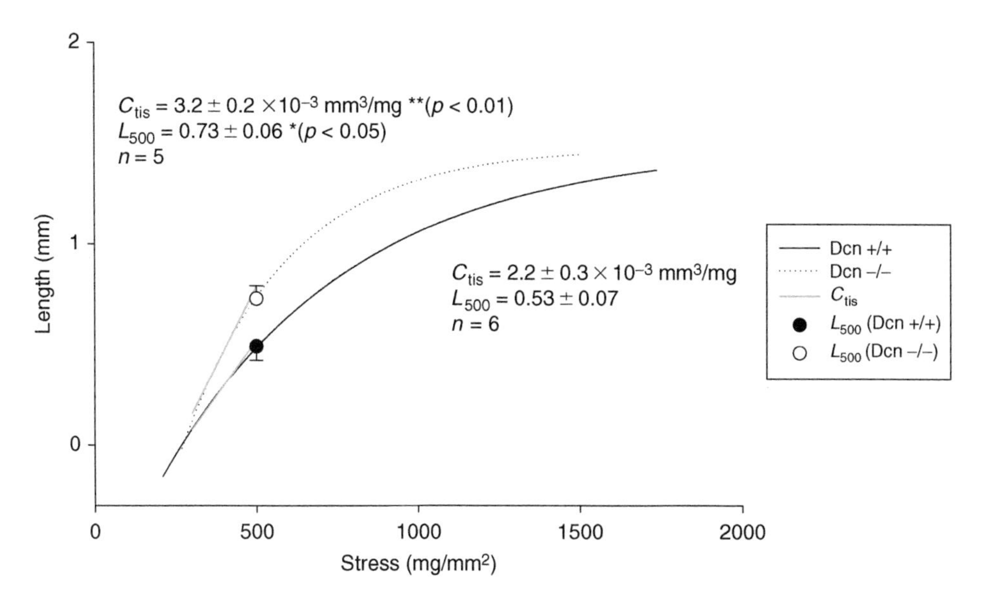

Figure 2 Stress–length curves during stepwise relaxation of parenchymal strips *in vitro*, showing increased compliance of the lung tissue in parenchymal strips from decorin-deficient mice. Compliance of the tissues (C_{tis}) was measured from the slope of the tangent to the curve between 300 and 500 mg/mm^2. L_{500} is the length of the strip above l_o at 500 mg/mm^2. *Dcn+/+*, wild type; *Dcn−/−*, decorin deficient. *, $p<0.05$ vs *Dcn+/+* **, $p<0.01$ vs *Dcn+/+*. Reproduced by permission from Ref. (39).

increase in versican mRNA. Biglycan protein was increased as was the expression of a large chondroitin/dermatan sulfate PG thought to be versican. We examined the effects of excessive mechanical strain on monolayers of adult human bronchial fibroblasts cultured on collagen-coated plates (44,45). After 24 h, both versican message and protein were increased in response to 30% stretch of the membrane on which cells were cultured. No change in either protein or message was observed for the small PGs, biglycan or lumican. We believe the change in versican in response to mechanical strain represents an adaptive mechanism to protect the cell from the traumatic effects of excessive mechanical stimuli. When the plasma membrane is subjected to mechanical strain it undergoes unfolding, elastic deformation, and/or stress failure (46). Versican is a large, hydrodynamic molecule with many GAG side chains that influence tissue viscoelastic properties (38) and is an important component of the pericellular matrix. Changes in versican in response to mechanical strain result in an altered pericellular layer with altered viscoelastic properties. This altered pericellular matrix may function as a type of "shock absorber," dispersing the effect of the physical force, and thereby protecting the cell from excessive mechanical strain and subsequent plasma membrane disruption. This modulation of the strain applied at the cell membrane would also potentially affect the strain transmitted via the cytoskeleton to the cell nucleus and the ensuing molecular response (Fig. 3).

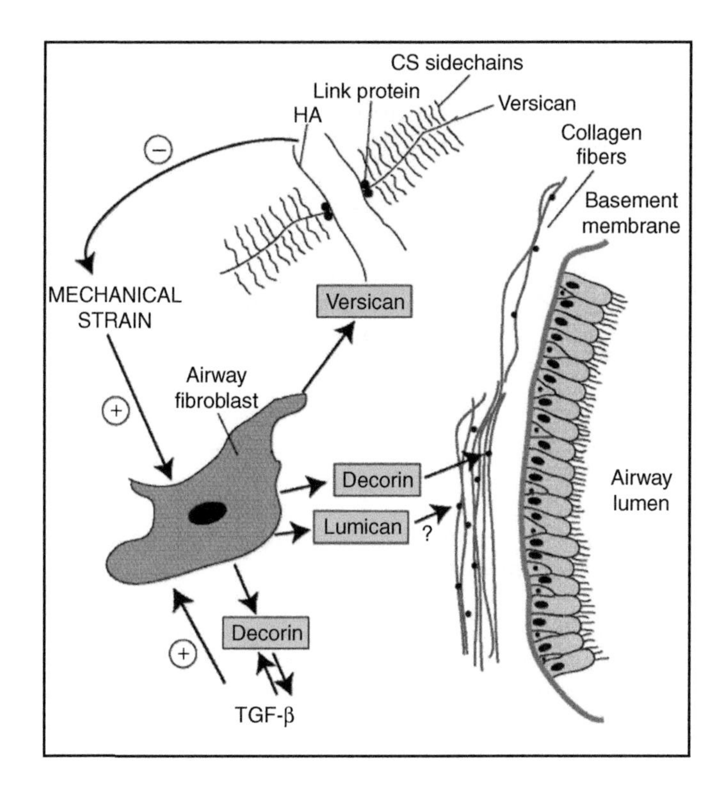

Figure 3 Proposed interactions between mechanical strain, proteoglycans (PGs), and transforming growth factor (TGF)-β in the extracellular environment of the airway wall. Excessive mechanical strain stimulates fibroblasts to increase PG secretion and deposition. The altered viscoelastic properties of the matrix subsequently modulate transmission of the mechanical signal to the airway structural cell and, thereby, protect the cell from mechanical strain-induced injury. In addition, increases in decorin may result in enhanced binding of TGF-β and thereby influence the effects of this cytokine on the fibroblast. Finally, changes in small PGs, such as decorin and lumican, may affect formation of collagen fibrils. CS, chondroitin sulfate; HA, hyaluronic acid.

VI. PGs and the Developing Lung

PGs influence lung development and branching morphogenesis through their effects on cell migration and the pericellular matrix. GAG synthesis and content have been shown to increase with lung development to a maximum and then fall abruptly at partuition (47). PG synthesis in the lung has been reported to be maximal at 80% gestation (48). HA seems to be particularly evident in early growing areas, whereas PGs are prevalent around parabronchi and CSPG is associated with collagen fibrils (49). Wang et al. (50) have shown, in the prenatal rat lung, that CSPG is present in relatively large amounts in the alveolar and airway ECM, whereas decorin is localized to developing airways and vessels, presumably related to collagen deposition. Syndecan, a cell surface PG, is developmentally regulated during lung morphogenesis; changes occur in its site of expression and its GAG structure during fetal maturation (51). In early postnatal lungs, CS, CSPG, and HSPG have been identified; immunoreactivity progressively diminishes with maturation (52). We have reported that HA content in the lung tissue rapidly diminishes after partuition, coincident with changes in lung mechanical properties (53). Hence, GAGs and PGs are altered during lung development and in the early postnatal stage, as alveolarization continues. These changes reflect the important role these molecules play in influencing the ability of cells to migrate through the pericellular matrix and the contribution of individual PG to the development of the various structural components of the lung.

VII. Role of PGs in Lung Disease

Lung diseases characterized by remodeling of the lung parenchyma or airway wall involve alterations in PGs. Changes in PGs have been best characterized in pulmonary fibrosis and, to a lesser extent, in asthma. Their roles in such diseases as emphysema, lung cancer, and ventilation-induced lung injury have not been as well defined.

A. Pulmonary Fibrosis

Pulmonary fibrosis is a chronic interstitial lung disease characterized by excess synthesis, deposition, and rearrangement of ECM molecules including collagen, elastic fibers, and PGs. PGs have been shown to be altered in pulmonary fibrosis of different etiologies. Bensadoun et al. (4,54) have shown, in histological studies in patients with both granulomatous and nongranulomatous lung fibrosis, that deposition of the large aggregating PG, versican, is increased. Versican forms part of the provisional matrix required for subsequent deposition of mature fibrous tissue. Westergren-Thorsson et al. (55) have recently reported that lung fibroblast clones from subjects with lung fibrosis produce different profiles of hyaluronin and decorin than fibroblast clones from normal controls. Animal models have been used extensively to study fibrotic lung disease; the most frequently employed model is that of bleomycin-induced lung fibrosis (56). In the bleomycin model, Westergren-Thorsson et al. (57) showed altered expression of the SLRPs, biglycan and decorin. Decorin message and protein decreased in rat lungs subsequent to bleomycin injury, biglycan message and protein increased. We have performed studies in the bleomycin model to further identify alterations in PG, the temporal sequence of these changes, and the effects on viscoelastic behavior of the tissues. We showed increases in all subclasses of PG studied, including the large, aggregating PG, versican; the basement membrane HSPG; and the SLRPs, biglycan and fibromodulin (14). The temporal pattern of increased PG profile was examined; increases in versican protein preceded increases in other of the PG proteins and occurred prior to the time point at which collagen deposition was enhanced (58). We also showed that changes in viscoelastic behavior of the tissues correlated with alterations in

PG (58). In subsequent studies (59), we isolated lung myofibroblasts from bleomycin-exposed rats and determined that these cells secreted increased amounts of TGF-β_1 and PG, as compared to fibroblasts from control animals. PG secretion was further enhanced by exogenous TGF-β and inhibited by anti-TGF-β antibodies and IFN-γ. The role of PG is compounded by the ability of decorin to bind TGF-β, reducing its bioavailability (60). A number of studies conducted in transgenic mice have now shown that gene transfer, or transgene overexpression of decorin, can modulate bleomycin-induced lung fibrosis (61–63).

B. Asthma

It is well established that airway wall remodeling is a key feature of asthma (64). There is thickening of the reticular basement membrane and increased matrix deposition in the subepithelial layer of the airways (65–68). Collagen and glycoproteins, such as fibronectin, account for some of these changes; PGs also contribute to the remodeling (6). Roberts (18) made the initial observation that, in patients dying of asthma, the airway wall showed prominent staining for biglycan, decorin, and versican. We have shown (6), in endobronchial biopsy specimens obtained from mild-asthmatic patients, that deposition of the PGs, versican, biglycan, and lumican, was increased in the subepithelial layer of the airway wall, as compared to that in airways obtained from normal volunteers. Reddington et al. (8) showed colocalization of decorin and TGF-β_1 in the subepithelial layer of the airway wall; however, the overall pattern of deposition was no different in biopsy specimens from asthmatic and control subjects. A recent report in mild-asthmatic subjects (69) showed a decrease in decorin in the lamina propria as compared to normal controls; conversely, biglycan was increased. Hence, the nature of PG deposition in the asthmatic airway wall is not entirely clear. More recent studies have addressed the question of PG deposition and asthma severity. In a study on fatal asthma, De Medeiros et al. (70) showed that in patients dying of fatal asthma, versican content in the internal area of both large and small airways was increased as compared to controls. In a study of patients with difficult or severe asthma (7), we used endobronchial biopsy to examine PG deposition within the airway wall. We showed that versican, biglycan, and lumican were all increased in the airway wall. Surprisingly, the pattern of deposition was different in asthmatics of severe versus moderate degree. While PG deposition in the subepithelial layer was relatively enhanced in both groups of asthmatics, as compared to normal controls, deposition within the smooth muscle layer was relatively greater in asthmatics with moderate clinical disease (Fig. 4).

These findings may have important implications for the functional effects of enhanced PG deposition on airway physiology and narrowing. Increased PG deposition in the subepithelial layer and subsequent thickening of the airway wall may lead to increased airway resistance because of decreases in luminal diameter, and to increased airway responsiveness because of the effects of a thickened airway wall on ASM shortening and subsequent airway narrowing (71). In a study of mild atopic asthmatics, PG deposition in the subepithelial layer was positively correlated with airway responsiveness (6). On the other hand, increased matrix

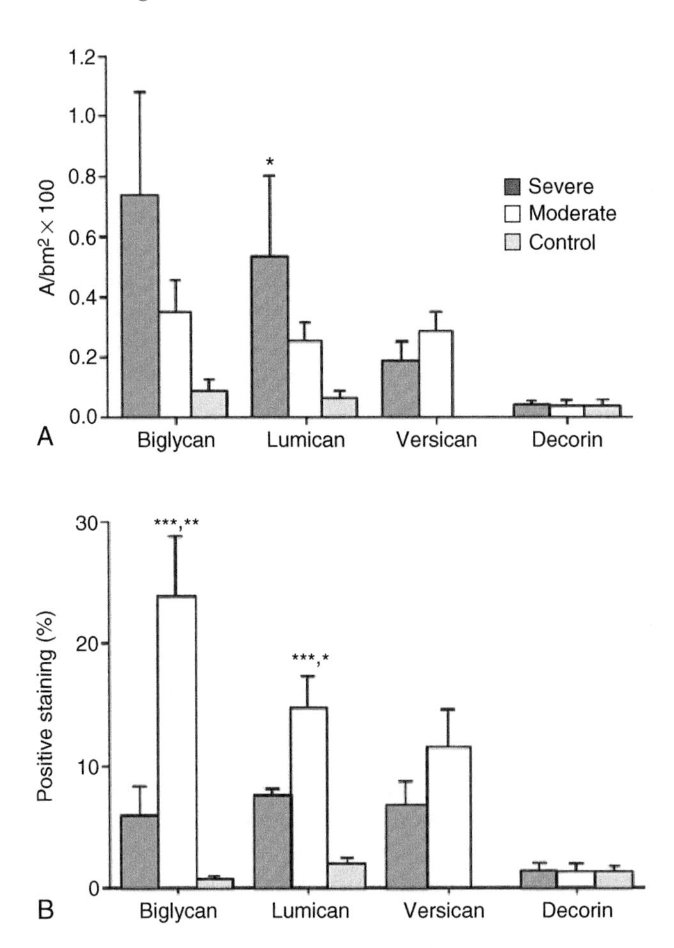

Figure 4 Changes in proteoglycans (PGs) deposition in the airway wall of patients with severe and moderate asthma and normal volunteers. Of the three groups, moderate asthmatics have the greatest amount of PG staining within the airway smooth muscle layer. (A) Area of positive staining in the subepithelial layer standardized for basement membrane length squared (A/bm^2) for biglycan, lumican, versican, and decorin in biopsy specimens from severe and moderate asthmatic patients and control subjects. Values are mean$\pm$standard error. *, $p<0.05$ vs control subjects. (B) Percentage area of smooth muscle layer staining positive for biglycan, lumican, versican, and decorin in biopsy specimens from severe and moderate asthmatic patients and control subjects. Values are mean$\pm$standard error. *, $p<0.05$; **, $p<0.01$ vs severe asthmatic patients. ***, $p<0.001$ vs control subjects. Reproduced by permission from Ref. 7.

deposition within the smooth muscle layer could potentially have a modulating effect on airway hyperresponsiveness (72–74). Excessive matrix could increase the impedence or resistance to ASM shortening, and thereby decreases the actual length change in the smooth muscle for a given degree of contractile stimulation. This could result in less-severe airway narrowing.

Some information is also available from *in vitro* studies. Fibroblasts obtained from endoscopic biopsies of airways of mild-asthmatic patients demonstrate increased production of versican, perlecan, and biglycan protein, as compared to cells isolated from normal patients (75). Additionally, we have shown that message for decorin is upregulated in these same cells (44). A recent study by Johnson et al. (30) reports that ASM cells obtained from asthmatic patients produce increased amounts of perlecan. Westergren-Thorrson et al. (75) showed in fibroblasts isolated from endoscopic biopsies that cells from asthmatic patients with the greatest degree of hyperresponsiveness produced larger amounts of PGs. Finally, we have shown that bronchial fibroblasts isolated from asthmatics patients show enhanced responses to mechanical strain, in terms of PG mRNA and protein secretion, as compared to fibroblasts from normal volunteers (44,45). As the asthmatic airway wall is subject to increased mechanical stimuli (76), altered PG metabolism in response to excessive mechanical strain in asthmatic airway fibroblasts is of particular significance.

A further issue is the effect of PGs on the structural cells present in the airway wall. Asthma is characterized by increased smooth muscle mass (64). PGs could potentially affect the rate of proliferation and/or apoptosis of these cells. Some information is available from *in vitro* studies. Hirst et al. (77) have shown that various ECM proteins, such as collagen and fibronectin, have the capacity to affect ASM proliferation. Johnson et al. (30) have reported that ECM proteins secreted by asthmatic ASM, which likely include PG, enhanced ASM proliferation in an autocrine fashion. We have obtained some preliminary data showing that human ASM cells grown on decorin matrices show decreased proliferation (28). Finally, the interaction between TGF-β and decorin may be important in the asthmatic airway wall (60). TGF-β is increased in the asthmatic airway wall (78) and plays an important role in asthmatic airway inflammation. Hence, the effect of changes in PG deposition in the airway wall in asthma could have a number of important consequences on asthmatic airway pathophysiology.

Data in animal models largely corroborate these findings. We have recently reported in the Brown Norway rat model of allergic asthma that repeated ovalbumin challenge leads to increased deposition of biglycan and decorin within the airway wall (79). Biglycan was deposited throughout the entire airway wall, whereas decorin largely spared the smooth muscle layer (Fig. 5). Similarly, Reinhardt et al. (80) showed increased decorin deposition in the airways of a mouse model of ovalbumin-induced asthma.

C. Emphysema

There is some information available on changes in PGs in clinical emphysema. Immunohistochemical study of lung tissue from patients with mild and severe emphysema showed decreased staining for decorin and biglycan, in the peribronchiolar area, and in the alveolar wall (81,82). HS PG was decreased in the alveolar region. Subsequent studies in fibroblasts isolated from patients with emphysema showed dysregulated decorin metabolism, suggesting an impaired

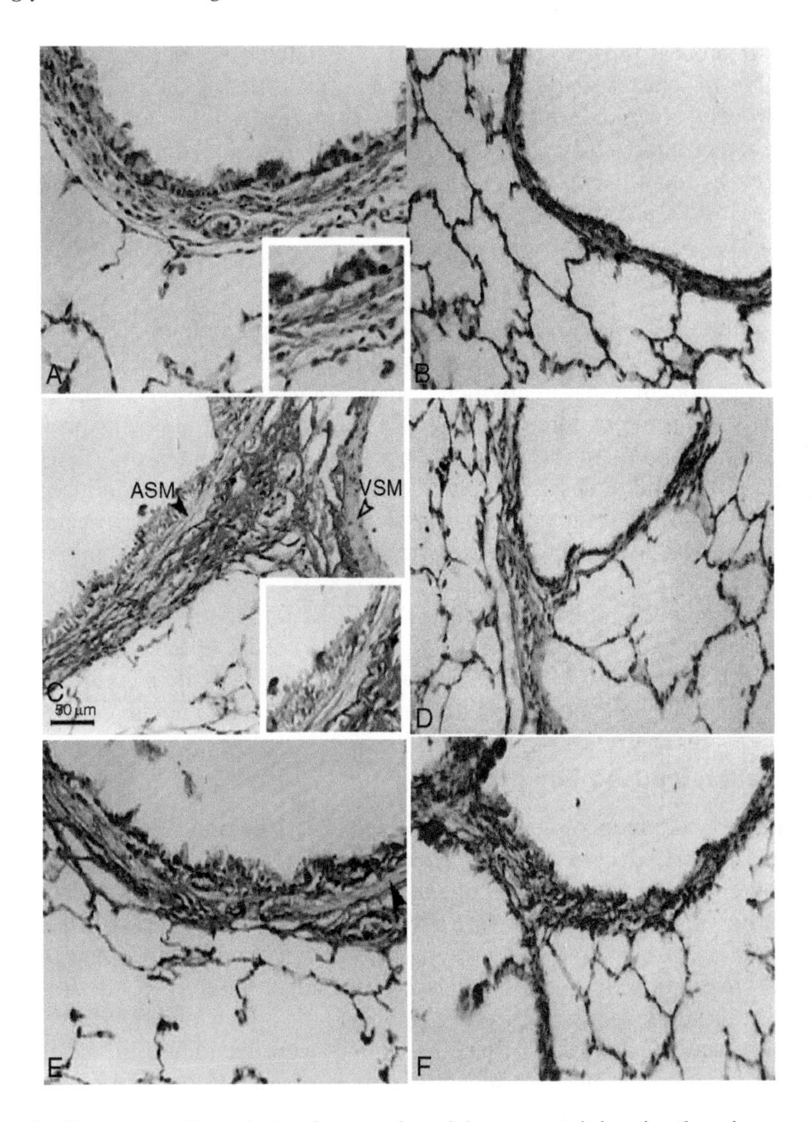

Figure 5 Representative photomicrographs of immunostaining in the airway wall of ovalbumin (OA)-challenged (A, C, E) and saline control Brown Norway rats (B, D, F) showing increased deposition of biglycan, decorin, and collagen type I in the airway wall of OA-sensitized and -challenged rats. Staining of large airways with primary antibody for biglycan (A, B), decorin (C, D), and collagen type I (E, F). Positive staining for biglycan was present within the smooth muscle layer (A, see inset) whereas decorin staining occurred around the smooth muscle bundles (C). Reproduced by permission from Ref. 79.

tissue repair process (83). Some data are also available in experimental models of emphysema. In elastase-induced emphysema in rodents, HSPG content was reduced in the alveolar wall; conversely, dermatin sulfate lung content was increased, again demonstrating dysregulation of PG metabolism (84,85). A further model of experimental emphysema has been described that is provoked by the agent β-D-xyloside, a compound which inhibits PG synthesis (86).

D. Pulmonary Edema

PGs could potentially play a role in the development of pulmonary edema because of their effects on tissue hydration and lung mechanical properties, and their function as an integral component of basement membranes. Some information is available in animal models of induced pulmonary edema (87). Alterations in the sieving properties of the basement membrane may enable enhanced transfer of solutes and fluids into the interstitial space (88). Studies in models of lesional pulmonary edema induced by neutrophil elastase showed fragmentation of basement membrane PG (perlecan), which was associated with loss of basement membrane integrity, increased microvascular permeability, and subsequent pulmonary edema (89,90). Hydraulic pulmonary edema led to weakening of PG interactions with other ECM components, and enhanced versican degradation (91). Degradation of hyaluronan occurs in the interstitial edema associated with alveolitis (92). Hence, alterations in matrix PG are a feature of cardiogenic and noncardiogenic models of experimental pulmonary edema.

E. Ventilation-Induced Lung Injury

Ventilation-induced lung injury is a well-recognized complication of mechanical ventilation, in part, due to alveolar overdistention (93,94). Recent studies have described alterations in matrix components in response to abnormal ventilatory regimens (95). Based on our data showing altered PG metabolism in response to excessive mechanical strain in *in vitro* systems (44), we postulated that excessive mechanical ventilation *in vivo* would similarly result in altered PG deposition. To test this hypothesis, we performed studies in which the effects of different ventilation regimes on tissue PGs and lung tissue mechanics were examined in an *in vivo* rat model (96). After 2 h of mechanical ventilation, versican, HSPG, and biglycan were all increased in rat lungs ventilated with large amplitude tidal volumes and zero positive end expiratory pressure. Versican and HSPG immunostaining became prominent in the alveolar wall and airspace; biglycan was identified in the airway wall. Pulmonary elastance and tissue resistance were significantly increased in rats receiving excessive ventilation, as compared to controls. These studies again provide evidence linking altered PG in the matrix with altered lung mechanics.

F. PGs and Lung Cancer

There is limited information available on the role of PGs in lung carcinoma. A number of studies have documented the antiproliferative action of decorin in other organ cancers (97–99); there is evidence that this effect occurs through downregulation of the epidermal growth factor (EGF) receptor (100). Whether decorin has a similar effect in lung cancer has not been established. Studies in A549 cells, a human lung carcinoma cell line, have shown that decorin upregulation leads to enhanced apoptosis in cancer, but not normal, cells (101). Gene expression studies in lung squamous cell and adenocarcinomas have shown decorin underexpression (102). Finally, in lymphangioleiomyomatosis, a fatal interstitial lung disease characterized by excessive smooth muscle proliferation, abnormal staining for versican, biglycan, and decorin is evident (103).

G. PGs and Other Lung Diseases

Involvement of PGs in other lung disease process has been even less well investigated. Hyperoxia-induced injury has been studied in newborn rats to give insight into hyperoxia-induced disease in the newborn. Increases in CSPG, hyaluronan, and biglycan have been identified in response to this insult (49,104). GAGs show an evolving profile during postresectional lung growth in the rat (105), mirroring processes described in the developing lung.

VIII. Conclusion

PGs form a key component of the lung ECM. They contribute to the structural integrity of the fiber network and are critical determinants of the mechanical properties of the lung tissues. They have important influences on various biological processes, including cell proliferation and responses to cytokines and chemokines, and participate in various autocrine pathways. Their involvement in different disease processes is currently being investigated; PGs potentially contribute to any disease in which the lung matrix is affected. Defining the role of PGs in disease biology is especially important as these molecules offer new, putative therapeutic targets.

References

1. Iozzo RV. Matrix proteoglycans: from molecular design to cellular function. Annu Rev Biochem 1998; 67:609–652.
2. van Kuppevelt TH, van Beuningen HM, Rutten TL, van den Brule AJ, Kuyper CM. Further characterization of a large proteoglycan in human lung alveoli. Eur J Cell Biol 1986; 39:386–390.
3. Roberts CR, Pare PD. Composition changes in human tracheal cartilage in growth and aging, including changes in proteoglycan structure. Am J Physiol Lung Cell Mol Physiol 1991; 261:L92–L101.
4. Bensadoun ES, Burke AK, Hogg JC, Roberts CR. Proteoglycans in granulomatous lung diseases. Eur Respir J 1997; 10:2731–2737.
5. Dolhnikoff M, Morin J, Roughley PJ, Ludwig MS. Expression of lumican in human lungs. Am J Respir Cell Mol Biol 1998; 19:582–587.
6. Huang J, Olivenstein R, Taha R, Hamid Q, Ludwig MS. Enhanced proteoglycan deposition in the airway wall of atopic asthmatics. Am J Respir Crit Care Med 1999; 160:725–729.

7. Pini L, Hamid Q, Shannon J, Lemelin L, Olivenstein R, Ernst P, Lemière C, Martin JG, Ludwig MS. Differences in proteoglycan deposition in the airways of moderate and severe asthmatics. Eur Respir J 2007; 29:71–77.
8. Redington AE, Roche WR, Holgate ST, Howarth PH. Co-localization of immunoreactive transforming growth factor-beta 1 and decorin in bronchial biopsies from asthmatic and normal subjects. J Pathol 1998; 186:410–415.
9. van Kuppevelt TH, Janssen HM, van Beuningen HM, Cheung KS, Schijen MM, Kuyper CM, Veerkamp JH. Isolation and characterization of a collagen fibril-associated dermatan sulphate proteoglycan from bovine lung. Biochim Biophys Acta 1987; 926:296–309.
10. Iozzo RV. The family of the small leucine-rich proteoglycans: key regulators of matrix assembly and cellular growth. Crit Rev Biochem Mol Biol 1997; 32:141–174.
11. Iozzo RV. The biology of the small leucine-rich proteoglycans. Functional network of interactive proteins. J Biol Chem 1999; 274:18843–18846.
12. Danielson KG, Baribault H, Holmes DF, Graham H, Kadler KE, Iozzo RV. Targeted disruption of decorin leads to abnormal collagen fibril morphology and skin fragility. J Cell Biol 1997; 136:729–743.
13. Corsi A, Xu T, Chen XD, Boyde A, Liang J, Mankani M, Somer B, Iozzo RV, Eichstetter I, Robey PG, Bianco P, Young MF. Phenotypic effects of biglycan deficiency are linked to collagen fibril abnormalities, are synergized by decorin deficiency, and mimic Ehlers-Danlos-like changes in bone and other connective tissues. J Bone Miner Res 2002; 17:1180–1189.
14. Venkatesan N, Ebihara T, Roughley PJ, Ludwig MS. Alterations in large and small proteoglycans in bleomycin-induced pulmonary fibrosis in rats. Am J Respir Crit Care Med 2000; 161:2066–2073.
15. Hedbom E, Heinegard D. Interaction of a 59-kDa connective tissue matrix protein with collagen I and collagen II. J Biol Chem 1989; 264:6898–6905.
16. Neame PJ, Kay CJ, McQuillan DJ, Beales MP, Hassell JR. Independent modulation of collagen fibrillogenesis by decorin and lumican. Cell Mol Life Sci 2000; 57:859–863.
17. Hedbom E, Heinegard D. Binding of fibromodulin and decorin to separate sites on fibrillar collagens. J Biol Chem 1993; 268:27307–27312.
18. Roberts CR. Is asthma a fibrotic disease? Chest 1995; 107:111S–117S.
19. van Kuppevelt TH, Cremers FP, Domen JG, van Beuningen HM, van den Brule AJ, Kuyper CM. Ultrastructural localization and characterization of proteoglycans in human lung alveoli. Eur J Cell Biol 1985; 36:74–80.
20. Murdoch AD, Dodge GR, Cohen I, Tuan RS, Iozzo RV. Primary structure of the human heparan sulfate proteoglycan from basement membrane HSPG2/perlecan). A chimeric molecule with multiple domains homologous to the low density lipoprotein receptor, laminin, neural cell adhesion molecules, and epidermal growth factor. J Biol Chem 1992; 267:8544–8557.
21. Maniscalco WM, Campbell MH. Alveolar type II cells synthesize hydrophobic cell-associated proteoglycans with multiple core proteins. Am J Physiol Lung Cell Mol Physiol 1992; 263:L348–L356.
22. Geng Y, McQuillan D, Roughley PJ. SLRP interaction can protect collagen fibrils from cleavage by collagenases. Matrix Biol 2006; 25:484–491.
23. Nietfeld JJ. Cytokines and proteoglycans. Experientia 1993; 49:456–469.

24. Ruoslahti E, Yamaguchi Y. Proteoglycans as modulators of growth factor activities. Cell 1991; 64:867–869.
25. Malmstrom J, Westergren-Thorsson G. Heparan sulfate upregulates platelet-derived growth factor receptors on human lung fibroblasts. Glycobiology 1998; 8:1149–1155.
26. Schaefer L, Babelova A, Kiss E, Hausser HJ, Baliova M, Krzyzankova M, Marxche G, Young MF, Mihalik D, Gotte M, Malle E, Schaefer RM, Grone HJ. The matrix component biglycan is proinflammatory and signals through Toll-like receptors 4 and 2 in macrophages. J Clin Invest 2005; 115:2223–2233.
27. Khoury J, Langleben D. Heparin-like molecules inhibit pulmonary vascular pericyte proliferation in vitro. Am J Physiol Lung Cell Mol Physiol 2000; 279:L252–L261.
28. Torregiani C, D'Antoni M, Michoud MC, Ferraro P, Martin JG, Ludwig MS. Culture of airway smooth muscle cells on decorin matrix reduces proliferation. Proc ATS 2006; 3:A261.
29. Tufvesson E, Westergren-Thorsson G. Biglycan and decorin induce morphological and cytoskeletal changes involving signalling by the small GTPases RhoA and Rac1 resulting in lung fibroblast migration. J Cell Sci 2003; 116:4857–4864.
30. Johnson PR, Burgess JK, Underwood PA, Au W, Poniris MH, Tamm M, Ge Q, Roth M, Black JL. Extracellular matrix proteins modulate asthmatic airway smooth muscle cell proliferation via an autocrine mechanism. J Allergy Clin Immunol 2004; 113:690–696.
31. Fung YC. Biomechanics. New York: Springer Verlag, 1993.
32. Yuan H, Kononov S, Cavalcante FS, Lutchen KR, Ingenito EP, Suki B. Effects of collagenase and elastase on the mechanical properties of lung tissue strips. J Appl Physiol 2000; 89:3–14.
33. Yuan H, Ingenito EP, Suki B. Dynamic properties of lung parenchyma: mechanical contributions of fiber network and interstitial cells. J Appl Physiol 1997; 83:1420–1431.
34. Bull HB. Tissue Elasticity. In: Redington J. W. ed., Washington: Waverly Press, 1957; 33–42.
35. Cavalcante FS, Ito S, Brewer K, Sakai H, Alencar AM, Almeida MP, Andrade JS, Jr, Majumdar A, Ingenito EP, Suki B. Mechanical interactions between collagen and proteoglycans: implications for the stability of lung tissue. J Appl Physiol 2005; 98:672–679.
36. Mijailovich SM, Stamenovic D, Fredberg JJ. Toward a kinetic theory of connective tissue micromechanics. J Appl Physiol 1993; 74:665–681.
37. Mijailovich SM, Stamenovic D, Brown R, Leith DE, Fredberg JJ. Dynamic moduli of rabbit lung tissue and pigeon ligamentum propatagiale undergoing uniaxial cyclic loading. J Appl Physiol 1994; 76:773–782.
38. Al Jamal R, Roughley PJ, Ludwig MS. Effect of glycosaminoglycan degradation on lung tissue viscoelasticity. Am J Physiol Lung Cell Mol Physiol 2001; 280:L306–L315.
39. Fust A, LeBellego F, Iozzo RV, Roughley PJ, Ludwig MS. Alterations in lung mechanics in decorin-deficient mice. Am J Physiol Lung Cell Mol Physiol 2005; 288:L159–L166.
40. Ingber DE. Tensegrity: the architectural basis of cellular mechanotransduction. Annu Rev Physiol 1997; 59:575–599.

41. Liu M, Tanswell AK, Post M. Mechanical force-induced signal transduction in lung cells. Am J Physiol Lung Cell Mol Physiol 1999; 277:L667–L683.
42. Xu J, Liu M, Liu J, Caniggia I, Post M. Mechanical strain induces constitutive and regulated secretion of glycosaminoglycans and proteoglycans in fetal lung cells. J Cell Sci 1996; 109:1605–1613.
43. Xu J, Liu M, Post M. Differential regulation of extracellular matrix molecules by mechanical strain of fetal lung cells. Am J Physiol Lung Cell Mol Physiol 1999; 276:L728–L735.
44. Ludwig MS, Ftouhi-Paquin N, Huang W, Page N, Chakir J, Hamid Q. Mechanical strain enhances proteoglycan message in fibroblasts from asthmatic subjects. Clin Exp Allergy 2004; 34:926–930.
45. Le Bellego F, Plante S, Chakir J, Hamid Q, Ludwig MS. Differences in MAP kinase phosphorylation in response to mechanical strain in asthmatic fibroblasts. Respir Res 2006; 7:68.
46. Vlahakis NE, Hubmayr RD. Invited review: plasma membrane stress failure in alveolar epithelial cells. J Appl Physiol 2000; 89:2490–2496.
47. Horwitz AL, Crystal RC. Content and synthesis of glycosaminoglycans in the developing lung. J Clin Invest 1975; 56:1312–1318.
48. Juul SE, Kinsella MG, Wight TN, Hodson WA. Alterations in nonhuman primate (M. nemestrina) lung proteoglycans during normal development and acute hyaline membrane disease. Am J Respir Cell Mol Biol 1993; 8:299–310.
49. Juul SE, Krueger RC, Jr., Scofield L, Hershenson MB, Schwartz NB. Hyperoxia alone causes changes in lung proteoglycans and hyaluronan in neonatal rat pups. Am J Respir Cell Mol Biol 1995; 13:629–638.
50. Wang Y, Sakamoto K, Khosla J, Sannes PL. Detection of chondroitin sulfates and decorin in developing fetal and neonatal rat lung. Am J Physiol Lung Cell Mol Physiol 2002; 282:L484–L490.
51. Brauker JH, Trautman MS, Bernfield M. Syndecan, a cell surface proteoglycan, exhibits a molecular polymorphism during lung development. Dev Biol 1991; 147:285–292.
52. Sannes PL, Burch KK, Khosla J, McCarthy KJ, Couchman JR. Immunohistochemical localization of chondroitin sulfate, chondroitin sulfate proteoglycan, heparan sulfate proteoglycan, entactin, and laminin in basement membranes of postnatal developing and adult rat lungs. Am J Respir Cell Mol Biol 1993; 8:245–251.
53. Tanaka R, Al Jamal R, Ludwig MS. Maturational changes in extracellular matrix and lung tissue mechanics. J Appl Physiol 2001; 91:2314–2321.
54. Bensadoun ES, Burke AK, Hogg JC, Roberts CR. Proteoglycan deposition in pulmonary fibrosis. Am J Respir Crit Care Med 1996; 154:1819–1828.
55. Westergren-Thorsson G, Sime P, Jordana M, Gauldie J, Sarnstrand B, Malmstrom A. Lung fibroblast clones from normal and fibrotic subjects differ in hyaluronan and decorin production and rate of proliferation. Int J Biochem Cell Biol 2004; 36:1573–1584.
56. Snider GL, Celli BR, Goldstein RH, O'Brien JJ, Lucey EC. Chronic interstitial pulmonary fibrosis produced in hamsters by endotracheal bleomycin. Lung volumes, volume-pressure relations, carbon monoxide uptake, and arterial blood gas studied. Am Rev Respir Dis 1978; 117:289–297.
57. Westergren-Thorsson G, Hernnas J, Sarnstrand B, Oldberg A, Heinegard D, Malmstrom A. Altered expression of small proteoglycans, collagen, and

transforming growth factor-beta 1 in developing bleomycin-induced pulmonary fibrosis in rats. J Clin Invest 1993; 92:632–637.
58. Ebihara T, Venkatesan N, Tanaka R, Ludwig MS. Changes in extracellular matrix and tissue viscoelasticity in bleomycin-induced lung fibrosis. Temporal aspects. Am J Respir Crit Care Med 2000; 162:1569–1576.
59. Venkatesan N, Roughley PJ, Ludwig MS. Proteoglycan expression in bleomycin lung fibroblasts: role of transforming growth factor-beta(1) and interferon-gamma. Am J Physiol Lung Cell Mol Physiol 2002; 283:L806–L814.
60. Yamaguchi Y, Mann DM, Ruoslahti E. Negative regulation of transforming growth factor-beta by the proteoglycan decorin. Nature 1990; 346:281–284.
61. Kolb M, Margetts PJ, Galt T, Sime PJ, Xing Z, Schmidt M, Gauldie J. Transient transgene expression of decorin in the lung reduces the fibrotic response to bleomycin. Am J Respir Crit Care Med 2001; 163:770–777.
62. Kolb M, Margetts PJ, Sime PJ, Gauldie J. Proteoglycans decorin and biglycan differentially modulate TGF-beta-mediated fibrotic responses in the lung. Am J Physiol Lung Cell Mol Physiol 2001; 280:L1327–L1334.
63. Giri SN, Hyde DM, Braun RK, Gaarde W, Harper JR, Pierschbacher MD. Antifibrotic effect of decorin in a bleomycin hamster model of lung fibrosis. Biochem Pharmacol 1997; 54:1205–1216.
64. Jeffery PK. Remodeling in asthma and chronic obstructive lung disease. Am J Respir Crit Care Med 2001; 164:S28–S38.
65. Roche WR, Beasley R, Williams JH, Holgate ST. Subepithelial fibrosis in the bronchi of asthmatics. Lancet 1989; 1:520–524.
66. Chu HW, Halliday JL, Martin RJ, Leung DY, Szefler SJ, Wenzel SE. Collagen deposition in large airways may not differentiate severe asthma from milder forms of the disease. Am J Respir Crit Care Med 1998; 158:1936–1944.
67. Laitinen A, Altraja A, Kampe M, Linden M, Virtanen I, Laitinen LA. Tenascin is increased in airway basement membrane of asthmatics and decreased by an inhaled steroid. Am J Respir Crit Care Med 1997; 156:951–958.
68. Chakir J, Shannon J, Molet S, Fukakusa M, Elias J, Laviolette M, Boulet LP, Hamid Q. Airway remodeling-associated mediators in moderate to severe asthma: effect of steroids on TGF-beta, IL-11, IL-17, and type I and type III collagen expression. J Allergy Clin Immunol 2003; 111:1293–1298.
69. de Kluijver J, Schrumpf JA, Evertse CE, Sont JK, Rabe KF, Hiemstra PS, Muaud T, Sterk PJ. Bronchial matrix and inflammation respond to inhaled steroids despite ongoing allergen exposure in asthma. Clin Exp Allergy 2005; 35:1361–1369.
70. de Medeiros MM, da Silva LF, dos Santos MA, Fernezlian S, Schrumpf JA, Roughley P, Hiemstra PS, Saldiva PHN, Mauad T, Dolhnikoff M. Airway proteoglycans are differentially altered in fatal asthma. J Pathol 2005; 207:102–110.
71. Moreno RH, Hogg JC, Pare PD. Mechanics of airway narrowing. Am Rev Respir Dis 1986; 133:1171–1180.
72. Macklem PT. A theoretical analysis of the effect of airway smooth muscle load on airway narrowing. Am J Respir Crit Care Med 1996; 153:83–89.
73. McParland BE, Macklem PT, Pare PD. Airway wall remodeling: friend or foe? J Appl Physiol 2003; 95:426–434.

74. Pare PD. Airway hyperresponsiveness in asthma: geometry is not everything! Am J Respir Crit Care Med 2003; 168:913–914.
75. Westergren-Thorsson G, Chakir J, Lafreniere-Allard MJ, Boulet LP, Tremblay GM. Correlation between airway responsiveness and proteoglycan production by bronchial fibroblasts from normal and asthmatic subjects. Int J Biochem Cell Biol 2002; 34:1256–1267.
76. Tschumperlin DJ, Drazen JM. Mechanical stimuli to airway remodeling. Am J Respir Crit Care Med 2001; 164:S90–S94.
77. Hirst SJ, Twort CH, Lee TH. Differential effects of extracellular matrix proteins on human airway smooth muscle cell proliferation and phenotype. Am J Respir Cell Mol Biol 2000; 23:335–344.
78. Minshall EM, Leung DY, Martin RJ, Song YL, Cameron L, Ernst P, Hamid Q. Eosinophil-associated TGF-beta1 mRNA expression and airways fibrosis in bronchial asthma. Am J Respir Cell Mol Biol 1997; 17:326–333.
79. Pini L, Torregiani C, Martin JG, Hamid Q, Ludwig MS. Airway remodeling in allergen-challenged Brown Norway rats: distribution of proteoglycans. Am J Physiol Lung Cell Mol Physiol 2006; 290:L1052–L1058.
80. Reinhardt AK, Bottoms SE, Laurent GJ, McAnulty RJ. Quantification of collagen and proteoglycan deposition in a murine model of airway remodelling. Respir Res 2005; 6:30.
81. Zandvoort A, Postma DS, Jonker MR, Noordhoek JA, Vos JT, van der Geld YM, Timens W. Altered expression of the Smad signalling pathway: implications for COPD pathogenesis. Eur Respir J 2006; 28: 533–541.
82. van Straaten JF, Coers W, Noordhoek JA, Huitema S, Flipsen JT, Kauffman HF, Timens W, Postma DS. Proteoglycan changes in the extracellular matrix of lung tissue from patients with pulmonary emphysema. Mod Pathol 1999; 12:697–705.
83. Noordhoek JA, Postma DS, Chong LL, Menkema L, Kauffman HF, Timens W, van Straaten JF, van der Geld YM. Different modulation of decorin production by lung fibroblasts from patients with mild and severe emphysema. COPD 2005; 2:17–25.
84. Karlinsky JB. Glycosaminoglycans in emphysematous and fibrotic hamster lungs. Am Rev Respir Dis 1982; 125:85–88.
85. van de Lest CH, Versteeg EM, Veerkamp JH, van Kuppevelt TH. Digestion of proteoglycans in porcine pancreatic elastase-induced emphysema in rats. Eur Respir J 1995; 8:238–245.
86. van Kuppevelt TH, van de Lest CH, Versteeg EM, Dekhuijzen PN, Veerkamp JH. Induction of emphysematous lesions in rat lung by beta-D-xyloside, an inhibitor of proteoglycan synthesis. Am J Respir Cell Mol Biol 1997; 16:75–84.
87. Miserocchi G, Negrini D, Passi A, De LG. Development of lung edema: interstitial fluid dynamics and molecular structure. News Physiol Sci 2001; 16:66–71.
88. Negrini D, Passi A, De LG, Miserochi G. Proteoglycan involvement during development of lesional pulmonary edema. Am J Physiol Lung Cell Mol Physiol 1998; 274:L203–L211.

89. Passi A, Negrini D, Albertini R, De LG, Miserocchi G. Involvement of lung interstitial proteoglycans in development of hydraulic- and elastase-induced edema. Am J Physiol Lung Cell Mol Physiol 1998; 275:L631–L635.
90. Negrini D, Passi A, De LG, Miserochi G. Proteoglycan involvement during development of lesional pulmonary edema. Am J Physiol Lung Cell Mol Physiol 1998; 274:L203–L211.
91. Negrini D, Passi A, De LG, Miserocchi G. Pulmonary interstitial pressure and proteoglycans during development of pulmonary edema. Am J Physiol Heart Circ Physiol 1996; 270:H2000–H2007.
92. Nettelbladt O, Tengblad A, Hallgren R. Lung accumulation of hyaluronan parallels pulmonary edema in experimental alveolitis. Am J Physiol Lung Cell Mol Physiol 1989; 257:L379–L384.
93. Vlahakis NE, Hubmayr RD. Cellular stress failure in ventilator-injured lungs. Am J Respir Crit Care Med 2005; 171:1328–1342.
94. Dos Santos CC, Slutsky AS. Invited review: mechanisms of ventilator-induced lung injury: a perspective. J Appl Physiol 2000; 89:1645–1655.
95. Berg JT, Fu Z, Breen EC, Tran HC, Mathieu-Costello O, West JB. High lung inflation increases mRNA levels of ECM components and growth factors in lung parenchyma. J Appl Physiol 1997; 83:120–128.
96. Al Jamal R, Ludwig MS. Changes in proteoglycans and lung tissue mechanics during excessive mechanical ventilation in rats. Am J Physiol Lung Cell Mol Physiol 2001; 281:L1078–L1087.
97. Grant DS, Yenisey C, Rose RW, Tootell M, Santra M, Iozzo RV. Decorin suppresses tumor cell-mediated angiogenesis. Oncogene 2002; 21:4765–4777.
98. Reed CC, Waterhouse A, Kirby S, Kay P, Owens RT, McQuillan DJ, Iozzo RV. Decorin prevents metastatic spreading of breast cancer. Oncogene 2005; 24:1104–1110.
99. Reed CC, Gauldie J, Iozzo RV. Suppression of tumorigenicity by adenovirus-mediated gene transfer of decorin. Oncogene 2002; 21:3688–3695.
100. Seidler DG, Goldoni S, Agnew C, Cardi C, Thakur ML, Owens RT, McQuillan DJ, Iozzo RV. Decorin protein core inhibits in vivo cancer growth and metabolism by hindering epidermal growth factor receptor function and triggering apoptosis via caspase-3 activation. J Biol Chem 2006; 281:26408–26418.
101. Tralhao JG, Schaefer L, Micegova M, Evaristo C, Schonherr E, Kayal S, Veiga-Fernandes H, Danel C, Iozzo RV, Kresse H, Lemarchand P. In vivo selective and distant killing of cancer cells using adenovirus-mediated decorin gene transfer. FASEB J 2003; 17:464–466.
102. Doniels-Silvers AL, Nimri CF, Stoner GD, Lubet RA, You M. Differential gene expression in human lung adenocarcinomas and squamous cell carcinomas. Clin Cancer Res 2002; 8:1127–1138.
103. Merrilees MJ, Hankin EJ, Black JL, Beaumont B. Matrix proteoglycans and remodelling of interstitial lung tissue in lymphangioleiomyomatosis. J Pathol 2004; 203:653–660.
104. Veness-Meehan KA, Rhodes DN, Stiles AD. Temporal and spatial expression of biglycan in chronic oxygen-induced lung injury. Am J Respir Cell Mol Biol 1994; 11:509–516.
105. Mueller MP, Thet LA. Changes in lung glycosaminoglycans during postresectional lung growth. J Appl Physiol 1987; 63:1033–1039.

Carbohydrate Chemistry, Biology and Medical Applications
Hari G. Garg, Mary K. Cowman and Charles A. Hales

DOI: 10.1016/B978-0-08-054816-6.00006-9

Chapter 6

Proteoglycans of the Intervertebral Disk

PETER J. ROUGHLEY

Genetics Unit, Shriners Hospital for Children and Department of Surgery, McGill University, Montreal, Quebec, H2X2P2 Canada

I. Introduction

The intervertebral disks are fibrocartilages that separate the vertebrae of the spine and permit the twisting and bending associated with spine mobility. They also provide the resistance to compression required for the spines of bipeds to counteract the force of gravity. At first sight, the intervertebral disks seem ill-equipped to deal with the constant mechanical insults to which they are exposed during daily life. They have no vascular system to provide the nutrients needed for cell metabolism, and are only sparsely populated with cells that are required to maintain the integrity of regions of the tissue that may be very remote. The solution to these problems resides in the unique structure of the extracellular matrix, which facilitates normal disk function and protects the cells from injury by mechanical trauma.

The disks are composed of two regions that are structurally and functionally distinct (1)—the outer annulus fibrosus (AF) and the central nucleus pulposus (NP). The AF consists of concentric lamellae rich in collagen, whereas the NP has a more gelatinous texture due to its lower collagen content and a high abundance of proteoglycan (Fig. 1). In large part, the collagen fibrils of the AF provide the resistance to bending and twisting, whereas the proteoglycan of the NP provides the osmotic properties needed to resist compressive loads (2). Disk structure is not constant with age, and following puberty the structure of the human NP begins to change from a gelatinous consistency to a more fibrous one. This change is associated with an increase in collagen content and a subsequent decline in proteoglycan content beginning early in adult life (3). The resulting decline in NP function is to some extent compensated by an increase in proteoglycan content

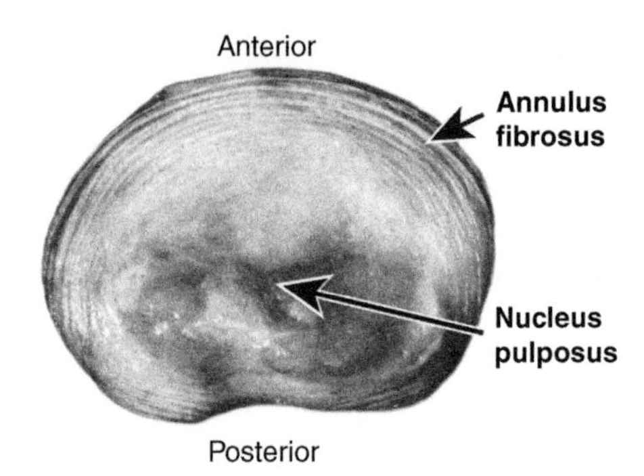

Figure 1 The intervertebral disk. A normal young adult human disk is shown with an outer lamellar annulus fibrosus and an inner gelatinous nucleus pulposus.

in the inner AF (4). The changes in NP structure in the juvenile are associated with the loss of notochordal cells that can influence disk cell metabolism (5). It is proposed that the age changes that have occurred by the young adult are a prelude to disk degeneration, which increases progressively throughout life (6).

II. Disk Proteoglycans

Disk proteoglycans may be divided into two distinct groups—those that are associated with the cell and those that reside in the extracellular matrix. The former group includes members of the syndecan (7) and glypican (8) families that are present at the cell surface, which are essential for cell signaling and cell–matrix interactions. While such proteoglycans are undoubtedly present on disk cells, there is little information concerning their identity. The extracellular proteoglycans include two major families—the hyalectans (9) that are characterized by their interaction with hyaluronan (HA) and the small leucine-rich repeat proteoglycans (SLRPs) (10) that commonly interact with collagen fibrils. Other proteoglycans are also present but in lower abundance, including perlecan (11), which may be involved in cell and matrix interactions. In contrast to the cell-associated proteoglycans, the proteoglycans of the disk extracellular matrix have been studied in considerable detail.

III. Aggrecan

In common with all cartilages, intervertebral disks are characterized by their high content of the proteoglycan aggrecan, a member of the hyalectan family. *In vivo*, aggrecan exists in the form of proteoglycan aggregates in association with HA and link protein (LP). The proteoglycan aggregates provide the osmotic properties needed to resist compressive loads (12). They may also be responsible for the avascular nature of the intervertebral disk, as high aggrecan contents are associated with the inhibition of endothelial cell migration (13). Each aggregate is composed of a central filament of HA with up to 100 aggrecan molecules radiating from it, with each interaction able to be stabilized by the presence of a LP. Proteoglycan aggregate structure is influenced by the length of the HA, the proportion of LP, and the degree of aggrecan processing. Proteoglycan aggregates may exist in two predominant molecular forms that differ in their LP content (14) and probably their functional characteristics. However, the structure of the proteoglycan aggregates is not constant throughout life, with decreases in HA length, aggrecan size, and possibly LP stabilization being evident (15).

Aggrecan is a modular proteoglycan with multiple functional domains. Its core protein consists of three globular regions (16), termed G1, G2, and G3 (Fig. 2), which are stabilized by disulfide bonds. The G1 and G2 regions are separated by a short interglobular domain (IGD), and the G2 and G3 regions are separated by a long glycosaminoglycan (GAG)-attachment region to which many keratan sulfate (KS) and chondroitin sulfate (CS) are attached. The G1 region is at the *N*-terminus of the core protein, and can be further subdivided into three functional domains, termed A, B1, and B2. The A domain is responsible for the interaction with LP, whereas the B-type domains are responsible for the interaction with HA (17). The G2 region also possesses two B-type domains, but does not appear to interact with HA (18), and at present its function is unknown. The G3 region resides at the C-terminus of the core protein and contains a variety of distinct structural domains. It is essential for normal posttranslational processing of the aggrecan core protein and subsequent aggrecan secretion (19).

The human aggrecan gene resides at chromosome 15q26 (20) and consists of 19 exons (21), with each exon encoding a distinct structural domain of the core

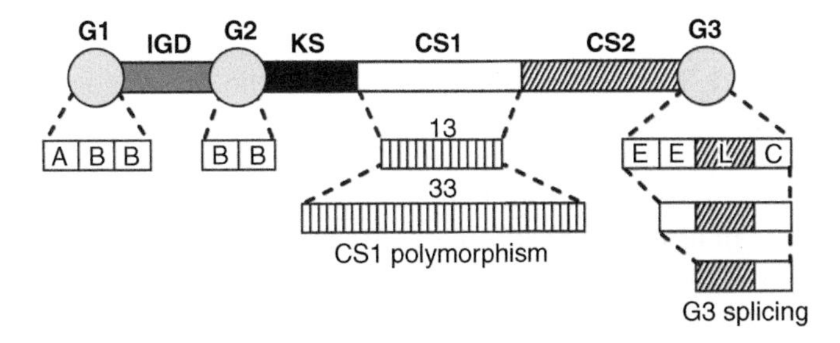

Figure 2 The structural domains of aggrecan. The aggrecan core protein is depicted with three disulfide-bonded globular regions (G1–3), an interglobular domain (IGD), and attachment regions for keratan sulfate (KS) and chondroitin sulfate (CS1 and CS2). The G1 region is composed of one domain (A) responsible for the interaction with LP and two domains (B) responsible for the interaction with hyaluronan. The G2 region also possesses two B-like domains. The CS1 domain exhibits length polymorphism due to a variable number of 19-amino acid tandem repeats, and individuals may possess between 13 and 33 repeats. The G3 region possesses up to four domains that show epidermal growth factor (E), lectin (L), and complement regulatory protein (C) homology. Its structure may vary because of alternative splicing of the E-like domains.

protein. Exons 3–6 encode the G1 region and exons 8–10 encode the G2 region. Most of the GAG-attachment region is encoded by exon 12. The G3 region of aggrecan is encoded by exons 13–18, with exons 13 and 14 each encoding an epidermal growth factor (EGF)-like domain, exons 15–17 encoding a lectin-like domain, and exon 18 encoding a CRP-like domain. However, the G3 region does not have a unique structure, as alternative slicing can result in the absence or presence of the EGF-like and the CRP-like domains (22). The presence of the lectin-like domain is essential for effective aggrecan secretion (19), but the functional significance of the other G3 domains is unclear though they could aid in matrix assembly following secretion. However, the continued presence of the G3 region on the aggrecan molecule does not appear to be essential in the mature disk matrix, as extracted aggrecan molecules are commonly devoid of the G3 region due to its proteolytic removal (23).

The GAG-attachment region is composed of three domains. The KS-attachment domain, which resides adjacent to the G2 region, consists of repeat motifs whose number varies between species (24). The following CS-attachment domain is divided into two subdomains, termed the CS1 and CS2 domains. The CS1 domain lies adjacent to the KS-attachment domain and also consists of repeat motifs whose number varies between species. The substitution of the KS and CS attachment sites within these domains may vary throughout life, and there is no unique aggrecan structure. The structure of the CS and KS chains on aggrecan can also vary throughout life due to changes in length and sulfation pattern (25,26). There does not, however, appear to be any major difference in GAG structure between the AF and NP in the same disk. The GAG-attachment region also possesses sites for the attachment of O-linked oligosaccharides, which may represent potential attachment sites for KS substitution (27). However, at least in the human, such substitution appears to be confined to the KS-attachment domain. The KS-attachment domain is devoid of CS, and the CS domains are devoid of KS (28). KS may also be present within the G1 region, the IGD and the G2 region, attached to either *O*-linked or *N*-linked oligosaccharides (29).

Aggrecan rarely exists in an intact form in disk extracellular matrix, but instead is subject to extracellular proteolytic processing. Each proteolytic cleavage of an intact aggrecan molecule results in one fragment that retains a G1 region and so remains bound to HA and a second fragment that remains free. These latter fragments contribute to the large quantity of nonaggregating proteoglycans present in the disk extracellular matrix (28). Such proteolysis continues throughout life, ultimately yielding proteoglycan aggregates that are enriched in aggrecan G1 regions, which remain resistant to proteolytic cleavage, and an increased abundance of nonaggregating proteoglycans. With time, continued proteolysis of the nonaggregating proteoglycans will generate fragments of a size small enough to be lost from the disk by diffusion but this is a slow process. Measurements of proteoglycan retention times have indicated that relatively intact aggrecan has a half-life of about 5 years, whereas the G1 regions have a half-life of about 21 years (30). The average half-life for the aggregating and nonaggregating proteoglycans is about 12 years in the normal disk, but is diminished to about 8 years in the degenerate disk. This is compatible with increased proteolysis being associated with disk degeneration. These half-lives are, however, considerably less than that of the collagen fibrils, which may have a half-life in excess of 100 years (31).

Aggrecanases and matrix metalloproteinases (MMPs) are associated with aggrecan proteolysis (32,33). The aggrecanases are of particular interest because of their selectivity for aggrecan. Five aggrecanase cleavage sites have been described in aggrecan (34), with one residing in the IGD domain and four in the CS2 domain. Increased activity of these metalloproteinases has been linked to the tissue destruction associated with intervertebral disk degeneration. A variety of MMPs have been associated with disk degeneration, including MMP1, 2, 3, 7, 9, 13, and 19 (35,36). This group of enzymes provides the ability to degrade both aggrecan and collagen. MMP19 is also of interest for its ability to degrade insulin-like growth factor, one of the growth factors associated with the stimulation of aggrecan synthesis (37). Among the MMPs, MMP3 (stromelysin) has received particular attention. Not only was it the first MMP shown to be secreted by human disk cells whose level of secretion was increased by cells of the degenerate disk (38), but a polymorphism in the promoter region of its gene has been described as a risk factor for an acceleration of disk degeneration (39). Of the aggrecanases, ADAMTS4 (aggrecanase-1) has been associated with the damage occurring in the herniated disk (40), but the lack of information to date does not exclude the participation of other aggrecanases such as ADAMTS5 (aggrecanase-2).

The increased production of metalloproteinases has been associated with adverse mechanical loading. The consequence of mechanical load depends on parameters such as magnitude, duration, and frequency of application (41,42). One general observation is that conditions that promote increased metalloproteinase expression also tend to downregulate aggrecan expression, so enhancing the detrimental effects of adverse loading. Appropriate mechanical loading is, however, essential for normal disk homeostasis as it is needed to promote aggrecan synthesis. The consequence of adverse mechanical loads is related not only to increased metalloproteinase secretion but also to the ability to activate secreted proenzyme (43). It is possible that the adverse effects of mechanical loads operate via stimulation of cytokine production by the disk cells, and interleukin-1 (IL-1) has been shown to increase metalloproteinase expression and decrease aggrecan expression by disk cells (44).

High concentrations of aggrecan are necessary to provide the osmotic properties necessary for the disk to resist compressive loads (2), with the GAGs producing the osmotic swelling pressure (Fig. 3). This functional ability is dependent on the aggrecan molecules possessing a high GAG content and having a size large enough for retention in the extracellular matrix. In articular cartilage, this latter property is dependent on the formation of large proteoglycan aggregates, as degradation products are rapidly lost from the tissue by diffusion. However, the unique organization of the intervertebral disk with its outer fibrous annulus permits the accumulation of large proteolytic degradation products of aggrecan irrespective of HA interaction. Any factor that results in a decrease in aggrecan GAG content in the extracellular matrix may result in overcompression of the tissue under load.

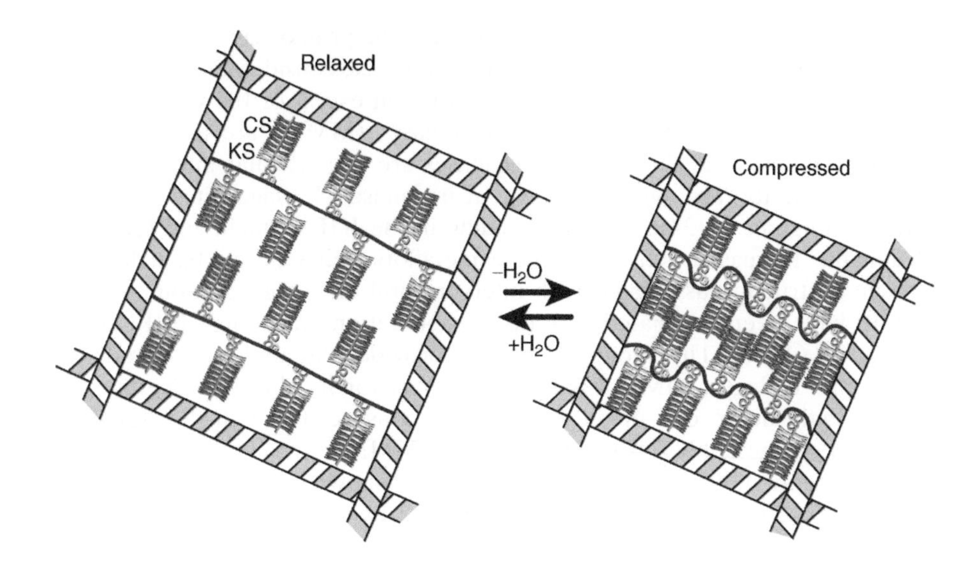

Figure 3 Aggrecan function. Aggrecan or its fragments if retained in the tissue exert an osmotic swelling pressure. In a relaxed state, the swelling of the aggrecan is resisted by the tensile forces generated in the surrounding collagen fibrils. Under load, water may be extruded from the tissue at the site of compression, bringing the aggrecan molecules into proximity and increasing their swelling potential. Upon removal of the load, water is drawn back as the aggrecan swells to return to the relaxed state. Aggrecan is depicted in the form of proteoglycan aggregates bound to hyaluronan, with each aggrecan molecule possessing numerous KS and CS side chains responsible for the swelling properties. Retaining collagen fibrils are depicted as banded lines.

This in turn may result in an adverse response by the disk cells, with increased proteinase secretion and subsequent tissue degeneration.

In the human, the aggrecan CS1 domain exhibits size polymorphism between individuals due to a variable number of repeats (Fig. 2) (45), resulting in the aggrecan molecules of different individuals having different numbers of CS chains. As the GAG-attachment region provides the high anionic charge density needed for the unique osmotic properties of aggrecan, it has been suggested that aggrecan function could be affected by size polymorphism within the CS1 domain. Individuals with the shortest core protein length would possess aggrecan with the lowest CS substitution, and such individuals might be at risk for premature loss of disk function and early tissue degeneration. Size polymorphism in the aggrecan CS1 domain has been associated with disk degeneration (46), though it is possible that the presence of one short aggrecan allele may be of little functional consequence and that two short alleles may be needed to be at risk (47).

Mutations in the aggrecan gene leading to chondrodyplasias have been described in the human, mouse, and chicken. In the human, a single base pair insertion in exon 12 causes a frameshift and results in a form of spondyloepiphyseal dysplasia (48). In the mouse, a 7-bp deletion in exon 5 causes a frameshift and results in a premature termination codon arising in exon 6 (49). In the chicken, a premature stop codon is present within exon 10 encoding the CS-attachment region (50), resulting in decreased message accumulation and underproduction of a truncated aggrecan. It is likely that the absence of a G3 region impairs secretion of the mutant aggrecan molecules. In the human, chondrodystrophic phenotypes have also been associated with the undersulfation of aggrecan due to gene defects in a sulfate transporter (51). These disorders illustrate the importance of aggrecan content and charge in embryonic skeletal development and growth.

IV. Versican

Versican is a ubiquitous member of the hyalectan family that is present in the intevertebral disk but at a much lower concentration than aggrecan. Its precise function in the mature disk is unclear. It is present throughout the disk, being localized between adjacent lamellae in the AF and spread diffusely in the NP (52). Versican is able to interact with HA and LP to form proteoglycan aggregates (53), but it is unclear at present whether it forms unique aggregates or mixed versican/aggrecan aggregates.

As with aggrecan, versican is a modular proteoglycan with multiple functional domains. Its core protein contains two terminal globular regions equivalent to the G1 and G3 regions of aggrecan separated by a large GAG-attachment domain (54). There is no region equivalent to the G2 region of the aggrecan. The G1 region possesses adjacent A and B domains analogous to those in aggrecan that are thought to function in a similar manner with respect to HA and LP interaction. The G3 region of versican possesses adjacent EGF-like, lectin-like, and CRP-like domains equivalent to aggrecan, but there is no evidence for alternative splicing unlike aggrecan (55). The GAG-attachment region of versican appears to contain only CS, though it is not clear whether under some circumstances this may be epimerized to DS. This CS-attachment region bears no structural similarity to that of aggrecan, and possesses many less potential sites for CS attachment. There is no report of the presence of KS in the GAG-attachment region, but it may be present within the G1 region (56). The versican present in the disk undergoes extensive proteolytic processing (56), and may contribute to the aggregated and nonaggregated populations of the disk proteoglycan. It is likely that metalloproteinases participate in versican degradation, and versican has been shown to be a substrate for the aggrecanases (57).

The human versican gene resides at chromosome 5q12–14 (58), and is encoded by 15 exons (59). Exons 2–6 encode the G1 region and exons 9–15 encode the G3 region. Exons 7 and 8 encode the CS-attachment region, and may undergo alternative splicing (60) to yield different versican isoforms. Four isoforms have been described—V_0 possessing the regions encoded by both exons 7 and 8, V_1 possessing only the region encoded by exon 8, V_2 possessing only the region encoded by exon 7, and V_3 possessing neither the region encoded by exon 7 or encoded by exon 8. The V_1 isoform is ubiquitously expressed and is the major form in the disk (56). There is currently no evidence for the expression of the other versican isoforms in the disk. Disruption of the versican gene in the hdf mouse results in early embryonic lethality (61). In the human, mutations in the versican gene have been associated with Wagner syndrome (62). These mutations are point substitutions within introns 7 or 8 that result in aberrant splicing and disruption of the GAG-attachment region. It is not clear how much of an effect the abnormal versican may have on intervertebral disk function.

V. Link Protein

LPs have a structure analogous to that of the G1 region of aggrecan and versican, possessing A, B1, and B2 domains and thus can be considered as members of the hyalectan family (63). The A domain has been shown to be responsible for interaction with the G1 region of aggrecan and presumably versican, whereas the B domains interact with HA (64). In human cartilage and intervertebral disk, LP can exist in three molecular forms, termed LP1, LP2, and LP3. LP1 and LP2 are intact LPs that differ by the presence of two or one *N*-linked oligosaccharide, respectively, in their *N*-terminal regions (65). LP3 is formed by proteolytic modification in the region between the two oligosaccharide chains. When the LPs are present in a proteoglycan aggregate, this site tends to be the only one that is accessible to most proteinases. MMPs have been implicated in such degradation (66), but the LPs are resistant to aggrecanases. Additional proteolysis does take place *in vivo* within the A domain to produce LP fragments. This type of processing has been mimicked *in vitro* by the action of free radicals.

LP serves several functions in the proteoglycan aggregates (Fig. 4). First, its interaction with the G1 region of aggrecan and HA stabilizes the proteoglycan aggregate and prevents its dissociation under physiological conditions. Second, it participates in a phenomenon termed "delayed aggregation", in which its interaction with newly synthesized aggrecan facilitates a conformational change in the G1 region to promote aggregate formation (67). Third, together with the G1 domain of aggrecan, it forms a protein coat covering the surface of HA, which helps protect the HA from hyaluronidases and free radical-mediated degradation (68). Finally, the *N*-terminal peptide of LP that is released by proteolysis can act as a growth factor to stimulate matrix production by the disk cells to maintain homeostasis (69). Proteolytic modification of LP takes place throughout life and commences at an early age in the disk. The modified LPs accumulate in the proteoglycan aggregates, but there is currently no evidence that such modified LPs are functionally impaired. However, LP content of the disk does declines with age (70), possibly as a result of proteolysis or diminished synthesis. This would be expected to impair aggregate function and stability and could contribute to the process of disk degeneration that progresses with age.

Mammals possess four LP genes (71), of which one is predominantly expressed in the intervertebral disk—the cartilage LP. Each LP gene is adjacent to a hyalectan proteoglycan gene, though it does not appear that adjacent gene products are necessarily coexpressed. The cartilage LP gene is adjacent to versican, and there is currently no evidence for the expression of the LP gene that resides

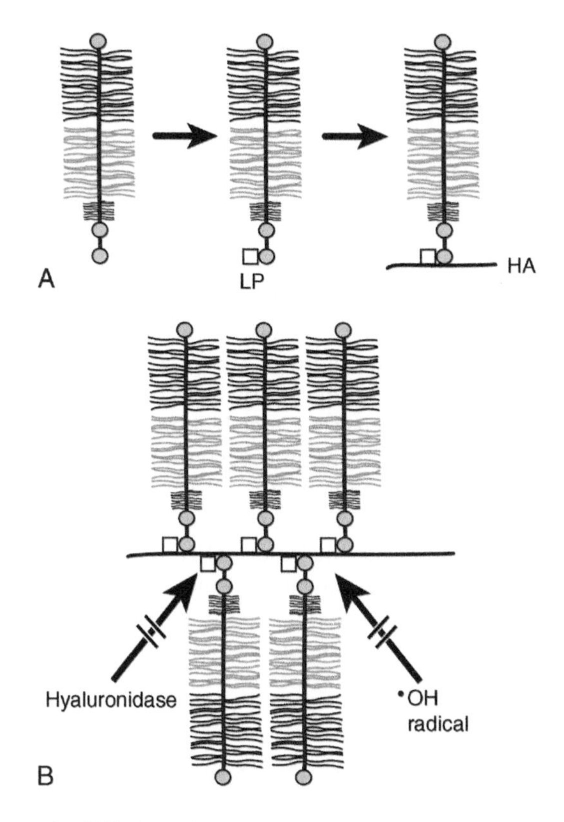

Figure 4 Link protein (LP) function. LPs are depicted in two functional roles. (A) LP interacts with nascent aggrecan to induce a conformational change in the aggrecan G1 region to facilitate interaction with hyaluronan (HA). (B) LP present in proteoglycan aggregates protects the HA core from degradation by preventing access of hyaluronidases or free radicals.

adjacent to aggrecan. The human cartilage LP gene resides at chromosome 5q13–14 (72) and possesses five exons, with exon 3 encoding the A domain and exons 4 and 5 encoding the B domains (73). The importance of LP in maintaining the structure of the proteoglycan aggregates has been demonstrated in knockout mice, which exhibit dwarfism and craniofacial abnormalities (74). To date, no human disorder has been directly linked to a mutation in LP, but its importance in maintaining intervertebral disk function is not in question.

VI. Hyaluronan

HA is a ubiquitous nonsulfated GAG characterized by its long length and unique mode of synthesis that occurs at the plasma membrane of the cell via a hyaluronan synthase (HAS) (75). Because of its mode of synthesis, HA is extruded directly into the extracellular space and can be retained as a cell coat (76). It is likely that proteoglycan aggregate formation initially occurs in the pericellular location. The mechanism whereby the proteoglycan aggregates are released from this environment and move to the more remote parts of the extracellular matrix is not clear. HA content in the disk increases with age and its size decreases (77). The increase in HA content may reflect an accumulation of partially degraded proteoglycan aggregates, and the decrease in size is probably a consequence of degradation as LP protection declines. HA degradation may have an adverse metabolic effect on the cells, as HA oligosaccharides can promote cell-mediated catabolism (78).

Mammals possess three HAS genes, termed HAS1, HAS2, and HAS3. While each of the three HASs presumably produces HA of an identical composition, there are differences in the chain length of the product and the ease with which it can be released from the cell surface (79). HAS2 is thought to be responsible for most HA synthesis in the skeleton, as mice lacking HAS2 gene expression show impaired skeletogenesis and die during mid-gestation (80). Mice lacking HAS1 or HAS3 expression show no skeletal phenotype. Recent studies on cartilage-specific HAS2 knockout mice verify that HAS2 gene expression is essential for intervertebral disk formation. This underlies the essential role of HA in normal disk function. It is quite possible that HA may have functions in the disk other than its role in proteoglycan aggregate formation, as it is present throughout the body including tissues in which members of the hyalectan family of proteoglycans are not prominent.

VII. Small Leucine-Rich Repeat Proteoglycans

As with all connective tissue, intervertebral disks contain a variety of small SLRPs, which on a weight basis are a minor component of the tissue but on a molar basis may rival the abundance of aggrecan. Decorin, biglycan, fibromodulin, and lumican are the major SLRPs present in disk where they help maintain the integrity of the tissue and modulate its metabolism. The SLRPs are not structurally related to the hyalectans, but belong to the large family of leucine-rich repeat proteins that are characterized by multiple adjacent domains bearing a common leucine-rich motif (81). They can be divided into several subfamilies based on the number of leucine-rich repeats, the type of GAG chain substituent, and their gene organization. In the case of decorin, biglycan, fibromodulin, and lumican, there are 10 leucine-rich repeats flanked by disulfide-bonded domains (Fig. 5). Decorin and biglycan are parts of one subfamily based on their substitution with dermatan sulfate (DS) chains and eight exon gene structure, whereas fibromodulin and lumican are part of a second subfamily based on their KS substitution and three exon gene structure. The genes for decorin and lumican reside close to one another at human chromosome 12q22 (82), while the gene for fibromodulin is at 1q32 (83) and that for biglycan is at Xq28 (84). The mature forms of fibromodulin and lumican present in the extracellular matrix correspond to removal of only the signal peptide, whereas for decorin and biglycan an additional amino acid sequence of 14 and 21 amino acids, respectively, are removed (85). These additional sequences have

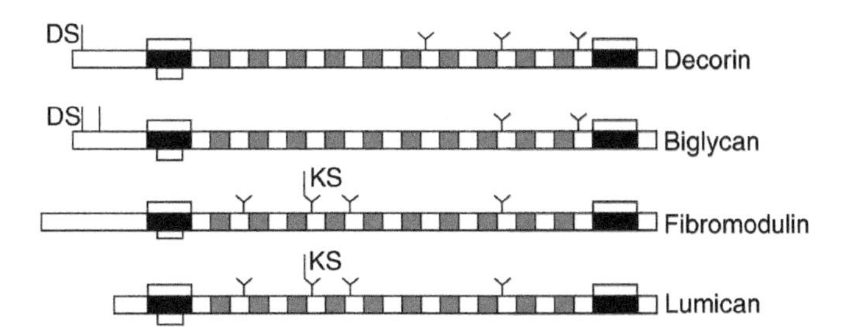

Figure 5 The structural domains of SLRPs. The core protein of decorin, biglycan, fibromodulin or lumican is depicted with two disulfide bonded domains (black boxes) flanking ten leucine-rich repeat domains (grey boxes). In the case of decorin or biglycan, one or two chondroitin/dermatan sulfate (DS) chains, respectively, reside in the amino terminal region. In the case of fibromodulin and lumican, keratan sulfate chains (KS) may reside on *N*-linked oligosaccharides located between the leucine-rich repeat domains. Decorin and biglycan also possess N-liked oligosaccharides but they are not further modified by KS substitution.

been considered as propeptides, though it is not clear whether their removal has any functional consequence.

The four SLRPs each possesses *N*-linked oligosaccharide chains within their central leucine-rich repeat region (86,87). In the case of fibromodulin and lumican, these *N*-linked oligosaccharides can be modified to KS (Fig. 5). Decorin and biglycan possess attachment sites for DS within the extreme *N*-terminus of the mature core proteins. Decorin has one such site at amino acid residue 4, whereas in biglycan has two sites at amino acids 5 and 10 in the human. Decorin and biglycan may also exist in nonglycanated forms devoid of DS chains, with the abundance of such forms accumulating with age (88). These nonglycanated forms appear to be the result of proteolysis within the *N*-terminal region of the core proteins. Nonglycanated forms of fibromodulin and lumican can also accumulate with age (89), due to the absence of KS synthesis. *In vitro*, decorin (90), biglycan (91), and fibromodulin (92) have been shown to be degraded by matrix metalloproteinases, and it is likely that such cleavage can occur *in vivo*. The structure and abundance of the SLRPs are not constant throughout life, but vary with both age and anatomical site.

The function of the SLRPs depends on both their core protein and GAG chains. The core proteins allow the SLRPs to interact with the fibrillar collagen that forms the framework of the tissue (93), and in so doing they help regulate fibril diameter during its formation and fibril–fibril interaction in the extracellular matrix. The continued interaction of decorin, fibromodulin, and lumican in the mature disk may limit access of the collagenases to their unique cleavage site on each collagen molecule, and in so doing may help protect the fibrils from proteolytic damage (94) and contribute to their long half-life (Fig. 6). Fibromodulin and lumican interact with the same region of the collagen molecule (95), whereas decorin interacts at a distinct site (96). Molecular modeling predicts that the SLRPs possess a "horse-shoe" conformation that is able to accommodate a

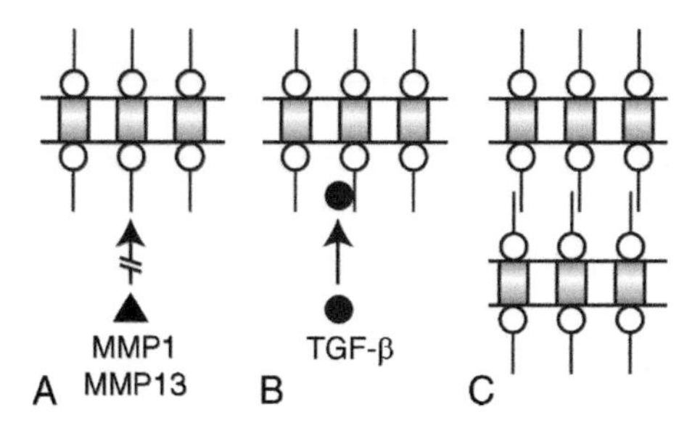

Figure 6 Small leucine-rich repeat proteoglycan (SLRP) function. SLRPs are depicted as globular core proteins (open circles) with an attached glycosaminoglycan chain, with the core proteins interacting with a collagen fibril (banded line). (A) The SLRP coat can impede the access of collagenases (MMP1 and MMP13) to the fibril surface and retard its degradation. (B) The dermatan sulfate chains of decorin may interact with transforming growth factor-β (TGF-β) sequestering the growth factor in the extracellular matrix. (C) The dermatan sulfate chains of decorin may self-associate to facilitate fibril–fibril interaction.

single-collagen molecule at the surface of the collagen fibrils within its concave face (97). However, X-ray diffraction analyses of decorin and biglycan crystals indicate that they exist as dimers with interlocking concave faces (98,99). It is not clear whether decorin dimers represent the functional form of the molecule in the tissue and how this impacts their interaction with other molecules.

Interaction of the SLRPs is not confined to the fibrillar collagens. Decorin, biglycan, fibromodulin, and lumican have been reported to interact with many other macromolecules, including types VI, XII, and XIV collagen, fibronectin, and elastin, all of which are present in the disk. Of particular note is the ability of biglycan to interact with type VI collagen and so participate in the formation of beaded filaments (100). This may be the principal role of biglycan in contrast to the fibrillar collagen binding of the other disk SLRPs. The SLRPs can also interact with growth factors, such as EGF, transforming growth factor-β, and tumor necrosis factor-α, and in doing so may influence disk cell metabolism. The GAG chains have been associated with the interaction with several growth factors, and enable the SLRPs to provide a sink for growth factor accumulation within the extracellular matrix so regulating growth factor access to the cells.

The level of SLRP synthesis varies with age and disk degeneration, though the precise consequence of such changes is unclear. The early stages of intervertebral disk degeneration are associated with increased SLRP production by the cells and accumulation in the extracellular matrix particularly in the AF (101). While this may help strengthen matrix architecture, it could also sequester growth factors responsible for matrix production, and hence participate in subsequent degeneration. In contrast, at later stages of disk degeneration, there is a progressive decline in SLRP production and levels particularly in the NP (101). Variations in SLRP expression have also been observed in disks from scoliotic spines, and may be associated with the development of the spinal curvature (102). It is clear that changes in SLRP expression can be influenced by adverse compressive loads, with the initial response being the upregulation of all SLRPs (103). It is also apparent that the cells of the AF and NP are different in the manner they respond to compressive loads.

Studies in knockout mice demonstrate that depletion in SLRP production can influence tissue properties. Absence of decorin results in lax, fragile skin, in which collagen fibril morphology is irregular (104). Absence of biglycan results in an osteoporosis-like phenotype, with a reduced growth rate and a decreased bone mass (105). Absence of lumican produces both skin laxity and corneal opacity, with an increased proportion of abnormally thick collagen fibrils (106). Absence of fibromodulin results in an abnormal collagen fibril organization in tendons (107). Thus, collagen fibril architecture is impaired in tissues in which SLRPs are deficient, but currently there is no information on how the intervertebral disks may be affected in these mice. Of these SLRP, only decorin has currently been linked to a human disorder, with a frameshift mutation being reported in congenital stromal dystrophy of the cornea (108). Impaired GAG synthesis can also have detrimental consequences, as deficiency in CS/DS substitution of decorin and biglycan has been associated with the progeriod form of Ehlers-Danlos syndrome (109). Again, there is no specific mention of how the intervertebral disks may be affected in these disorders.

VIII. Perlecan

Perlecan was originally described as a heparan sulfate (HS) proteoglycan of basement membranes. It is present in the basement membranes of all human tissues tested, and has also been shown to be present in the extracellular matrix of cartilaginous tissues that are devoid of basement membranes (11). With respect to cartilages, it has been shown to be present in the vertebral growth plate and cartilaginous end plates of the intervertebral disk, the cartilaginous primordia of the spine and limbs, and mature articular cartilage (110,111).

The human perlecan gene resides at 1p36–35 (112) and was originally considered to consist of 94 exons (113). However, analysis of human chromosome 1 indicates the presence of 97 exons. The perlecan core protein consists of five distinct domains (114). Domain I is unique; domain II contains repeats having homology to those present in the low-density lipoprotein receptor; domain III contains regions similar to those in the short arm of the laminin A chain; domain IV contains repeats similar to those in the neural cell adhesion molecule; and domain V contains EGF-like repeats. Structural variation can exist within the core protein due to alternative spicing in domain IV (115). GAG-attachment sites have been described in both the N-terminal domain I (116) and the C-terminal domain V (117), with three such sites being present in domain I and two in domain V. Both domains may be substituted with either HS or CS. In the case of cartilage, perlecan may be present in forms that are devoid of GAG, possess HS or CS, or both HS and CS (111). In cartilage, perlecan does not exist as the intact molecule but has undergone extensive proteolysis. The absence of GAG, the type of GAG substitution, the core protein variation, and its degree of proteolysis could influence the functional properties of perlecan.

Perlecan can interact with a variety of basement membrane and extracellular matrix components, including laminin, type IV collagen, nidogen, fibulin, fibronectin, and PRELP (proline/arginine-rich end leucine-rich repeat protein) (118,119). In doing so, perlecan may help maintain the structural organization of the tissues in which it is present. The HS chains and core protein also interact with fibroblast growth factors and their receptors (120). This interaction is important for the correct presentation of the FGFs to their receptors, and also protects them from proteolytic degradation so increasing their biological half-life. Thus, perlecan may also influence cell metabolism.

The essential role of perlecan in cartilage and skeletal development has been demonstrated in knockout mice (121), which survive to birth, but display severe skeletal defects with short axial and limb bones, and striking abnormalities in the growth plates of their long-bones. Mutations in the human perlecan gene have been identified in dyssegmental dysplasia (122) and the milder Schwartz-Jampel syndrome (chondrodystrophic myotonia) (123), which are characterized by their skeletal defects. Presumably, these defects include the intervertebral disks, but this has yet to be studied in detail.

IX. Conclusion

The intervertebral disk proteoglycans are essential for normal tissue organization and metabolism, and perturbation in their abundance or structure can have dire consequences on tissue formation in the embryo, growth in the juvenile, and function in the adult. In the adult, detrimental changes in proteoglycan content and structure are associated with disk degeneration.

Acknowledgments

The author thanks the Shriners of North America, the Canadian Institutes of Health Research, and the Arthritis Society of Canada for financial support.

References

1. Oegema TR. Biochemistry of the intervertebral disc. Clin Sports Med 1993; 12:419–439.
2. Urban JPG, Roberts S, Ralphs JR. The nucleus of the intervertebral disc from development to degeneration. Am Zool 2000; 40:53–61.
3. Antoniou J, Steffen T, Nelson F, Winterbottom N, Hollander AP, Poole RA, Aebi M, Alini M. The human lumbar intervertebral disc. Evidence for changes in the biosynthesis and denaturation of the extracellular matrix with growth, maturation, ageing, and degeneration. J Clin Invest 1996; 98:996–1003.
4. Roughley PJ. Biology of intervertebral disc aging and degeneration: involvement of the extracellular matrix. Spine 2004; 29:2691–2699.

5. Aguiar DJ, Johnson SL, Oegema TR. Notochordal cells interact with nucleus pulposus cells: regulation of proteoglycan synthesis. Exp Cell Res 1999; 246:129–137.
6. Roberts S, Urban JPG. Degeneration of intervertebral disc. Arthritis Res Ther 2003; 5:120–130.
7. Woods A. Syndecans: transmembrane modulators of adhesion and matrix assembly. J Clin Invest 2001; 107:935–941.
8. Fransson L. Glypicans. Int J Biochem Cell Biol 2003; 35:125–129.
9. Margolis RU, Margolis RK. Aggrecan-versican-neurocan family of proteoglycans. Meth Enzymol 1994; 245:105–128.
10. Iozzo RV. The family of the small leucine-rich proteoglycans: key regulators of matrix assembly and cellular growth. Crit Rev Biochem Mol Biol 1997; 32:141–174.
11. Iozzo RV, Cohen IR, Grässel S, Murdoch AD. The biology of perlecan: the multifaceted heparan sulphate proteoglycan of basement membranes and pericellular matrices. Biochem J 1994; 302:625–639.
12. Hascall VC. Proteoglycans: the chondroitin sulfate/keratan sulfate proteoglycan of cartilage. ISI Atlas Sci Biochem 1988; 1:189–198.
13. Johnson WEB, Caterson B, Eisenstein SM, Roberts S. Human intervertebral disc aggrecan inhibits endothelial cell adhesion and cell migration *in vitro*. Spine 2005; 30:1139–1147.
14. Buckwalter JA, Pita JC, Muller FJ, Nessler J. Structural differences between two populations of articular cartilage proteoglycan aggregates. J Orthop Res 1994; 12:144–148.
15. Buckwalter JA, Roughley PJ, Rosenberg LC. Age-related changes in cartilage proteoglycans: quantitative electron microscopic studies. Microsc Res Tech 1994; 28:398–408.
16. Doege KJ, Sasaki M, Kimura T, Yamada Y. Complete coding sequence and deduced primary structure of the human cartilage large aggregating proteoglycan, aggrecan. Human-specific repeats, and additional alternatively spliced forms. J Biol Chem 1991; 266:894–902.
17. Watanabe H, Cheung SC, Itano N, Kimata K, Yamada Y. Identification of hyaluronan-binding domains of aggrecan. J Biol Chem 1997; 272: 28057–28065.
18. Fosang AJ, Hardingham TE. Isolation of the *N*-terminal globular protein domains from cartilage proeoglycans. Identification of G2 domain and its lack of interaction with hyaluronate and link protein. Biochem J 1989; 261:801–809.
19. Zheng J, Luo W, Tanzer ML. Aggrecan synthesis and secretion—a paradigm for molecular and cellular coordination of multiglobular protein folding and intracellular trafficking. J Biol Chem 1998; 273:12999–13006.
20. Korenberg JR, Chen XN, Doege K, Grover J, Roughley PJ. Assignment of the human aggrecan gene (AGC1) to 15q26 using fluorescence in situ hybridization analysis. Genomics 1993; 16:546–548.
21. Valhmu WB, Palmer GD, Rivers PA, Ebara S, Cheng J-F, Fischer S, Ratcliffe A. Structure of the human aggrecan gene: exon-intron organization and association with the protein domains. Biochem J 1995; 309:535–542.
22. Fülöp C, Walcz E, Valyon M, Glant TT. Expression of alternatively spliced epidermal growth factor-like domains in aggrecans of different species. Evidence for a novel module. J Biol Chem 1993; 268:17377–17383.

23. Dudhia J, Davidson CM, Wells TM, Hardingham TE, Bayliss MT. Studies on the G3 domain of aggrecan from human cartilage. Ann NY Acad Sci 1996; 785:245–247.
24. Barry FP, Neame PJ, Sasse J, Pearson D. Length variation in the keratan sulfate domain of mammalian aggrecan. Matrix 1994; 14:323–328.
25. Roughley PJ, White RJ. Age-related changes in the structure of the proteoglycan subunits from human articular cartilage. J Biol Chem 1980; 255:217–224.
26. Brown GM, Huckerby TN, Bayliss MT, Nieduszynski IA. Human aggrecan keratan sulfate undergoes structural changes during adolescent development. J Biol Chem 1998; 273:26408–26414.
27. Santer V, White RJ, Roughley PJ. O-linked oligosaccharides of human articular cartilage proteoglycans. Biochim Biophys Acta 1982; 716:277–282.
28. Roughley PJ, Melching LI, Heathfield TF, Pearce RH, Mort JS. The structure and degradation of aggrecan in human intervertebral disc. Eur Spine J 2006; 15(Suppl 3):326–332.
29. Barry FP, Rosenberg LC, Gaw JU, Koob TJ, Neame PJ. *N*- and *O*-linked keratan sulfate on the hyaluronan binding region of aggrecan from mature and immature bovine cartilage. J Biol Chem 1995; 270:20516–20524.
30. Sivan SS, Tsitron E, Wachtel E, Roughley PJ, Sakkee N, Van der Ham F, DeGroot J, Roberts S, Maroudas A. Aggrecan turnover in human intervertebral disc as determined by the racemization of aspartic acid. J Biol Chem 2006; 281:13009–13014.
31. Sivan SS, Tsitron E, Wachtel E, Roughley P, Sakkee N, van der HF, Degroot J, Maroudas A. Age-related accumulation of pentosidine in aggrecan and collagen from normal and degenerate human intervertebral discs. Biochem J 2006; 399:29–35.
32. Sztrolovics R, Alini M, Roughley PJ, Mort JS. Aggrecan degradation in human intervertebral disc and articular cartilage. Biochem J 1997; 326:235–241.
33. Roberts S, Caterson B, Menage J, Evans EH, Jaffray DC, Eisenstein SM. Matrix metalloproteinases and aggrecanase—their role in disorders of the human intervertebral disc. Spine 2000; 25:3005–3013.
34. Tortorella MD, Pratta M, Liu RQ, Austin J, Ross OH, Abbaszade I, Burn T, Arner E. Sites of aggrecan cleavage by recombinant human aggrecanase-1 (ADAMTS-4). J Biol Chem 2000; 275:18566–18573.
35. Le Maitre CL, Freemont AJ, Hoyland JA. Localization of degradative enzymes and their inhibitors in the degenerate human intervertebral disc. J Pathol 2004; 204:47–54.
36. Le Maitre C, Freemont A, Hoyland J. Human disc degeneration is associated with increased MMP 7 expression. Biotech Histochem 2006; 81:125–131.
37. Gruber HE, Ingram JA, Hanley EN, Jr. Immunolocalization of MMP-19 in the human intervertebral disc: implications for disc aging and degeneration. Biotech Histochem 2005; 80:157–162.
38. Nemoto O, Yamagishi M, Yamada H, Kikuchi T, Takaishi H. Matrix metalloproteinase-3 production by human degenerated intervertebral disc. J Spinal Disord 1997; 10:493–498.
39. Takahashi M, Haro H, Wakabayashi Y, Kawa-uchi T, Komori H, Shinomiya K. The association of degeneration of the intervertebral disc with 5a/6a

polymorphism in the promoter of the human matrix metalloproteinase-3 gene. J Bone Joint Surg 2001; 83B:491–495.
40. Hatano E, Fujita T, Ueda Y, Okuda T, Katsuda S, Okada Y, Matsumoto T. Expression of ADAMTS-4 (aggrecanase-1) and possible involvement in regression of lumbar disc herniation. Spine 2006; 31:1426–1432.
41. MacLean JL, Lee CR, Alini M, Iatridis JC. Anabolic and catabolic mRNA levels of the intervertebral disc vary with the magnitude and frequency of in vivo dynamic compression. J Orthop Res 2004; 22:1193–1200.
42. MacLean JJ, Lee CR, Alini M, Iatridis JC. The effects of short-term load duration on anabolic and catabolic gene expression in the rat tail intervertebral disc. J Orthop Res 2005; 23:1120–1127.
43. Hsieh AH, Lotz JC. Prolonged spinal loading induces matrix metalloproteinase-2 activation in intervertebral discs. Spine 2003; 28:1781–1788.
44. Le Maitre CL, Freemont AJ, Hoyland JA. The role of interleukin-1 in the pathogenesis of human intervertebral disc degeneration. Arthritis Res Ther 2005; 7:R732–R745.
45. Doege KJ, Coulter SN, Meek LM, Maslen K, Wood JG. A human-specific polymorphism in the coding region of the aggrecan gene. Variable number of tandem repeats produce a range of core protein sizes in the general population. J Biol Chem 1997; 272:13974–13979.
46. Kawaguchi Y, Osada R, Kanamori M, Ishihara H, Ohmori K, Matsui H, Kimura T. Association between an aggrecan gene polymorphism and lumbar disc degeneration. Spine 1999; 24:2456–2460.
47. Roughley P, Martens D, Rantakokko J, Alini M, Mwale F, Antoniou J. The involvement of aggrecan polymorphism in degeneration of human intervertebral disc and articular cartilage. Eur Cell Mater 2006; 11:1–7.
48. Gleghorn L, Ramesar R, Beighton P, Wallis G. A mutation in the variable repeat region of the aggrecan gene (*AGC1*) causes a form of spondyloepiphyseal dysplasia associated with severe, premature osteoarthritis. Am J Hum Genet 2005; 77:484–490.
49. Watanabe H, Kimata K, Line S, Strong D, Gao L, Kozak CA, Yamada Y. Mouse cartilage matrix deficiency (*cmd*) caused by a 7 bp deletion in the aggrecan gene. Nat Genet 1994; 7:154–157.
50. Li H, Schwartz NB, Vertel BM. cDNA cloning of chick cartilage chondroitin sulfate (aggrecan) core protein and identification of a stop codon in the aggrecan gene associated with the chondrodystrophy, nanomelia. J Biol Chem 1993; 268:23504–23511.
51. Superti-Furga A, Rossi A, Steinmann B, Gitzelmann R. A chondrodysplasia family produced by mutations in the *diastrophic dysplasia sulfate transporter* gene: genotype/phenotype correlations. Am J Med Genet 1996; 63:144–147.
52. Melrose J, Ghosh P, Taylor TKF. A comparative analysis of the differential spatial and temporal distributions of the large (aggrecan, versican) and small (decorin, biglycan, fibromodulin) proteoglycans of the intervertebral disc. J Anat 2001; 198:3–15.
53. Matsumoto K, Shionyu M, Go M, Shimizu K, Shinomura T, Kimata K, Watanabe H. Distinct interaction of versican/PG-M with hyaluronan and link protein. J Biol Chem 2003; 278:41205–41212.

54. Zimmermann DR, Ruoslahti E. Multiple domains of the large fibroblast proteoglycan, versican. EMBO J 1989; 8:2975–2981.
55. Grover J, Roughley PJ. Versican gene expression in human articular cartilage and comparison of mRNA splicing variation with aggrecan. Biochem J 1993; 291:361–367.
56. Sztrolovics R, Grover J, CS-Szabo G, Shi SL, Zhang YP, Mort JS, Roughley PJ. The characterization of versican and its message in human articular cartilage and intervertebral disc. J Orthop Res 2002; 20:257–266.
57. Sandy JD, Westling J, Kenagy RD, Iruela-Arispe ML, Verscharen C, Rodriguez-Mazaneque JC, Zimmermann DR, Lemire JM, Fischer JW, Wight TN, Clowes AW. Versican V1 proteolysis in human aorta *in vivo* occurs at the Glu^{441}-Ala^{442} bond, a site that is cleaved by recombinant ADAMTS-1 and ADAMTS-4. J Biol Chem 2001; 276:13372–13378.
58. Iozzo RV, Naso MF, Cannizzaro LA, Wasmuth JJ, McPherson JD. Mapping of the versican proteoglycan gene (CSPG2) to the long arm of human chromosome 5 (5q12–5q14). Genomics 1992; 14:845–851.
59. Naso MF, Zimmermann DR, Iozzo RV. Characterization of the complete genomic structure of the human versican gene and functional analysis of its promoter. J Biol Chem 1994; 269:32999–33008.
60. Dours-Zimmermann MT, Zimmermann DR. A novel glycosaminoglycan attachment domain identified in two alternative splice variants of human versican. J Biol Chem 1994; 269:32992–32998.
61. Mjaatvedt CH, Yamamura H, Capehart AA, Turner D, Markwald RR. The Cspg2 gene, disrupted in the hdf mutant, is required for right cardiac chamber and endocardial cushion formation. Dev Biol 1998; 202:56–66.
62. Mukhopadhyay A, Nikopoulos K, Maugeri A, de Brouwer AP, van Nouhuys CE, Boon CJ, Perveen R, Zegers HA, Wittebol-Post D, van den Biesen PR, van der Velde-Visser SD, Brunner HG, Black GC, Hoyng CB, Cremers FP. Erosive vitreoretinopathy and wagner disease are caused by intronic mutations in CSPG2/Versican that result in an imbalance of splice variants. Invest Ophthalmol Vis Sci 2006; 47:3565–3572.
63. Neame PJ, Barry FP. The link proteins. Experientia 1993; 49:393–402.
64. Grover J, Roughley PJ. The expression of functional link protein in a baculovirus system: analysis of mutants lacking the A, B and B′ domains. Biochem J 1994; 300:317–324.
65. Roughley PJ, Poole AR, Mort JS. The heterogeneity of link proteins isolated from human articular cartilage proteoglycan aggregates. J Biol Chem 1982; 257:11908–11914.
66. Nguyen Q, Murphy G, Hughes CE, Mort JS, Roughley PJ. Matrix metalloproteinases cleave at two distinct sites on human cartilage link protein. Biochem J 1993; 295:595–598.
67. Melching LI, Roughley PJ. Studies on the interaction of newly secreted proteoglycan subunits with hyaluronate in human articular cartilage. Biochim Biophys Acta Gen Subj 1990; 1035:20–28.
68. Rodriguez E, Roughley P. Link protein can retard the degradation of hyaluronan in proteoglycan aggregates. Osteoarthr Cartil 2006; 14:823–829.
69. Mwale F, Demers CN, Petit A, Roughley P, Poole AR, Steffen T, Aebi M, Antoniou J. A synthetic peptide of link protein stimulates the biosynthesis

of collagens II, IX and proteoglycan by cells of the intervertebral disc. J Cell Biochem 2003; 88:1202–1213.
70. Pearce RH, Mathieson JM, Mort JS, Roughley PJ. Effect of age on the abundance and fragmentation of link protein of the human intervertebral disc. J Orthop Res 1989; 7:861–867.
71. Spicer AP, Joo A, Bowling RA, Jr. A hyaluronan binding link protein gene family whose members are physically linked adjacent to chrondroitin sulfate proteoglycan core protein genes. The missing links. J Biol Chem 2003; 278:21083–21091.
72. Osborne-Lawrence SL, Sinclair AK, Hicks RC, Lacey SW, Eddy RL, Jr., Byers MG, Shows TB, Duby AD. Complete amino acid sequence of human cartilage link protein (*CRTL1*) deduced from cDNA clones and chromosomal assignment of the gene. Genomics 1990; 8:562–567.
73. Dudhia J, Bayliss MT, Hardingham TE. Human link protein gene: structure and transcription pattern in chondrocytes. Biochem J 1994; 303:329–333.
74. Watanabe H, Yamada Y. Mice lacking link protein develop dwarfism and craniofacial abnormalities. Nat Genet 1999; 21:225–229.
75. Weigel PH, Hascall VC, Tammi M. Hyaluronan synthases. J Biol Chem 1997; 272:13997–14000.
76. Inkinen RI, Lammi MJ, Agren U, Tammi R, Puustjärvi K, Tammi MI. Hyaluronan distribution in the human and canine intervertebral disc and cartilage endplate. Histochem J 1999; 31:579–587.
77. Hardingham TE, Adams P. A method for the determination of hyaluronate in the presence of other glycosaminoglycans and its application to human intervertebral disc. Biochem J 1976; 159:143–147.
78. Knudson W, Casey B, Nishida Y, Eger W, Kuettner KE, Knudson CB. Hyaluronan oligosaccharides perturb cartilage matrix homeostasis and induce chondrocytic chondrolysis. Arthritis Rheum 2000; 43:1165–1174.
79. Itano N, Sawai T, Yoshida M, Lenas P, Yamada Y, Imagawa M, Shinomura T, Hamaguchi M, Yoshida Y, Ohnuki Y, Miyauchi S, Spicer AP, McDonald JA, Kimata K. Three isoforms of mammalian hyaluronan synthases have distinct enzymatic properties. J Biol Chem 1999; 274:25085–25092.
80. Camenisch TD, Spicer AP, Brehm-Gibson T, Biesterfeldt J, Augustine ML, Calabro A, Jr., Kubalak S, Klewer SE, McDonald JA. Disruption of hyaluronan synthase-2 abrogates normal cardiac morphogenesis and hyaluronan-mediated transformation of epithelium to mesenchyme. J Clin Invest 2000; 106:349–360.
81. Hocking AM, Shinomura T, McQuillan DJ. Leucine-rich repeat glycoproteins of the extracellular matrix. Matrix Biol 1998; 17:1–19.
82. Grover J, Chen XN, Korenberg JR, Roughley PJ. The human lumican gene—organization, chromosomal, location, and expression in articular cartilage. J Biol Chem 1995; 270:21942–21949.
83. Sztrolovics R, Chen X-N, Grover J, Roughley PJ, Korenberg JR. Localization of the human fibromodulin gene (FMOD) to chromosome 1q32 and completion of the cDNA sequence. Genomics 1994; 23:715–717.
84. McBride OW, Fisher LW, Young MF. Localization of PGI (biglycan, BGN) and PGII (decorin, DCN, PG-40) genes on human chromosomes Xq13-qter and 12q, respectively. Genomics 1990; 6:219–225.

85. Roughley PJ, White RJ, Mort JS. Presence of pro-forms of decorin and biglycan in human articular cartilage. Biochem J 1996; 318:779–784.
86. Neame PJ, Choi HU, Rosenberg LC. The primary structure of the core protein of the small, leucine-rich proteoglycan (PG I) from bovine articular cartilage. J Biol Chem 1989; 264:8653–8661.
87. Plaas AHK, Neame PJ, Nivens CM, Reiss L. Identification of the keratan sulfate attachment sites on bovine fibromodulin. J Biol Chem 1990; 265:20634–20640.
88. Roughley PJ, White RJ, Magny M-C, Liu J, Pearce RH, Mort JS. Non-proteoglycan forms of biglycan increase with age in human articular cartilage. Biochem J 1993; 295:421–426.
89. Sztrolovics R, Alini M, Mort JS, Roughley PJ. Age-related changes in fibromodulin and lumican in human intervertebral discs. Spine 1999; 24:1765–1771.
90. Imai K, Hiramatsu A, Fukushima D, Pierschbacher MD, Okada Y. Degradation of decorin by matrix metalloproteinases: identification of the cleavage sites, kinetic analyses and transforming growth factor-b1 release. Biochem J 1997; 322:809–814.
91. Monfort J, Tardif G, Reboul P, Mineau F, Roughley P, Pelletier JP, Martel-Pelletier J. Degradation of small leucine-rich repeat proteoglycans by matrix metalloprotease-13: identification of a new biglycan cleavage site. Arthritis Res Ther 2006; 8:R26–R34.
92. Heathfield TF, Onnerfjord P, Dahlberg L, Heinegard D. Cleavage of fibromodulin in cartilage explants involves removal of the N-terminal tyrosine sulfate-rich region by proteolysis at a site that is sensitive to matrix metalloproteinase-13. J Biol Chem 2004; 279:6286–6295.
93. Svensson L, Heinegård D, Oldberg Å. Decorin-binding sites for collagen type I are mainly located in leucine-rich repeats 4–5. J Biol Chem 1995; 270:20712–20716.
94. Geng Y, McQuillan D, Roughley PJ. SLRP interaction can protect collagen fibrils from cleavage by collagenases. Matrix Biol 2006; 25:484–491.
95. Svensson L, Närlid I, Oldberg Å. Fibromodulin and lumican bind to the same region on collagen type I fibrils. FEBS Lett 2000; 470:178–182.
96. Hedbom E, Heinegård D. Binding of fibromodulin and decorin to separate sites on fibrillar collagens. J Biol Chem 1993; 268:27307–27312.
97. Scott JE. Proteodermatan and proteokeratan sulfate (decorin, lumican/fibromodulin) proteins are horseshoe shaped. Implications for their interactions with collagen. Biochemistry 1996; 35:8795–8799.
98. Scott PG, McEwan PA, Dodd CM, Bergmann EM, Bishop PN, Bella J. Crystal structure of the dimeric protein core of decorin, the archetypal small leucine-rich repeat proteoglycan. Proc Natl Acad Sci USA 2004; 101:15633–15638.
99. Scott PG, Dodd CM, Bergmann EM, Sheehan JK, Bishop PN. Crystal structure of the biglycan dimer and evidence that dimerization is essential for folding and stability of class I small leucine-rich repeat proteoglycans. J Biol Chem 2006; 281:13324–13332.
100. Feng H, Danfelter M, Stromqvist B, Heinegard D. Extracellular matrix in disc degeneration. J Bone Joint Surg 2006; 88A(Suppl 2):25–29.

101. Cs-Szabo G, Ragasa-San Juan D, Turumella V, Masuda K, Thonar EJ, An HS. Changes in mRNA and protein levels of proteoglycans of the anulus fibrosus and nucleus pulposus during intervertebral disc degeneration. Spine 2002; 27:2212–2219.
102. Bertram H, Steck E, Zimmermann G, Chen BH, Carstens C, Nerlich A, Richter W. Accelerated intervertebral disc degeneration in scoliosis versus physiological ageing develops against a background of enhanced anabolic gene expression. Biochem Biophys Res Commun 2006; 342:963–972.
103. Chen J, Yan W, Setton LA. Static compression induces zonal-specific changes in gene expression for extracellular matrix and cytoskeletal proteins in intervertebral disc cells in vitro. Matrix Biol 2004; 22:573–583.
104. Danielson KG, Baribault H, Holmes DF, Graham H, Kadler KE, Iozzo RV. Targeted disruption of decorin leads to abnormal collagen fibril morphology and skin fragility. J Cell Biol 1997; 136:729–743.
105. Xu T, Bianco P, Fisher LW, Longenecker G, Smith E, Goldstein S, Bonadio J, Boskey A, Heegaard AM, Sommer B, Satomura K, Dominguez P, Zhao C, Kulkarni AB, Gehron Robey P, Young MF. Targeted disruption of the biglycan gene leads to an osteoporosis-like phenotype in mice. Nat Genet 1998; 20:78–82.
106. Chakravarti S, Magnuson T, Lass JH, Jepsen KJ, LaMantia C, Carroll H. Lumican regulates collagen fibril assembly: skin fragility and corneal opacity in the absence of lumican. J Cell Biol 1998; 141:1277–1286.
107. Svensson L, Aszódi A, Reinholt FP, Fässler R, Heinegård D, Oldberg Å. Fibromodulin-null mice have abnormal collagen fibrils, tissue organization, and altered lumican deposition in tendon. J Biol Chem 1999; 274:9636–9647.
108. Bredrup C, Knappskog PM, Majewski J, Rodahl E, Boman H. Congenital stromal dystrophy of the cornea caused by a mutation in the decorin gene. Invest Ophthalmol Vis Sci 2005; 46:420–426.
109. Seidler DG, Faiyaz-Ul-Haque M, Hansen U, Yip GW, Zaidi SH, Teebi AS, Kiesel L, Gotte M. Defective glycosylation of decorin and biglycan, altered collagen structure, and abnormal phenotype of the skin fibroblasts of an Ehlers-Danlos syndrome patient carrying the novel Arg270Cys substitution in galactosyltransferase I (beta4GalT-7). J Mol Med 2006; 84:583–594.
110. Melrose J, Smith S, Ghosh P, Whitelock J. Perlecan, the multidomain heparan sulfate proteoglycan of basement membranes, is also a prominent component of the cartilaginous primordia in the developing human fetal spine. J Histochem Cytochem 2003; 51:1331–1341.
111. Melrose J, Roughley P, Knox S, Smith S, Lord M, Whitelock J. The structure, location, and function of perlecan, a prominent pericellular proteoglycan of fetal, postnatal, and mature hyaline cartilages. J Biol Chem 2006; 281:36905–36914.
112. Kallunki P, Eddy RL, Byers MG, Kestila M, Shows TB, Tryggvason K. Cloning of human heparan sulfate proteoglycan core protein, assignment of the gene (HSPG2) to 1p36.1–p35 and identification of a BamHI restriction fragment length polymorphism. Genomics 1991; 11:389–396.
113. Cohen IR, Grässel S, Murdoch AD, Iozzo RV. Structural characterization of the complete human perlecan gene and its promoter. Proc Natl Acad Sci USA 1993; 90:10404–10408.

114. Murdoch AD, Dodge GR, Cohen I, Tuan RS, Iozzo RV. Primary structure of the human heparan sulfate proteoglycan from basement membrane (HSPG2/perlecan). A chimeric molecule with multiple domains homologous to the low density lipoprotein receptor, laminin, neural cell adhesion molecules, and epidermal growth factor. J Biol Chem 1992; 267:8544–8557.
115. Noonan DM, Hassell JR. Perlecan, the large low-density proteoglycan of basement membranes: structure and variant forms. Kidney Int 1993; 43:53–60.
116. Dolan M, Horchar T, Rigatti B, Hassell JR. Identification of sites in domain I of perlecan that regulate heparan sulfate synthesis. J Biol Chem 1997; 272:4316–4322.
117. Tapanadechopone P, Hassell JR, Rigatti B, Couchman JR. Localization of glycosaminoglycan substitution sites on domain V of mouse perlecan. Biochem Biophys Res Commun 1999; 265:680–690.
118. Hopf M, Göhring W, Kohfeldt E, Yamada Y, Timpl R. Recombinant domain IV of perlecan binds to nidogens, laminin-nidogen complex, fibronectin, fibulin-2 and heparin. Eur J Biochem 1999; 259:917–925.
119. Bengtsson E, Mörgelin M, Sasaki T, Timpl R, Heinegård D, Aspberg A. The leucine-rich repeat protein PRELP binds perlecan and collagens and may function as a basement membrane anchor. J Biol Chem 2002; 277:15061–15068.
120. Chang Z, Meyer K, Rapraeger AC, Friedl A. Differential ability of heparan sulfate proteoglycans to assemble the fibroblast growth factor receptor complex *in situ*. FASEB J 2000; 14:137–144.
121. Arikava-Hirasawa E, Watanabe H, Takami H, Hassell JR, Yamada Y. Perlecan is essential for cartilage and cephalic development. Nat Genet 1999; 23:354–358.
122. Arikawa-Hirasawa E, Wilcox WR, Le AH, Silverman N, Govindraj P, Hassell JR, Yamada Y. Dyssegmental dysplasia, Silverman-Handmaker type, is caused by functional null mutations of the perlecan gene. Nat Genet 2001; 27:431–434.
123. Arikawa-Hirasawa E, Le AH, Nishino I, Nonaka I, Ho NC, Francomano CA, Govindraj P, Hassell JR, Devaney JM, Spranger J, Stevenson RE, Iannaccone S, Dalakas MC, Yamada Y. Structural and functional mutations of the perlecan gene cause Schwartz-Jampel syndrome, with myotonic myopathy and chondrodysplasia. Am J Hum Genet 2002; 70:1368–1375.

Carbohydrate Chemistry, Biology and Medical Applications
Hari G. Garg, Mary K. Cowman and Charles A. Hales

DOI: 10.1016/B978-0-08-054816-6.00007-0

Chapter 7

Small Leucine-Rich Repeat Proteoglycans of Skin

PAUL G. SCOTT

Department of Biochemistry, University of Alberta, Edmonton, Alberta T6G 2H7 Canada

I. Introduction

On a weight basis, collagen is the major component of the extracellular matrix of dermis, accounting for about 90% of the protein and about 80% of the dry weight of the tissue (1). The remainder is made up of smaller amounts of other proteins (e.g., elastin and dermatopontin), glycoproteins (such as fibronectin and tenascin), glycosaminoglycans (GAGs; such as hyaluronan), and proteoglycans. As in other fibrous connective tissues, it is the collagen that gives dermis its tensile strength. The GAGs and proteoglycans (glycoconjugate macromolecules that consist of a core protein to which is attached at least one sulfated GAG chain), on the other hand, confer turgor and the related physical properties of resilience and resistance to compression. Proteoglycans and GAGs also regulate the behavior of the resident cells in many different ways that are still being elucidated. Therefore, although they are quantitatively minor components, the proteoglycans and GAGs have a major impact on the physiology and pathology of the skin.

II. GAGs and Proteoglycans of Skin

The content of GAG in human skin is about 2–3 mg/g fresh tissue (2). The major GAGs, present in similar amounts, are hyaluronan (HA) (previously called hyaluronic acid) and the eponymous dermatan sulfate (DS) (3). There are much smaller amounts of chondroitin 4 sulfate (C4S), chondroitin 6 sulfate (C6S), keratan sulfate (KS), and heparan sulfate (HS). Only HA normally exists in the tissue as a free GAG; the others are all covalently attached to protein as components of various proteoglycans.

Proteoglycans are found in the epidermis, within the basal lamina below the epidermis, and in the dermis. Some are associated with cell membranes, while others are associated with collagen fibrils or with elastin fibers in the dermis, or are in the interfibrillar space. Membrane and basal lamina proteoglycans are generally substituted with HS (e.g., the glypicans) or with both HS and CS (e.g., the syndecans) (4). Most of the CS in the extracellular matrix of skin is present on versican: a large proteoglycan belonging to the lectican gene family that also includes aggrecan, the major proteoglycan of cartilage (5). The most abundant proteoglycan in normal, uninjured skin is decorin, carrying most of the DS that is found in this organ.

III. Small Leucine-Rich Repeat Proteins/Proteoglycans

The small leucine-rich repeat proteins/proteoglycans (SLRPs) are components of the extracellular matrix. They are a subfamily of the much larger family of leucine-rich repeat proteins (LRPs), of which nearly 5000 are known (6). The leucine-rich repeat (LRR) is a sequence motif of the general form LxxLxLxxNxL/I or LxxLxLxxCxxL/I (7), where L is normally leucine, "x" is any residue, N is asparagine, C is cysteine, and I is isoleucine. The complete LRR averages 24 residues in length and includes one of the above conserved segments nested within more variable sequences. All LRPs are characterized by the presence of 2–52 copies (tandem repeats) of the LRR. In extracellular LRPs, this LRR domain is flanked by disulphides.

At least 12 different extracellular matrix macromolecules are recognized as SLRPs (Fig. 1). They are organized into three main classes, according to overall sequence similarity, number of LRRs, and pattern of near-*N*-terminal cysteine residues (8). Two of the SLRPs discussed here: decorin and biglycan, belong to Class I, and two: fibromodulin and lumican, belong to Class II. ECM2, chondroadherin, and nyctalopin may also be SLRPs, although they each show distinct sequence and compositional characteristics. Additional proteins might eventually be classified as SLRPs, as more sophisticated bioinformatic analyses are applied to the genome or as more structural information becomes available.

The SLRPs are mostly glycosylated with *N*- and/or *O*-linked oligosaccharides. Some, like decorin and biglycan, are also "full-time" proteoglycans (i.e., normally have one or more sulfated GAG chains attached to their protein cores), while others, like fibromodulin and lumican, are "part-time" proteoglycans, carrying *N*-linked oligosaccharides that can be extended with sulfated disaccharides to become GAG (KS) chains. The predominant SLRP in human skin is decorin, present at about 3.5 mg/g dry weight of dermis (9). There is also about 0.5 mg/g of

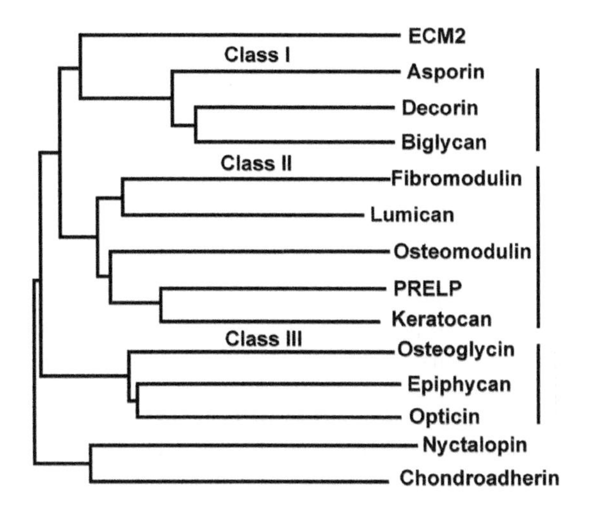

Figure 1 Dendrogram to illustrate the relationships between the different small leucine-rich repeat proteins/proteoglycans (SLRPs). SLRPs belonging to Classes I and II have 12 leucine-rich repeats while those in Class III have 7. The spacing of *N*-terminal cysteines differs in each class. ECM2 is considered to be "Class I related," while nyctalopin and chondroadherin each have distinct sequence and structural characteristics.

biglycan (9) and smaller amounts of lumican (10), fibromodulin (11), and possibly also other SLRPs.

IV. Decorin: The Major SLRPs of Skin

The covalent structure of decorin is illustrated in Fig. 2. This proteoglycan consists of a protein core of a single polypeptide chain of 329 or 330 amino acids (depending on species), close to the *N*-terminus of which is usually attached one sulfated GAG chain (12). In chicken, there is a second form of decorin that carries two GAG chains (13).

A. Glycosaminoglycan

In decorin from mammalian skin, the single sulfated GAG is a high (60–70%) L-iduronate-containing DS. L-iduronate is the C5 epimer of D-glucuronic acid from which it is converted during elongation of the GAG chain (14). The repeating disaccharide unit of skin DS is therefore either L-iduronosyl-α-1,3-*N*-acetyl-L-galactosamine-4-sulfate (60–70%) or D-glucuronosyl-β-1,3-*N*-acetyl-D-galactosamine-4-sulfate (30–40%); both occurring within the same chain so that all DS chains are actually copolymers. The linkage between galactosamine and uronic acid in either type of disaccharide is β-1,4. The presence of L-iduronate is believed to confer on the DS chain-increased flexibility compared to CS (15). The L-iduronate is occasionally sulfated at C2 or C3 (18), and the hexosamine

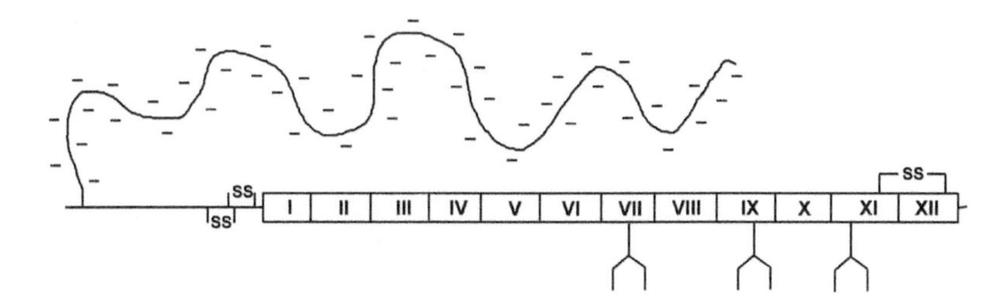

Figure 2 Covalent structures of skin small leucine-rich repeat proteoglycans. The protein cores of decorin, biglycan, lumican, and fibromodulin all consist largely of 12 leucine-rich repeats, shown as boxes. The first 21 amino acids of the decorin protein core constitute a flexible tether to which the single GAG chain (shown decorated with negative charges) is attached (at serine 4). The two GAG chains of biglycan are also attached near to the *N*-terminus of its protein core (at serines 5 and 10). There are three *N*-linked oligosaccharides on the decorin core (shown as forks), and two on that of biglycan (on LRRs IX and XI). The two near-*N*-terminal disulphides in decorin and biglycan form a cystine knot. A third disulphide connects LRRs XI and XII. Lumican and fibromodulin also have 12 LRRs of similar sizes to those of decorin, and most probably have the same arrangement of disulphides. See text for other details.

may also be sulfated at C6 in decorin from some tissues such as sclera (16). In articular cartilage and bone, the GAG chain on decorin has very little or no L-iduronate, in which case it is classified as CS (17). Both DS and CS chains are heterogeneous in size, even if derived from a purified ("homogeneous") population of proteoglycan molecules. Moreover, the sizes of the GAG chains vary between different connective tissues and with the rate of turnover of the tissue (e.g., during wound healing—see the text below). The usual range for skin is between 35 and 70 disaccharides, corresponding to a molar mass of about 15–30 kDa.

DS and CS chains connect to the proteoglycan protein cores by a tetrasaccharide (D-glucuronosyl-β-1,3-D-galactosyl-β-1,3-D-galactosyl-β-1,4-D-xylose). The single GAG chain of decorin is attached, via an *O*-glycoside link from this xylose, to residue 4 (serine) of the mature protein core (12).

B. N-Linked Oligosaccharides

There are three potential sites for *N*-glycosylation of the protein core of decorin (i.e., asparagine residues within an asparagine-X-serine or asparagine-X-threonine sequence), all of which can be substituted with *N*-linked oligosaccharides of the complex (*N*-acetyllactosamine) type (18,19). One of these sites, however, is often not fully substituted, so that some decorin protein cores carry only two *N*-linked oligosaccharides (20). The *N*-linked oligosaccharides of decorin appear to be necessary for the solubility of the protein core because their removal results in protein-driven self-aggregation (19).

C. Protein Core

1. Leucine-Rich Repeats

The central portion of the decorin polypeptide chain consists of 12 tandem LRRs that range in length from 21 to 30 residues, arranged in four groups of 3, each with 1 short (S) repeat and 2 long (T) repeats. The consensus sequences of the S and T repeats are, respectively, xxaPzxLPxxLxxLxLxxNxI and zzxxaxxxxFxxaxxLxxLxLxxNxI, where "x" indicates a variable residue; "z" is frequently a gap; "a" is Val, Leu, or Ile; and I is Ile or Leu. These STT "supermotifs" probably represent ancestral genes that were replicated during the evolution of the SLRPs (8).

2. Ear Repeat

A distinguishing feature of the protein cores of the SLRPs is the presence toward the C-terminal end of the sequence of one very long LRR, stabilized by a disulphide bridge. This LRR has been termed the "ear repeat" because, in the crystal structures of decorin (21) and biglycan (22), several of its amino acid residues are seen to project out from the body of the protein core as a flexible loop (the "ear"). Sequence alignments suggest that this feature is shared by all SLRPs (21).

3. Disulphides

The amino acid sequence of decorin includes six cysteine residues paired to form three disulphides (23). The crystal structure of decorin (21) revealed that the four cysteines on the *N*-terminal side of the LRR domain contained within a Cx_3CxCx_6C motif form a cystine knot, with the first cysteine connected to the third and the second to the fourth. The third disulphide located toward the C-terminal end of the LRR domain connects LRRs 11 and 12, providing a stable scaffold anchoring the ends of the flexible "ear" loop. Collectively, these three disulphides, and the tertiary structure they maintain, are critical for function because chemical reduction (23) leads to irreversible loss of the ability of decorin to bind to collagen (see below).

4. Flexible N-Terminal Tether for the GAG Chain

No electron density corresponding to residues 1–21 was observed in the crystal structure of the protein core of bovine skin decorin (21). These residues are therefore disordered and presumably serve as a flexible tether for the attachment of the single DS chain.

5. Structure and Stabilization of the Decorin Protein Core

Figure 3 shows a cartoon representation of the topology and secondary structure of the decorin protein core. The overall shape, like that of other LRPs (24), is an arch, or bent "solenoid", in which the polypeptide chain follows a right-handed helical path. The inner concave face of the arch is made up of 14 β-strands: the first 2 being antiparallel and forming a β-hairpin that is stabilized by the 2 disulphides. All the other β-strands are parallel and connected on the convex aspect of the protein by β-turns and loops to short segments of either β-sheet, polyproline II helix, 3–10 helix, α-helix, or irregular secondary structure. The highly conserved leucines

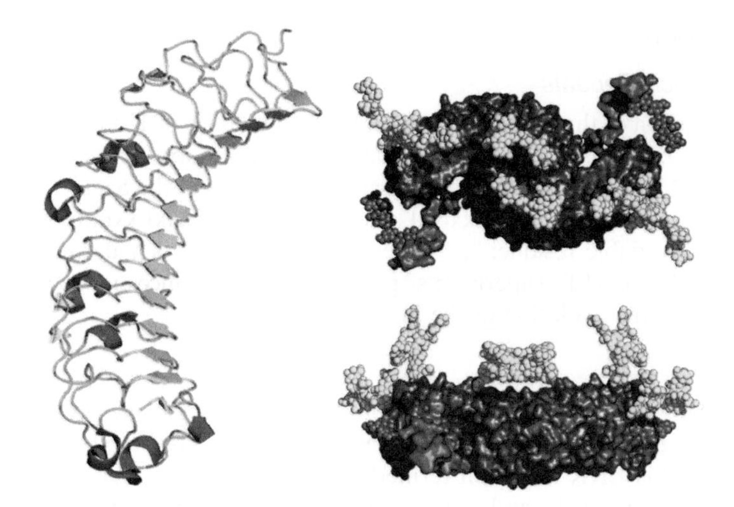

Figure 3 Three-dimensional structure of decorin. On the left is shown a picture of the decorin protein core based on the highest resolution crystal structure (PDB accession code 1XKU). The *N*-terminus is at the top. β-strands are shown in yellow, helices (α, 3–10, and polyproline II) in red, loops and turns in green. All other Class I and Class II SLRPs are believed to have very similar secondary and tertiary structures to that shown here for decorin. On the right are two orthogonal views of the decorin dimer. The protein core of the dimer is shown as a smooth surface in slate blue. The *N*-terminal 21 amino acids are shown as a smooth surface in marine blue. Spheres are used to represent the N-linked oligosaccharides (yellow) and the hexasaccharide (marine blue) connecting the GAG chain to serine residue 4 of the protein core. The protein core dimer structure is derived from the crystal structure (PDB accession code 1XEC). The other features shown are based on a combination of small-angle X-ray scattering data and molecular modeling (P.G.S., unpublished work).

and other strongly hydrophobic amino acid side chains are all buried within the hydrophobic core of the protein. The conserved asparagines of the LRRs are also found within the hydrophobic core, where their side chain amides form two "ladders" of four hydrogen bonds each, linking LRRs II through VI and VIII through XII. Pairs of salt bridges that connect adjacent LRRs across the concave β-sheet face provide a third source of stabilization energy for the native structure.

V. Biglycan

The second Class I SLRP found in skin is biglycan, whose amino acid sequence is 55% identical (and an additional 29% similar) to that of decorin from the same species. Biglycan was named for the fact that its protein core has two sites (serines at positions 5 and 10) for substitution with GAG chains (25), although both sites may not always be fully used (26,27). Biglycan also has two sites, rather than the three on decorin from most species, for substitution with N-linked oligosaccharides.

A. Tissue Distribution and Localization

Biglycan was originally detected in bone (25) and subsequently purified from articular cartilage (28). It is also a minor component in many other fibrous connective tissues including skin where it is associated with elastin fibers (29,30). It may also facilitate the assembly of type VI collagen, a minor beaded filament-forming collagen in dermis, into hexagonal networks (31).

B. Structure of the Protein Core and the Importance of Dimerization

Given the similarities in their amino acid sequences, it is not surprising that the three-dimensional structures of the protein cores of biglycan and decorin are closely superimposable (22). Moreover, both biglycan and decorin exist in solution as stable homodimers (22,32). Their crystal structures reveal the same antiparallel symmetrical arrangement of monomers within these dimers, with the two concave β-sheet faces in contact over about two-thirds of their surfaces. Dimerization is likely to be important for the structures and functions of these SLRPs and possibly also for asporin: the third Class I SLRP. We have recently provided evidence that dimerization is essential for the stability of native decorin and biglycan, so that folded monomers of these proteins may not exist under physiological conditions (22). Occlusion of much of the parallel β-sheet face of each decorin or biglycan protomer by dimerization must normally prevent this part of the protein from interacting with other proteins (33).

VI. Lumican and Fibromodulin

Lumican was originally discovered in the cornea (34), and later named for its putative role in controlling corneal transparency (10,35). Fibromodulin was first extracted from articular cartilage and named for its effects on collagen fibril formation *in vitro* (11). Both lumican and fibromodulin are only minor components of skin but are important to its structure, as attested to by the changes seen in the skins of null mice (36).

Like the Class I SLRPs (decorin, biglycan, and asporin), the Class II SLRPs also have 12 LRRs. The spacing of the *N*-terminal cysteines is slightly different: Cx_3CxCx_9C in Class II compared to Cx_3CxCx_6C in Class I, but the pairing of cysteines is expected to be the same, with the first connected to the third and the second to the fourth, forming a cystine knot. The fifth and sixth cysteines are also expected to form a disulphide and to connect, as in decorin and biglycan, LRRs 11 and 12. As seen earlier for decorin (12), reduction of the disulphides in fibromodulin destroys its ability to bind to collagen (37).

On the *N*-terminal side of the first cysteine in lumican and fibromodulin are sequences of 18 residues and about 56 residues, respectively, that are, for the most part, very poorly conserved compared to those within the main LRR domain. By analogy with the *N*-terminal regions of decorin and biglycan, these sequences are predicted to lack regular secondary structure. They do, however, include several tyrosine residues—two in lumican and up to nine in fibromodulin, that can be sulfated (38), possibly facilitating electrostatic interactions with, as yet unidentified, positively charged protein ligands.

Although the LRR domains of lumican and fibromodulin share only about 30% sequence identity with those of decorin or biglycan, they are of almost identical lengths. Moreover, the key structure-determining hydrophobic residues and asparagines are highly conserved in all four SLRPs. Consequently, the overall topology of the Class II SLRP protein cores is expected to be very similar to that of decorin and biglycan. Unlike the latter, however, lumican and fibromodulin are normally monomeric in solution (P.G. Scott and C.M. Dodd, unpublished data).

Lumican has four potential sites for substitution with *N*-linked oligosaccharides while fibromodulin has five. The extension of one branch of each of these oligosaccharides with *N*-acetyllactosamine disaccharides and the sulfation of both glucosamine and galactose converts them into KS chains (39.40).

VII. Functions of Skin SLRPs

The biological activities that have been ascribed to the SLRPs fall into two general categories: (1) organization and physical properties of the extracellular matrix and (2) effects on cell behavior such as attachment, migration, and proliferation. Although there are exceptions, activities belonging to the first category are mostly due to the sulfated GAG chain(s) (and possibly the sulfated tyrosines in the case of fibromodulin and lumican), and those activities belonging to the second category are mostly due to the protein cores.

A. Collagen Fibril and Fiber Organization

All four SLRPs discussed here are believed to affect the formation of collagen fibrils and/or their organization into fibers and fiber bundles. These insights came originally from studies on the localization of SLRPs within tissues and on their physical interaction with soluble collagen or with collagen fibrils *in vitro*. More recently, confirmation of the importance of the SLRPs for normal tissue architecture has come from studying mice in which the genes have been knocked out, either singly or in combination (41).

Decorin binds to the fibril-forming collagens (e.g., types I and III in skin) and to the beaded filament-forming collagen (e.g., type VI). Biglycan also promotes the assembly of type VI collagen into hexagonal networks. All these interactions can be readily demonstrated *in vitro* (14,31,42). It is, however, from much earlier work on the localization of GAG chains that many important ideas about the function of the proteoglycans in fibrous connective tissues developed.

1. *Periodic Arrangement of SLRPs on Collagen Fibrils*

Cationic metal-containing stains bind to the negatively charged sulfate groups of GAGs, rendering them visible in the electron microscope. The copper-containing stain Alcian Blue and its derivatives Cuprolinic Blue and Cupromeronic Blue, developed by Scott (43,44), have been used to produce many striking images showing that sulfated GAGs are specifically and periodically associated with particular locations ("bands") on the collagen fibrils (44). In tendon and skin, these are the D and E bands, both within the gap zone of the collagen fibrils; in cornea, these are the A and C bands, both within the overlap zone. We now know that these different tissue-specific localizations reflect the presence of distinct small proteoglycans, predominantly DS-containing decorin in tendon and skin and KS-containing fibromodulin and lumican in cornea. An example of such an image for skin is shown in Fig. 4.

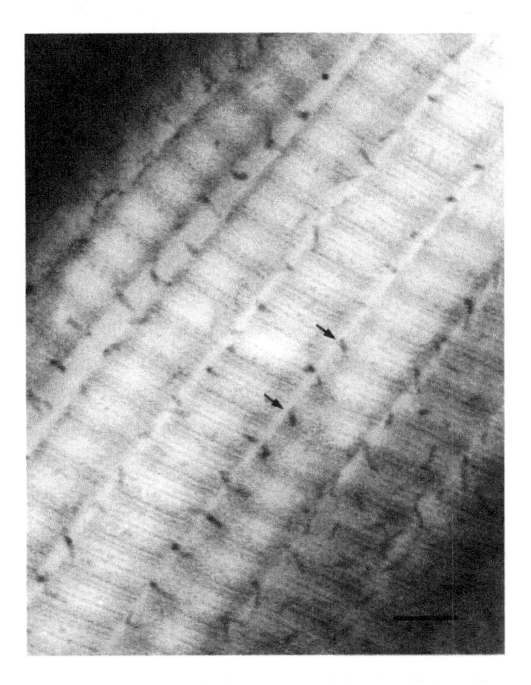

Figure 4 Longitudinal section of a skin collagen fiber, stained with Cuprolinic blue and uranyl acetate, showing densely packed, banded collagen fibrils with decorin DS chains attached periodically at D or E bands (arrowed). The scale bar represents 100 nm.

2. GAG Chains of SLRPs as Tissue Organizers

In electron microscopic images of some tissues, the densely stained GAG filaments appear to span the gaps between parallel collagen fibrils and possibly to link them together. These observations, and other data on the interaction of SLRPs with collagen *in vitro*, have led to the notion that the small proteoglycans are important in controlling the formation of collagen fibrils and/or their organization into collagen fibers and fiber bundles.

3. Possible Interaction Between GAG Chains

Early reports that DS prepared by proteolytic digestion is able to bind to itself (45,46), and later modeling and NMR studies on isolated GAGs, led to the suggestion that the connections between collagen fibrils seen in some electron micrographs were mediated by noncovalent (van der Waal's) interactions between the hydrophobic faces of the monosaccharides of helically structured GAG chains (47). Highly purified DS peptides, retaining only two or three amino acids attached to the xylose residue of the linkage region, however, were shown in our own work not to self-associate by light scattering (48), nor, by others (49), to bind to the DS chains of intact decorin by gel-chromatography. On the other hand, tryptic peptides retaining positively charged amino acids did interact in this latter test system. The self-association of DS previously reported may, therefore, have been a consequence of incomplete proteolysis of the decorin protein core (or incomplete purification of the GAG peptides), and appears not to be a property of DS per se. More recently, it has become apparent that the interfibrillar GAG filaments seen in the electron microscope may be an artifact of preparation, because freeze-drying solutions of decorin generates complexes that are aggregated through the DS chains (21,32).

4. *Electrostatic Repulsion Between Gags May Facilitate Collagen Fiber Formation*

In a hydrated connective tissue such as skin, DS chains probably do not bind directly to one another, rather they are far more likely to exert a strong mutual repulsion as a consequence of their high density of negative charges. This does not, of course, preclude a role for decorin in connective tissue organization based on its physical and chemical properties. In the decorin null mouse, collagen fibrils are irregular in outline and form anastomoses and random fusions (50). Thus, the decorin that is normally present appears to define and delimit the surfaces of the collagen fibrils. The parallel orientation and association of collagen fibrils into fibers and fiber-bundles is impaired in the periodontal ligament of decorin deficient mice (51) and in postburn hypertrophic scars (52) (Fig. 5). The latter are deficient in decorin compared to normal skin or mature scars (9). Skin collagen defects are also seen in variant forms of Ehlers-Danlos syndrome, where mutations in a galactosyltransferase gene lead to the secretion of GAG-deficient decorin (53).

All these observations suggest that it is the DS chain on decorin that is important in the organization of connective tissues, with the protein core, which has itself been detected at the D-band by immunohistochemistry (54), serving to attach the proteoglycan to specific sites on the collagen fibril. In the fibrous connective tissues that are richest in decorin, most collagen fibrils are organized into parallel bundles (fibers) visible in the light microscope. This alignment is probably facilitated by the antiadhesive properties of the highly hydrated anionic GAG chains, allowing collagen fibrils to slide past one another and to orient along lines of tension. Conversely, a lack of decorin may impede the organization or reorganization of collagen fibrils.

5. *Biglycan and Tissue Organization*

Biglycan was originally described as pericellular in developing tissues (55). It is not able to compete effectively with decorin for binding to collagen fibrils (56), but may nevertheless attach to collagen in the absence of decorin (57,58). The deletion

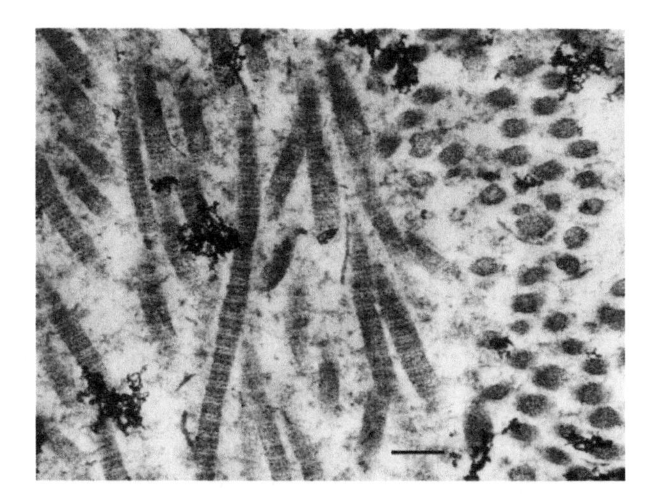

Figure 5 Section of human postburn hypertrophic scar tissue, stained with Cuprolinic blue and uranyl acetate, showing thin collagen fibrils embedded in an abundant interfibrillar matrix. Clumps of densely stained amorphous materials are probably biglycan and/or versican. Note the paucity of periodically attached decorin DS chains on the collagen fibrils (contrast with Fig. 4). The scale bar represents 100 nm.

of the biglycan gene leads most obviously to defects in skeletal development and bone formation (59) but the skin and other soft connective tissues are involved because collagen fibril abnormalities are seen there also (60). Deletion of both decorin and biglycan causes much more severe osteopenia and soft tissue abnormalities than deletion of either gene alone (60). The skin in these double mutant mice is especially fragile and is reminiscent of the rare human progeroid variant of Ehlers-Danlos syndrome (61,62). Type VI collagen is found in skin as thin beaded filaments and as hexagonal networks whose formation can be promoted *in vitro* by biglycan (31). This activity of biglycan is ascribed to the two GAG chains that are present on each protein monomer.

6. Lumican and Fibromodulin in Tissue Organization

Mice deficient in lumican display skin laxity and fragility resembling certain types of Ehlers-Danlos syndrome (63), while the consequences of a deficiency of fibromodulin are mainly apparent in the tendon where it is found at relatively high concentration (64). The effects of a combined deficiency of lumican and fibromodulin suggest that, while both these Class II SLRPs attach to the same (A and C band) sites on collagen fibrils, albeit with different affinities (65), they function at different stages in fibril and fiber development (66).

B. Effects of SLRPs on Cell Behavior

The SLRPs are antiproliferative for several types of cell [reviewed in Ref. (67)]. This particular biological property, which has probably been most extensively investigated in relation to decorin, has attracted considerable interest, largely because of its implications for tumor growth, invasion, and metastasis, and prospects for control of these pathological processes. Decorin expression is often very low in carcinoma cells but is upregulated in the connective tissue stroma surrounding tumors (68). Induction of decorin expression in carcinoma cells reduces their proliferative and metastatic potential (69) and lumican expression in melanoma cells has similar effects (70).

1. *Transforming Growth Factor-β and SLRPs*

Yamaguchi and Ruoslahti (71) first noted that overexpression of decorin in Chinese hamster ovary cells reduced their rate of proliferation and density at saturation. This effect was subsequently attributed to the direct binding and inactivation by decorin of the transforming growth factor-β (TGF-β) that was considered to be an autocrine stimulator of cell proliferation in these cultures (72). Biglycan was also reported to bind to TGF-β in the solid-phase competition assay system used in this study.

The list of SLRPs that bind TGF-β was later extended to include fibromodulin (73). These latter authors found, however, that a high-affinity TGF-β binding site was present on only 1 SLRP molecule in 10. Furthermore, the DS or CS chain(s) normally present on decorin and biglycan inhibited interaction with TGF-β. The use in these investigations of fusion proteins expressed in *Escherichia coli* (which does not carry out posttranslational modifications such as disulphide bond formation or glycosylation) is of concern, as no evidence was presented that the SLRPs tested were in their native, folded conformations. Nor had Yamaguchi and coworkers (71,72) presented such evidence in relation to the recombinant proteins that they had studied earlier. Moreover, it was subsequently reported that decorin enhanced the binding of TGF-β to receptors on osteoblastic MC3T3-E1 cells and increased its bioactivity (74), and did not affect the TGF-β-mediated inhibition of the proliferation of U937 monocytic cells, even in 100,000-fold molar excess (75). The latter authors did, however, report that TGF-β attached to collagen-bound decorin, and it had been earlier shown that TGF-β accumulates in the pericellular matrix secreted by fibroblasts (76). These and other observations, such as the release of active TGF-β by proteolysis with plasmin or thrombin (76), or with matrix metalloproteinases (MMPs) (77), have given rise to the concept that the extracellular matrix acts as a reservoir (or sink) for growth factors such as TGF-β.

In some experimental systems, administration of decorin may downregulate the production of TGF-β (78,79), providing a possible explanation for an effect on its biological activity that is independent of either direct binding to decorin or of competition for cell surface receptors. Notwithstanding the uncertainty surrounding the mechanism(s), there have been some spectacular successes in the use of decorin in experimental animals to treat TGF-β-mediated fibrotic processes such as glomerular nephritis (80), gliotic scarring (81), bleomycin-induced lung fibrosis (82,83), and intra-articular fibrous adhesions (84). TGF-β released from certain tumors is believed to contribute to their evasion of T-cell-mediated immunity, and local overexpression of decorin has been shown to abrogate the growth of neurogliomas in rats (79). Curiously, although biglycan has been reported also to bind TGF-β *in vitro*, it seems not to have the same beneficial therapeutic effect as decorin on TGF-β-mediated fibrosis *in vivo*, and may even have the opposite effect (83).

2. TGF-β-Independent effects of SLRPs on Cell Behavior

Decorin can upregulate p21, a potent inhibitor of cyclin-dependent kinases, leading to inhibition of cell division (85,86). It can also stimulate phosphorylation of the epidermal growth factor (EGF) receptor and thus activate the mitogen-activated protein kinase signal pathway in sqamous carcinoma cells (87). Stable expression of decorin leads to the sustained downregulation of EGF receptor kinase activity and can block the growth of tumor xenografts *in vivo* (88).

3. Effects on Angiogenesis

The development of new blood vessels, that is angiogenesis, is necessary to support wound healing. On the other hand, inappropriate vascularization is a component of many pathological processes (e.g., diabetic retinopathy and rheumatoid arthritis), and supports the growth of solid tumors (89). Antiangiogenic factors have therefore long been proposed as therapeutics, and the SLRPs, decorin in particular, which has been shown to downregulate the expression of vascular endothelial growth factor, might possibly be useful in this application (90). In some experimental systems, however, decorin has been reported to have proangiogenic effects (91,92) so that its role in angiogenesis is complex and at the present time incompletely understood.

VIII. Involvement of SLRPs in Skin Pathology

There have been many descriptions of changes in the structures or in the absolute or relative amounts of SLRPs, associated with different pathologies. It is not always obvious, however, whether these are the cause of the pathology or merely secondary to it, but in either case they are likely to have deleterious effects on the affected connective tissue. Perhaps, surprisingly, there are relatively few mutations or polymorphisms in the protein sequences of the SLRPs that have been directly linked to disease, all of those so far described occurring in the eye (6,93). Although most interactions between SLRPs and other macromolecules can be rationalized as playing some physiological or otherwise beneficial role, some interactions, such as those between decorin or biglycan and lipoproteins within atherosclerotic plaques (94,95) or between spirochetes and decorin (96), are obviously damaging to the host.

A. Aberrant Wound Healing

Deep second- and third-degree burns often heal with hypertrophic scarring, a condition that presents some months after the original injury as red, raised, inelastic and frequently itchy masses of scar tissue. These scars are uncomfortable, cosmetically troublesome, and often interfere with motion. Molecular and cellular aspects of postburn hypertrophic scarring are reviewed in Ref. (97).

1. GAGs and Proteoglycans in Hypertrophic Scars

There are abnormalities in the amounts and nature of the GAGs in hypertrophic scars, which has been recognized for more than 30 years. Chondroitin 4 sulfate, which is barely detectable in normal skin, can be readily demonstrated in hypertrophic scars (98), where it is especially elevated in the characteristic nodular structures (99). DS, in contrast, is absent from all but the more normal appearing parallel-fibered areas of the scars (100).

Swann et al. (101) later extracted proteoglycans from human skin and normal and hypertrophic scars with the powerful denaturing agent guanidinium chloride (to disrupt the collagen), and separated and studied them using techniques that had earlier been successfully applied to the proteoglycans of cartilage. The major conclusions were that the hypertrophic scars contained more proteoglycan than did normal scars and that this included a higher proportion of low-density DS proteoglycan. Subsequent work identified in these extracts a small DS proteoglycan with the N-terminal sequence that we now know to belong to decorin (102). Two small protein cores of 21.5 and 17 kDa were found for this DS-PG, both much shorter than that expected for intact decorin (~42 kDa). On the other hand, the GAG chains were longer (~23.5 kDa) than those from normal skin (~20 kDa). It was suggested that the smaller protein cores might represent alternately spliced decorin variants, but other evidence (9) suggests that they are more likely to be proteolytic fragments.

2. Decorin Is Deficient in Hypertrophic Scars

Garg et al. (103) isolated L-iduronate-containing small proteoglycans from skin and postburn scar tissues. Although the two small proteoglycans were not separated, evidence was presented for an increased ratio of biglycan to decorin in the scar tissues compared to normal skin. In our own work, the absolute amounts of decorin, biglycan, and versican in human skin and in postburn hypertrophic and mature scars were measured using quantitative inhibition ELISA methods (9). The amounts of decorin in the hypertrophic scars were found to be about 25% of those in normal skin or in mature scars, while biglycan and versican were each about sixfold higher. The longer DS chains previously noted were found to belong to both decorin and biglycan.

In normally healing incisional wounds in humans and experimental animals, decorin appears within about a week, at the same time as new collagen fibrils (104,105). In contrast, in healing second- and third-degree burn wounds, many of which become hypertrophic, decorin expression appears to be largely suppressed for 12 months or longer (106). Given its importance for collagen fibril and fiber structure and organization, it is reasonable to propose that the paucity of decorin is at least partly to blame for the derangement of the connective tissue in hypertrophic scars. The longer DS chains on the decorin that is present in the scars might also contribute to the increased separation between collagen fibrils, as noted in the chemically inflamed skin of mice (107). The increased amounts of biglycan (which is presumably associated with the collagen fibrils in the absence of decorin) and of versican (which itself does not attach to collagen but most probably occupies the enlarged interfibrillar spaces) are probably also important contributors to the pathology. The 2.4-fold increase in total GAG content, which is disproportionately high compared to the 12% increase in water content (9), probably accounts in large part for the greater turgor of active hypertrophic scars.

3. *Role of TGF-β in Hypertrophic Scarring*

Excessive amounts and/or uncontrolled activity of TGF-β have been implicated as a causative factor in many forms of fibroproliferative disease (108), including postburn hypertrophic scarring (109). Circulating levels of TGF-β are about twofold higher in recovering burn patients than in normal subjects and remain elevated for several months after injury (110). The profibrotic effects of TGF-β on cultured fibroblasts include the stimulation of the synthesis of collagen, fibronectin, and GAG (111,112); the downregulation of MMPs; and the upregulation of tissue inhibitors of matrix metalloproteinases (TIMPs) (113). TGF-β also downregulates the synthesis of decorin and upregulates the synthesis of biglycan and versican by human skin fibroblasts (114,115).

4. *Distinct Phenotype of Fibroblasts in Healing Burn Wounds*

The abnormal chemical composition of the hypertrophic scar extracellular matrix might be explained as resulting from the known effects of TGF-β on the metabolism of fibroblasts. However, when fibroblasts are cultured from postburn hypertrophic scars, they are found to have a permanently altered phenotype compared to those from uninjured skin of the same patients, often making more collagen, and always less collagenase (116), less nitric oxide (117), and less decorin with a longer DS chain (114). Some form of cell selection must therefore be operating, and the simplest explanation is probably that the fibroblasts in healing burn wounds are derived largely from the deep dermis or superficial fascia where they survived the original injury. Support for this suggestion comes also from the observation that fibroblasts from reticular dermis synthesize less decorin than do those from papillary dermis (118).

B. Ehlers-Danlos and Progeroid Syndromes

In 1987, Kresse et al. (119) described a young male patient of pronounced progeroid appearance and signs of Ehlers-Danlos syndrome, whose fibroblasts secreted reduced amounts of the intact (proteoglycan form) of decorin, together with decorin protein core lacking a GAG chain. This patient suffered from developmental and connective tissue abnormalities, including osteopenia, hypermobile joints, loose skin, and impaired wound healing. The primary defect was subsequently shown to be a deficiency of galactosyltransferase I, the enzyme that catalyzes the second glycosylation step in the biosynthesis of GAG chains (120). Specific point mutations within this gene (*B4GALT7*) have since been defined in this patient and in other patients diagnosed with Ehlers-Danlos syndrome (121).

A deficiency of decorin core protein has been described in the skins of two Japanese patients (122,123). The cause of the decorin deficiency in the first patient appears not to have been further investigated but in the second patient, (123) the coding region of the cDNA was shown to be normal while the levels of decorin mRNA were markedly reduced. These observations, together with the lack of responsiveness of the patient's fibroblasts to stimulation of decorin expression by interleukin-1β, suggested a mutation in a decorin gene regulatory element. In this particular female patient, the presenting symptom was a large nonhealing skin ulcer resulting from a mastectomy 2 years earlier. Synthesis of decorin by skin fibroblasts would thus appear to be essential for normal wound healing, a conclusion that is also supported by the observation that healing of incisional or excisional wounds in the decorin null mouse is significantly delayed (124).

C. Lyme Disease

Lyme disease is caused by the spirochete *Borrelia burgdorferi.* Ticks deposit this microorganism in the dermis where it attaches to the decorin that coats the surfaces of collagen fibrils (96). Two decorin-binding adhesins (DbpA and DbpB—both cell surface lipoproteins) have been identified (125,126). DbpA binds with higher affinity than DbpB, probably through an electrostatic mechanism involving specific (positively charged) lysine residues interacting with the negatively charged DS chains (127). This interaction facilitates the colonization by *B. burgdorferi* of the skin (where it proliferates), and may also be important at later stages in the biology of the spirochete, after it disseminates from the skin and infects other connective tissues. This organism may use decorin to evade the humoral immune system (128).

IX. Summary and Conclusions

Skin contains at least four small proteoglycans: decorin, biglycan, lumican, and fibromodulin (in decreasing order of abundance). All have protein cores made up largely of 12 leucine-rich repeats and that probably share the same arch-shaped topology, although decorin and biglycan are dimers while lumican and fibromodulin are most probably monomers. Decorin and biglycan are substituted with DS chains, attached by a flexible "tether" close to the *N*-terminus. All four proteoglycans also carry *N*-linked oligosaccharides that in lumican and fibromodulin can be extended with sulfated *N*-acetyllactosamine disaccharides to become KS chains. The protein cores attach the small proteoglycans to specific sites on the surfaces of collagen fibrils where, primarily through their GAG chains, they serve to define the boundaries of the fibrils and to prevent their random fusion. The DS chain of decorin also appears to facilitate the alignment of collagen fibrils into parallel bundles (i.e., fibers). Many other biological activities, mainly mediated by the protein cores, have been proposed for these small proteoglycans, including control of cell proliferation in general and of angiogenesis in particular. Reduced amounts of decorin, and/or its replacement by biglycan and versican, are associated with disorganized and mechanically dysfunctional connective tissue in hypertrophic scars and certain variant forms of Ehlers-Danlos syndrome. Better understanding of the structures and activities of the SLRPs may lead eventually to novel therapies for fibrotic and metastatic disease.

References

1. Pinnell SR, Murad S. Collagen. In: Goldsmith LA, ed. Biochemistry and Physiology of the Skin. New York: Oxford University Press, 1983; 385–410.
2. Varma RS, Varma R. Glycosaminoglycans and proteoglycans of skin. In: Varma RS, Varma R (eds.) Glycosaminoglycans and Proteoglycans in Physiological and Pathological Processes of Body Systems. Basel: Karger, 1982; 151–164.
3. Shetlar MR, Shetlar CL, Kischer CW. Glycosaminoglycans in granulation tissue and hypertrophic scars. Burns 1981; 8:27–31.
4. Delehedde M, Lyon M, Sergeant N, Rahmoune H, Fernig DG. Proteoglycans: pericellular and cell surface multireceptors that integrate external stimuli in the mammary gland. J Mammary Gland Biol Neoplasia 2001; 6:253–273.
5. Aspberg A, Miura R, Bourdoulous S, Shimonaka M, Heinegard D, Schachner M, Ruoslahti E, Yamaguchi Y. The C-type lectin domains of lecticans, a family of aggregating chondroitin sulfate proteoglycans, bind tenascin-R by protein-protein interactions independent of carbohydrate moiety. Proc Natl Acad Sci USA 1997; 94:10116–10121.
6. Matsushima N, Tachi N, Kuroki Y, Enkhbayar P, Osaki M, Kamiya M, Kretsinger RH. Structural analysis of leucine-rich-repeat variants in proteins associated with human diseases. Cell Mol Life Sci 2005; 62:2771–2791.
7. Kobe B, Deisenhofer J. The leucine-rich repeat: a versatile binding motif. Trends Biochem Sci 1994; 19:415–421.
8. Matsushima N, Ohyanagi T, Tanaka T, Kretsinger RH. Super-motifs and evolution of tandem leucine-rich repeats within the small proteoglycans–biglycan, decorin, lumican, fibromodulin, PRELP, keratocan, osteoadherin, epiphycan, and osteoglycin. Proteins 2000; 38:210–225.
9. Scott PG, Dodd CM, Tredget EE, Ghahary A, Rahemtulla F. Chemical characterization and quantification of proteoglycans in human post-burn hypertrophic and mature scars. Clin Sci (Lond) 1996; 90:417–425.
10. Chakravarti S, Stallings RL, SundarRaj N, Cornuet PK, Hassell JR. Primary structure of human lumican (keratan sulfate proteoglycan) and localization of the gene (LUM) to chromosome 12q21.3-q22. Genomics 1995; 27:481–488.
11. Oldberg A, Antonsson P, Lindblom K, Heinegard D. A collagen-binding 59-kd protein (fibromodulin) is structurally related to the small interstitial-proteoglycans PG-S1 and PG-S2 (decorin). EMBO J 1989; 8:2601–2604.
12. Chopra RK, Pearson CH, Pringle GA, Fackre DS, Scott PG. Dermatan sulphate is located on serine-4 of bovine skin proteodermatan sulphate. Demonstration that most molecules possess only one glycosaminoglycan chain and comparison of amino acid sequences around glycosylation sites in different proteoglycans. Biochem J 1985; 232:277–279.

13. Blaschke UK, Hedbom E, Bruckner P. Distinct isoforms of chicken decorin contain either one or two dermatan sulfate chains. J Biol Chem 1996; 271:30347–30353.
14. Malmstrom A, Fransson LA. Biosynthesis of dermatan sulfate. I. Formation of L-iduronic acid residues. J Biol Chem 1975; 250:3419–3425.
15. Ferro DR, Provasoli A, Ragazzi M, Casu B, Torri G, Bossennec V, Perly B, Sinay P, Petitou M, Choay J. Conformer populations of L-iduronic acid residues in glycosaminoglycan sequences. Carbohydr Res 1990; 195:157–167.
16. Fransson LA. Periodate oxidation of L-iduronic acid residues in dermatan sulphate. Carbohydr Res 1974; 36:339–348.
17. Cheng F, Heinegard D, Malmstrom A, Schmidtchen A, Yoshida K, Fransson LA. Patterns of uronosyl epimerization and 4-/6-O-sulphation in chondroitin/dermatan sulphate from decorin and biglycan of various bovine tissues. Glycobiology 1994; 4:685–696.
18. Scott PG, Dodd CM. The *N*-linked oligosaccharides of bovine skin proteodermatan sulphate. Biochem Soc Trans 1989; 17:1031–1032.
19. Scott PG, Dodd CM. Self-aggregation of bovine skin proteodermatan sulphate promoted by removal of the three *N*-linked oligosaccharides. Connect Tissue Res 1990; 24:225–235.
20. Glossl J, Beck M, Kresse H. Biosynthesis of proteodermatan sulfate in cultured human fibroblasts. J Biol Chem 1984; 259:14144–14150.
21. Scott PG, McEwan PA, Dodd CM, Bergmann EM, Bishop PN, Bella J. Crystal structure of the dimeric protein core of decorin, the archetypal small leucine-rich repeat proteoglycan. Proc Natl Acad Sci U S A 2004; 101:15633–15638.
22. Scott PG, Dodd CM, Bergmann EM, Sheehan JK, Bishop PN. Crystal structure of the biglycan dimer and evidence that dimerization is essential for folding and stability of class I small leucine-rich repeat proteoglycans. J Biol Chem 2006; 281:13324–13332.
23. Scott PG, Winterbottom N, Dodd CM, Edwards E, Pearson CH. A role for disulphide bridges in the protein core in the interaction of proteodermatan sulphate and collagen. Biochem Biophys Res Commun 1986; 138:1348–1354.
24. Kobe B, Kajava AV. The leucine-rich repeat as a protein recognition motif. Curr Opin Struct Biol 2001; 11:725–732.
25. Fisher LW, Termine JD, Young MF. Deduced protein sequence of bone small proteoglycan I (biglycan) shows homology with proteoglycan II (decorin) and several nonconnective tissue proteins in a variety of species. J Biol Chem 1989; 264:4571–4576.
26. Johnstone B, Markopoulos M, Neame P, Caterson B. Identification and characterization of glycanated and non-glycanated forms of biglycan and decorin in the human intervertebral disc. Biochem J 1993; 292:661–666.
27. Roughley PJ, White RJ, Magny MC, Liu J, Pearce RH, Mort JS. Non-proteoglycan forms of biglycan increase with age in human articular cartilage. Biochem J 1993; 295:421–426.
28. Neame PJ, Choi HU, Rosenberg LC. The primary structure of the core protein of the small, leucine-rich proteoglycan (PG I) from bovine articular cartilage. J Biol Chem 1989; 264:8653–8661.

29. Baccarani-Contri M, Vincenzi D, Cicchetti F, Mori G, Pasquali-Ronchetti I. Immunocytochemical localization of proteoglycans within normal elastin fibers. Eur J Cell Biol 1990; 53:305–312.
30. Reinboth B, Hanssen E, Cleary EG, Gibson MA. Molecular interactions of biglycan and decorin with elastic fiber components: biglycan forms a ternary complex with tropoelastin and microfibril-associated glycoprotein 1. J Biol Chem 2002; 277:3950–3957.
31. Wiberg C, Heinegard D, Wenglen C, Timpl R, Morgelin M. Biglycan organizes collagen VI into hexagonal-like networks resembling tissue structures. J Biol Chem 2002; 277:49120–49126.
32. Scott PG, Grossmann JG, Dodd CM, Sheehan JK, Bishop PN. Light and X-ray scattering show decorin to be a dimer in solution. J Biol Chem 2003; 278:18353–19359.
33. McEwan PA, Scott PG, Bishop PN, Bella J. Structural correlations in the family of small leucine-rich repeat proteins and proteoglycans. J Struct Biol 2006; 155:294–305.
34. Hassell JR, Newsome DA, Krachmer JH, Rodrigues MM. Macular corneal dystrophy: failure to synthesize a mature keratan sulfate proteoglycan. Proc Natl Acad Sci U S A 1980; 77:3705–3709.
35. Blochberger TC, Vergnes JP, Hempel J, Hassell JR. cDNA to chick lumican (corneal keratan sulfate proteoglycan) reveals homology to the small interstitial proteoglycan gene family and expression in muscle and intestine. J Biol Chem 1992; 267:347–352.
36. Chakravarti S. Functions of lumican and fibromodulin: lessons from knockout mice. Glycoconj J 2002; 19:287–293.
37. Scott PG, Nakano T, Dodd CM. Isolation and characterization of small proteoglycans from different zones of the porcine knee meniscus. Biochim Biophys Acta 1997; 1336:254–262.
38. Onnerfjord P, Heathfield TF, Heinegard D. Identification of tyrosine sulfation in extracellular leucine-rich repeat proteins using mass spectrometry. J Biol Chem 2004; 279:26–33.
39. Plaas AH, Wong-Palms S. Biosynthetic mechanisms for the addition of polylactosamine to chondrocyte fibromodulin. J Biol Chem 1993; 268: 26634–26644.
40. Plaas AH, Neame PJ, Nivens CM, Reiss L. Identification of the keratan sulfate attachment sites on bovine fibromodulin. J Biol Chem 1990; 265:20634–20640.
41. Ameye L, Young MF. Mice deficient in small leucine-rich proteoglycans: novel in vivo models for osteoporosis, osteoarthritis, Ehlers-Danlos syndrome, muscular dystrophy, and corneal diseases. Glycobiology 2002; 12:107–116R.
42. Vogel KG. Tendon structure and response to changing mechanical load. J Musculoskelet Neuronal Interact 2003; 3:323–325.
43. Scott JE, Quintarelli G, Dellovo MC. The chemical and histochemical properties of Alcian Blue. I. The mechanism of Alcian Blue staining. Histochemie 1964; 4:73–85.
44. Scott JE. Collagen–proteoglycan interactions. Localization of proteoglycans in tendon by electron microscopy. Biochem J 1980; 187:887–891.

45. Fransson LA. Interaction between dermatan sulphate chains. I. Affinity chromatography of copolymeric galactosaminioglycans on dermatan sulphate-substituted agarose. Biochim Biophys Acta 1976; 437:106–115.
46. Fransson LA, Coster L. Interaction between dermatan sulphate chains. II. Structural studies on aggregating glycan chains and oligosaccharides with affinity for dermatan sulphate-substituted agarose. Biochim Biophys Acta 1979; 582:132–144.
47. Scott JE, Heatley F, Wood B. Comparison of secondary structures in water of chondroitin-4-sulfate and dermatan sulfate: implications in the formation of tertiary structures. Biochemistry 1995; 34:15467–15474.
48. Zangrando D, Gupta R, Jamieson AM, Blackwell J, Scott PG. Light scattering studies of bovine skin proteodermatan sulfate. Biopolymers 1989; 28:1295–1308.
49. Bittner K, Liszio C, Blumberg P, Schonherr E, Kresse H. Modulation of collagen gel contraction by decorin. Biochem J 1996; 314:159–166.
50. Danielson KG, Baribault H, Holmes DF, Graham H, Kadler KE, Iozzo RV. Targeted disruption of decorin leads to abnormal collagen fibril morphology and skin fragility. J Cell Biol 1997; 136:729–743.
51. Hakkinen L, Strassburger S, Kahari VM, Scott PG, Eichstetter I, Lozzo RV, Larjava H. A role for decorin in the structural organization of periodontal ligament. Lab Invest 2000; 80:1869–1880.
52. Linares HA, Kischer CW, Dobrkovsky M, Larson DL. The histiotypic organization of the hypertrophic scar in humans. J Invest Dermatol 1972; 59:323–331.
53. Quentin-Hoffmann E, Harrach B, Robenek H, Kresse H. Genetic defects in proteoglycan biosynthesis. Padiatr Padol 1993; 28:37–41.
54. Pringle GA, Dodd CM. Immunoelectron microscopic localization of the core protein of decorin near the d and e bands of tendon collagen fibrils by use of monoclonal antibodies. J Histochem Cytochem 1990; 38:1405–1411.
55. Bianco P, Fisher LW, Young MF, Termine JD, Robey PG. Expression and localization of the two small proteoglycans biglycan and decorin in developing human skeletal and non-skeletal tissues. J Histochem Cytochem 1990; 38:1549–1563.
56. Brown DC, Vogel KG. Characteristics of the in vitro interaction of a small proteoglycan (PG II) of bovine tendon with type I collagen. Matrix 1989; 9:468–478.
57. Schonherr E, Witsch-Prehm P, Harrach B, Robenek H, Rauterberg J, Kresse H. Interaction of biglycan with type I collagen. J Biol Chem 1995; 270:2776–2783.
58. Scott PG, Nakano T, Dodd CM. Small proteoglycans from different regions of the fibrocartilaginous temporomandibular joint disc. Biochim Biophys Acta 1995; 1244:121–128.
59. Xu T, Bianco P, Fisher LW, Longenecker G, Smith E, Goldstein S, Bonadio J, Boskey A, Heegaard AM, Sommer B, Satomura K, Dominguez P, Zhao C, Kulkarni A, Robey PG, Young MF. Targeted disruption of the biglycan gene leads to an osteoporosis-like phenotype in mice. Nat Genet 1998; 20:78–82.
60. Corsi A, Xu T, Chen XD, Boyde A, Liang J, Mankani M, Sommer B, Iozzo RV, Eichstetter I, Robey PG, Bianco P, Young MF. Phenotypic effects of biglycan deficiency are linked to collagen fibril abnormalities, are synergized by decorin

deficiency, and mimic Ehlers-Danlos-like changes in bone and other connective tissues. J Bone Miner Res 2002; 17:1180–1189.
61. Beavan LA, Quentin-Hoffmann E, Schonherr E, Snigula F, Leroy JG, Kresse H. Deficient expression of decorin in infantile progeroid patients. J Biol Chem 1993; 268:9856–9862.
62. Wu J, Utani A, Endo H, Shinkai H. Deficiency of the decorin core protein in the variant form of Ehlers-Danlos syndrome with chronic skin ulcer. J Dermatol Sci 2001; 27:95–103.
63. Chakravarti S, Magnuson T, Lass JH, Jepsen KJ, LaMantia C, Carroll H. Lumican regulates collagen fibril assembly: skin fragility and corneal opacity in the absence of lumican. J Cell Biol 1998; 141:1277–1286.
64. Svensson L, Aszodi A, Reinholt FP, Fassler R, Heinegard D, Oldberg A. Fibromodulin-null mice have abnormal collagen fibrils, tissue organization, and altered lumican deposition in tendon. J Biol Chem 1999; 274: 9636–9647.
65. Svensson L, Narlid I, Oldberg A. Fibromodulin and lumican bind to the same region on collagen type I fibrils. FEBS Lett 2000; 470:178–182.
66. Ezura Y, Chakravarti S, Oldberg A, Chervoneva I, Birk DE. Differential expression of lumican and fibromodulin regulate collagen fibrillogenesis in developing mouse tendons. J Cell Biol 2000; 151:779–788.
67. Kresse H, Schonherr E. Proteoglycans of the extracellular matrix and growth control. J Cell Physiol 2001; 189:266–274.
68. Iozzo RV, Cohen I. Altered proteoglycan gene expression and the tumor stroma. Experientia 1993; 49:447–455.
69. Santra M, Skorski T, Calabretta B, Lattime EC, Iozzo RV. De novo decorin gene expression suppresses the malignant phenotype in human colon cancer cells. Proc Natl Acad Sci U S A 1995; 92:7016–7020.
70. Vuillermoz B, Khoruzhenko A, D'Onofrio MF, Ramont L, Venteo L, Perreau C, Antonicelli F, Maquart FX, Wegrowski Y. The small leucine-rich proteoglycan lumican inhibits melanoma progression. Exp Cell Res 2004; 296:294–306.
71. Yamaguchi Y, Ruoslahti E. Expression of human proteoglycan in Chinese hamster ovary cells inhibits cell proliferation. Nature 1988; 336:244–246.
72. Yamaguchi Y, Mann DM, Ruoslahti E. Negative regulation of transforming growth factor-beta by the proteoglycan decorin. Nature 1990; 346:281–284.
73. Hildebrand A, Romaris M, Rasmussen LM, Heinegard D, Twardzik DR, Border WA, Ruoslahti E. Interaction of the small interstitial proteoglycans biglycan, decorin and fibromodulin with transforming growth factor beta. Biochem J 1994; 302:527–534.
74. Takeuchi Y, Kodama Y, Matsumoto T. Bone matrix decorin binds transforming growth factor-beta and enhances its bioactivity. J Biol Chem 1994; 269:32634–32638.
75. Hausser H, Groning A, Hasilik A, Schonherr E, Kresse H. Selective inactivity of TGF-beta/decorin complexes. FEBS Lett 1994; 353:243–245.
76. Taipale J, Koli K, Keski-Oja J. Release of transforming growth factor-beta 1 from the pericellular matrix of cultured fibroblasts and fibrosarcoma cells by plasmin and thrombin. J Biol Chem 1992; 267:25378–25384.
77. Imai K, Hiramatsu A, Fukushima D, Pierschbacher MD, Okada Y. Degradation of decorin by matrix metalloproteinases: identification of the cleavage

sites, kinetic analyses and transforming growth factor-beta1 release. Biochem J 1997; 322:809–814.
78. Costacurta A, Priante G, D'Angelo A, Chieco-Bianchi L, Cantaro S. Decorin transfection in human mesangial cells downregulates genes playing a role in the progression of fibrosis. J Clin Lab Anal 2002; 16:178–186.
79. Stander M, Naumann U, Dumitrescu L, Heneka M, Loschmann P, Gulbins E, Dichgans J, Weller M. Decorin gene transfer-mediated suppression of TGF-beta synthesis abrogates experimental malignant glioma growth in vivo. Gene Ther 1998; 5:1187–1194.
80. Border WA, Noble NA, Yamamoto T, Harper JR, Yamaguchi Y, Pierschbacher MD, Ruoslahti E. Natural inhibitor of transforming growth factor-beta protects against scarring in experimental kidney disease. Nature 1992; 360:361–364.
81. Logan A, Baird A, Berry M. Decorin attenuates gliotic scar formation in the rat cerebral hemisphere. Exp Neurol 1999; 159:504–510.
82. Kolb M, Margetts PJ, Galt T, Sime PJ, Xing Z, Schmidt M, Gauldie J. Transient transgene expression of decorin in the lung reduces the fibrotic response to bleomycin. Am J Respir Crit Care Med 2001; 163:770–777.
83. Kolb M, Margetts PJ, Sime PJ, Gauldie J. Proteoglycans decorin and biglycan differentially modulate TGF-beta-mediated fibrotic responses in the lung. Am J Physiol Lung Cell Mol Physiol 2001; 280:L1327–L1334.
84. Fukui N, Fukuda A, Kojima K, Nakajima K, Oda H, Nakamura K. Suppression of fibrous adhesion by proteoglycan decorin. J Orthop Res 2001; 19:456–462.
85. De Luca A, Santra M, Baldi A, Giordano A, Iozzo RV. Decorin-induced growth suppression is associated with up-regulation of p21, an inhibitor of cyclin-dependent kinases. J Biol Chem 1996; 271:18961–18965.
86. Santra M, Mann DM, Mercer EW, Skorski T, Calabretta B, Iozzo RV. Ectopic expression of decorin protein core causes a generalized growth suppression in neoplastic cells of various histogenetic origin and requires endogenous p21, an inhibitor of cyclin-dependent kinases. J Clin Invest 1997; 100:149–157.
87. Moscatello DK, Santra M, Mann DM, McQuillan DJ, Wong AJ, Iozzo RV. Decorin suppresses tumor cell growth by activating the epidermal growth factor receptor. J Clin Invest 1998; 101:406–412.
88. Csordas G, Santra M, Reed CC, Eichstetter I, McQuillan DJ, Gross D, Nugent MA, Hajnoczky G, Iozzo RV. Sustained down-regulation of the epidermal growth factor receptor by decorin. A mechanism for controlling tumor growth in vivo. J Biol Chem 2000; 275:32879–32887.
89. Brown LF, Guidi AJ, Schnitt SJ, Van De Water L, Iruela-Arispe ML, Yeo TK, Tognazzi K, Dvorak HF. Vascular stroma formation in carcinoma in situ, invasive carcinoma, and metastatic carcinoma of the breast. Clin Cancer Res 1999; 5:1041–1056.
90. Grant DS, Yenisey C, Rose RW, Tootell M, Santra M, Iozzo RV. Decorin suppresses tumor cell-mediated angiogenesis. Oncogene 2002; 21:4765–4777.
91. Schonherr E, O'Connell BC, Schittny J, Robenek H, Fastermann D, Fisher LW, Plenz G, Vischer P, Young MF, Kresse H. Paracrine or virus-mediated induction of decorin expression by endothelial cells contributes to tube formation and prevention of apoptosis in collagen lattices. Eur J Cell Biol 1999; 78:44–55.

92. Schonherr E, Levkau B, Schaefer L, Kresse H, Walsh K. Decorin-mediated signal transduction in endothelial cells. Involvement of Akt/protein kinase B in up-regulation of p21(WAF1/CIP1) but not p27(KIP1). J Biol Chem 2001; 276:40687–40692.
93. Majava M, Bishop PN, Hagg P, Scott PG, Rice A, Inglehearn C, Hammond CJ, Spector TD, Ala-Kokko L, Mannikko M. Novel mutations in the small leucine-rich repeat protein/proteoglycan (SLRP) genes in high myopia. Hum Mutat 2007; 28:336–344.
94. Pentikainen MO, Oorni K, Lassila R, Kovanen PT. The proteoglycan decorin links low density lipoproteins with collagen type I. J Biol Chem 1997; 272:7633–7638.
95. Olin KL, Potter-Perigo S, Barrett PH, Wight TN, Chait A. Biglycan, a vascular proteoglycan, binds differently to HDL2 and HDL3: role of apoE. Arterioscler Thromb Vasc Biol 2001; 21:129–135.
96. Guo BP, Norris SJ, Rosenberg LC, Hook M. Adherence of Borrelia burgdorferi to the proteoglycan decorin. Infect Immun 1995; 63:3467–3472.
97. Scott PG, Ghahary A, Wang JF, Tredget EE. Molecular and cellular basis of hypertrophic scarring. In: Herndon DN, ed. Total Burn Care. London: Saunders, 2007; 596–607.
98. Shetlar MR, Shetlar CL, Chien SF, Linares HA, Dobrkovsky M, Larson DL. The hypertrophic scar. Hexosamine containing components of burn scars. Proc Soc Exp Biol Med 1972; 139:544–547.
99. Shetlar MR, Shetlar CL, Linares HA. The hypertrophic scar: location of glycosaminoglycans within scars. Burns 1977; 4:14–19.
100. Alexander SA, Donoff RB. The histochemistry of glycosaminoglycans within hypertrophic scars. J Surg Res 1980; 28:171–181.
101. Swann DA, Garg HG, Jung W, Hermann H. Studies on human scar tissue proteoglycans. J Invest Dermatol 1985; 84:527–531.
102. Swann DA, Garg HG, Hendry CJ, Hermann H, Siebert E, Sotman S, Stafford W. Isolation and partial characterization of dermatan sulfate proteoglycans from human post-burn scar tissues. Coll Relat Res 1988; 8:295–313.
103. Garg HG, Siebert JW, Garg A, Neame PJ. Inseparable iduronic acid-containing proteoglycan PG(IdoA) preparations of human skin and post-burn scar tissues: evidence for elevated levels of PG(IdoA)-I in hypertrophic scar by N-terminal sequencing. Carbohydr Res 1996; 284:223–228.
104. Yeo TK, Brown L, Dvorak HF. Alterations in proteoglycan synthesis common to healing wounds and tumors. Am J Pathol 1991; 138:1437–1450.
105. Oksala O, Salo T, Tammi R, Hakkinen L, Jalkanen M, Inki P, Larjava H. Expression of proteoglycans and hyaluronan during wound healing. J Histochem Cytochem 1995; 43:125–135.
106. Sayani K, Dodd CM, Nedelec B, Shen YJ, Ghahary A, Tredget EE, Scott PG. Delayed appearance of decorin in healing burn scars. Histopathology 2000; 36:262–272.
107. Kuwaba K, Kobayashi M, Nomura Y, Irie S, Koyama Y. Elongated dermatan sulphate in post-inflammatory healing skin distributes among collagen fibrils separated by enlarged interfibrillar gaps. Biochem J 2001; 358:157–163.
108. Leask A, Abraham DJ. TGF-beta signaling and the fibrotic response. FASEB J 2004; 18:816–827.

109. Wang R, Ghahary A, Shen Q, Scott PG, Roy K, Tredget EE. Hypertrophic scar tissue and fibroblasts produce more TGF-β1 mRNA and protein than normal skin and cells. Wound Repair Regen 2000; 8:128–137.
110. Tredget EE, Shankowsky HA, Pannu R, Nedelec B, Iwashina T, Ghahary A, Taerum TV, Scott PG. Transforming growth factor-beta in thermally injured patients with hypertrophic scars: effects of interferon alpha-2b. Plast Reconstr Surg 1998; 102:1317–1328.
111. Ignotz RA, Massague J. Transforming growth factor-beta stimulates the expression of fibronectin and collagen and their incorporation into the extracellular matrix. J Biol Chem 1986; 261:4337–4345.
112. Sporn MB, Roberts AB, Wakefield LM, de Crombrugghe B. Some recent advances in the chemistry and biology of transforming growth factor-beta. J Cell Biol 1987; 105:1039–1045.
113. Overall CM, Wrana JL, Sodek J. Transforming growth factor-beta regulation of collagenase, 72 kDa-progelatinase, TIMP and PAI-1 expression in rat bone cell populations and human fibroblasts. Connect Tissue Res 1989; 20:289–294.
114. Scott PG, Dodd CM, Ghahary A, Shen YJ, Tredget EE. Fibroblasts from post-burn hypertrophic scar tissue synthesize less decorin than normal dermal fibroblasts. Clin Sci (Lond) 1998; 94:541–547.
115. Kahari VM, Larjava H, Uitto J. Differential regulation of extracellular matrix proteoglycan (PG) gene expression. Transforming growth factor-beta 1 up-regulates biglycan (PGI), and versican (large fibroblast PG) but down-regulates decorin (PGII) mRNA levels in human fibroblasts in culture. J Biol Chem 1991; 266:10608–10615.
116. Arakawa M, Hatamochi A, Mori Y, Mori K, Ueki H, Moriguchi T. Reduced collagenase gene expression in fibroblasts from hypertrophic scar tissue. Br J Dermatol 1996; 134:863–868.
117. Wang R, Ghahary A, Shen YJ, Scott PG, Tredget EE. Human dermal fibroblasts produce nitric oxide and express both constitutive and inducible nitric oxide synthase isoforms. J Invest Dermatol 1996; 106:419–427.
118. Schonherr E, Beavan LA, Hausser H, Kresse H, Culp LA. Differences in decorin expression by papillary and reticular fibroblasts in vivo and in vitro. Biochem J 1993; 290:893–899.
119. Kresse H, Rosthoj S, Quentin E, Hollmann J, Glossl J, Okada S, Tonnesen T. Glycosaminoglycan-free small proteoglycan core protein is secreted by fibroblasts from a patient with a syndrome resembling progeroid. Am J Hum Genet 1987; 41:436–453.
120. Quentin E, Gladen A, Roden L, Kresse H. A genetic defect in the biosynthesis of dermatan sulfate proteoglycan: galactosyltransferase I deficiency in fibroblasts from a patient with a progeroid syndrome. Proc Natl Acad Sci U S A 1990; 87:1342–1346.
121. Seidler DG, Faiyaz-Ul-Haque M, Hansen U, Yip GW, Zaidi SH, Teebi AS, Kiesel L, Gotte M. Defective glycosylation of decorin and biglycan, altered collagen structure, and abnormal phenotype of the skin fibroblasts of an Ehlers-Danlos syndrome patient carrying the novel Arg270Cys substitution in galactosyltransferase I (beta4GalT-7). J Mol Med 2006; 84:583–594.

122. Wu J, Utani A, Endo H, Shinkai H. Deficiency of the decorin core protein in the variant form of Ehlers-Danlos syndrome with chronic skin ulcer. J Dermatol Sci 2001; 27:95–103.
123. Fushimi H, Kameyama M, Shinkai H. Deficiency of the core proteins of dermatan sulphate proteoglycans in a variant form of Ehlers-Danlos syndrome. J Intern Med 1989; 226:409–416.
124. Jarvelainen H, Puolakkainen P, Pakkanen S, Brown EL, Hook M, Iozzo RV, Sage EH, Wight TN. A role for decorin in cutaneous wound healing and angiogenesis. Wound Repair Regen 2006; 14:443–452.
125. Guo BP, Brown EL, Dorward DW, Rosenberg LC, Hook M. Decorin-binding adhesins from Borrelia burgdorferi. Mol Microbiol 1998; 30:711–723.
126. Hagman KE, Lahdenne P, Popova TG, Porcella SF, Akins DR, Radolf JD, Norgard MV. Decorin-binding protein of Borrelia burgdorferi is encoded within a two-gene operon and is protective in the murine model of Lyme borreliosis. Infect Immun 1998; 66:2674–2683.
127. Brown EL, Guo BP, O'Neal P, Hook M. Adherence of Borrelia burgdorferi. Identification of critical lysine residues in DbpA required for decorin binding. J Biol Chem 1999; 274:26272–26278.
128. Liang FT, Brown EL, Wang T, Iozzo RV, Fikrig E. Protective niche for Borrelia burgdorferi to evade humoral immunity. Am J Pathol 2004; 165:977–985.

Carbohydrate Chemistry, Biology and Medical Applications
Hari G. Garg, Mary K. Cowman and Charles A. Hales

DOI: 10.1016/B978-0-08-054816-6.00008-2

Chapter 8

Functional Glycosaminoglycans in the Eye

MASAHIRO ZAKO* AND MASAHIKO YONEDA†

**Department of Ophthalmology, Aichi Medical University, Nagakute, Aichi 480-1195, Japan*
†Biochemistry and Molecular Biology Laboratory, Aichi Prefectural College of Nursing and Health, Nagoya, Aichi 463-8502, Japan

I. Introduction

A large number of investigations have revealed limitless functional contributions of glycosaminoglycans, the most characteristic carbohydrates, to ocular systems. Here, we focus on the roles of glycosaminoglycans in ocular pathogenic conditions and the clinical applications of glycosaminoglycans for eye diseases, and attempt to review them for both health professionals and general readers. We deal with macular corneal dystrophy, glaucoma, cataract, diabetic retinopathy, retinal detachment/proliferative vitreoretinopathy, myopia, thyroid eye disease, and pseudoexfoliation syndrome in the eye disease section, and heparin, hyaluronan, and chondroitin sulfate in the clinical application section. Figure 1 shows a schematic diagram of the eye in cross section indicating each ocular component. The candidate glycosaminoglycans involved in the ocular components of each eye disease described in this chapter are summarized in both the legend for Fig. 1 and Table 1.

II. Glycosaminoglycans in Eye Diseases

A. Macular Corneal Dystrophy

Macular corneal dystrophy is a rare dystrophy that is characterized by abnormal deposits in the corneal stroma, keratocytes, Descemet's membrane, and

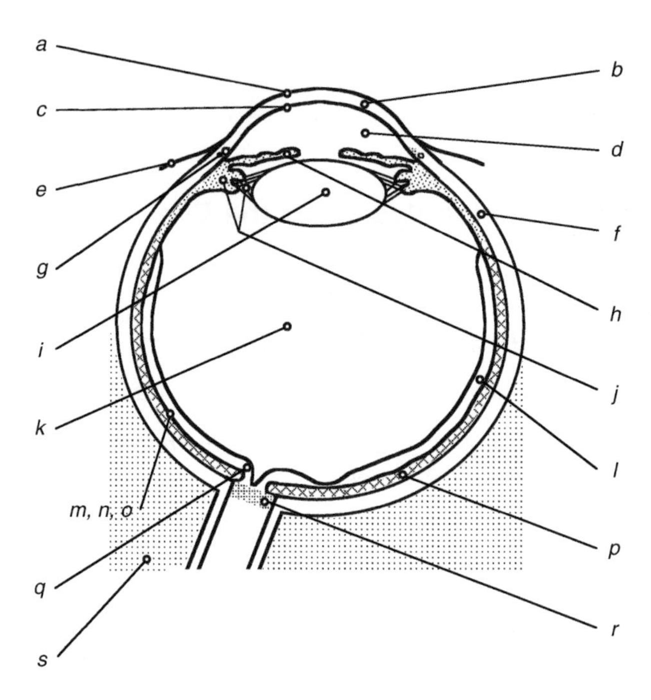

Figure 1 Schematic diagram of the eye in horizontal section indicating each ocular component. *a*, corneal epithelium; *b*, keratocyte; *c*, corneal endothelium; *d*, aqueous humor; *e*, conjunctiva; *f*, sclera; *g*, trabecular meshwork; *h*, iris; *i*, lens; *j*, ciliary zonule and body; *k*, vitreous; *l*, retina; *m*, interphotoreceptor matrix; *n*, retinal pigment epithelium; *o*, Bruch's membrane; *p*, choroid; *q*, optic nerve head; *r*, lamina cribrosa; *s*, extraocular muscles and tissues. The candidate glycosaminoglycans involved in the ocular components of each eye disease described in this chapter are as follows: macular corneal dystrophy (*b*, *c*: KS, CS/DS, HA), glaucoma (*d*: HA; *g*: CS/DS, HS, HA; *q*, *r*: CS, HS, HA), cataract (*i*: CS/DS, HS, HA), diabetic retinopathy (*k*: HA; *l*: HS), retinal detachment/proliferative vitreoretinopathy (*k*, *l*, *m*, *n*: CS/DS, HS, HA), myopia (*f*, *p*: CS), thyroid eye disease (*s*: CS, HA), pseudoexfoliation syndrome (*c*, *d*, *g*, *h*, *i*, *j*: KS, CS/DS, HA). KS, keratan sulfate; CS/DS, chondroitin sulfate/dermatan sulfate; HS, heparan sulfate; HA, hyaluronan.

endothelium, which is accompanied by progressive clouding. Corneal cells and organ cultures from patients with macular corneal dystrophy show diminished synthesis of keratan sulfate proteoglycans (1,2). Nakazawa et al. indicated that the associated error in the synthesis of corneal keratan sulfate in macular corneal dystrophy is caused by failure of specific sulfotransferases involved in sulfation (3).

Macular corneal dystrophy can be classified into three immunophenotypes, types I, IA, and II, according to the serum level of sulfated keratan sulfate and immunoreactivity of the corneal tissue. In macular corneal dystrophy type I, the keratan sulfate level is low in both the serum and corneal tissue. In macular

Table 1 Glycosaminoglycans Associated with Eye Diseases

KS	Macular corneal dystrophy, pseudoexfoliation syndrome
CS/DS	Macular corneal dystrophy, glaucoma, cataract, retinal detachment/proliferative vitreoretinopathy, myopia, thyroid eye disease, pseudoexfoliation syndrome
HS	Glaucoma, cataract, diabetic retinopathy, retinal detachment/proliferative vitreoretinopathy
HA	Macular corneal dystrophy, glaucoma, cataract, diabetic retinopathy, retinal detachment/proliferative vitreoretinopathy, thyroid eye disease, pseudoexfoliation syndrome

KS, keratan sulfate; CS/DS, chondroitin sulfate/dermatan sulfate; HS, heparan sulfate; HA, hyaluronan.

corneal dystrophy type IA, the serum keratan sulfate level is low, but keratan sulfate accumulated within the keratocytes reacts with 5D4, a monoclonal antibody that recognizes a sulfated epitope on the keratan sulfate chain. In macular corneal dystrophy type II, the serum keratan sulfate level is often normal, but accumulated corneal keratan sulfate reacts with the 5D4 antibody. As an example, a quantitative analysis revealed that the antigenic keratan sulfate content of the cornea of a macular corneal dystrophy type I patient was at least 800 times lower than that in normal control subjects (4). The key molecules involved in the disease are modification enzymes of keratan sulfate. For example, the activity of GlcNAc6ST was found to be decreased in a cornea with macular corneal dystrophy (5), and recent studies have identified mutations in a carbohydrate sulfotransferase gene encoding corneal *N*-acetylglucosamine 6-*O*-sulfotransferase on chromosome 16q22 as one of the causes of macular corneal dystrophy (6–9).

The keratan and chondroitin/dermatan sulfate levels in normal human corneas and corneas affected by macular corneal dystrophies types I and II were compared (10). The results revealed that the keratan sulfate chain size was reduced and chain sulfation was absent in type I, and that sulfation of both GlcNAc and Gal was significantly reduced in type II. The chondroitin/dermatan sulfate chain sizes were also decreased in all diseased corneas, and the contents of 4- and 6-sulfated disaccharides were proportionally increased. The keratan sulfate chain concentrations were reduced in both types I and II, whereas the chondroitin/dermatan sulfate chain concentrations were increased in both types. Hyaluronan, which is not normally present in healthy adult corneas, was also detected in both disease subtypes.

Macular corneal dystrophy types I and II have also been characterized histochemically. In normal corneas, high levels of sulfated keratan sulfate were detected in the stroma, Bowman's layer, and Descemet's membrane with low levels in the keratocytes, epithelium, and endothelium. Furthermore, in normal corneas, negligible levels of labeling for *N*-acetyllactosamine (unsulfated keratan sulfate) were detected. In macular corneal dystrophy type I corneas, sulfated keratan sulfate was not detected anywhere, but a specific distribution of *N*-acetyllactosamine (unsulfated keratan sulfate) was evident with heavily labeled deposits in the stroma, keratocytes, endothelium, and disrupted posterior region of Descemet's membrane (11). In macular corneal dystrophy type II corneas, the midstroma contained 30% less sulfur than normal corneas (12).

B. Glaucoma

Glaucoma is a type of optic neuropathy associated with characteristic optic disk damage, which may result in certain visual field loss patterns, at least partly in response to suboptimal intraocular pressure. There are three representative ocular lesions containing glycosaminoglycans, namely the aqueous humor, trabecular meshwork, and optic nerve head/lamina cribrosa, which are involved in glaucoma.

1. Aqueous Humor

The concentration of hyaluronan was found to be significantly lower in the aqueous humor of primary open-angle glaucoma patients than in that of nonglaucomatous control patients. Specifically, the median hyaluronan concentrations in the primary open-angle glaucoma and nonglaucoma groups were 298.4 and 545.1 μg/liter, respectively (13). This change in the aqueous humor glycosaminoglycans may be important for prednisolone-induced ocular hypertension. Interestingly, the hyaluronan concentration in rabbit aqueous humor was significantly lower in eyes treated with topical prednisolone than in control eyes, and returned to the normal level after withdrawal of the prednisolone (14).

2. Trabecular Meshwork

Hyaluronan is present in the part of the trabecular meshwork that surrounds the collector channels and blood vessels in the sclera of normal eyes, and may regulate the outflow pathways (15). The hyaluronan contents in the tissues of primary open-angle glaucoma and age-matched normal eyes were examined. Quantitative biochemical profiles of the glycosaminoglycans in the trabecular meshwork revealed depletion of hyaluronan and accumulation of chondroitin sulfates in the primary open-angle glaucoma trabecular meshwork (16). Similarly, the primary open-angle glaucoma juxtacanalicular tissue was depleted of hyaluronan and showed accumulation of chondroitin sulfates that may increase the outflow resistance and, consequently, increase intraocular pressure in patients with primary open-angle glaucoma (17).

An examination of the effects of chronic and acute loss of glycosaminoglycans caused by Streptomyces hyaluronidase and chondroitinase ABC from the aqueous outflow pathway on the intraocular pressure and outflow facility in monkey eyes revealed that the intraocular pressure and outflow facility remained unchanged following acute and chronic intracameral chondroitinase ABC and hyaluronidase treatment in monkeys (18). However, the anterior chambers in dogs with normotensive and glaucomatous eyes treated with bovine testicular hyaluronidase showed different results (19). In normotensive eyes, hyaluronidase significantly increased the rate of constant-pressure perfusion, and hyaluronidase abolished the staining of colloidal iron in the trabecular meshwork. Treatment of glaucomatous eyes with hyaluronidase did not significantly increase the rate of constant-pressure perfusion, and the trabecular meshworks remained stained with colloidal iron after the hyaluronidase treatment, suggesting that some glycosaminoglycans were resistant to the actions of this enzyme.

Glycosaminoglycans may be involved in the occurrence of secondary glaucoma. Topical administration of dexamethasone resulted in an increase in the total chondroitin sulfate content in the rabbit aqueous outflow pathway, whereas the total hyaluronan content was decreased (20). Similarly, in human trabecular cells, there was a significant decrease in hyaluronan synthesis following treatment with dexamethasone compared to untreated control cells, although many other components of the extracellular matrix, such as fibronectin and elastin, increased after the dexamethasone treatment (21). However, the rabbit corneoscleral trabeculum in steroid-induced glaucoma showed increased levels of ground substance and cytoplasmic organelles, as well as great sensitivity of the ground substance to hyaluronidase (22).

The proteoglycan localizations in the trabecular tissue of eyes with neovascular glaucoma have been examined. Chondroitin sulfate and dermatan sulfate proteoglycans were present in association with collagen fibrils, while heparan sulfate proteoglycans were present in association with the basal lamina of both the vascular endothelial cells and the trabecular cells (23). The large accumulation of basal lamina-like material with heparan sulfate proteoglycans may be one of the causes of the intraocular pressure increase in goniodysgenetic (developmental) glaucoma (24).

The effects of antiglaucoma drugs (epinephrine, timolol, and pilocarpine) on the glycosaminoglycans of the rabbit trabecular meshwork were examined by organ culture. Treatment with epinephrine significantly reduced hyaluronan and increased chondroitin sulfate in the radiolabeled precursor incorporation rate, but the other drugs had little effect (25).

The aqueous humor is characterized by a high ascorbic acid concentration, and addition of ascorbic acid to the medium of human trabecular meshwork cells in culture resulted in a significant dose-dependent stimulation of hyaluronan synthesis and secretion (26).

3. *Optic Nerve Head/Lamina Cribrosa*

The optic nerve head is the portion of the optic nerve observed in the fundus formed by the meeting of all the retinal nerve fibers. The lamina cribrosa is a thin sieve-like membrane composed of neuroglia and connective tissue that bridges the posterior scleral foramen and is continuous with the choroid and the deepest third of the sclera. Fiber bundles of the optic nerve pass through the perforations of the lamina cribrosa toward the optic chiasm. Because glycosaminoglycans and proteoglycans compose the optic nerve head and lamina cribrosa, they may be involved in the occurrence of glaucoma.

Chondroitin and dermatan sulfate-containing proteoglycans exist throughout the support tissues of the human optic nerve head (27,28). Rodents also show evidence of chondroitin/dermatan sulfate proteoglycans in all the connective tissue structures of the optic nerve head (29). Heparan sulfate proteoglycans are localized to the margins of the collagenous laminar plates of the scleral lamina cribrosa and along the margins of the optic nerve septa and the pia mater (30). There is also a specific keratan sulfate proteoglycan associated with astrocytes in the rat optic nerve, as identified by the TED15 monoclonal antibody (31), but is not reported for human.

The sulfated proteoglycans in the human lamina cribrosa showed age-related changes (32). In the eyes of older individuals, the chondroitin/dermatan sulfate and heparan sulfate proteoglycan filaments were found to be shorter than those in younger persons. A mild decline in the diameter of the filaments with aging was also noted.

There are some differences in the glycosaminoglycans and matrix metalloproteinases between normal and glaucomatous optic nerve heads. In normal adult eyes, hyaluronan was found surrounding the myelin sheaths in the retrolaminar nerve, while hyaluronan staining was virtually absent around the myelin sheaths of the retrolaminar nerve in eyes with primary open-angle glaucoma (33). When laser-induced glaucomatous eyes were compared with normal eyes, accumulation and enlargement of collagen-associated proteoglycan filaments were observed, accompanied by the destruction of collagenous beams (34). Accumulation of chondroitin sulfate proteoglycans was most evident, and prominent filamentous heparan sulfate proteoglycans were noted in thickened astrocytic and vascular basal laminae. There was also increased immunostaining for matrix metalloproteinases in the glaucomatous optic nerve head that suggests increased expression of these proteins in glaucoma and thereby implies a role in the tissue remodeling and degenerative changes seen in glaucomatous optic nerve heads (35).

Recently, increased levels of autoantibodies recognizing glycosaminoglycans of the optic nerve head were identified in the serum of some patients with glaucoma (36). These autoantibodies may increase the susceptibility of the optic nerve head to damage in the patients by changing the functional properties of the lamina cribrosa, its vasculature, or both. Furthermore, the levels of serum autoantibodies were higher in patients with normal-tension glaucoma than in patients with primary open-angle glaucoma (37).

C. Cataract

The lens comprises three parts: the capsule, lens epithelium, and lens fibers. Any opacity in the lens that causes it to lose its transparency and/or scatter light is called a cataract. Heparan sulfates and chondroitin sulfates are essential for lens differentiation and homeostasis, and abnormal expressions of these glycosaminoglycans may be involved in the occurrence of a cataract.

Lens epithelial cells synthesize and secrete one or more high-molecular-weight glycoconjugates that contain heparan sulfate (38). Analysis of lens capsules revealed the presence of heparan sulfate in the lens epithelium and capsule (39,40). Heparan sulfate proteoglycans in the lens capsule colocalize with and bind to fibroblast growth factors (FGFs) (41,42). The heparan sulfate chains of perlecan in the lens capsule have an essential function in lens formation, as revealed by a gene-targeting study that involved loss of attachment sites for heparan sulfate side chains (43). An investigation of intact chick lens revealed the presence of chondroitin sulfate proteoglycans with a broad molecular size distribution of the chondroitin sulfate chains (44).

Children with congenital or infantile cataracts were found to excrete fragments of glycosaminoglycans (heparan and chondroitin sulfates) in their urine, and showed accumulation of these glycosaminoglycan fragments in their blood and lenses (45). Quantitative electron microscopy analyses of the three main types of glycosaminoglycans (heparan, chondroitin, and dermatan sulfates) in young and senile (cataractous) lens capsules demonstrated reduced amounts of glycosaminoglycans in the senile cataractous lens capsules (46). In addition, examination of capsules containing epithelial lens cells from normal and cataractous mice (Nakano strain) revealed that the proteoglycan synthesis was specifically decreased in the cataractous lens (47). Furthermore, glycosaminoglycan analyses showed increased synthesis of hyaluronan and decreased synthesis of heparan sulfate in the cataract capsules.

Thus, the expression levels of the glycosaminoglycan chains of proteoglycans in the lens may be important for normal lens cell biology, and the disruption induced by many factors may finally result in the formation of a cataract.

D. Diabetic Retinopathy

Diabetic retinopathy is a progressive dysfunction of the retinal vasculature caused by chronic hyperglycemia. The effects of advanced glycation end products (AGEs), which are increased in the vitreous of diabetic patients, on the photolysis of hyaluronan have been examined. Exposure to light decreased the molecular weight of hyaluronan, and the addition of AGEs promoted this change (48). The photosensitizer activities of AGEs may be associated with accelerated depolymerization of hyaluronan in diabetic patients. Microvascular leakage in diabetic retinopathy in association with the expression of heparan sulfate proteoglycans in retinal microvessels has been investigated (49). The results revealed that the increased microvascular permeability in human diabetic retinopathy was not associated with changes in the expression of heparan sulfate proteoglycans, suggesting that the mechanism underlying retinal leakage is different from that of diabetic glomerular capillary leakage, in which loss of heparan sulfate proteoglycans has been observed. On the other hand, there is a report that both the synthesis of heparan sulfate proteoglycans and the mRNA expression of perlecan were decreased in the retina of streptozotocin-diabetic rats compared with normal rats (50). This decrease in heparan sulfate proteoglycan synthesis may account for the decrease in retinal basement membrane anionic sites and increased capillary permeability that occur in diabetes, but further investigations are necessary to elucidate the functions of glycosaminoglycans in diabetic retinopathy.

Many case-control studies have demonstrated an association between diabetes and cataract. Incubation of lens capsules with glucose *in vitro* resulted in changes in the mechanical and thermal properties of type IV collagen consistent with increased cross-linking, suggesting that some glucose-mediated covalent cross-linking of type IV collagen occurs in lens capsules, which may be involved in the formation of cataract in diabetes mellitus (51).

E. Retinal Detachment/Proliferative Vitreoretinopathy

Retinal detachment is a condition in which fluid exists in the subretinal space and causes separation of the neural retina from the underlying retinal pigment epithelium. Proliferative vitreoretinopathy is the proliferation of avascular fibrocellular retinal membranes associated with rhegmatogenous retinal detachment, a retina detached as a result of a retinal break or tear, which causes severe damage to the neural retina and retinal pigment epithelium. Contact between the retinal pigment epithelium and the vitreous fluid through retinal breakage may change the content of glycosaminoglycans in the fluid.

The glycosaminoglycans in the subretinal fluid of rhegmatogenous retinal detachment were characterized (52). The results revealed that hyaluronan alone (HA type) was present in 50% of the eyes. A combination of chondroitin sulfate (chSA) and hyaluronan (chSA type) was present in 15% of the eyes. A combination of dermatan sulfate (DS) and hyaluronan (DS type) was present in 35% of the eyes. Retinal detachment with a demarcation line resulted in subretinal strand formation in the DS-type eyes, while no such formation was seen in the chSA type. Vitreous haze was observed in one eye of the DS type. All eyes with grade C proliferative vitreoretinopathy were the DS type. The eyes with reoperated surgeries were the DS type. The presence of DS may indicate an advanced condition of retinal detachment.

The glycosaminoglycans in normal vitreous were identified as hyaluronan (92%) and chondroitin sulfate (8%). In contrast, up to 18% of the total glycosaminoglycans in pathological samples were identified as chondroitin sulfate (53). In pathological vitreous, two fractions of glycosaminoglycans representing about 10% were identified as undersulfated chondroitin and heparan sulfate. The hydrodynamic size of hyaluronan differed between normal and pathological samples, and the hyaluronan in samples from patients with detached retinas showed a small hydrodynamic size.

Subretinal fluid from patients with rhegmatogenous retinal detachment showed hyaluronan in 70% of the eyes examined (54). Other samples showed no hyaluronan but hyaluronidase activity. The hyaluronidase activity in the subretinal fluid increased with the duration of the retinal detachment.

Part of the mechanism of the above vitreous glycosaminoglycan transformation in retinal detachment may be explained by modification of blood components.

Hyaluronan represented 91% of the total glycosaminoglycans synthesized by normal vitreous. Vitreous fibrosis was induced by intravitreal injection of monocytes and lymphocytes (55). In the fibrotic vitreous, the synthesis of hyaluronan was decreased to 30%, whereas the synthesis of chondroitin sulfate was increased to 47% of the total newly synthesized glycosaminoglycans. Similarly, vitreous fibrosis was induced by intravitreal injection of erythrocytes (56). In the fibrotic vitreous, the synthesis of hyaluronan was decreased to 26%, whereas the synthesis of chondroitin sulfate increased to 59% of the total newly synthesized glycosaminoglycans. These results suggest that blood cells may alter the synthesis of glycosaminoglycans in induced fibrotic vitreous.

The carbohydrates of the posterior vitreoretinal juncture were examined by electron microscopy (57). Globular material of intermediate electron density was found in the basement membrane of the retina and on collagen fibrils in the vitreous cortex after cetylpyridinium chloride fixation and disappeared after Streptomyces hyaluronidase digestion. These observations suggest that the globular material is hyaluronan that is more labile along the basement membrane than toward the inner vitreous cortex. A fine filamentous network may be formed by the oligosaccharide chains associated with vitreous proteins as part of the vitreoretinal juncture layer.

Vidaurri-Leal et al. reported the following implications for retinal pigment epithelial cells in the pathogenesis of proliferative vitreoretinopathy (58). Normal human retinal pigment epithelial cells in confluent monolayers maintained a hexagonal to oval shape. However, when these monolayers were overlaid with autologous vitreous, the retinal pigment epithelial cells became elongated and migrated into the vitreous gel as bipolar fibrocyte-like cells. Retinal pigment epithelial cells overlaid with hyaluronate did not change their morphological features. When retinal pigment epithelial monolayers were overlaid with a collagen gel, the cells changed from their normal epithelial shape to bipolar fibrocyte-like cells that later migrated into the collagen gel. Thus, exposure of retinal pigment epithelial cells to vitreous in the presence of collagen may play a role in the pathogenesis of proliferative vitreoretinopathy.

F. Myopia

Myopia is a condition in which the eye is too long or the refractive power is too great to bring objects at a distance clearly into focus. Recent studies in animal models have shown that the development of and recovery from induced myopia is associated with visually guided changes in scleral glycosaminoglycan synthesis.

The initial studies on myopia were performed using chick eyes. A chick model of myopia, designated monocular occlusion, was used to evaluate the biochemical changes in the sclera (59). There was a 34% increase in glycosaminoglycans and a 20.7% decrease in cell density within the posterior sclera of myopic eyes. The $^{35}SO_4$ incorporation was significantly increased in the posterior sclera of myopic eyes, and the increased and accumulated scleral proteoglycan was identified as aggrecan, which may increase the volume of the extracellular matrix in the posterior sclera for the ocular enlargement.

The proteoglycan synthesis in chick sclera was modulated by a vision-dependent mechanism. Proteoglycan synthesis was measured in chick sclera at the onset of form-deprivation myopia, as well as in the period immediately following the removal of the occluder (60). After 24 h of form-deprivation, proteoglycan synthesis was 33% higher in myopic eyes than in paired control eyes. The rate of proteoglycan synthesis further increased to 83% higher than the control rate and remained elevated throughout the period of deprivation. Removal of the occluder resulted in a rapid drop in the rate of proteoglycan synthesis to the control level within 24 h.

A chick model of myopia was modified by using spectacle lenses, which may mimic similar situations in humans. It was found that chick eye growth compensated for the defocusing imposed by the spectacle lenses, in that the eyes elongated in response to a hyperopic defocus imposed by negative lenses and slowed their elongation in response to a myopic defocus imposed by positive lenses. The compensatory modulation of the eye length involved changes in the synthesis of glycosaminoglycans in the sclera, with synthesis increasing in eyes in response to negative spectacles lenses and decreasing in eyes in response to positive lenses (61). In addition, changes in the synthesis of glycosaminoglycans in the choroid were correlated with changes in choroidal thickness, because eyes with positive lenses developed thicker choroids and these choroids synthesized more glycosaminoglycans than choroids from eyes with negative lenses. Changes in scleral glycosaminoglycan synthesis accompanied the lens-induced changes in the length of the eye. Furthermore, changes in the thickness of the choroid were also associated with changes in the synthesis of glycosaminoglycans.

A study on myopia caused by monocular deprivation was similarly performed using mammalian eyes (62). Axial myopia was induced in tree shrews by monocular deprivation imposed with a translucent diffuser. In comparison to control eyes, the deprived eyes became myopic and elongated. Sulfated glycosaminoglycan levels were significantly lower in the deprived eyes than in the control eyes at the posterior pole (−15.6%), at the nasal equatorial region (−18.1%), and in the rest of the sclera (−11.6%). These findings suggest that the deprived sclera contained fewer proteoglycans or that the proteoglycans were less glycosylated or less sulfated.

Scleral glycosaminoglycan synthesis was monitored as an indicator of remodeling in myopic and recovering tree shrew sclerae (63). Myopia was induced by monocular deprivation of pattern vision. The animals were allowed to recover from the induced myopia by removal of the occluder. Eyes developing myopia showed a significant reduction in scleral glycosaminoglycan synthesis, particularly in the region of the posterior pole. In recovering eyes, significant changes in glycosaminoglycan synthesis were apparent after 24 h of recovery. After 3 days of recovery, the level of glycosaminoglycan synthesis was significantly elevated, and then returned to the level in the contralateral control eye after 9 days of recovery. Regulatory changes in scleral metabolism can be rapidly evoked by changes in the visual conditions, and active remodeling is associated with changes in eye size during both myopia development and recovery.

The differences in glycosaminoglycan synthesis between the scleral layer of the chick eye during myopia development and recovery were examined (64). Glycosaminoglycan synthesis in the fibrous scleral layers of myopic and recovering eyes did not differ significantly from those in the contralateral control eyes. In contrast, glycosaminoglycan synthesis was significantly elevated relative to the control eyes in the cartilaginous scleral layer of eyes developing myopia, while there was a significant decrease in synthesis in the cartilaginous layer of recovering eyes. The fibrous scleral layer of the chick eye does not display the characteristic differential patterns of glycosaminoglycan synthesis observed in mammalian sclera during myopia development and recovery. However, the cartilaginous layer of the chick sclera does display differential glycosaminoglycan expressions, although the direction of the regulation is opposite to that found in the fibrous sclera of mammals.

Recently, part of the mechanism for eye growth in myopia has been determined. Induction of myopia leads to decreased glycosaminoglycan synthesis, increased levels of matrix metalloproteinase-2, and decreased amounts of tissue inhibitor of matrix metalloproteinase-2 in the fibrous sclera of both chicks and tree shrews (65,66), while transforming growth factor (TGF)-β-2 regulates the visual eye growth in the final steps (67).

G. Thyroid Eye Disease

Thyroid eye disease is an immunological disorder that affects the orbital muscles and fat. Hyperthyroidism is observed with orbitopathy at some point in most patients, although the two are commonly synchronous. Histological examination of the retroocular connective tissues in thyroid eye disease reveals lymphocytic infiltration and accumulation of glycosaminoglycans produced locally by fibroblasts, which contribute to the pathogenesis of ophthalmopathy.

Effective parameters involving glycosaminoglycans to indicate the activity of thyroid eye disease have been reported. The concentrations of glycosaminoglycans were determined in patients with thyroid eye disease and control subjects (68). The orbital extracellular matrix glycosaminoglycans exhibited a significant increase in the tissue fractions containing chondroitin sulfate A and hyaluronan in patients with thyroid eye disease in comparison to those from control subjects. Patients with increased glycosaminoglycan concentrations responded well to steroids and/or orbital irradiation. Therefore, glycosaminoglycans are candidates for activity markers in patients with thyroid eye disease.

Transmission electron microscopy analysis revealed a marked expansion of the endomysial space in thyroid eye disease extraocular muscle biopsies compared with that in control biopsies (69). An increased number of collagen fibers with hyaluronan were detected by immunogold staining, although the serum levels of hyaluronan and urinary glycosaminoglycans were not found to be sensitive indicators for the presence of these molecules within the extraocular muscles (69). Imai et al. further confirmed that the local accumulation of glycosaminoglycans in thyroid eye disease was not associated with the serum hyaluronan concentration (70).

On the other hand, Kahaly et al. reported that urinary glycosaminoglycan excretion is an effective parameter for the activity of thyroid eye disease (71). Urinary glycosaminoglycan excretion was quantified in patients with thyroid eye disease and control subjects. In comparison with the control subjects, a significant elevation of urinary glycosaminoglycan excretion was found in patients with ophthalmopathy, whereas patients with thyroid eye disease and no ophthalmopathy and patients with toxic nodular goiter exhibited no markedly increased values. In particular, patients with active untreated ophthalmopathy showed a twofold increase in urinary glycosaminoglycan excretion on average. In contrast, high values were not detected in patients with inactive ophthalmopathy and the elevated values decreased after treatment, consistent with the clinical findings.

Humoral and cell-mediated immunity responses are related to glycosaminoglycan synthesis by retrobulbar fibroblasts in patients with thyroid eye disease. Previous studies have produced the following results. First, an ELISA using hyaluronan as the antigen detected isotype IgG antibodies in the sera of thyroid eye disease patients and healthy control subjects. In comparison with the control subjects, significantly higher hyaluronan antibody levels were found in the thyroid eye disease patients. Furthermore, when hyaluronan synthesis was measured in retrobulbar fibroblasts from the control subjects and patients after coculture with lymphocytes, the patient lymphocytes showed a marked ability to increase the hyaluronan concentration compared to the control lymphocytes (72,73). The hyaluronan concentration after incubation of patient retrobulbar fibroblasts with autologous lymphocytes was markedly more elevated than the intrinsic hyaluronan production of control retrobulbar fibroblasts. Second, thyroid eye disease is characterized by infiltration of mast cells into the orbit (74). The effects of mast cell coculture on human orbital fibroblasts have been determined. HMC-1, an established human mast cell line, activated human orbital fibroblasts to produce increased levels of hyaluronan, as was evidenced by a twofold increase in [^{3}H]-glucosamine incorporation into macromolecules upon coculture. Some molecules stimulated the accumulation of glycosaminoglycans in cultured human retroocular fibroblasts. Thus, these molecules released from lymphocytes, macrophages, or other cells infiltrating the retroocular space may play roles in the pathogenesis of thyroid eye disease.

Some of the released molecules have been identified. Interleukin (IL)-1, which is produced by macrophages and fibroblasts within the thyroid eye disease orbit, stimulated glycosaminoglycan synthesis by normal orbital fibroblasts (75). IL-1 and TGF-β significantly stimulated glycosaminoglycan accumulation by retroocular connective tissue in a dose- and time-dependent fashion (76,77). Although retroocular tissue fibroblasts synthesized hyaluronan, as well as large and small chondroitin sulfate proteoglycans, both insulin-like growth factor (IGF)-I and platelet-derived growth factor (PDGF) increased the synthesis of hyaluronan and proteoglycans in a dose-dependent manner (78). Furthermore, IGF-1 predominantly stimulated the secretion of small chondroitin sulfate proteoglycans, while PDGF increased that of large chondroitin sulfate proteoglycans. Recombinant interferon-γ stimulated glycosaminoglycan accumulation in retroocular fibroblast cultures (79). In contrast, interferon-γ had no consistent effect on macromolecular accumulation in dermal fibroblast cultures derived from the pretibium or areas ordinarily noninvolved in thyroid eye disease. Therefore, retroocular fibroblasts may be uniquely targeted for one action of interferon-γ.

Cyclic AMP (cAMP) stimulates glycosaminoglycan synthesis by retroocular tissue fibroblasts (80). Antithyrotropin (TSH)-receptor antibodies are involved in the pathogenesis of thyroid eye disease, and TSH-receptor antibodies increase cAMP as a second messenger in thyroid cells. The effects of dibutyryl cAMP on glycosaminoglycan synthesis by retroocular tissue fibroblasts have been examined. Retroocular tissue fibroblasts mainly synthesized hyaluronan, large chondroitin sulfate proteoglycans, and small chondroitin sulfate proteoglycans as glycosaminoglycans in cell culture. Dibutyryl cAMP increased hyaluronan and proteoglycan syntheses by retroocular tissue fibroblasts, and particularly stimulated the secretion of the large proteoglycans.

On the other hand, some molecules inhibit glycosaminoglycan synthesis. Pentoxifylline, an analogue of methylxanthine theobromine, inhibits the proliferation and biosynthetic activities of fibroblasts (81). Fibroblasts from the extraocular muscles of patients with thyroid eye disease and normal extraocular muscles were cultured *in vitro* in the presence or absence of pentoxifylline. In these fibroblast cultures, exposure to pentoxifylline resulted in a dose-dependent inhibition of glycosaminoglycan synthesis in all the fibroblasts. Therefore, pentoxifylline may be useful for the treatment of thyroid eye disease. In addition, treatment of fibroblasts with an IL-1-receptor antagonist or soluble IL-1 receptor significantly inhibited IL-1-stimulated glycosaminoglycan synthesis, indicating that such treatments may be useful for the prevention of thyroid eye disease (75).

Interestingly, fibroblasts derived from retroocular connective tissue and skin in thyroid eye disease exhibited different hormonal regulation. Specifically, skin fibroblasts responded to T3 (100 nmol/liter) and dexamethasone (100 nmol/liter) with 27% and 55% inhibition of glycosaminoglycan accumulation, respectively, whereas retroocular fibroblasts responded to the two hormones with 12% and 8% inhibition, respectively (82).

H. Pseudoexfoliation Syndrome

Pseudoexfoliation syndrome is a degenerative systemic disorder that is primarily characterized by deposits of distinct fibrillar material on the surfaces lining the anterior and posterior chambers of the eye, although these are most easily visualized on the lens capsule. The disease is often associated with cataract and severe high-intraocular pressure glaucoma that result in rapid deterioration of the optic nerve.

The concentrations of hyaluronan and galactosaminoglycans have been measured in the aqueous humor of patients with pseudoexfoliation syndrome and healthy subjects. The results revealed that the hyaluronan levels were significantly higher (three- to eightfold) in the patients than in the healthy subjects, but there was no significant alteration in the galactosaminoglycan concentration (83). Similarly, the aqueous humor hyaluronan levels in both patients with pseudoexfoliation syndrome and those with exfoliative glaucoma were significantly higher than in the control subjects (84).

The hyaluronan in the component of pseudoexfoliation material from human donor eyes has been analyzed histochemically. Hyaluronan was found to coat the fibrillar exfoliation material on the lens, zonules, iris epithelium, and ciliary body (85). The major component of the pseudoexfoliation material on the posterior surface of the iris was histochemically verified as chondroitin sulfate, while the minor component was hyaluronan (86). A polarization microscopic analysis demonstrated the presence of more sulfated glycosaminoglycans in exfoliation syndrome than in control subjects.

The components of the precapsular pseudoexfoliation deposits of anterior lens capsules were investigated by immunofluorescence and electron microscopic immunogold techniques, and extensive labeling of the pseudoexfoliation material for chondroitin sulfate was detected (87). On the other hand, the pseudoexfoliation syndrome material deposited on the surface of the anterior lens capsule revealed immunoreactivity for keratan sulfate and dermatan sulfate proteoglycans (88). Thus, the overproduction and/or abnormal metabolism of glycosaminoglycans identified in the pseudoexfoliation material may indicate important roles for proteoglycans in the pathogenic pathway of pseudoexfoliation syndrome.

III. Glycosaminoglycans as Clinical Applications

A. Heparin

1. Decreased Postoperative Inflammatory Response for Cataract Surgery

The effect of intraocular infusion of enoxaparin, a low-molecular-weight heparin, on the postoperative inflammatory response was evaluated for pediatric cataract surgery. The number of cells and degree of flare were minimal in the group with enoxaparin in the infusion bottle, and the total number of postoperative inflammation-related complications was also lower in this group without any enoxaparin-related complications (89). Similarly, addition of heparin sodium to the irrigation solution decreased the levels of postoperative inflammatory and fibrinoid reactions and related complications, such as synechiae, pupil irregularity, and intraocular lens decentration, after pediatric cataract surgery (90). Lens aspiration using intracameral heparin combined with primary posterior capsulorhexis and optic capture of a heparin-coated intraocular lens is a useful technique for preventing secondary visual axis opacification in pediatric cataract patients (91). The influence of heparin sodium in the irrigation solution on postoperative inflammation and cellular reactions on the anterior surface of a hydrophilic intraocular lens was also evaluated for senile cataract surgery (92). Heparin sodium added to the standard irrigation solution reduced disturbances of the blood–aqueous barrier in the early postoperative period. Therefore, infusion of heparin during surgery may minimize the postoperative inflammatory response and decrease the number of postoperative inflammation-related complications.

2. Reduction of Posterior Capsular Opacification

Topical heparin eyedrops were effective for long-term reduction of fibrotic posterior capsule opacification after extracapsular cataract extraction with intraocular lens implantation (93). Implantation of a heparin drug delivery system into the posterior chamber of experimental animals maintained a significantly higher heparin level in the aqueous humor for a long period of time, suggesting the potential for effective prevention of posterior capsular opacification after phacoemulsification of the lens with no toxic or side effects (94).

Although patients with pseudoexfoliation syndrome generally showed higher blood–aqueous barrier permeability than the control subjects, the patients implanted with heparin surface-modified intraocular lenses showed decreased permeability compared with the control subjects after surgery (95). Correspondingly, eyes with exfoliation syndrome exhibited a reduced incidence of posterior capsule opacification after implantation of heparin surface-modified intraocular lenses (96).

3. Reduction of Fibrin Formation

Heparin may prevent postoperative fibrin formation in eyes undergoing surgery for complications of proliferative diabetic retinopathy, proliferative vitreoretinopathy, and glaucoma filtration surgery.

The inhibitory effect of infusion of low-molecular-weight heparin sodium during lensectomy, vitrectomy, and retinotomy on intraocular fibrin formation was demonstrated in a rabbit model (97). Similarly, the preventative effect of heparin on postoperative intraocular fibrin clot formation was evaluated in the rabbit after vitrectomy and cyclocryotherapy. A single anterior chamber injection of heparin supplemented in the infusion solution or a single intravenous injection each resulted in a statistically significant reduction in postoperative intraocular fibrin formation (98). No ocular bleeding complications developed postoperatively.

A single collagen shield soaked in heparin achieved anterior chamber anticoagulant levels and resulted in fibrin inhibition during a 6-h study period, while a subconjunctival heparin injection did not alter the baseline aqueous anticoagulant activity (99). No complications related to collagen shield heparin delivery were encountered.

4. Reduction of Proliferative Vitreoretinopathy

Asaria et al. reported a significant reduction in the incidence of postoperative proliferative vitreoretinopathy in patients receiving 5-fluorouracil and low-molecular-weight heparin therapy and in the reoperation rate resulting from proliferative vitreoretinopathy (100). However, Charteris et al. reported that a combined perioperative infusion of 5-fluorouracil and low-molecular-weight heparin did not significantly increase the success rate of vitreoretinal surgery for established proliferative vitreoretinopathy (101). Furthermore, Williams et al. revealed no significant reduction in the reproliferation rate in proliferative vitreoretinopathy patients prospectively receiving high-molecular-weight heparin and dexamethasone (102).

On the other hand, the maximum tolerated dose of enoxaparin, a low-molecular-weight heparin, during vitrectomy for rhegmatogenous retinal detachment with proliferative vitreoretinopathy and severe diabetic retinopathy was determined (103). The study was able to achieve the 6.0 IU/ml maximum dose in the infusion fluid, and enoxaparin dose escalation did not result in a dose-dependent increase in acute side effects.

5. Reduction of Infection

Heparin surface modification and heparin treatments of lenses may reduce the incidence of postoperative endophthalmitis and intraocular inflammation. Significantly fewer *Staphylococcus epidermidis* attached to heparin surface-modified intraocular lenses and to regular poly(methyl methacrylate) intraocular lenses treated with heparin than to untreated poly(methyl methacrylate) intraocular lenses (104). When heparin was added to the medium, the numbers of *Pseudomonas aeruginosa* adhering to the contact lenses were significantly lower than those adhering to the control lenses (105).

B. Hyaluronan/Hyaluronidase

1. Intraocular Injection of Hyaluronidase

The use of intraocular sodium hyaluronan is complicated by a postoperative rise in intraocular pressure. This rise in intraocular pressure is thought to stem from "clogging" of the trabecular meshwork by the large molecules of hyaluronan. Hein et al. demonstrated the potential usefulness of hyaluronidase and documented the lack of harmful side effects histopathologically (106). Knepper et al. compared two enzymes, testicular hyaluronidase and Streptomyces hyaluronidase, that degrade hyaluronan (107), and concluded that Streptomyces hyaluronidase is more effective than testicular hyaluronidase for decreasing aqueous outflow resistance and that hyaluronan is an important glycosaminoglycan contributor to aqueous outflow resistance in the normal rabbit eye.

Recently, the efficacy of intravitreous ovine hyaluronidase for the management of vitreous hemorrhage was evaluated clinically (108). Ovine hyaluronidase at 55 IU showed statistically significant efficacy in both hemorrhage clearance and improvement in the best corrected visual acuity, suggesting a therapeutic utility of ovine hyaluronidase in the management of vitreous hemorrhage.

2. Precorneal Residence of Hyaluronan

Sodium hyaluronan solutions have been advocated for the management of a variety of dry eye states, and sodium hyaluronan could be used as an additive in various drug-release systems for the eye.

Quantitative gamma scintigraphy was used to evaluate the precorneal residence times of 0.2% and 0.3% sodium hyaluronate solutions and a polymer-free solution of buffered saline in patients with keratoconjunctivitis sicca and a group of normal volunteers (109). The mean values for the sodium hyaluronate solutions were significantly longer than those for buffered saline. Gurny et al. reported that there was no statistically significant difference between the quantities of 0.125% sodium hyaluronate and phosphate buffer solutions remaining in the precorneal space at 20 min (110). However, a comparison of 0.250% sodium hyaluronate with the phosphate buffer solution revealed a statistically significant difference in the amount remaining in the precorneal space after the same interval. In fact, 53% of the 0.250% sodium hyaluronate solution remained on the cornea, compared with just 30% of the 0.125% sodium hyaluronate solution and 18.3% of the phosphate buffer solution. Therefore, a sodium hyaluronate solution of 0.250% may have a prolonged residence time on the precorneal surface.

3. Corneal Epithelium Protection for Dry Eye Syndrome and Other Corneal Disorders

Dry eye syndrome, which affects approximately 10–20% of the adult population, is a clinical condition of ocular discomfort caused by deficient tear production and/or excessive tear evaporation. Artificial tears are often effective for relieving the symptoms of mild and moderate dry eye syndrome by replenishing the deficient tear volume. Sodium hyaluronate has been proposed as a component of artificial tears because of its viscoelastic rheology.

The efficacy of eyedrops containing sodium hyaluronate was examined in the treatment of dry eye syndrome, and the presence of sodium hyaluronate was found to reduce the symptoms of ocular irritation and lengthen the noninvasive breakup time in subjects with dry eye syndrome more effectively than saline, in terms of the peak effect and duration of action (111). Similarly, many clinical studies have shown that sodium hyaluronate effectively improved ocular surface conditions associated with dry eye syndrome (112–116).

Basic studies have been undertaken to try and elucidate the mechanism of hyaluronan for corneal epithelial protection. Hyaluronan facilitates corneal epithelial wound healing in nondiabetic and diabetic rats, although the healing rate in diabetic rats was slower than that in normal control rats (117). Rabbit corneas were used to examine the efficacy of sodium hyaluronate by comparing its effects on the rate of epithelial healing, and sodium hyaluronate concentrations of 0.1% and 0.5% were found to significantly accelerate the recovery time of iodine vapor-induced corneal erosions (118). Topical application of 1% sodium hyaluronan enhanced the formation of hemidesmosomes in the basement membrane of rabbit corneas during the early healing phase in *n*-heptanol-induced corneal wounds (119). Recently, hyaluronan was shown to have antioxidant properties and it tended to reduce the toxic effects of preservatives (120). Therefore, hyaluronan may be useful not only for dry eye syndrome but also for ocular surface disorders involving oxidative stress and ophthalmic drug therapies to improve ocular tolerance.

4. Corneal Endothelium Protection

Sodium hyaluronate is usually used for cataract surgery because the corneal endothelial damage induced by mechanical trauma, such as phacoemulsification and aspiration, was found to be significantly reduced by preinjection of sodium hyaluronate into the anterior chamber (121,122). The influences of hyaluronan on free radical formation, corneal endothelium damage, and inflammation parameters during phacoemulsification were investigated in the rabbit eye (123). Hyaluronan decreased free radical formation by about 58–60% during phacoemulsification, reduced mean corneal thickness modifications by about 76–80%, and decreased corneal endothelial cell loss by about 54–61%.

5. Implications for Increased Intraocular Pressure

Aqueous exchange via all commercially available ophthalmic viscosurgical devices was reported to cause a postoperative increase in intraocular pressure in rabbits, although there was considerable variation in the maximum intraocular pressure values obtained and the times when these values occurred (124). Injection of hyaluronan caused a rapid increase in the intraocular pressure, while injection of bovine testicular hyaluronidase significantly decreased the high intraocular pressure (125). When injected in the absence of hyaluronan, however, the decrease in intraocular pressure mediated by hyaluronidase was not significant. Interestingly, the intraocular pressure following intracameral injection of hyaluronan was proportional to the polymer size of hyaluronan in rabbits (126). The elevation of intraocular pressure following injection of hyaluronan during ophthalmic surgery may be avoided by rapid fragmentation of the large molecular size hyaluronan polymers.

Hyaluronan-induced hypertension in rats was significantly decreased by the application of one drop of brimonidine (0.2%), latanoprost (0.005%), or timolol (0.5%), suggesting that intracameral administration of hyaluronan could be a model for ocular hypertension (127).

C. Chondroitin Sulfate/Chondroitinase

The effect of chondroitin sulfate on the metabolism of the trabecular meshwork was studied by replacing the aqueous humor in the eye of albino rabbits with chondroitin sulfate solution (128). Intracameral injections caused an elevation of intraocular pressure. Light microscopy revealed that the trabecular beams were compact and sclerotic and the intertrabecular spaces were narrower. Chamber angle tissues incorporated more radioactive precursors into glycosaminoglycans than control tissues. Therefore, the glycosaminoglycan metabolism of the trabecular meshwork may be modified by chondroitin sulfate.

Chondroitinase ABC was introduced into the anterior chamber of cynomolgus monkeys (129). Following the injection, the intraocular pressure was decreased in the experimental eyes. Structurally, the intertrabecular spaces appeared wider and marked ballooning of the juxtacanalicular tissue was observed. The outer trabecular beams and inner wall of Schlemm's canal were greatly disorganized. Considerable loss of the juxtacanalicular tissue was noted. These observations suggest that chondroitinase ABC digested the trabecular glycosaminoglycans, causing disorganization of the trabecular beams and triggering intraocular pressure reduction.

In the experimental environment around porcine corneas, Hagenah and Bohnke concluded that corneal cryopreservation in the presence of chondroitin sulfate produced higher corneal endothelial cell densities than preservation in conventional cryopreservation medium alone (130). However, combining FGF and chondroitin sulfate in the culture medium appeared to be disadvantageous. FGF and chondroitin sulfate were used as supplements in organ culture medium to compare the regeneration ability of corneal endothelium with scattered damage (131). After 1 week of culture, the cell density in the group supplemented with chondroitin sulfate alone was higher than that in the control group, whereas the cell density in the group supplemented with chondroitin sulfate plus FGF was lower than that of the control group and the morphology was even worse.

IV. Conclusion

The eye is one of the most ideal organs for defining the functions of glycosaminoglycans, because the greater part of the globe is composed of glycosaminoglycans and the organ is easily assessed functionally and structurally. In fact, glycosaminoglycans have been shown to play key roles in ocular pathogenic conditions and treatments for eye diseases, as described in this chapter. Recently, clinical trials of intravitreous hyaluronidase injection for the management of vitreous hemorrhage were performed, and showed statistically significant efficacy of the therapy without any serious adverse events (108). We believe that this is an epoch-making event in ophthalmic clinical records, and further studies on the functions of glycosaminoglycans will make it possible to develop new therapies for ocular diseases.

References

1. Hassell JR, Newsome DA, Krachmer JH, Rodrigues MM. Macular corneal dystrophy: failure to synthesize a mature keratan sulfate proteoglycan. Proc Natl Acad Sci USA 1980; 77:3705–3709.
2. Klintworth GK, Smith CF. Abnormalities of proteoglycans and glycoproteins synthesized by corneal organ cultures derived from patients with macular corneal dystrophy. Lab Invest 1983; 48:603–612.
3. Nakazawa K, Hassell JR, Hascall VC, Lohmander LS, Newsome DA, Krachmer J. Defective processing of keratan sulfate in macular corneal dystrophy. J Biol Chem 1984; 259:13751–13757.
4. Edward DP, Thonar EJ, Srinivasan M, Yue BJ, Tso MO. Macular dystrophy of the cornea. A systemic disorder of keratan sulfate metabolism. Ophthalmology 1990; 97:1194–1200.
5. Hasegawa N, Torii T, Kato T, Miyajima H, Furuhata A, Nakayasu K, Kanai A, Habuchi O. Decreased GlcNAc 6-*O*-sulfotransferase activity in the cornea with macular corneal dystrophy. Invest Ophthalmol Vis Sci 2000; 41:3670–3677.
6. Akama TO, Nishida K, Nakayama J, Watanabe H, Ozaki K, Nakamura T, Dota A, Kawasaki S, Inoue Y, Maeda N, Yamamoto S, Fujiwara T, Thonar EJ, Shimomura Y, Kinoshita S, Tanigami A, Fukuda MN. Macular corneal dystrophy

type I and type II are caused by distinct mutations in a new sulphotransferase gene. Nat Genet 2000; 26:237–241.

7. El-Ashry MF, Abd El-Aziz MM, Wilkins S, Cheetham ME, Wilkie SE, Hardcastle AJ, Halford S, Bayoumi AY, Ficker LA, Tuft S, Bhattacharya SS, Ebenezer ND. Identification of novel mutations in the carbohydrate sulfotransferase gene (CHST6) causing macular corneal dystrophy. Invest Ophthalmol Vis Sci 2002; 43:377–382.
8. Niel F, Ellies P, Dighiero P, Soria J, Sabbagh C, San C, Renard G, Delpech M, Valleix S. Truncating mutations in the carbohydrate sulfotransferase 6 gene (CHST6) result in macular corneal dystrophy. Invest Ophthalmol Vis Sci 2003; 44:2949–2953.
9. Iida-Hasegawa N, Furuhata A, Hayatsu H, Murakami A, Fujiki K, Nakayasu K, Kanai A. Mutations in the CHST6 gene in patients with macular corneal dystrophy: immunohistochemical evidence of heterogeneity. Invest Ophthalmol Vis Sci 2003; 44:3272–3277.
10. Plaas AH, West LA, Thonar EJ, Karcioglu ZA, Smith CJ, Klintworth GK, Hascall VC. Altered fine structures of corneal and skeletal keratan sulfate and chondroitin/dermatan sulfate in macular corneal dystrophy. J Biol Chem 2001; 276:39788–39796.
11. Lewis D, Davies Y, Nieduszynski IA, Lawrence F, Quantock AJ, Bonshek R, Fullwood NJ. Ultrastructural localization of sulfated and unsulfated keratan sulfate in normal and macular corneal dystrophy type I. Glycobiology 2000; 10:305–312.
12. Quantock AJ, Fullwood NJ, Thonar EJ, Waltman SR, Capel MS, Ito M, Verity SM, Schanzlin DJ. Macular corneal dystrophy type II: multiple studies on a cornea with low levels of sulphated keratan sulphate. Eye 1997; 11:57–67.
13. Navajas EV, Martins JR, Melo LA, Jr, Saraiva VS, Dietrich CP, Nader HB, Belfort R, Jr. Concentration of hyaluronic acid in primary open-angle glaucoma aqueous humor. Exp Eye Res 2005; 80:853–857.
14. Laurent UB. Reduction of the hyaluronate concentration in rabbit aqueous humour by topical prednisolone. Acta Ophthalmol (Copenh) 1983; 61:751–755.
15. Gong H, Underhill CB, Freddo TF. Hyaluronan in the bovine ocular anterior segment, with emphasis on the outflow pathways. Invest Ophthalmol Vis Sci 1994; 35:4328–4332.
16. Knepper PA, Goossens W, Hvizd M, Palmberg PF. Glycosaminoglycans of the human trabecular meshwork in primary open-angle glaucoma. Invest Ophthalmol Vis Sci 1996; 37:1360–1367.
17. Knepper PA, Goossens W, Palmberg PF. Glycosaminoglycan stratification of the juxtacanalicular tissue in normal and primary open-angle glaucoma. Invest Ophthalmol Vis Sci 1996; 37:2414–2425.
18. Hubbard WC, Johnson M, Gong H, Gabelt BT, Peterson JA, Sawhney R, Freddo T, Kaufman PL. Intraocular pressure and outflow facility are unchanged following acute and chronic intracameral chondroitinase ABC and hyaluronidase in monkeys. Exp Eye Res 1997; 65:177–190.
19. Gum GG, Samuelson DA, Gelatt KN. Effect of hyaluronidase on aqueous outflow resistance in normotensive and glaucomatous eyes of dogs. Am J Vet Res 1992; 53:767–770.

20. Knepper PA, Collins JA, Frederick R. Effects of dexamethasone, progesterone, and testosterone on IOP and GAGs in the rabbit eye. Invest Ophthalmol Vis Sci 1985; 26:1093–1100.
21. Engelbrecht-Schnur S, Siegner A, Prehm P, Lutjen-Drecoll E. Dexamethasone treatment decreases hyaluronan-formation by primate trabecular meshwork cells in vitro. Exp Eye Res 1997; 64:539–543.
22. Francois J, Benozzi G, Victoria-Troncoso V, Bohyn W. Ultrastructural and morphometric study of corticosteroid glaucoma in rabbits. Ophthalmic Res 1984; 16:168–178.
23. Kubota T, Tawara A, Khalil A, Honda M, Inomata H. Distribution of proteoglycans in the trabecular tissue of eyes with neovascular glaucoma. Ger J Ophthalmol 1996; 5:392–398.
24. Tawara A, Inomata H. Distribution and characterization of sulfated proteoglycans in the trabecular tissue of goniodysgenetic glaucoma. Am J Ophthalmol 1994; 117:741–755.
25. Yoneyama J. Effects of antiglaucoma agents on glycosaminoglycans in organ-cultured rabbit trabecular meshwork. Ophthalmologica 1994; 208:278–283.
26. Schachtschabel DO, Binninger E. Stimulatory effects of ascorbic acid on hyaluronic acid synthesis of in vitro cultured normal and glaucomatous trabecular meshwork cells of the human eye. Z Gerontol 1993; 26:243–246.
27. Morrison JC, Rask P, Johnson EC, Deppmeier L. Chondroitin sulfate proteoglycan distribution in the primate optic nerve head. Invest Ophthalmol Vis Sci 1994; 35:838–845.
28. Sawaguchi S, Yue BY, Fukuchi T, Iwata K, Kaiya T. Sulfated proteoglycans in the human lamina cribrosa. Invest Ophthalmol Vis Sci 1992; 33:2388–2398.
29. Morrison J, Farrell S, Johnson E, Deppmeier L, Moore CG, Grossmann E. Structure and composition of the rodent lamina cribrosa. Exp Eye Res 1995; 60:127–135.
30. Morrison JC, Jerdan JA, L'Hernault NL, Quigley HA. The extracellular matrix composition of the monkey optic nerve head. Invest Ophthalmol Vis Sci 1988; 29:1141–1150.
31. Geisert EE, Williams RC, Bidanset DJ. A CNS specific proteoglycan associated with astrocytes in rat optic nerve. Brain Res 1992; 571:165–168.
32. Sawaguchi S, Yue BY, Fukuchi T, Iwata K, Kaiya T. Age-related changes of sulfated proteoglycans in the human lamina cribrosa. Curr Eye Res 1993; 12:685–692.
33. Gong H, Ye W, Freddo TF, Hernandez MR. Hyaluronic acid in the normal and glaucomatous optic nerve. Exp Eye Res 1997; 64:587–595.
34. Fukuchi T, Sawaguchi S, Yue BY, Iwata K, Hara H, Kaiya T. Sulfated proteoglycans in the lamina cribrosa of normal monkey eyes and monkey eyes with laser-induced glaucoma. Exp Eye Res 1994; 58:231–243.
35. Yan X, Tezel G, Wax MB, Edward DP. Matrix metalloproteinases and tumor necrosis factor alpha in glaucomatous optic nerve head. Arch Ophthalmol 2000; 118:666–673.
36. Tezel G, Edward DP, Wax MB. Serum autoantibodies to optic nerve head glycosaminoglycans in patients with glaucoma. Arch Ophthalmol 1999; 117:917–924.
37. Tomita G. The optic nerve head in normal-tension glaucoma. Curr Opin Ophthalmol 2000; 11:116–120.

38. Heathcote JG, Orkin RW. Biosynthesis of sulphated macromolecules by rabbit lens epithelium. I. Identification of the major macromolecules synthesized by lens epithelial cells in vitro. J Cell Biol 1984; 99:852–860.
39. Kennedy A, Frank RN, Mancini MA. In vitro production of glycosaminoglycans by retinal microvessel cells and lens epithelium. Invest Ophthalmol Vis Sci 1986; 27:746–754.
40. Mohan PS, Spiro RG. Characterization of heparan sulfate proteoglycan from calf lens capsule and proteoglycans synthesized by cultured lens epithelial cells. Comparison with other basement membrane proteoglycans. J Biol Chem 1991; 266:8567–8575.
41. Lovicu FJ, McAvoy JW. Localization of acidic fibroblast growth factor, basic fibroblast growth factor, and heparan sulphate proteoglycan in rat lens: implications for lens polarity and growth patterns. Invest Ophthalmol Vis Sci 1993; 34:3355–3365.
42. Schulz MW, Chamberlain CG, McAvoy JW. Binding of FGF-1 and FGF-2 to heparan sulphate proteoglycans of the mammalian lens capsule. Growth Factors 1997; 14:1–13.
43. Rossi M, Morita H, Sormunen R, Airenne S, Kreivi M, Wang L, Fukai N, Olsen BR, Tryggvason K, Soininen R. Heparan sulfate chains of perlecan are indispensable in the lens capsule but not in the kidney. EMBO J 2003; 22:236–245.
44. Nakazawa K, Takeuchi N, Iwata S. Turnover of proteoglycans by chick lens epithelial cells in cell culture and intact lens. J Biochem (Tokyo) 1989; 106:784–793.
45. Sulochana KN, Ramakrishnan S, Vasanthi SB, Madhavan HN, Arunagiri K, Punitham R. First report of congenital or infantile cataract in deranged proteoglycan metabolism with released xylose. Br J Ophthalmol 1997; 81:319–323.
46. Winkler J, Wirbelauer C, Frank V, Laqua H. Quantitative distribution of glycosaminoglycans in young and senile (cataractous) anterior lens capsules. Exp Eye Res 2001; 72:311–318.
47. Nakazawa K, Takehana M, Iwata S. Biosynthesis of proteoglycans by lens epithelial cells of cataractous mouse (Nakano strain). Exp Eye Res 1985; 40:609–618.
48. Katsumura C, Sugiyama T, Nakamura K, Obayashi H, Hasegawa G, Oku H, Ikeda T. Effects of advanced glycation end products on hyaluronan photolysis: a new mechanism of diabetic vitreopathy. Ophthalmic Res 2004; 36:327–331.
49. Witmer AN, van den Born J, Vrensen GF, Schlingemann RO. Vascular localization of heparan sulfate proteoglycans in retinas of patients with diabetes mellitus and in VEGF-induced retinopathy using domain-specific antibodies. Curr Eye Res 2001; 22:190–197.
50. Bollineni JS, Alluru I, Reddi AS. Heparan sulfate proteoglycan synthesis and its expression are decreased in the retina of diabetic rats. Curr Eye Res 1997; 16:127–130.
51. Bailey AJ, Sims TJ, Avery NC, Miles CA. Chemistry of collagen cross-links: glucose-mediated covalent cross-linking of type-IV collagen in lens capsules. Biochem J 1993; 296:489–496.

52. Hara A, Nakagomi Y. Analysis of glycosaminoglycans of subretinal fluid in rhegmatogenous retinal detachment–preliminary report. Jpn J Ophthalmol 1995; 39:137–142.
53. Theocharis DA, Feretis E, Papageorgacopoulou N. Glycosaminoglycans in the vitreous body of patients with retinal detachment. Biochem Int 1991; 25:397–407.
54. Hayasaka S, Shiono T, Hara S, Mizuno K. Lysosomal hyaluronidase in the subretinal fluid of patients with rhegmatogenous retinal detachments. Am J Ophthalmol 1982; 94:58–63.
55. Katakami C, Appel A, Raymond LA, Lipman MJ, Kao WW. Synthesis of chondroitin sulfate by fibrotic vitreous induced by monocytes and lymphocytes. Exp Eye Res 1985; 41:509–518.
56. Katakami C, Raymond LA, Lipman MJ, Appel A, Kao WW. Change in the synthesis of glycosaminoglycans by fibrotic vitreous induced by erythrocytes. Biochim Biophys Acta 1986; 880:40–45.
57. Rhodes RH. An ultrastructural study of the complex carbohydrates of the mouse posterior vitreoretinal juncture. Invest Ophthalmol Vis Sci 1982; 22:460–477.
58. Vidaurri-Leal J, Hohman R, Glaser BM. Effect of vitreous on morphologic characteristics of retinal pigment epithelial cells. A new approach to the study of proliferative vitreoretinopathy. Arch Ophthalmol 1984; 102:1220–1223.
59. Rada JA, Thoft RA, Hassell JR. Increased aggrecan (cartilage proteoglycan) production in the sclera of myopic chicks. Dev Biol 1991; 147:303–312.
60. Rada JA, McFarland AL, Cornuet PK, Hassell JR. Proteoglycan synthesis by scleral chondrocytes is modulated by a vision dependent mechanism. Curr Eye Res 1992; 11:767–782.
61. Nickla DL, Wildsoet C, Wallman J. Compensation for spectacle lenses involves changes in proteoglycan synthesis in both the sclera and choroid. Curr Eye Res 1997; 16:320–326.
62. Norton TT, Rada JA. Reduced extracellular matrix in mammalian sclera with induced myopia. Vision Res 1995; 35:1271–1281.
63. McBrien NA, Lawlor P, Gentle A. Scleral remodeling during the development of and recovery from axial myopia in the tree shrew. Invest Ophthalmol Vis Sci 2000; 41:3713–3719.
64. Gentle A, Truong HT, McBrien NA. Glycosaminoglycan synthesis in the separate layers of the chick sclera during myopic eye growth: comparison with mammals. Curr Eye Res 2001; 23:179–184.
65. Rada JA, Perry CA, Slover ML, Achen VR. Gelatinase A and TIMP-2 expression in the fibrous sclera of myopic and recovering chick eyes. Invest Ophthalmol Vis Sci 1999; 40:3091–3099.
66. Siegwart JT, Jr, Norton TT. Selective regulation of MMP and TIMP mRNA levels in tree shrew sclera during minus lens compensation and recovery. Invest Ophthalmol Vis Sci 2005; 46:3484–3492.
67. Schippert R, Brand C, Schaeffel F, Feldkaemper MP. Changes in scleral MMP-2, TIMP-2 and TGFbeta-2 mRNA expression after imposed myopic and hyperopic defocus in chickens. Exp Eye Res 2006; 82:710–719.
68. Kahaly G, Forster G, Hansen C. Glycosaminoglycans in thyroid eye disease. Thyroid 1998; 8:429–432.

69. Pappa A, Jackson P, Stone J, Munro P, Fells P, Pennock C, Lightman S. An ultrastructural and systemic analysis of glycosaminoglycans in thyroid-associated ophthalmopathy. Eye 1998; 12:237–244.
70. Imai Y, Odajima R, Shimizu T, Shishiba Y. Serum hyaluronan concentration determined by radiometric assay in patients with pretibial myxedema and Graves' ophthalmopathy. Endocrinol Jpn 1990; 37:749–752.
71. Kahaly G, Schuler M, Sewell AC, Bernhard G, Beyer J, Krause U. Urinary glycosaminoglycans in Graves' ophthalmopathy. Clin Endocrinol (Oxf) 1990; 33:35–44.
72. Kahaly G, Hansen CE, Stover C, Beyer J, Otto E. Glycosaminoglycan antibodies in endocrine ophthalmopathy. Horm Metab Res 1993; 25:637–639.
73. Kahaly G, Stover C, Beyer J, Otto E. In vitro synthesis of glycosaminoglycans in endocrine ophthalmopathy. Acta Endocrinol (Copenh) 1992; 127: 397–402.
74. Smith TJ, Parikh SJ. HMC-1 mast cells activate human orbital fibroblasts in coculture: evidence for up-regulation of prostaglandin E2 and hyaluronan synthesis. Endocrinology 1999; 140:3518–3525.
75. Tan GH, Dutton CM, Bahn RS. Interleukin-1 (IL-1) receptor antagonist and soluble IL-1 receptor inhibit IL-1-induced glycosaminoglycan production in cultured human orbital fibroblasts from patients with Graves' ophthalmopathy. J Clin Endocrinol Metab 1996; 81:449–452.
76. Korducki JM, Loftus SJ, Bahn RS. Stimulation of glycosaminoglycan production in cultured human retroocular fibroblasts. Invest Ophthalmol Vis Sci 1992; 33:2037–2042.
77. Imai Y, Ibaraki K, Odajima R, Shishiba Y. Analysis of proteoglycan synthesis by retro-ocular tissue fibroblasts under the influence of interleukin 1 beta and transforming growth factor-beta. Eur J Endocrinol 1994; 131:630–638.
78. Imai Y, Odajima R, Inoue Y, Shishiba Y. Effect of growth factors on hyaluronan and proteoglycan synthesis by retroocular tissue fibroblasts of Graves' ophthalmopathy in culture. Acta Endocrinol (Copenh) 1992; 126:541–552.
79. Smith TJ, Bahn RS, Gorman CA, Cheavens M. Stimulation of glycosaminoglycan accumulation by interferon gamma in cultured human retroocular fibroblasts. J Clin Endocrinol Metab 1991; 72:1169–1171.
80. Imai Y, Ibaraki K, Odajima R, Shishiba Y. Effects of dibutyryl cyclic AMP on hyaluronan and proteoglycan synthesis by retroocular tissue fibroblasts in culture. Endocr J 1994; 41:645–654.
81. Chang CC, Chang TC, Kao SC, Kuo YF, Chien LF. Pentoxifylline inhibits the proliferation and glycosaminoglycan synthesis of cultured fibroblasts derived from patients with Graves' ophthalmopathy and pretibial myxoedema. Acta Endocrinol (Copenh) 1993; 129:322–327.
82. Smith TJ, Bahn RS, Gorman CA. Hormonal regulation of hyaluronate synthesis in cultured human fibroblasts: evidence for differences between retroocular and dermal fibroblasts. J Clin Endocrinol Metab 1989; 69:1019–1023.
83. Lamari F, Katsimpris J, Gartaganis S, Karamanos NK. Profiling of the eye aqueous humor in exfoliation syndrome by high-performance liquid chromatographic analysis of hyaluronan and galactosaminoglycans. J Chromatogr B Biomed Sci Appl 1998; 709:173–178.

84. Gartaganis SP, Georgakopoulos CD, Exarchou AM, Mela EK, Lamari F, Karamanos NK. Increased aqueous humor basic fibroblast growth factor and hyaluronan levels in relation to the exfoliation syndrome and exfoliative glaucoma. Acta Ophthalmol Scand 2001; 79:572–575.
85. Fitzsimmons TD, Fagerholm P, Wallin O. Hyaluronan in the exfoliation syndrome. Acta Ophthalmol Scand 1997; 75:257–260.
86. Baba H. Histochemical and polarization optical investigation for glycosaminoglycans in exfoliation syndrome. Graefes Arch Clin Exp Ophthalmol 1983; 221:106–109.
87. Schlotzer-Schrehardt U, Dorfler S, Naumann GO. Immunohistochemical localization of basement membrane components in pseudoexfoliation material of the lens capsule. Curr Eye Res 1992; 11:343–355.
88. Winkler J, Lunsdorf H, Wirbelauer C, Reinhardt DP, Laqua H. Immunohistochemical and charge-specific localization of anionic constituents in pseudoexfoliation deposits on the central anterior lens capsule from individuals with pseudoexfoliation syndrome. Graefes Arch Clin Exp Ophthalmol 2001; 239:952–960.
89. Rumelt S, Stolovich C, Segal ZI, Rehany U. Intraoperative enoxaparin minimizes inflammatory reaction after pediatric cataract surgery. Am J Ophthalmol 2006; 141:433–437.
90. Bayramlar H, Totan Y, Borazan M. Heparin in the intraocular irrigating solution in pediatric cataract surgery. J Cataract Refract Surg 2004; 30:2163–2169.
91. Dada T, Dada VK, Sharma N, Vajpayee RB. Primary posterior capsulorhexis with optic capture and intracameral heparin in paediatric cataract surgery. Clin Experiment Ophthalmol 2000; 28:361–363.
92. Kruger A, Amon M, Abela-Formanek C, Schild G, Kolodjaschna J, Schauersberger J. Effect of heparin in the irrigation solution on postoperative inflammation and cellular reaction on the intraocular lens surface. J Cataract Refract Surg 2002; 28:87–92.
93. Mastropasqua L, Lobefalo L, Ciancaglini M, Ballone E, Gallenga PE. Heparin eyedrops to prevent posterior capsule opacification. J Cataract Refract Surg 1997; 23:440–446.
94. Xie L, Sun J, Yao Z. Heparin drug delivery system for prevention of posterior capsular opacification in rabbit eyes. Graefes Arch Clin Exp Ophthalmol 2003; 241:309–313.
95. Ravalico G, Tognetto D, Baccara F. Heparin-surface-modified intraocular lens implantation in eyes with pseudoexfoliation syndrome. J Cataract Refract Surg 1994; 20:543–549.
96. Zetterstrom C. Incidence of posterior capsule opacification in eyes with exfoliation syndrome and heparin-surface-modified intraocular lenses. J Cataract Refract Surg 1993; 19:344–347.
97. Iverson DA, Katsura H, Hartzer MK, Blumenkranz MS. Inhibition of intraocular fibrin formation following infusion of low-molecular-weight heparin during vitrectomy. Arch Ophthalmol 1991; 109:405–409.
98. Johnson RN, Balyeat E, Stern WH. Heparin prophylaxis for intraocular fibrin. Ophthalmology 1987; 94:597–601.
99. Murray TG, Stern WH, Chin DH, MacGowan-Smith EA. Collagen shield heparin delivery for prevention of postoperative fibrin. Arch Ophthalmol 1990; 108:104–106.

100. Asaria RH, Kon CH, Bunce C, Charteris DG, Wong D, Khaw PT, Aylward GW. Adjuvant 5-fluorouracil and heparin prevents proliferative vitreoretinopathy: results from a randomized, double-blind, controlled clinical trial. Ophthalmology 2001; 108:1179–1183.
101. Charteris DG, Aylward GW, Wong D, Groenewald C, Asaria RH, Bunce C. PVR Study Group. A randomized controlled trial of combined 5-fluorouracil and low-molecular-weight heparin in management of established proliferative vitreoretinopathy. Ophthalmology 2004; 111:2240–2245.
102. Williams RG, Chang S, Comaratta MR, Simoni G. Does the presence of heparin and dexamethasone in the vitrectomy infusate reduce reproliferation in proliferative vitreoretinopathy? Graefes Arch Clin Exp Ophthalmol 1996; 234:496–503.
103. Lane RG, Jumper JM, Nasir MA, MacCumber MW, McCuen BW 2nd. A prospective, open-label, dose-escalating study of low molecular weight heparin during repeat vitrectomy for PVR and severe diabetic retinopathy. Graefes Arch Clin Exp Ophthalmol 2005; 243:701–705.
104. Abu el-Asrar AM, Shibl AM, Tabbara KF, al-Kharashi SA. Heparin and heparin-surface-modification reduce Staphylococcus epidermidis adhesion to intraocular lenses. Int Ophthalmol 1997; 21:71–74.
105. Duran JA, Malvar A, Rodriguez-Ares MT, Garcia-Riestra C. Heparin inhibits Pseudomonas adherence to soft contact lenses. Eye 1993; 7:152–154.
106. Hein SR, Keates RH, Weber PA. Elimination of sodium hyaluronate-induced decrease in outflow facility with hyaluronidase. Ophthalmic Surg 1986; 17:731–734.
107. Knepper PA, Farbman AI, Telser AG. Exogenous hyaluronidases and degradation of hyaluronic acid in the rabbit eye. Invest Ophthalmol Vis Sci 1984; 25:286–293.
108. Kuppermann BD, Thomas EL, de Smet MD, Grillone LR. Vitrase for Vitreous Hemorrhage Study Groups. Pooled efficacy results from two multinational randomized controlled clinical trials of a single intravitreous injection of highly purified ovine hyaluronidase (Vitrase) for the management of vitreous hemorrhage. Am J Ophthalmol 2005; 140:573–584.
109. Snibson GR, Greaves JL, Soper ND, Prydal JI, Wilson CG, Bron AJ. Precorneal residence times of sodium hyaluronate solutions studied by quantitative gamma scintigraphy. Eye 1990; 4:594–602.
110. Gurny R, Ryser JE, Tabatabay C, Martenet M, Edman P, Camber O. Precorneal residence time in humans of sodium hyaluronate as measured by gamma scintigraphy. Graefes Arch Clin Exp Ophthalmol 1990; 228:510–512.
111. Johnson ME, Murphy PJ, Boulton M. Effectiveness of sodium hyaluronate eyedrops in the treatment of dry eye. Graefes Arch Clin Exp Ophthalmol 2006; 244:109–112.
112. Sand BB, Marner K, Norn MS. Sodium hyaluronate in the treatment of keratoconjunctivitis sicca. A double masked clinical trial. Acta Ophthalmol (Copenh) 1989; 67:181–183.
113. Yokoi N, Komuro A, Nishida K, Kinoshita S. Effectiveness of hyaluronan on corneal epithelial barrier function in dry eye. Br J Ophthalmol 1997; 81:533–536.

114. Iester M, Orsoni GJ, Gamba G, Taffara M, Mangiafico P, Giuffrida S, Rolando M. Improvement of the ocular surface using hypotonic 0.4% hyaluronic acid drops in keratoconjunctivitis sicca. Eye 2000; 14:892–898.
115. McDonald CC, Kaye SB, Figueiredo FC, Macintosh G, Lockett C. A randomised, crossover, multicentre study to compare the performance of 0.1% (w/v) sodium hyaluronate with 1.4% (w/v) polyvinyl alcohol in the alleviation of symptoms associated with dry eye syndrome. Eye 2002; 16:601–607.
116. Aragona P, Papa V, Micali A, Santocono M, Milazzo G. Long term treatment with sodium hyaluronate-containing artificial tears reduces ocular surface damage in patients with dry eye. Br J Ophthalmol 2002; 86:181–184.
117. Nakamura M, Sato N, Chikama TI, Hasegawa Y, Nishida T. Hyaluronan facilitates corneal epithelial wound healing in diabetic rats. Exp Eye Res 1997; 64:1043–1050.
118. Shimmura S, Ono M, Shinozaki K, Toda I, Takamura E, Mashima Y, Tsubota K. Sodium hyaluronate eyedrops in the treatment of dry eyes. Br J Ophthalmol 1995; 79:1007–1011.
119. Chung JH, Kim WK, Lee JS, Pae YS, Kim HJ. Effect of topical Na-hyaluronan on hemidesmosome formation in n-heptanol-induced corneal injury. Ophthalmic Res 1998; 30:96–100.
120. Debbasch C, De La Salle SB, Brignole F, Rat P, Warnet JM, Baudouin C. Cytoprotective effects of hyaluronic acid and Carbomer 934P in ocular surface epithelial cells. Invest Ophthalmol Vis Sci 2002; 43:3409–3415.
121. Glasser DB, Matsuda M, Edelhauser HF. A comparison of the efficacy and toxicity of and intraocular pressure response to viscous solutions in the anterior chamber. Arch Ophthalmol 1986; 104:1819–1824.
122. Miyauchi S, Horie K, Morita M, Nagahara M, Shimizu K. Protective efficacy of sodium hyaluronate on the corneal endothelium against the damage induced by sonication. J Ocul Pharmacol Ther 1996; 12:27–34.
123. Camillieri G, Nastasi A, Gulino P, Bucolo C, Drago F. Effects of hyaluronan on free-radical formation, corneal endothelium damage, and inflammation parameters after phacoemulsification in rabbits. J Ocul Pharmacol Ther 2004; 20:151–157.
124. Torngren L, Lundgren B, Madsen K. Intraocular pressure development in the rabbit eye after aqueous exchange with ophthalmic viscosurgical devices. J Cataract Refract Surg 2000; 26:1247–1252.
125. Harooni M, Freilich JM, Abelson M, Refojo M. Efficacy of hyaluronidase in reducing increases in intraocular pressure related to the use of viscoelastic substances. Arch Ophthalmol 1998; 116:1218–1221.
126. Equi RA, Jumper M, Cha C, Stern R, Schwartz DM. Hyaluronan polymer size modulates intraocular pressure. J Ocul Pharmacol Ther 1997; 13:289–295.
127. Benozzi J, Nahum LP, Campanelli JL, Rosenstein RE. Effect of hyaluronic acid on intraocular pressure in rats. Invest Ophthalmol Vis Sci 2002; 43:2196–2200.
128. Yue BY, Lin CC, Fei PF, Tso MO. Effects of chondroitin sulfate on metabolism of trabecular meshwork. Exp Eye Res 1984; 38:35–44.
129. Sawaguchi S, Yue BY, Yeh P, Tso MO. Effects of intracameral injection of chondroitinase ABC in vivo. Arch Ophthalmol 1992; 110:110–117.

130. Hagenah M, Bohnke M. Corneal cryopreservation with chondroitin sulfate. Cryobiology 1993; 30:396–406.
131. Lin CP, Bohnke M, Draeger J. Effects of fibroblast growth factor and chondroitin sulfate on predamaged corneal endothelium. An organ culture study. Ophthalmic Res 1990; 22:173–177.

Carbohydrate Chemistry, Biology and Medical Applications
Hari G. Garg, Mary K. Cowman and Charles A. Hales

DOI: 10.1016/B978-0-08-054816-6.00009-4

Chapter 9

Biological Function of Glycosaminoglycans

MICHAEL ROTH*†, ELENI PAPAKONSTANTINOU‡ AND GEORGE KARAKIULAKIS‡

**Department of Internal Medicine, Pneumology, Pulmonary Cell Research, University Hospital Basel, CH-4031 Basel, Switzerland*
†The Woolcock Institute for Medical Research, Molecular Medicine, Camperdown, NSW-2050, Australia
‡Department of Pharmacology, School of Medicine, Aristotle University, GR-54124 Thessaloniki, Greece

I. Introduction

The extracellular matrix (ECM) comprises essentially of molecules such as collagens, elastin, glycosaminoglycans (GAGs), and proteoglycans (1). ECM is subject to a continuous remodeling, in an organ-, tissue-, and cell-type specific manner. The daily turnover of the ECM in the healthy human lung is >10% (2) and is critical for the maintenance of the structural and functional integrity of the lung (1). Most of the ECM is synthesized by the constituent cells of a tissue, such as fibroblasts and epithelial cells, which also produce most of the degrading enzymes such as matrix metalloproteases (MMPs). However, ECM-degrading enzymes can also be produced by tissue infiltrating immune cells during inflammation, wound repair, or tumor genesis (2).

GAGs represent a major component of the ECM that also undergo significant alterations in content, synthesis, and distribution during neonatal organ growth (3), acute injury (4), development of fibro-degenerative and thrombotic diseases, or age-related tissue degeneration (5–13). GAGs either are an integral part of the ECM or are located directly on the cellular membrane where they can function as protein receptors or activators. The turnover of these local GAGs and their composition is essential for organ function and homeostasis. Specific GAGs such as heparan sulfate, heparin, and hyaluronic acid have been used for years for the treatment of blood clotting disorders (6,7) and joint lubrication (8–10). However, most of the GAGs have not been extensively investigated for possible therapeutic use. In this respect, it is of interest that the biological function may depend not only on the structure of a GAG but also on the length of its polysaccharide chains, as it has been shown for hyaluronic acid (11–13).

Four structurally distinct groups of GAGs have been assigned: chondroitin/dermatan sulfate, heparan sulfate/heparin, keratan sulfate, and hyaluronic acid. All GAGs are linear polysaccharides that consist of repeated amino sugars (*N*-acetyl-D-glucosamine, *N*-acetyl-D-galactosamine) and uronic acids (D-glucuronic, L-iduronic acid), only in keratan sulfate, the uronic acids are replaced by galactose. In order to understand the biological function of GAGs, one must be aware that GAGs are not encoded by genes, as they are assembled of the above described amino sugars and uronic acids, by specific synthases, which are regulated by tissue or cell-type local microenvironments. Moreover, GAGs are not synthesized by the action of a single synthase, but result from the as yet poorly understood interactions of several independently regulated synthases (14–20). Furthermore, there is evidence that during the assembly of GAGs, the addition of sulfate residues to a polysaccharide chain determines the function of the GAGs (21). For instance, sulfate residues are essential for the activity of heparan, chondroitin, and dermatan sulfates. Specific GAG chains seem to be directed by a serine–glycine motif on a protein core to form proteoglycans (22). Thus, the polysaccharide chains encode information that is not stored in the DNA (23,24). The enormous possibilities with which polysaccharide chains can be assembled to form a single GAG molecule result in an incredible diversity and it would be immature to assume that the synthesis of GAG chains is random. Therefore, this overview on the biological function of GAGs must be regarded as an attempt to deal with this intriguing and highly prospective field. Because the function of GAGs is different among species, this chapter is focused on human cell culture models or human tissue, and only few animal studies have been included.

II. Biological Function of GAGs

A. Chondroitin Sulfates

Three different types of chondroitin sulfate (A, B, and C) have been described, and classified by the location of the sulfate group. Chondroitin sulfate A and C are the most abundant GAGs in joints and contribute to the lubrication of the joint gaps (8,9,12,25,26). The function of chondroitin sulfates depends on the location of their sulfate groups, which determines their structure and their binding characteristics. Chondroitin sulfate carrying proteoglycans are aggrecan, bamacan, betaglycan, brevican, neurocan, serglycin, syndecan, and a CD44 isoform (27–29).

Chondroitin sulfates activate ECM-degrading enzymes, such as MMPs, and are therefore involved in the degradation of aggrecan by MMP-2 (30,31), MMP-13

(32), ADAMTS4 (32), and ADAMTS7B (33). MMP-2 is activated by direct binding of chondroitin sulfate to the C-domain of pro-MMP-2, which is essential to present the inactive enzyme to its activator, the membrane type 3 matrix metalloproteinase (30). Chondroitin sulfate is therefore a key regulator for protein activation and degradation and may explain its involvement in chondrodysplasias (34) and also affects the function of perlecan, though it is yet unclear whether the location of perlecan in the cartilage dictates the link to chondroitin sulfate or vice versa (35). In breast cancer cells, increased numbers of chondroitin sulfate chains linked to P-selectin enhanced the attachment of cancer cells to vascular endothelial cells forcing tissue invasion of tumor cells (36). Increased levels of chondroitin sulfates were also found in melanoma (37) and ovarian carcinoma (38) and enhanced the mitogenic effect of platelet-derived growth factor-BB on fibrosarcoma cells (39). Chondroitin sulfates have a role in tissue remodeling, tumor progression, and metastasis (40).

Chondroitin sulfates have been used in clinical trials of osteoarthritis, where they improved knee mobility and reduced pain, by providing better lubrication (41,42). This effect is claimed to be achieved not only by local injection but also by increased dietary intake of chondroitin sulfates (43,44). However, the effect was only achieved when the compounds were taken regularly over a period of at least 6 months (43,44), and therefore, oral administration must be approached with caution, also taking into account bioavailability aspects. In a much smaller study, chondroitin sulfates were administered to patients with osteoarthritis who also suffered from psoriasis, and the outcome showed reduced skin thickness, less skin inflammation, and flaking (45).

B. Dermatan Sulfate

Dermatan sulfate binds to a variety of proteoglycans, affects the function of growth factors, modifies the action of the heparin cofactors (46–48), and influences proliferation in a cell-type specific manner. Dermatan sulfate also occurs as a circulating GAG in the blood and binds to several proteins that are part of the coagulation cascade, including activated protein C, heparin cofactor II, fibrinogen, fibronectin, apolipoprotein B, kininogen, trypsin inhibitor A, and factor H. All these proteins are sensitive to serine protease, but the role of dermatan sulfate binding in the degradation is unclear (49). Binding of dermatan sulfate to activated protein C and heparin cofactor II accounts for its antithrombotic activity (50). Because heparin cofactor is highly expressed by several tumor types, dermatan sulfate has been employed as a carrier for anticancer drugs (51).

Dermatan sulfate comprises a part of the structure of the main proteoglycan of the skin decorin, which consists of a small core protein (40 kDa) with one dermatan sulfate chain attached. Decorin, and therefore dermatan sulfate, regulates the assembly of collagen within the ECM. The protein core of decorin binds to collagen fibers, while the length of the dermatan sulfate chain regulates the distance between the collagen fibers, and the length of the dermatan sulfate chain is assumed to reduce with age (52). Dermatan sulfate bridges the gap between collagen fibers stabilizing the ECM toward mechanical stress (53). Similarly, dermatan sulfate coordinates the alignment of opticin along collagen fibers resulting in stabilized vitreous gels (54).

Dermatan sulfate is necessary for thrombopoietin-induced megakaryocytopoiesis (55) and triggers the maturation of mesenchymal and neuronal stem cells (27,55–57). In this context, decorin and versican are rich in dermatan sulfate and neutralize phospholipase A2, which is a proinflammatory and profibrotic factor (58). Dermatan sulfate also regulates the differentiation of bone forming cells (59), which involves the blockade of NF-κB inhibitors (60) leading to the expression of endothelial and intercellular adhesion molecule-1 (61). Therefore, dermatan sulfate is the only chondroitin sulfate that modifies the action of transcription factors without being attached to a proteoglycan.

In contrast to other chondroitin sulfates, dermatan sulfate functions as a docking molecule for a range of human pathogenic microorganisms. Dermatan sulfate mediated the adhesion of *Penicilliunum marneffi* to the host's ECM (62), of pneumococci to nasopharyngeal epithelial cells (63), and of spirochetes to fibronectin (64). Dermatan sulfate also mediated infection by *Propionibacterium acnes* (65) and of spirochetes by inactivating α-defensin (66). A supportive role in viral infection has been reported for *Herpes* simplex infection acting as a docking device (67). However, a supportive role of dermatan sulfate in viral infection depends on the species or strain of the microorganism, as dermatan sulfate reduced adherence of *Chlamydia trachomatis* (68). A summary of the functions of dermatan sulfate is provided in Fig. 1.

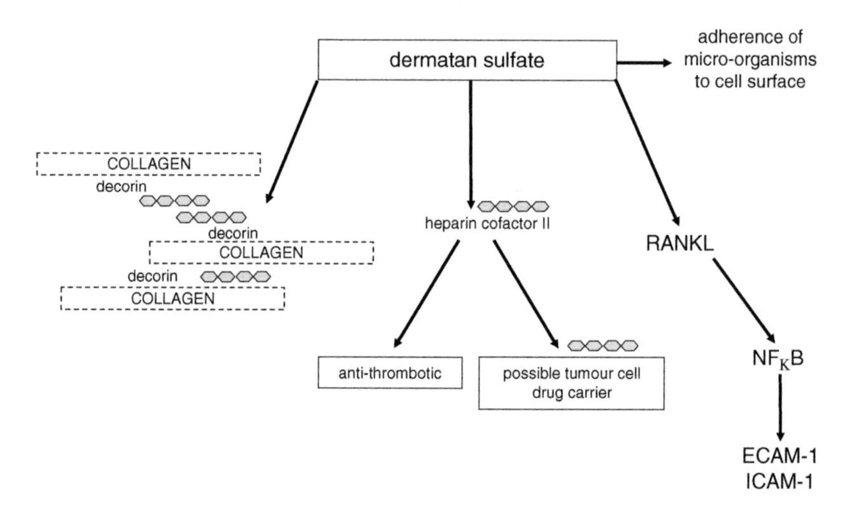

Figure 1 Biological functions of dermatan sulfate (chondroitin sulfate B) that have been described in human cells or tissues.

C. Heparin and Heparan Sulfate

Heparin and heparan sulfate are the best known GAGs as they have been used since decades to treat diseases that are associated with increased blood clotting (69–72). The molecular mass of heparin ranges from 6 to 50 kDa, but can reach up to 100 kDa, and those used therapeutically have a molecular weight of 12–15 kDa.

Importantly, endothelial cells produce a highly sulfated heparin that is capable of binding antithrombin III directly on the vascular wall, preventing adherence of platelets and clotting factors to the vessel wall (73). The composition of heparin can vary to a large extend and 32 variants of the disaccharide units can occur, depending on the location of the sulfate groups (70). The difference between heparin and heparan sulfate is defined as that heparin have more than 70% of *N*-sulfonation (74), while heparan sulfate has less than 30% *N*-sulfonation (75). One heparin unit forms a complex with antithrombin, which increases the binding capacity for thrombin and factor Xa by approximately 1000-fold (76,77). Thus, patients in long-term therapy with heparin must be monitored for thrombin and blood clotting in order to prevent uncontrolled bleeding and vessel leakage (78).

In regard to vessel integrity and tissue remodeling, as well as tumor development, the interaction of heparins with specific isoforms of the vascular endothelium-derived growth factor (VEGF) and its receptors is of interest. Heparin binds and activates VEGF-A145 and -A165, but not -A121; it binds to VEGF-B167, but not to -B186. It binds to VEGF-E, -F1, and F2 but not to VEGF-C or -D (79). The heparin's capacity to bind VEGF isoforms depends on the number of sulfate groups, and the localization of VEGF in the tissue is assumed to be directed by heparins (80), but this hypothesis has recently been challenged (81). The effect of heparins on the function of VEGF is highly controversial. In some studies, heparins enhanced the VEGF function, which is of specific interest as heparin is used in patients with neoplasia (82), while in other conditions heparins counteracted VEGF actions (83,84). Fibronectin needs to be first link to heparin that modifies its conformation and enables it to bind to VEGF (85). This controversy demonstrates the danger in extrapolating conclusions based on data obtained in transformed or genetically modified human cells as well as in animal models to a human disease, a problem that is often encountered in basic science and is unfortunately widely ignored.

Heparins initiate the binding of the fibroblast growth factor (FGF) to its receptor (86,87). Moreover, the tricomposite becomes internalized in a cell and is transported into the nucleus, suggesting that heparins may influence FGF-induced intracellular signaling (88). Heparins affect interleukin-11 signaling by upregulating the Erk1/2 mitogen-activated protein kinase activity (89), while activation of the heparin-binding epidermal growth factor (HB-EGF)-like growth factor inhibits cytokine-induced activation of NF-κB (90) and regulates the endocytosis of superoxide dismutase by endothelial cells (91).

Proteoglycans that bind heparin are syndecan-1 to -4 with multiple binding sites (92). During inflammation, syndecans protect cathepsin G and neutrophil elastase from degradation, thereby controlling protease and antiprotease activity during remodeling and wound repair (93). Heparin also binds to agrin, betaglycan, glypican-1 to -6, and perlecan (94,95). It can bind directly to collagen XVIII and to the receptor for hyaluronic acid, CD44 (94,95). The latter effect enables FGF, VEGF, and HB-EGF to signal through CD44, in a possibly cell-type specific pattern (Fig. 2).

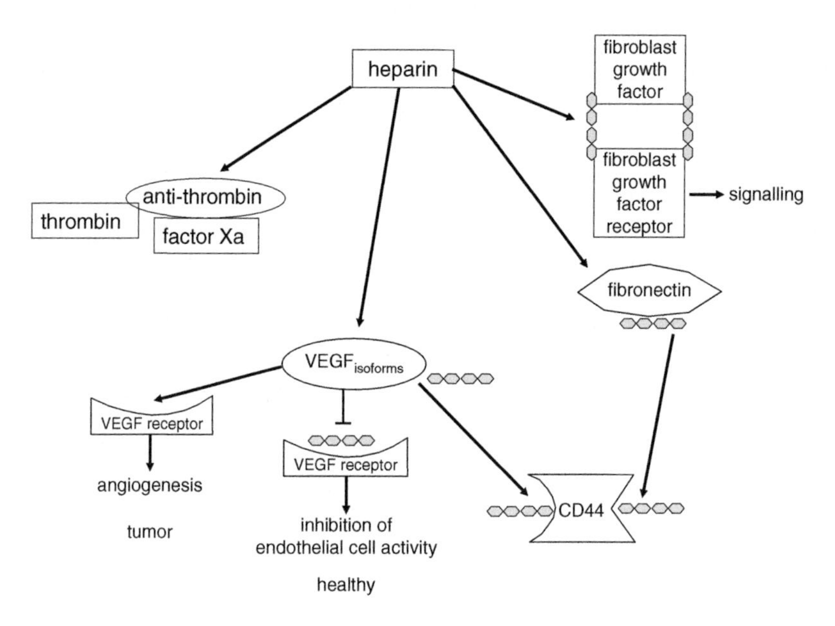

Figure 2 The various biological effects that have been described for heparins and heparan sulfates in human cells or tissues.

D. Hyaluronic Acid

The function of hyaluronic acid was initially confined to the maintenance and stability of the ECM (96). However, the action of hyaluronic acid varies with its size, which determines its function in a cell-type specific manner (97–101). Hyaluronic acid represents more than 50% of the ECM in the skin. High molecular weight hyaluronic acid (>1,000 kDa) controls tissue water content, ECM lubrication, structural integrity, free oxygen radicals, and distribution of plasma proteins (96,100,101). The synthesis of hyaluronic acid is achieved by hyaluronan synthase-1 to -3 (102,103). The stability of hyaluronic acid varies with its microenvironment, as its half-life is less than 10 min in blood, up to 12 h in the skin, and extends to months in the vitreous gel of the eye (100,101). Hyaluronic acid is the only GAG with a function of its breakdown molecules, as small hyaluronic acid molecules and fragments stimulated the maturation of dentritic cells and the synthesis of proinflammatory IL-1β, IL-12, and TNF-α (103–105). The latter effect seems to be restricted to an interaction of hyaluronic acid fragments with the Toll-like receptor 4 (104,105). The observation that bacterial spreading in the

ECM is facilitated by small hyaluronic acid molecules implies a function in host defense, but has not been proven (106). Increased levels of hyaluronic acid were associated with colon inflammation, psoriasis, osteoarthritis, rheumatoid arthritis, and scleroderma (101,107,108), as well as with viral infections (109). Hyaluronic acid is produced by endothelial cells and binds to its receptor, CD44, expressed by activated T and B cells, inducing the attachment of the two cell types (110). This interaction is controlled by the expression of specific hyaluronic acids subtypes or by modifications of CD44 on platelets (111). High molecular weight hyaluronic acid and CD44 are essential for scarless embryonic wound repair (112).

Hyaluronic acid also binds to the receptor for hyaluronic acid-mediated motility (RHAMM) (113), which controls the GAGs effect on cell migration, proliferation, and motility (114). The role of CD44 or RHAMM in activating T and B cells remains unclear. In the context of inflammation, TSG-6 is yet another hyaluronic acid-binding protein, which is synthesized by various cell types during inflammation and by growth factors (115). Local injection of high molecular weight hyaluronic acid restores the physical properties of the synovial fluid and is employed to treat osteoarthritis (116). The effects of hyaluronic acid are show in Fig. 3.

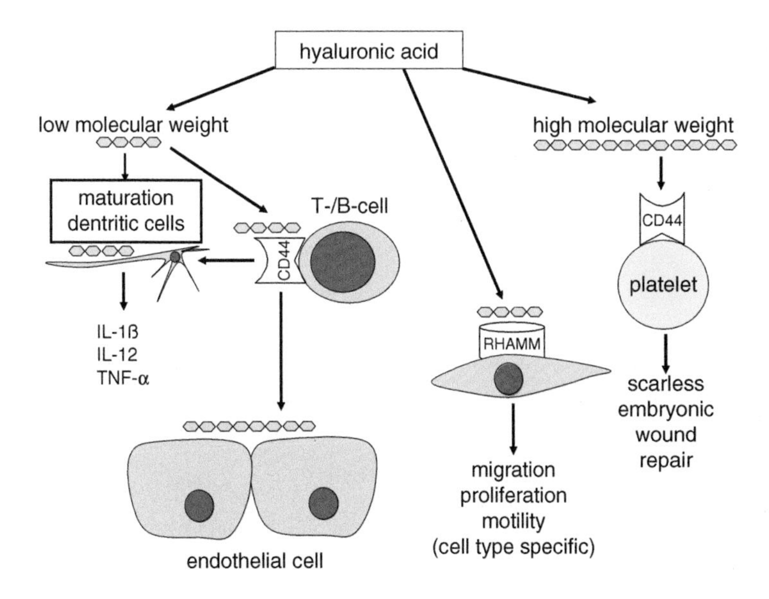

Figure 3 The biological functions of low and high molecular weight forms of hyaluronic acid described in human cells and tissues.

E. Keratan Sulfate

Keratan sulfate is highly expressed in the cornea, bones, brain, and the ECM. Keratan sulfate chains are formed by the action of glycosyl transferases that link the two saccharids in an alternate pattern (117). Sulfation of keratan sulfate occurs while a chain is elongated in a cell-type or organ-specific pattern (118–121). Keratan sulfates isolated from bone-derived ECM have the highest content of sulfate groups, and sulfation regulates the organ or cell-type specific function of this GAGs (120,121). The binding of keratan sulfate to a protein core is tissue-specific, as it is linked with the N-group to an asparagine residue in the cornea (122) and by the O-group to a serine in the ECM (123). In the ECM, keratan sulfate is linked to aggrecan and increases in length and sulfation with age (122,123). During reproduction the content of keratan sulfate in the endometrium increases at the time point of egg implantation (124).

In the cornea, keratan sulfate is linked to the leucine-rich repeats of the small leucine-rich proteoglycans keratocan, lumican, and mimecan (125,126). In the ECM, keratan sulfate is linked to aggrecans by the O-group in vertebrates (127). However, species-specific variations in this sequence exist and the degree of sulfation varies with the binding site (128). MUC1, a product of the endometrial epithelial cells, changes its function when linked to keratan sulfate (129). The function of keratan sulfate chains bound to CD44 (130) and to keratin is unknown (131). In neural tissues, keratan sulfate is bound to aggrecan (132) and was found in specific proteoglycans of the nervous system including ABAKAN, claustrin, PG-1000, phosphocan, and SV2 (133–135).

Animal models suggest that keratan sulfate together with dermatan sulfate regulates the water content in the cornea (136). In mice, macrophages do not adhere to any surface that is coated with keratan sulfate-enriched luminican (125). High content of keratan sulfate also inhibits growth of neuronal cells (137), and when found with aggrecan reduced the severity of osteoarthritis (138). In conclusion, keratan sulfate might be a key factor for treatment of age-related degenerative diseases (Fig. 4).

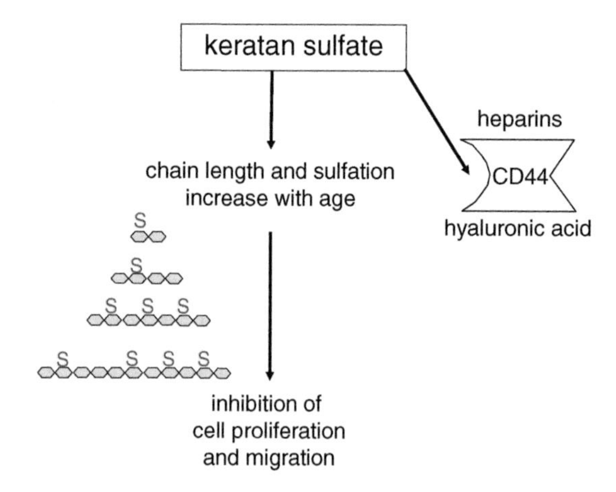

Figure 4 Possible biological functions of keratan sulfate and its suggested role for cell maturation and tissue degeneration.

III. Concluding Remarks

Even though heparin was discovered in 1916 at John Hopkins University and a purified product (Connaught Laboratories in Canada) was first used in humans as an anticoagulant in 1935 (139), and despite the fact that the structure of hyaluronic acid was reported back in 1950 (140), it was not until the 1970s that the molecular structure of most GAGs was unraveled and not before the 1980s that the involvement of GAGs in tissue structure (high viscosity, low compressibility, rigidity) and in many biological functions such as cell recognition, adhesion, migration, proliferation, organogenesis, control of reproduction, differentiation, growth, protein folding, metabolism, and transport was recognized. Advances in glycobiology demonstrated that GAGs, mostly as cell surface-bound polysaccharides, appear to be involved in nearly all levels of cell biology and pathogenesis. It was not surprising, therefore, that by mid 1990s, research in carbohydrates was considered as one of the hottest topics (141). However, the available evidence has not provided clear cut associations for most GAGs, between specific structure and biological function, in accordance to the classical physiological and pharmacological principle of "structure–action" interrelation. Deriving clear cut conclusions is also hampered by the fact the function of GAGs may also vary on their size or the level of sulfation. Furthermore, the characterization of GAGs is difficult and tricky because they are constantly modified, not only *in vivo* but also after their isolation. Glycobiology, in our opinion, has not yet lived up to its expectations and is far from achieving its full potential in unraveling the biological function of these extremely versatile molecules. On contrary, we believe that GAGs have the potential to provide alternative targets or molecules for drug development.

References

1. Dunsmore SE, Rannels DE. Extracellular matrix biology in the lung. Am J Physiol 1996; 270:L3–L27.
2. McAnulty RJ, Laurent GJ. Pathogenesis of lung fibrosis and potential new therapeutic strategies. Exp Nephrol 1995; 3:96–107.
3. Schmid KL, Grundboek-Jusko Kimura JA, Tschopp FA, Zollinger R, Binette JP, Lewis W, Hayashi S. The distribution of the glycosaminoglycans in the anatomic components of the lung and the changes in the concentration of these macromolecules during development and ageing. Biochim Biophys Acta 1982; 716:178–187.
4. Cantor JO, Bray BA, Ryan SF, Mandl I, Turino GM. Glycosaminoglycan and collagen synthesis in N-nitroso-N-methylurethane induced pulmonary fibrosis. Proc Soc Exp Biol Med 1980; 164:1–8.
5. Cantor JO, Osman O, Cerreta JM, Mandl I, Turino GM. Glycosaminoglycan synthesis in explants derived from bleomycin-treated fibrotic hamster lungs. Proc Soc Exp Biol Med 1983; 173:362–366.
6. Kumar N, Bentolila A, Domb AJ. Structure and biological activity of heparinoid. Mini Rev Med Chem 2005; 5:441–447.

7. Rabenstein DL. Heparin and heparan sulphate: structure and function. Nat Prod Rep 2002; 19:312–331.
8. Blewis ME, Nugent-Derfus GE, Schmidt TA, Schumacher BL, Sah RL. A model of synovial fluid lubricant composition in normal and injured joints. Eur Cell Mater 2007; 13:26–39.
9. Wang F, Garza LA, Kang S, Varani J, Orringer JS, Fisher GJ, Voorhees JJ. In vivo stimulation of de novo collagen production caused by cross-linked hyaluronic acid dermal filler injections in photodamaged human skin. Arch Dermatol 2007; 143:155–163.
10. Ito T, Fraser IP, Yeo Y, Highley CB, Bellas E, Kohane DS. Anti-inflammatory function of an in situ cross-linkable conjugate hydrogel of hyaluronic acid and dexamethasone. Biomaterials 2007; 28:1778–1786.
11. Vitanzo PC Jr, Sennett BJ. Hyaluronans: is clinical effectiveness dependent on molecular weight? Am J Orthop 2006; 35:421–428.
12. Kato Y, Nakamura S, Nishimura M. Beneficial actions of hyaluronan (HA) on arthritic joints: effects of molecular weight of HA on elasticity of cartilage matrix. Biorheology 2006; 43:347–354.
13. Joddar B, Ramamurthi A. Fragment size- and dose-specific effects of hyaluronan on matrix synthesis by vascular smooth muscle cells. Biomaterials 2006; 27:2994–3004.
14. Berninsone PM, Hirschberg CB. Nucleotide sugar transporters of the Golgi apparatus. Curr Opin Struct Biol 2000; 10:542–547.
15. Sugahara K, Kitagawa H. Recent advances in the study of the biosynthesis and functions of d-glycosaminoglycans. Curr Opin Struct Biol 2000; 10:518–527.
16. Itano N, Kimata K. Mammalian hyaluronan synthases. IUBMB Life 2002; 54:195–199.
17. DeAngelis PL. Hyaluronan synthases: fascinating glycosyltransferases from vertebrates, bacterial pathogens, and algal viruses. Cell Mol Life Sci 1999; 56:670–682.
18. Kusche-Gullberg M, Kjellen L. Sulfotransferases in glycosaminoglycan biosynthesis. Curr Opin Struct Biol 2003; 13:605–611.
19. Silbert JE, Sugumaran G. Biosynthesis of chondroitin/dermatan sulfate. IUBMB Life 2002; 54:177–186.
20. Funderburgh JL. Keratan sulfate biosynthesis. IUBMB Life 2002; 54:187–194.
21. Sugahara K, Yamada S, Yoshida K, de Waard P, Vliegenthart JF. A novel sulfated structure in the carbohydrate-protein linkage region isolated from porcine intestinal heparin. J Biol Chem 1992; 267:1528–1533.
22. Prydz K, Dalen KT. Synthesis and sorting of proteoglycans. J Cell Sci 2000; 113:193–205.
23. Gama CI, Hsieh-Wilson LC. Chemical approaches to deciphering the glycosaminoglycan code. Curr Opin Chem Biol 2005; 9:609–619.
24. Sasisekharan R, Raman R, Prabhakar V. Glycomics approach to structure–function relationships of glycosaminoglycans. Annu Rev Biomed Eng 2006; 8:181–231.
25. Fujita S, Iizuka T, Dauber W. Localization of keratan sulphate and chondroitin-6-sulphate on the anteriorly, displaced human temporomandibular joint disc–histological and, immunohistochemical analysis. J Oral Rehabil 2001; 28:962–970.

26. Bayliss MT, Davidson C, Woodhouse SM, Osborne DJ. Chondroitin sulphation in human joint tissues varies with age, zone and topography. Acta Orthop Scand Suppl 1995; 266:22–25.
27. Bandtlow CE, Zimmermann DR. Proteoglycans in the developing brain: new conceptual insights for old proteins. Physiol Rev 2000; 80:1267–1290.
28. Rhodes KE, Fawcett JW. Chondroitin sulphate proteoglycans: preventing plasticity or protecting the CNS? J Anat 2004; 204:33–48.
29. Kawashima H, Hirose M, Hirose J, Nagakubo D, Plaas AH, Miyasaka M. Binding of a large chondroitin sulphate/dermatan sulphate proteoglycan, versican, to L-selectin, P-selectin, and CD44. J Biol Chem 2000; 275:35448–35456.
30. Iida J, Wilhelmson KL, Ng J, Lee P, Morrison C, Tam E, Overall CM, McCarthy JB. Cell surface chondroitin sulfate glycosaminoglycan in melanoma: role in the activation of pro-MMP-2 (pro-gelatinase A). Biochem J 2007; 403:553–563.
31. Rodriguez E, Roland SK, Plaas A, Roughley PJ. The glycosaminoglycan attachment regions of human aggrecan. J Biol Chem 2006; 28127: 18444–18450.
32. Miwa HE, Gerken TA, Huynh TD, Flory DM, Hering TM. Mammalian expression of full-length bovine aggrecan and link protein: formation of recombinant proteoglycan aggregates and analysis of proteolytic cleavage by ADAMTS-4 and MMP-13. Biochim Biophys Acta 2006; 1760:472–486.
33. Somerville RP, Longpre JM, Apel ED, Lewis RM, Wang LW, Sanes JR, Leduc R, Apte SS. ADAMTS7B, the full-length product of the ADAMTS7 gene, is a chondroitin sulphate proteoglycan containing a mucin domain. J Biol Chem 2004; 279:35159–35175.
34. Schwartz NB, Domowicz M. Chondrodysplasias due to proteoglycan defects. Glycobiology 2002; 124:57R–68R.
35. Melrose J, Roughley P, Knox S, Smith S, Lord M, Whitelock J. The structure, location, and function of perlecan, a prominent pericellular proteoglycan of fetal, postnatal, and mature hyaline cartilages. J Biol Chem 2006; 281:36905–36914.
36. Monzavi-Karbassi B, Stanley JS, Hennings L, Jousheghany F, Artaud C, Shaaf S, Kieber-Emmons T. Chondroitin sulfate glycosaminoglycans as major P-selectin ligands on metastatic breast cancer cell lines. Int J Cancer 2007; 120:1179–1191.
37. Kiewe P, Bechrakis NE, Schmittel A, Ruf P, Lindhofer H, Thiel E, Nagorsen D. Increased chondroitin sulphate proteoglycan expression (B5 immunoreactivity) in metastases of uveal melanoma. Ann Oncol 2006; 17:1830–1834.
38. Pothacharoen P, Siriaunkgul S, Ong-Chai S, Supabandhu J, Kumja P, Wanaphirak C, Sugahara K, Hardingham T, Kongtawelert P. Raised serum chondroitin sulfate epitope level in ovarian epithelial cancer. J Biochem 2006; 140:517–524.
39. Fthenou E, Zafiropoulos A, Tsatsakis A, Stathopoulos A, Karamanos NK, Tzanakakis GN. Chondroitin sulfate A chains enhance platelet derived growth factor-mediated signalling in fibrosarcoma cells. Int J Biochem Cell Biol 2006; 38:2141–2150.
40. Wu XZ. Sulfated oligosaccharides and tumor: promoter or inhibitor? Panminerva Med 2006; 48:27–31.

41. Derrett-Smith E, Beynon HL. Supplements and injections for joint disease. Br J Hosp Med (Lond) 2006; 67:290–293.
42. Volpi N. Therapeutic applications of glycosaminoglycans. Curr Med Chem 2006; 13:1799–1810.
43. Das A Jr, Hammad TA. Efficacy of a combination of FCHG49 glucosamine hydrochloride, TRH122 low molecular weight sodium chondroitin sulfate and manganese ascorbate in the management of knee osteoarthritis. Osteoarthritis Cartilage 2000; 85:343–350.
44. Uebelhart D, Malaise M, Marcolongo R, DeVathaire F, Piperno M, Mailleux E, Fioravanti A, Matoso L, Vignon E. Intermittent treatment of knee osteoarthritis with oral chondroitin sulfate: a one-year, randomized, double-blind, multicenter study versus placebo. Osteoarthr Cartil 2004; 124:269–276.
45. Verges J, Montell E, Herrero M, Perna C, Cuevas J, Perez M, Moller I. Clinical and histopathological improvement of psoriasis with oral chondroitin sulfate: a serendipitous finding. Dermatol Online J 2005; 11:31.
46. Kinsella MG, Bressler SL, Wight TN. The regulated synthesis of versican, decorin, and biglycan: extracellular matrix proteoglycans that influence cellular phenotype. Crit Rev Eukaryot Gene Expr 2004; 143:203–234.
47. Sugahara K, Mikami T, Uyama T, Mizuguchi S, Nomura K, Kitagawa H. Recent advances in the structural biology of chondroitin sulfate and dermatan sulfate. Curr Opin Struct Biol 2003; 13:612–620.
48. Villena J, Brandan E. Dermatan sulfate exerts an enhanced growth factor response on skeletal muscle satellite cell proliferation and migration. J Cell Physiol 2004; 198:169–178.
49. Saito A, Munakata H. Analysis of plasma proteins that bind to glycosaminoglycans. Biochim Biophys Acta 2007; 1770:241–624.
50. Du HY, Ji SL, Song HF, Ye QN, Cao JC. The relationship between the structure of dermatan sulfate derivatives and their antithrombotic activities. Thromb Res 2007; 119:377–384.
51. Ranney D, Antich P, Dadey E, Mason R, Kulkarni P, Singh O, Chen H, Constantanescu A, Parkey R. Dermatan carriers for neovascular transport targeting, deep tumor penetration and improved therapy. J Control Release 2005; 109:222–235.
52. Nomura Y. Structural change in decorin with skin aging. Connect Tissue Res 2006; 47:249–255.
53. Henninger HB, Maas SA, Underwood CJ, Whitaker RT, Weiss JA. Spatial distribution and orientation of dermatan sulfate in human medial collateral ligament. J Struct Biol 2007; 158:33–45.
54. Hindson VJ, Gallagher JT, Halfter W, Bishop PN. Opticin binds to heparan and chondroitin sulfate proteoglycans. Invest Ophthalmol Vis Sci 2005; 46:4417–4423.
55. Kashiwakura I, Teramachi T, Kakizaki I, Takagi Y, Takahashi TA, Takagaki K. The effects of glycosaminoglycans on thrombopoietin-induced megakaryocytopoiesis. Haematologica 2006; 91:445–451.
56. Prante C, Bieback K, Funke C, Schon S, Kern S, Kuhn J, Gastens M, Kleesiek K, Gotting C. The formation of extracellular matrix during chondrogenic differentiation of mesenchymal stem cells correlates with increased levels of xylosyltransferase I. Stem Cells 2006; 24:2252–2261.

57. Ida M, Shuo T, Hirano K, Tokita Y, Nakanishi K, Matsui F, Aono S, Fujita H, Fujiwara Y, Kaji T, Oohira A. Identification and functions of chondroitin sulfate in the milieu of neural stem cells. J Biol Chem 2006; 281:5982–5991.
58. Fuentes L, Hernandez M, Nieto ML, Sanchez Crespo M. Biological effects of group IIA secreted phosholipase A(2). FEBS Lett 2002; 531:7–11.
59. Waddington RJ, Roberts HC, Sugars RV, Schonherr E. Differential roles for small leucine-rich proteoglycans in bone formation. Eur Cell Mater 2003; 6:12–21.
60. Huntington JA. Mechanisms of glycosaminoglycan activation of the serpins in hemostasis. J Thromb Haemost 2003; 1:1535–1549.
61. Theoleyre S, Kwan Tat S, Vusio P, Blanchard F, Gallagher J, Ricard-Blum S, Fortun Y, Padrines M, Redini F, Heymann D. Characterization of osteoprotegerin binding to glycosaminoglycans by surface plasmon resonance: role in the interactions with receptor activator of nuclear factor kappaB ligand (RANKL) and RANK. Biochem Biophys Res Commun 2006; 347:460–467.
62. Penc SF, Pomahac B, Eriksson E, Detmar M, Gallo RL. Dermatan sulfate activates nuclear factor-kappab and induces endothelial and circulating intercellular adhesion molecule-1. J Clin Invest 1999; 103:1329–1335.
63. Srinoulprasert Y, Kongtawelert P, Chaiyaroj SC. Chondroitin sulfate B and heparin mediate adhesion of Penicillium marneffei conidia to host extracellular matrices. Microb Pathog 2006; 40:126–132.
64. Tonnaer EL, Hafmans TG, Van Kuppevelt TH, Sanders EA, Verweij PE, Curfs JH. Involvement of glycosaminoglycans in the attachment of pneumococci to nasopharyngeal epithelial cells. Microbes Infect 2006; 8: 316–322.
65. Lodes MJ, Secrist H, Benson DR, Jen S, Shanebeck KD, Guderian J, Maisonneuve JF, Bhatia A, Persing D, Patrick S, Skeiky YA. Variable expression of immunoreactive surface proteins of Propionibacterium acnes. Microbiology 2006; 152:3667–3681.
66. Fischer JR, LeBlanc KT, Leong JM. Fibronectin binding protein BBK32 of the Lyme disease spirochete promotes bacterial attachment to glycosaminoglycans. Infect Immun 2006; 74:435–441.
67. Schmidtchen A, Frick IM, Bjorck L. Dermatan sulphate is released by proteinases of common pathogenic bacteria and inactivates antibacterial alpha-defensin. Mol Microbiol 2001; 39:708–713.
68. Zaretzky FR, Pearce-Pratt R, Phillips DM. Sulfated polyanions block Chlamydia trachomatis infection of cervix-derived human epithelia. Infect Immun 1995; 63:3520–3526.
69. Banfield BW, Leduc Y, Esford L, Visalli RJ, Brandt CR, Tufaro F. Evidence for an interaction of herpes simplex virus with chondroitin sulphate proteoglycans during infection. Virology 1995; 208:531–539.
70. Riley JF, Shepherd DM, Stroud SW. The function of heparin. Nature 1955; 176:1123.
71. Nugent MA. Heparin sequencing brings structure to the function of complex oligosaccharides. Proc Natl Acad Sci USA 2000; 97:10301–10303.
72. Iozzo RV. Series introduction. Heparan sulfate proteoglycans: intricate molecules with intriguing functions. J Clin Invest 2001; 108:165–167.

73. Merry CLR, Lyon M, Deakin JA, Hopwood JJ, Gallagher JT. Highly sensitive sequencing of the sulfated domains of Heparan sulfate. J Biol Chem 1999; 274:18455–18462.
74. Roden L, Ananth S, Campbell P, Curenton T, Ekborg G, Manzella S, Pillion D, Meezan E. Heparin–an introduction. Adv Exp Med Biol 1992; 313:1–20.
75. Fransson LA, Carlstedt I, Coster L, Malmstrom A. The functions of the heparan sulphate proteoglycans. Ciba Found Symp 1986; 124:125–142.
76. Wallace A, Rovelli G, Hofsteenge J, Stone SR. Effect of heparin on the interaction between thrombin and hirudin. Eur J Biochem 1987; 169: 373–376.
77. Teitel JM, Rosenberg RD. Protection of factor Xa from neutralization by the heparin-antithrombin complex. J Clin Invest 1983; 71:1383–1391.
78. Leonardi MJ, McGory ML, Ko CY. The rate of bleeding complications after pharmacologic deep venous thrombosis prophylaxis: a systematic review of 33 randomized controlled trials. Arch Surg 2006; 141:790–797.
79. Soker S, Goldstaub D, Svahn CM, Vlodavsky I, Levi BZ, Neufeld G. Variations in the size and sulfation of heparin modulate the effect of heparin on the binding of VEGF165 to its receptors. Biochem Biophys Res Commun 1994; 203:1339–1347.
80. Springer ML, Banfi A, Ye J, von Degenfeld G, Kraft PE, Saini SA, Kapasi NK, Blau HM. Localization of vascular response to VEGF is not dependent on heparin binding. FASEB J 2007; 21:2074–2085.
81. Jakobsson L, Kreuger J, Holmborn K, Lundin L, Eriksson I, Kjellen L, Claesson-Welsh L. Heparan sulfate in trans potentiates VEGFR-mediated angiogenesis. Dev Cell 2006; 10:625–634.
82. Pyda M, Korybalska K, Ksiazek K, Grajek S, Lanocha M, Lesiak M, Wisniewska-Elnur J, Olasinska A, Breborowicz A, Cieslinski A, Witowski J. Effect of heparin on blood vascular endothelial growth factor levels in patients with ST-elevation acute myocardial infarction undergoing primary percutaneous coronary intervention. Am J Cardiol 2006; 98:902–905.
83. Moy AB, Blackwell K, Wu MH, Granger HJ. Growth factor- and heparin-dependent regulation of constitutive and agonist-mediated human endothelial barrier function. Am J Physiol Heart Circ Physiol 2006; 291: H2126–H2135.
84. Wijelath ES, Rahman S, Namekata M, Murray J, Nishimura T, Mostafavi-Pour Z, Patel Y, Suda Y, Humphries MJ, Sobel M. Heparin-II domain of fibronectin is a vascular endothelial growth factor-binding domain: enhancement of VEGF biological activity by a singular growth factor/matrix protein synergism. Circ Res 2006; 99:853–860.
85. Mitsi M, Hong Z, Costello CE. Nugent MA, Heparin-mediated conformational changes in fibronectin expose vascular endothelial growth factor binding sites. Biochemistry 2006; 45:10319–10328.
86. Harmer NJ, Insights into the role of heparan sulphate in fibroblast growth factor signalling. Biochem Soc Trans 2006; 34:442–445.
87. Jackson RA, Nurcombe V, Cool SM. Coordinated fibroblast growth factor and heparan sulfate regulation of osteogenesis. Gene 2006; 379:79–91.
88. Wiedlocha A, Sorensen V. Signaling, internalization, and intracellular activity of fibroblast growth factor. Curr Top Microbiol Immunol 2004; 286:45–79.

89. Rajgopal R, Butcher M, Weitz JI, Shaughnessy SG. Heparin synergistically enhances interleukin-11 signaling through up-regulation of the MAPK pathway. J Biol Chem 2006; 281:20780–20787.
90. Mehta VB, Besner GE. Heparin-binding epidermal growth factor-like growth factor inhibits cytokine-induced NF-kappa B activation and nitric oxide production via activation of the phosphatidylinositol 3-kinase pathway. J Immunol 2005; 175(3):1911–1918.
91. Chu Y, Piper R, Richardson S, Watanabe Y, Patel P, Heistad DD. Endocytosis of extracellular superoxide dismutase into endothelial cells: role of the heparin-binding domain. Arterioscler Thromb Vasc Biol 2006; 26:1985–1990.
92. Tkachenko E, Rhodes JM, Simons M. Syndecans: new kids on the signaling block. Circ Res 2005; 96:488–500.
93. Kainulainen V, Wang H, Schick C, Bernfield M. Syndecans, heparan sulfate proteoglycans, maintain the proteolytic balance of acute wound fluids. J Biol Chem 1998; 273:11563–11569.
94. Rabenstein DL. Heparin and heparan sulfate: structure and function. Nat Pro Rep 2002; 19:312–331.
95. Iozzo RV, San Antonio JD. Heparan sulfate proteoglycans: heavy hitters in the angiogenesis arena. J Clin Invest 2001; 108:349–355.
96. Papakonstantinou E, Karakiulakis G, Roth M, Block LH. Platelet-derived growth factor stimulates the secretion of hyaluronic acid by proliferating human vascular smooth muscle cells. Proc Natl Acad Sci USA 1995; 92:9881–9885.
97. Papakonstantinou E, Roth M, Eickelberg O, Mirtsou-Fidani V, Mora M, Karakiulakis G. Hyaluronic acid isolated and purified from human atherosclerotic aortas is involved in the progression of atheromatosis. Biochem Biophyss Newslet 1997; 43:52–54.
98. Papakonstantinou E, Karakiulakis G, Eickelberg O, Perruchoud AP, Block LH, Roth M. A 340-kDa hyaluronic acid secreted by human vascular smooth muscle cells regulates their proliferation and migration. Glycobiol 1998; 8:821–830.
99. Papakonstantinou E, Roth M, Block LH, Mirtsou-Fidani V, Argiridis P, Karakiulakis G. The differential expression of hyaluronic acid in the layers of human atheromatic aortas is associated with vascular smooth muscle cell proliferation and invasion. Atheroscl 1998; 138:79–89.
100. Laurent TC, Fraser JR. Hyaluronan. FASEB J 1992; 6:2397–2404.
101. Mohamadzadeh M, DeGrendele H, Arizpe H, Estess P, Siegelman M. Proinflammatory stimuli regulate endothelial hyaluronan expression and CD44/HA-dependent primary adhesion. J Clin Invest 1998; 101:97–108.
102. Adamia S, Maxwell CA, Pilarski LM. Hyaluronan and hyaluronan synthases: potential therapeutic targets in cancer. Curr Drug Targets Cardiovasc Haematol Disord 2005; 5:3–14.
103. Termeer CC, Hennies J, Voith U, Ahrens T, Weiss JM, Prehm P, Simon JC. Oligosaccharides of hyaluronan are potent activators of dendritic cells. J Immunol 2000; 165:1863–1870.
104. Termeer C, Benedix F, Sleeman J, Fieber C, Voith U, Ahrens T, Miyake K, Freudenberg M, Galanos C, Simon JC. Oligosaccharides of hyaluronan activate dendritic cells via toll-like receptor 4. J Exp Med 2002; 195:99–111.

105. Taylor KR, Trowbridge JM, Rudisill JA, Termeer CC, Simon JC, Gallo RL. Hyaluronan fragments stimulate endothelial recognition of injury through TLR4. J Biol Chem 2004; 279:17079–17084.
106. Itano N, Kimata K. Mammalian hyaluronan synthases. IUBMB Life 2002 Hynes WL, Walton SL. Hyaluronidases of Gram-positive bacteria. FEMS Microbiol Lett 2000; 183:201–207.
107. de la Motte CA, Hascall VC, Drazba J, Bandyopadhyay SK. Strong SA, mononuclear leukocytes bind to specific hyaluronan structures on colon mucosal smooth muscle cells treated with polyinosinic acid:polycytidylic acid:inter-alpha-trypsin inhibitor is crucial to structure and function. Am J Pathol 2003; 163:121–133.
108. Majors AK, Austin RC, de la Motte CA, Pyeritz RE, Hascall VC, Kessler SP, Sen G, Strong SA. Endoplasmic reticulum stress induces hyaluronan deposition and leukocyte adhesion. J Biol Chem 2003; 278:47223–47231.
109. Siegelman MH, DeGrendele HC, Estess P. Activation and interaction of CD44 and hyaluronan in immunological systems. J Leukoc Biol 1999; 66:315–321.
110. Johnson P, Maiti A, Brown KL, Li R. A role for the cell adhesion molecule CD44 and sulfation in leukocyte-endothelial cell adhesion during an inflammatory response? Biochem Pharmacol 2000; 59:455–465.
111. Koshiishi I, Shizari M, Underhill CB. CD44 can mediate the adhesion of platelets to hyaluronan. Blood 1994; 84:390–396.
112. Turley EA, Noble PW, Bourguignon LY. Signaling properties of hyaluronan receptors. J Biol Chem 2002; 277:4589–4592.
113. Toole BP. Hyaluronan: from extracellular glue to pericellular cue. Nat Rev Cancer 2004; 4:528–539.
114. Milner CM, Day AJ. TSG-6: a multifunctional protein associated with inflammation. J Cell Sci 2003; 116:1863–1873.
115. Lesley J, Gal I, Mahoney DJ, Cordell MR, Rugg MS, Hyman R, Day A, Mikecz K. TSG-6 modulates the interaction between hyaluronan and cell surface CD44. J Biol Chem 2004; 279:25745–25754.
116. Gossec L, Dougados M. Intra-articular treatments in osteoarthritis: from the symptomatic to the structure modifying. Ann Rheum Dis 2004; 63: 478–482.
117. Christner JE, Distler JJ, Jourdian GW. Biosynthesis of keratan sulfate: purification and properties of a galactosyltransferase from bovine cornea. Arch Biochem Biophys 1979; 192:548–558.
118. Degroote S, Lo-Guidice JM, Strecker G, Ducourouble MP, Roussel P, Lamblin G. Characterization of an N-acetylglucosamine-6-O-sulfotransferase from human respiratory mucosa active on mucin carbohydrate chains. J Biol Chem 19972; 72:29493–29501.
119. Uchimura K, Muramatsu H, Kadomatsu K, Fan QW, Kurosawa N, Mitsuoka C, Kannagi R, Habuchi O, Muramatsu T. Molecular cloning and characterization of an N-acetylglucosamine-6-O-sulfotransferase. J Biol Chem 1998; 273:22577–22583.
120. Krusius T, Finne J, Margolis RK, Margolis RU. Identification of an O-glycosidic mannose-linked sialylated tetrasaccharide and keratan sulfate oligosaccharides in the chondroitin sulfate proteoglycan of brain. J Biol Chem 1986; 261:8237–8242.

121. Nieduszynski IA, Huckerby TN, Dickenson JM, Brown GM, Tai GH, Morris HG, Eady S. There are two major types of skeletal keratan sulphates. Biochem J 1990; 271:243–245.
122. Praus R, Brettschneider I. Glycosaminoglycans in embryonic and postnatal human cornea. Ophthal Res 1975; 7:452–458.
123. Brown GM, Huckerby TN, Bayliss MT, Nieduszynski IA. Human aggrecan keratan sulfate undergoes structural changes during adolescent development. J Biol Chem 1998; 273:26408–26414.
124. Graham RA, Li TC, Cooke ID, Aplin JD. Keratan sulphate as a secretory product of human endometrium: cyclic expression in normal women. Hum Reprod 1994; 9:926–9230.
125. Funderburgh JL, Mitschler RR, Funderburgh ML, Roth MR, Chapes SK, Conrad GW. Macrophage receptors for lumican. A corneal keratan sulfate proteoglycan. Invest Ophthalmol Vis Sci 1997; 38:1159–1167.
126. Corpuz LM, Funderburgh JL, Funderburgh ML, Bottomley GW, Prakash S, Conrad GW. Molecular cloning and tissue distribution of keratocan. Bovine corneal keratan sulfate proteoglycan 37A. J Biol Chem 1996; 271:9759–9763.
127. Barry FP, Rosenberg LC, Gaw JU, Koob TJ, Neame PJ. N- and O-linked keratan sulfate on the hyaluronan binding region of aggrecan from mature and immature bovine cartilage. J Biol Chem 1995; 270:20516–20524.
128. Flannery CR, Little CB, Caterson B. Molecular cloning and sequence analysis of the aggrecan interglobular domain from porcine, equine, bovine and ovine cartilage: comparison of proteinase-susceptible regions and sites of keratan sulfate substitution. Matrix Biol 1998; 16:507–511.
129. Aplin JD, Hey NA, Graham RA. Human endometrial MUC1 carries keratan sulfate: characteristic glycoforms in the luminal epithelium at receptivity. Glycobiology 1998; 8:269–276.
130. Tuhkanen AL, Tammi M, Tammi R. CD44 substituted with heparan sulfate and endo-beta- galactosidase-sensitive oligosaccharides: a major proteoglycan in adult human epidermis. J Invest Dermatol 1997; 109:213–218.
131. Schafer IA, Sorrell JM. Human keratinocytes contain keratin filaments that are glycosylated with keratan sulfate. Exp Cell Res 1993; 207:213–219.
132. Domowicz M, Li H, Hennig A, Henry J, Vertel BM, Schwartz NB. The biochemically and immunologically distinct CSPG of notochord is a product of the aggrecan gene. Dev Biol 1995; 171:655–664.
133. Seo H, Geisert EE. A keratan sulfate proteoglycan marks the boundaries in the cortical barrel fields of the adult rat. Neurosci Lett 1995; 197:13–16.
134. Carlson SS, Iwata M, Wight TN. A chondroitin sulfate/keratan sulfate proteoglycan, PG-1000, forms complexes which are concentrated in the reticular laminae of electric organ basement membranes. Matrix Biol 1996; 15:281–292.
135. Burg MA, Cole GJ. Claustrin, an antiadhesive neural keratan sulfate proteoglycan, is structurally related to MAP1B. J Neurobiol 1994; 25:1–22.
136. Chakravarti S, Magnuson T, Lass JH, Jepsen KJ, LaMantia C, Carroll H. Lumican regulates collagen fibril assembly: skin fragility and corneal opacity in the absence of lumican. J Cell Biol 1998; 141:1277–1286.
137. Olsson L, Stigson M, Perris R, Sorrell JM, Lofberg J. Distribution of keratan sulphate and chondroitin sulphate in wild type and white mutant axolotl embryos during neural crest cell migration. Pigment Cell Res 1996; 9:5–17.

138. Guerassimov A, Zhang Y, Cartman A, Rosenberg LC, Esdaile J, Fitzcharles MA, Poole AR. Immune responses to cartilage link protein and the G1 domain of proteoglycan aggrecan in patients with osteoarthritis. Arthritis Rheum 1999; 42:527–533.
139. http://www.healthheritageresearch.com/Heparin-Conntact9608.html
140. Kaye MA, Stacey M. Observations on the chemistry of hyaluronic acid. Process Biochem 1950; 2:13.
141. Dwek RA. Glycobiology: more functions for oligosaccharides. Science 1995; 269:1234–1235.

Carbohydrate Chemistry, Biology and Medical Applications
Hari G. Garg, Mary K. Cowman and Charles A. Hales

DOI: 10.1016/B978-0-08-054816-6.00010-0

Chapter 10

Physiological, Pathophysiological and Therapeutic Roles of Heparin and Heparan Sulfate

JIN XIE,* SARAVANABABU MURUGESAN[†,1] AND ROBERT J. LINHARDT*[,†,‡]

**Department of Chemistry and Chemical Biology, Rensselaer Polytechnic Institute, Troy, NY 12180, USA*
[†]Department of Chemical and Biological Engineering, Rensselaer Polytechnic Institute, Troy, NY 12180, USA
[‡]Department of Biology, Rensselaer Polytechnic Institute, Troy, NY 12180, USA

I. Introduction

Heparin and heparan sulfate belong to the glycosaminoglycan (GAG) family of carbohydrates. They are linear acidic complex polysaccharides found on the cell surface and in the extracellular matrix (1). Heparin and heparan sulfate GAGs are biosynthesized as proteoglycans (PGs) with multiple GAG chains linked to a variety of core proteins. Heparin PGs are found exclusively in the granules of subsets of mast cells, whereas heparan sulfate PGs have a much greater distribution in the body, being associated with stromal matrices, basement membranes, and almost all cell surfaces. Heparin and heparin sulfate PGs interact with cell surface binding proteins and are internalized by receptor-mediated endocytosis through their GAG chains (2).

Heparin and heparan sulfate are the most intensively studied GAGs as a result of their anticoagulant properties. However, in last 1–2 decades, it has become obvious that heparin and heparan sulfate not only have anticoagulant activities but also exhibit a number of diverse biological functions including ones regulating cell growth and differentiation, inflammatory processes, host defense

---

[1]Present address: Development Engineering, Process Research and Development, Bristol-Myers Squibb, New Brunswick, NJ 08903.

and viral infection mechanisms, cell–cell and cell–matrix interactions, lipid transport, and clearance/metabolism. These functions result from the direct interactions between heparin and heparan sulfate and heparin-binding proteins (3). It was proposed as early as 1979 that in the classic lock-and-key model, heparin can be viewed as "a bag of skeleton keys that could fit many locks" (4). These interactions play important roles in the normal physiological and pathological processes, thereby moving heparin's utility beyond its use as an anticoagulant drug, offering therapeutic benefit for a wide range of diseases.

In this chapter, we review the physiological, pathophysiological, and therapeutic roles of heparin and heparan sulfate in different biological processes.

II. Structure of Heparin and Heparan Sulfate

Heparin and heparan sulfate are relatively large anionic polysaccharides having an average molecular weight of approximately 12,000 (corresponding to ~20 disaccharide units), polydispersity of approximately 1.2–1.4, and a molecular weight range of 5,000–40,000 (5). Heparin is composed of primarily of 2-*O*-sulfo-α-L-iduronic acid and 2-deoxy-2-sulfamino-6-*O*-sulfo-α-D-glucose, while heparan sulfate consists primarily of β-D-glucuronic acid and 2-acetamido-2-deoxy-α-D-glucose as major saccharide repeating units. These saccharide units are joined through 1 → 4 glycosidic linkages. Figure 1 shows the major disaccharide sequence and variable sequence of heparin and heparan sulfate.

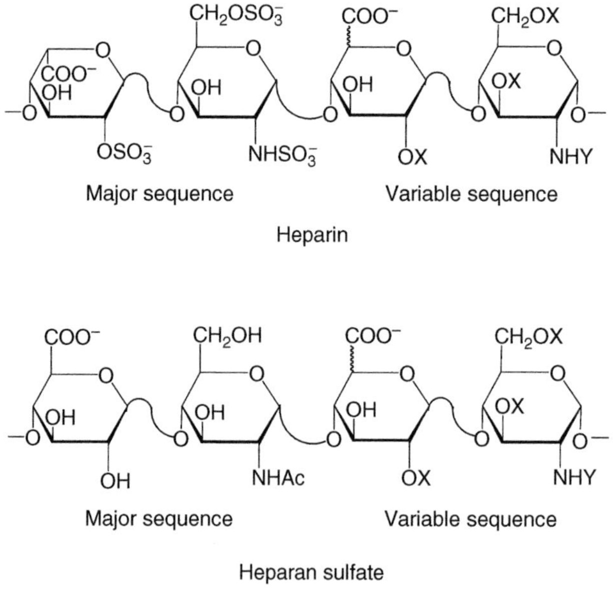

Figure 1 Major and minor disaccharide repeating units in heparin and heparan sulfate (X = H or SO_3^-, Y = Ac, SO_3^-, or H).

While heparin and heparan sulfate PGs are biosynthesized linked to different core proteins, distinction between heparin and heparan sulfate GAGs is difficult because both structural and functional criteria are inadequate to separate these two forms. A number of differential modifications of individual disaccharide units lead the microheterogeneity of heparin and heparan sulfate. The amino group of the glucosamine residue can be substituted by acetyl or sulfo group or unsubstituted. The 3- or 6-OH group of the glucosamine residue can be substituted by sulfo group or unsubstituted. The 2-OH group of the uronic acid residue can also be substituted with a sulfo group or unsubstituted. In heparin, the uronic acid residues typically consist of 80–90% L-iduronic acid and 10–20% D-glucuronic acid, while the opposite ratio is typically observed in heparan sulfate (3). This differential modification of saccharides results in 48 possible disaccharide units (6).

Heparin and heparan sulfate are highly charged and have an average net charge of −40 to −75 per chain. This charge is maintained through a wide range of pH values as the pK_a of its anionic groups are approximately 3.3 (carboxyl) and <1.5 (*O*-sulfo and *N*-sulfo). This polyanionic character leads heparin and heparan sulfate to interact ionically with many proteins. Heparin and heparan sulfate can also bind to proteins such as brain natriuretic peptide (BNP) through hydrogen binding while ionic interactions, contribute only a small portion of the free energy (7). Hydrophobic interaction may also play a minor role in heparin's interaction with antithrombin III (AT III), where hydrophobic interactions are reported between hydrophobic amino acid side chains in AT III and the *N*-acetyl group in the AT III-binding pentasaccharide sequence in porcine mucosal heparin (8).

III. Blood Coagulation Processes

Heparin is widely used as an anticoagulant drug based on its ability to accelerate the rate at which AT III inhibits serine proteases in the blood coagulation cascade. The interaction of heparin with AT III is the first well-studied heparin–protein interaction (5). In the simplest terms, blood coagulation is a process that changes the circulating soluble blood protein, fibrinogen, into an insoluble fibrin gel. The gel plugs leaks in blood vessels and stops the loss of blood. This biochemical process requires coagulation factors—calcium and phospholipids. Coagulation involves a cascade initiated by the intrinsic and extrinsic activation of coagulation factors (i.e., factor X) that will produce the activated serine protease (i.e., factor Xa). Having generated factor Xa, the final steps are the activation of prothrombin (factor II) to form thrombin (factor IIa) that acts to convert soluble fibrinogen to fibrin, which is cross-linked, completing the basic clotting pathway. AT III is a member of the serine protease inhibitor (Serpin) superfamily that regulates the proteolytic activities of the procoagulant proteases (i.e., factor Xa, thrombin) of both the intrinsic and extrinsic blood coagulation cascade. It acts by forming a stable 1:1 enzyme–inhibitor complex with the target enzyme, blocking its active site, thus preventing blood from clotting. Hemostasis is regulated by a dynamic balance between anticoagulant and procoagulant factors in the blood and blood vessels.

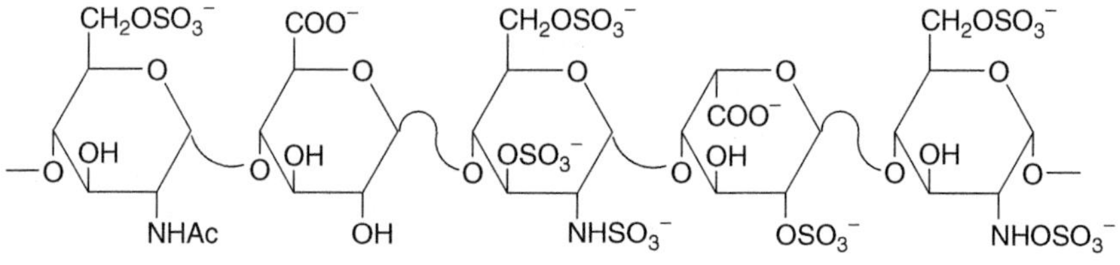

Figure 2 The heparin and heparan sulfate pentasaccharide sequence binding to AT III.

Heparin and the closely related molecule heparan sulfate act as cofactors in AT III inhibition process by affording a 1000-fold increase in the rate at which AT III inhibits coagulation enzymes (9). In the absence of heparin, AT III is a relatively ineffective inhibitor of factor Xa and thrombin (k_{assoc} [factor Xa] = 2.6×10^3 $M^{-1} s^{-1}$; k_{assoc} [thrombin] = 1×10^4 $M^{-1} s^{-1}$) (10). Heparin accelerates the interaction between AT III and target proteases by two distinct mechanisms. First, long-chain heparin is able to act as a "template," to which both AT III and procoagulant enzymes bind, bringing them into proximity. Heparin has a sequence-specific pentasaccharide (Fig. 2) that binds AT III with high affinity (K_d ~ 10–20 nM) and this binding induces a conformational change in the Serpin. The long-chain heparin also provide a site for thrombin to bind adjacent to AT III through a positively charged exosite on thrombin near its active site. The interaction between AT III and thrombin is thereby promoted due to heparin bridging the proteins in a ternary complex. The minimum heparin chain length, for the acceleration of the reaction of AT III and thrombin, is approximately 16 saccharides. AT III, to which a heparin chain having only five specific saccharide residues (Fig. 2) is bound, can inactivate factor Xa. In both cases, AT III is bound to the active site of the serine proteases as an acyl–enzyme intermediate. As a result of the cleavage, the affinity of AT III for the heparin chain is markedly diminished, causing heparin to dissociate in an unaltered form, free to catalyze further reactions between AT III and its target enzymes. The acyl–enzyme intermediate hydrolyzes very slowly to yield the unaltered serine proteases and the cleaved AT III, which is no longer an active Serprin and has lost its heparin-binding affinity (3).

The conformational changes in the Serpin, induced by the specific pentasaccharide sequence within heparin, promote the strong binding between AT III and heparin, and also leads to an exposure of the reactive center loop (RCL), the region of the Serprin responsible for primary interaction with the target enzyme (factor Xa). Thus, the reactivity of AT III toward serine protease is enhanced. The crystal structure of an AT III–pentasaccharide complex was used to identify residues in AT III involved in the interaction as well as important functional groups within the pentasaccharide. The conformational rearrangement of AT III most notably happens around the heparin-binding site. The main differences are found in the stretches 5–48, 108–199, 203–215, 218–223, 324–329, 353–362, 379–386, and 414–419. The less-well-ordered N-terminal residues 12–16 shift sideways, widening the cleft where the pentasaccharide binds. The N-terminus of the

A-helix rearranges, allowing the primary amines of Arg-46 and Arg-47 to move by 17 and 8 Å, respectively, to hydrogen bond with the sulfate residues of the pentasaccharide, whereas residues 44 and 45 make way for the cofactor. The D-helix tilts by some 10°, whereas at its N-terminus, residues 113–118 coil to form the two-turn P-helix at right angles with the D-helix. At the C-terminus, the D-helix is extended by one-and-a-half turns, moving Arg-132, Lys-133, and Lys-136 toward the pentasaccharide-binding site. These residues do not come within hydrogen bonding distance of the pentasaccharide. However, they would be able to interact with additional saccharide units of full-length heparin (11). The conformational change is also involved in the activation of AT III toward factor Xa. AT III has a mobile reactive site loop that is initially exposed as a substrate for coagulant proteases. On cleavage by the protease, the loop becomes inserted as a sixth strand in the central β-sheet (the A-sheet) of the molecule. The closing of the A-sheet with an accompanying expulsion of the partially inserted residues P14 and P15 of the reactive site loop occurs by an allosteric mechanism as it is more than 30 Å distant from the pentasaccharide. In this way, the protease is believed to be irreversibly trapped as a reaction intermediate covalently bound to the serprin. Johnson et al. (12) were able to crystallize the pentasaccharide-activated AT III and factor Xa complex only after mutating three residues on a remote surface of AT III to engender the formation of crystal contacts, which, due to electrostatic repulsion, would not otherwise have been formed. This crystal structure is consistent with the exosite-dependent recognition mechanism. Several of the residues seen to participate in exosite contacts have previously been identified by mutagenesis studies. Three residues on factor Xa were shown to contribute significantly to the rate of inhibition by AT in the presence of the pentasaccharide, Glu37, Glu39, and Arg150. Site-directed mutation study showed that substitution of Arg150 by Ala reduces the rate of inhibition by nearly tenfold. Both crystal structure and the mutagenesis study investigate the importance of the P insertion of AT III, with the deletion of Arg399 reducing the pentasaccharide-activated rate of inhibition by over twofold (12). Figure 3 indicates the heparin-binding mechanism of AT III.

The role of individual saccharide residues of the heparin pentasaccharide in the allosteric activation of AT III has also been determined by studying the effect of truncating the pentasaccharide residue, at either its reducing or its nonreducing end. These studies established that the three saccharide residues on the nonreducing end of the pentasaccharide sequence are capable of fully activating AT III. While the reducing end residues are not essential for the activation, they stabilize the activated conformation (13).

IV. Cell Growth and Differentiation

Heparan sulfate PGs play an essential role in regulating cell recognition and cell growth by its interactions with signaling molecules such as fibroblast growth factors (FGFs), vascular endothelial growth factor (VEGF), hepatocyte growth factor (HGF), transforming growth factor-β (TGF-β), and platelet-derived growth factor (PDGF).

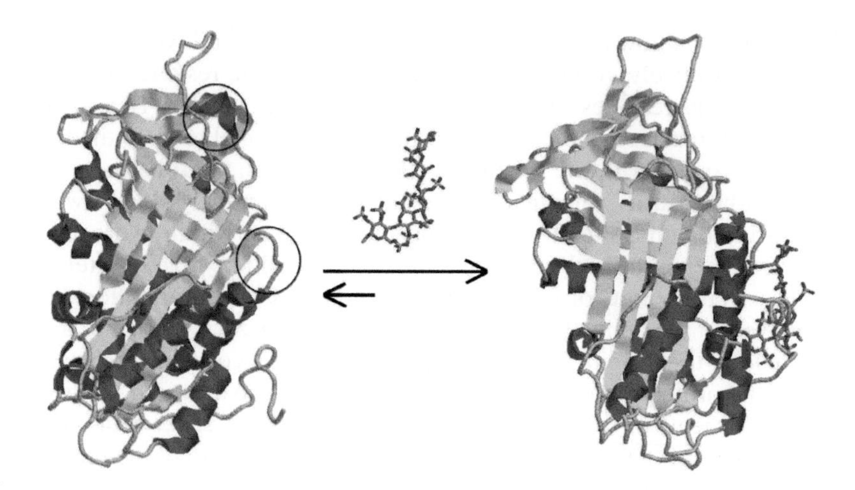

Figure 3 The heparin-binding mechanism of AT III. β-sheet A facing A-helix in the back, D-helix to the right, the reactive center loop (RCL) at the top, and Arg399 displayed in spacefill. Heparin pentasaccharide showed in stick. In native AT, the N-terminal region of the RCL is incorporated as strand 4 in β-sheet A, which constrains the RCL and the P1 Arg393 side chain.

FGFs are polypeptides with a conserved core of approximately 120-amino acid residues and more variable N- and C-terminal domains. They have been demonstrated to induce neovascularization *in vivo* and to be implicated in the growth of new blood vessels during wound healing and chick embryo development. *In vitro*, FGF induces cell proliferation, migration, and production of proteases in endothelial cells by using a dual-receptor system (14). The primary component of this system is a signal-transducing FGF receptors (FGFRs), which is composed of an extracellular ligand-binding portion consisting of three immunoglobulin-like domains (D1–D3), a single transmembrane helix, and a cytoplasmic portion with protein tyrosine kinase activity. The second component of this receptor system consists of heparan sulfate PGs of the cell surface required for FGF to bind to and activate FGFR (15). Receptor dimerization is an obligatory event in FGF signaling. FGF is a monomer and is unable by itself to induce FGFR activation. It has to function in concert with either soluble or cell surface-bound heparan sulfate to promote FGFR dimerization.

Acidic fibroblast growth factor (FGF1) and basic fibroblast growth factor (FGF2) have been extensively studied for their interactions with heparin and heparan sulfate. Pellegrini et al. (16) described an asymmetrical structure of the FGFR2 ectodomain in a dimeric form that is induced by simultaneous binding to FGF1 and a heparin decasaccharide. The complex is assembled around a central heparin molecule linking two FGF1 ligands into a dimer that bridges between two receptor chains. The asymmetric heparin binding involves contacts with both FGF1 molecules but only one receptor chain (Fig. 4A) (16). Robinson et al. have expanded this study and observed that FGF1–FGFR2–heparin 2:2:1 ternary complexes formed spontaneously in solution. They found that octasaccharides were the shortest heparin fragments that formed 2:2:1 complexes and induced FGF1

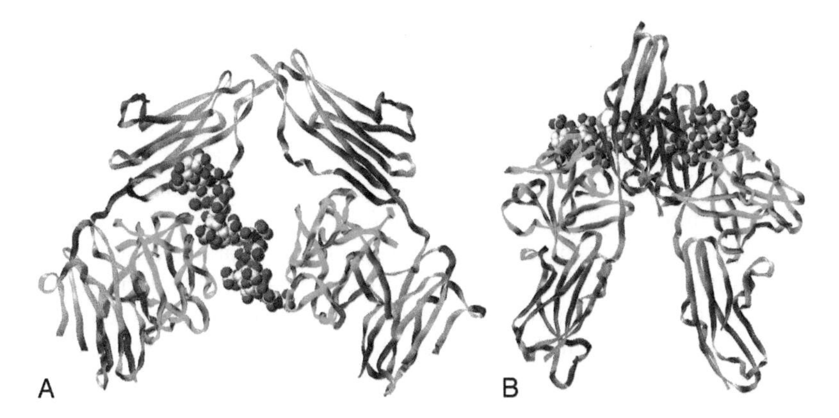

Figure 4 (A) Crystal structure of FGF2-FGFR1-heparin 2:2:2 ternary complex. (B) Crystal structure of FGF1-FGFR2-heparin 2:2:1 ternary complex. FGF2 and FGF1 are shown in green ribbon, FGFR1 and FGFR2 are shown as gold ribbons, and heparin oligosaccharides are shown as a red (oxygen), white (carbon), blue (nitrogen) and yellow (sulfur) space filling model.

mitogenesis in heparan sulfate-deficient FGFR2-transfected cell lines (17). A cooperative mechanism was also proposed for the asymmetrical ternary complex formation. In the presence of heparin, free FGF1 will dimerize to form 2:1 FGF1–heparin structure which is a prerequisite for subsequent FGFR2 dimerization. A crystal structure of a dimeric ternary complex of FGF2, FGFR1 and a heparin decasaccharide (2:2:2) has also been reported (Fig. 4B) (18). According to this model, heparin interacts through its nonreducing end with both FGF2 and FGFR1 and promotes the formation of a stable 1:1:1 FGF2–FGFR1–heparin ternary complex (19). A second 1:1:1 FGF2–FGFR1–heparin ternary complex is recruited to the first complex via direct FGFR1–FGFR1 contacts, secondary interactions between FGF2 in one ternary complex and FGFR1 in the other ternary complex, and indirect heparin-mediated FGFR1–FGFR1 contacts. In the absence of heparin, the direct receptor–receptor contacts and secondary ligand–receptor interactions are not sufficient for appreciable dimerization. Heparin hexasaccharides are sufficient to promote receptor dimerization. Less was known about the interaction between heparin and other FGF species.

HGF is a member of the plasminogen-related growth factor family and contains 697 residues in its mature processed form. It activates the dimerization of the Met tyrosine kinase receptor and results in the phosphorylation of the kinase. The phosphotyrosines recruit further downstream signaling molecules. The activation of Met also triggers the mitogen-activated protein kinase pathway (20). There are at least two possible ways for the dimerization of Met by HGF. The first involves two Met receptors binding to two separate and nonequivalent binding sites on one HGF molecule to form a 1:2 ligand–receptor signaling complex. The second model suggests a 2:2 signaling complex formed either by association of two binary ligand–receptor complexes or by ligand dimerization. Heparan sulfate increases the mitogenic potency of HGF and causes the oligomerization of HGF, thus concentrating the ligand at the cell surface and making interaction with Met more energetically favorable. Tetrasaccharides of heparan sulfate have been shown to be of minimal chain length that shows activity *in vivo*; however, hexasaccharides were shown to be more capable of Met kinase activation. The precise role of heparin and heparan sulfate in the Met–HGF complex formation is limited to the NK1 isoform that contains the N-terminal domain and followed kringle domain of HGF. There are several models for the role of heparin and heparan sulfate in the binding of NK1 isoform and Met. In the first model, NK1 undergoes a conformational change upon the interaction with heparin and heparan sulfate, which enables the ligand to dimerize and bind to Met in a stable manner (21). The second model proposes that heparin and heparan sulfate stabilize the ligand's dimer. Although in the absence of heparin and heparan sulfate, NK1 can form the dimer, heparin and heparan sulfate cement the interaction and shift the equilibrium to the active, dimeric form of ligand (22).

Vascular endothelial growth factor (VEGF), an endothelial cell mitogen, is one of the major regulators of angiogenesis. It has three major isoforms, VEGF 120, VEGF 164, and VEGF 188, but only two, VEGF 164 and VEGF 188, interact with heparin and heparan sulfate. Heparin is known to affect the interaction between VEGF 164 and VEGF receptors, interfering with VEGF 164-mediated activities. The crystal structure obtained by Leahy and coworkers showed that the 6- to 7-mer of heparin is required for efficient VEGF heparin binding, then a 22-mer of heparin is required to promote optimal interaction of VEGF and its receptors (23).

In summary, heparan sulfate PG localizes a number of different growth factors at the cell surface or in the extracellular matrix and promotes their activities of cell growth and differentiation.

V. Inflammation

It has been recognized since 1920s that heparin can inhibit certain aspects of the allergic inflammatory response. Inflammation is the first response of the immune system to infection or irritation and may be referred to as the innate cascade. An essential feature of any inflammatory response is the rapid recruitment of leukocytes from the blood to the site of inflammation, usually through postcapillary venules (24). During this process, leukocytes have to migrate through the blood-vessel wall and enter tissues through a multistep process known as extravasation or diapedesis. Leukocytes first attach to and roll on the inflamed endothelium, followed by the endothelial cell-bound chemokine activation. The activated leukocytes then can attach to the endothelium and be degraded by the subendothelial basement membrane. Leukocytes migrate into the target tissue along chemokine gradients. After leukocytes enter the target tissue, they can perform various immune activities such as pathogen elimination and tissue repair. Recent studies have indicated that heparin and heparan sulfate are required for several stages of inflammation.

In the first adhesion step of the leukocytes extravasation process, selectins (including L-selectin, E-selectin, and P-selectin), along with heparan sulfate PGs, play an important role. Selectins are a family of transmembrane glycoproteins found on the endothelium, platelets, and leukocytes. Heparan sulfate interacts with L- and P-selectins but not with E-selectin. L-selectin is important in the infiltration of leukocytes into the kidney (25). Studies show that collagen XVIII, a basement membrane heparan sulfate PG, is involved in leukocyte recruitment. Collagen XVIII provides a link between initial rolling of inflammatory cells through L-selectin followed by the induction of chemokine-induced, integrin-dependent adhesion (26). L-selectin only binds a subset of renal heparan sulfate GAG chains attached to the collagen XVIII protein core. This collagen XVIII PG is predominantly present in modularly tubular and vascular basement membranes (27). L-selectin binding to heparan sulfate is critically dependent on *O*-sulfo groups as well as the on length of the GAG chain. Unsubstituted amino groups do not play a role in L-selectin binding but these amino groups somehow influence the biosynthesis of L-selectin-binding sites on heparan sulfate PGs. Heparan sulfate *O*-sulfo groups, however, are critical for L-selectin binding. The L-selectin-binding domain is located in domains of alternating *N*-acetyl and *N*-sulfo residues. The presence of iduronate residues inhibits the L-selectin binding, suggesting that regions with a high level of 6-*O*-sulfo groups and GlcA content are ideal for L-selectin binding. Decreasing the sulfation of heparan sulfate chains reduces the binding of L-selectin.

P-selectin also binds to heparan sulfate and heparin, but binding generally occurs with weaker affinity than for the interaction with L-selectin. P-selectin interacts with P-selectin glycoprotein ligand expressed on a majority of leukocytes, to mediate tethering (initial attachment), rolling, and weak adhesion of leukocytes on the activated endothelial cells. P-selectin also mediates the aggregation of the activated platelets to leukocytes (28). P-selectin binds to A375 cells, a cell line of a human malignant melanoma, that express heparan sulfate PGs on cell surfaces. Inhibitors of both PG biosynthesis and heparinases reduce adhesion of A375 cells to P-selectin. Heparin, but not other GAGs, abolished the P-selectin binding to these cells (29). The molecular mechanism and the exact oligosaccharide structure of heparan sulfate responsible for its interaction with P-selectin are not yet clear.

In addition to the regulations of leukocyte rolling process by binding to selectins, heparin and heparan sulfate also have been implicated in chemokine and cytokine regulation. Heparan sulfate regulates the gradients of chemokines and cytokines produced by endothelial cells that have been stimulated by proinflammatory factors, such as interleukin (IL)1β and tumor necrosis factor (TNF)α. Chemokines are a large family of soluble proteins involved in leukocyte activation through G–protein-coupled receptor's signaling (30). One of the most well-studied interactions is that of heparan sulfate and the CXL chemokine IL-8, which is a member of the interleukin family. Heparan sulfate binding to IL-8 activates the transportation of the chemokine across the endothelial cell layer (31). *In vivo* studies showed that endothelial cells deficient in highly sulfated heparan sulfate are much less efficient in mediating chemokine (IL-8) movement across the endothelial cell layer (32). The interaction of chemokines with heparan sulfate can also protect chemokines from proteolysis and induce their oligomerization. The resulting higher order chemokine oligomers are required for maximal activity. CC chemokine ligand 5 (CCL5) that has been mutated to remove its heparan sulfate-binding site can function as a dominant-negative inhibitor by forming nonfunctional heterodimers with wild-type CCL5. The resultant heterodimers are unable to produce higher order oligomers, induced by heparan sulfate (33). Heparan sulfate also plays an essential role in immobilizing chemokines on the luminal surface of endothelial cells establishing chemokine gradients on the vascular endothelium. The stromal cell-derived factor-1α (SDF-1α) is a proinflammatory mediator, a potent chemoattractant for a variety of cells, like monocytes and T cells, and also a potent inhibitor of the cellular entry of HIV (34). Heparan sulfate is involved in the binding and localization of SDF-1α to the cell surface (35). Heparin dodecasaccharide or tetradecasaccharide is required for binding to SDF-1α (36). Without chemokines binding to heparan sulfate, leukocytes are unable to form stable, integrin-mediated interactions with the endothelium and cannot migrate in a directional manner through the blood-vessel wall. Endothelium-bound chemokines, but not soluble chemokines, induced a rapid extension of inactive lymphocyte function-associated antigen 1 (37). The extended form of this integrin can then interact with endothelial cell intercellular adhesion molecule 1 and trigger stable cell adhesion.

These findings indicate multiple functions of heparan sulfate during inflammatory reactions and suggest that heparan sulfate participates in each of the main steps of leukocyte extravasation, including facilitating L-selectin-dependent cell rolling, modulating chemokines to recruit different leukocyte subsets across the endothelium during inflammatory responses.

VI. Host Defense and Viral Infection Mechanisms

Heparan sulfate is present at the surface of cells through its covalent binding to a cell membrane-attached core proteins. A large number of viruses interact with heparan sulfate that facilitates the attachment to host cells and subsequent infection.

Malaria is a serious, sometimes fatal disease caused by a parasite, *Plasmodium* sporozoite, and is transmitted by the bite of an infected mosquito. More than 40% of the people in the world live in areas at risk for malaria. *Plasmodium* sporozoites traverse the cytosol of several cells before invading a hepatocyte (38). The invasion of liver cells by these sporozoites has been ascribed to the interaction of circumsporozoite (CS) protein, a 40–45 kDa antigen (39) that covers the plasma membrane of sporozoites, with GAG side chains of PGs located on the surface of the hepatocytes (40,41). Both heparin and heparan sulfate interact with CS protein and help in the binding of the parasite to liver cells. Surface plasmon resonance (SPR) experiments indicate that heparin has fast on-rate (k_a) and slow off-rate (k_d) kinetics for CS protein with a K_d of 41 nM. Isothermal titration calorimetry shows that the minimum sequence within the heparin polymer capable of a strong interaction with the CS protein is a decasaccharide. High concentrations of exogenous heparin or heparan sulfate can inhibit the binding of CS protein to HepG2 cell, a human hepatoma cell line. However, at low concentrations, heparin or heparan sulfate can also act as a cross-linking agent effectively promoting a formation of multimeric CS complexes and thus enhancing the CS binding to the target cells (42).

Hepatitis C virus (HCV) is the major cause of posttransfusion and community-acquired hepatitis in the world. The majority of HCV-infected individuals develop chronic hepatitis that may progress to liver cirrhosis and hepatocellular carcinoma. The HCV structural proteins comprise the core protein and the two envelope glycoproteins E1 and E2 (43). Several lines of evidence have demonstrated that the HCV envelope proteins may play a crucial role in the initiation of infection by mediating virus–host cell membrane interaction. E2 is thought to initiate viral attachment, whereas E1 may be involved in virus–cell membrane fusion (44–46). A comparative structural analysis of the E2 protein of various HCV isolates demonstrated that positively charged amino acid residues are highly conserved in the N-terminus of E2 hypervariable region 1 (47), thereby suggesting the negatively charged cell surface GAG, heparan sulfate, as an HCV receptor. Heparin directly interacts with E2 and binding of E2 is specifically inhibited by highly sulfated heparan sulfate and heparin. Partial enzymatic degradation of cellular heparan sulfate resulted in a marked reduction of E2 binding (48). These studies suggest that binding of the HCV envelope glycoproteins to a specific heparan sulfate structure on host cells is an important step for the initiation of viral infection and that this interaction represents an important target of antiviral host immune responses in HCV infection *in vivo*. The level of heparan sulfate sulfation appears to play a crucial role for E2 binding. A high total sulfate/disaccharide ratio and the presence of trisulfated disaccharide sequence, →4) 2-*O*-sulfo-α-L-iduronic acid (1 → 4) 2-oxy-2-sulfoamino-6-*O*-sulfo-α-D-glucose (1→, are important for efficient heparan sulfate–E2 protein binding. These results also suggested that the mechanism of heparan sulfate–E2 protein-binding ranges from simple charge effects to highly specific receptor-like interactions. Recent studies demonstrated that the interaction of the viral envelope is mediated by both envelope glycoproteins E1 and E2. Similar to findings for envelope glycoprotein E2, highly sulfated heparan sulfate played an important role in mediating binding of E1. Although recombinant-soluble E1 had a lower affinity for immobilized heparin in SPR assays than E2 and E1 exhibited a significantly different cellular binding profile from E2, the functional differences of E1 and E2 envelope glycoproteins may not be directly extrapolated to the *in vivo* situation, where E1 is likely to be in a different conformation as a complex with E2 (49). Barth et al. also have studied the functional relevance of HCV–heparan sulfate binding. Competition experiments suggested that highly sulfated heparan sulfate is an important molecule for binding of the viral envelope to the cell surface and may contribute to viral entry in concert with other cell surface molecules such as the tetraspanin CD81 and scavenger receptor SR-BI.

There are also many other viruses in which the binding to heparin or heparan sulfate is important for the initiation of infection. For example, it has been known for a long time that removal of cell surface heparan sulfate or the use of soluble heparin and heparan sulfate as competing agents reduced both HIV attachment and infection on several cell lines, including CD4-positive HeLa cells, macrophages, and T-cell lines. Many studies showed this is caused by the binding of heparan sulfate to the viral surface glycoprotein, gp120. The *O*-sulfation, particularly 6-*O*-sulfation, and *N*-substitution (acetylation or sulfation) were essential for the heparan sulfate–gp120 interaction (50). Vives et al. found that the cell surface molecule CD4 induced conformational change of gp120, dramatically increasing binding to heparan sulfate. These results suggest an implication of heparan sulfate in the late stages of the virus–cell attachment process (51). The herpes simplex virus (HSV) uses heparan sulfate PG to target (52) and infect (53) cells through the binding of heparan sulfate with viral coat glycoproteins gC and gB (54).

VII. Cell–Cell and Cell–Matrix Interactions

Extracellular matrix (ECM) is composed of various glycoproteins and PGs. It provides support and anchorage for cells and regulates the intercellular communications. Individual ECM components influence cell arrangements and activities through binding to transmembrane receptors, thus initiating intracellular signaling and altering gene expression and other downstream events. Controlling ECM composition and organization modulates cell behaviors and serves important roles during both physiological and pathological processes (55). Heparan sulfate PGs are widely distributed at the cell surface and in ECM. Interactions of heparan sulfate PG with a number of matrix proteins are important in regulating cell behavior and fibril formation during development and pathophysiological events. The heparan sulfate chain has various important biological properties and influences cell behavior through interactions with a variety of matrix proteins (56).

In animals, the most abundant glycoproteins in the ECM are collagens. Collagen V, a quantitatively minor component of connective tissues, plays a crucial role in matrix organization. It is a fibrillar collagen that generally copolymerizes with collagen I to form heterotypic fibrils. It also has the property of forming fibrils that show preferential pericellular localization. Its predominant molecular form is the heterotrimer $(\alpha1(V))_2\alpha2(V)$. Collagen V binds heparin through a 12-kDa fragment of the $\alpha1(V)$ chain referred to as HepV. The recombinant HepV fragment supports heparin-dependent cell adhesion (57). Five basic residues (Lys^{905}, Arg^{909}, Arg^{912}, Arg^{918}, and Arg^{921}) of the Lys^{905}–Arg^{921} sequence of the $\alpha1(V)$ chain participate in the binding of HepV to heparin (58). SPR experiment showed HepV bound to heparin and heparan sulfate with a similar affinity ($K_D \sim 18$ and 36 nM, respectively) in a cation-dependent manner, whereas the native collagen V heterotrimer $(\alpha1(V))_2\alpha2(V)$ bound to heparin and heparan sulfate with even higher affinity ($K_D \sim 5.6$ and 2.0 nM, respectively). Heat and chemical denaturation strongly decreased the binding of the native collagen V to heparin. These indicate that the collagen triple helix plays a major role in stabilizing the interaction with heparin. Heparin 2-*O*-sulfo and 6-*O*-sulfo groups were shown to be crucial for the HepV binding. Octasaccharide of heparin or decasaccharide of heparan sulfate is required for HepV binding. Complexes formed between HepV and heparin/heparan sulfate in the present of cations (Ca^{2+} or Mg^{2+}) exhibited a lower dissociation rate, which suggested that cations stabilize the complexes (56). Although collagen V is present in normal tissues as trimeric molecules, HepV could be released in pathophysiological situations where extensive ECM remodeling occurs. Interactions mediated by heparin and heparan sulfate could modulate biological activities of monomeric HepV and trimeric collagen V.

Along with ECM glycoproteins, PGs also communicate extracellular information to the cell. Transmembrane heparan sulfate PGs of the syndecan family bind to ECM proteins and cooperate with heterodimeric integrin receptors to regulate cell adhesive activities. Syndecan-4, a member of this family, is upregulated during tissue repair (59) through its binding to plasma fibronectin (60) and tenascin-C (61) via its heparan sulfate side chains. Fibronectin is a ubiquitously expressed, multifunctional extracellular glycoprotein that promotes cell adhesion. It forms a provisional matrix with covalently cross-linked fibrin at sites of tissue injury. This fibrin–fibronectin matrix coordinates the activities of cells involved in wound repair (62). Syndecan-4 (heparan sulfate PG) binds to fibronectin resulting in optimal cell response to fibrin–fibronectin matrices. Soluble heparin prevents cell spreading and organization of actin stress fibers. Syndecan-4 also functions cooperatively with integrins in fibroblast binding to fibronectin. Integrin binding to the fibronectin cell-binding domain together with heparan sulfate side chain of syndecan-4 interacting with heparin-binding site on fibronectin are essential for cells to spread and organize their cytoskeleton. Syndecan-4 is also required for tenascin-C modulation of cell behavior. Tenascin-C is an ECM protein that regulates cell response to fribronectin with the provisional matrix and has modulatory effects on cell–ECM interactions. With a fibrin–fibronectin matrix, tenascin-C impacts the ability of fibroblasts to deposit and contract the matrix by affecting the morphology and signaling pathways of adherent cells (63). Tenascin-C acts in a syndecan-4—dependent manner to suppress fibronectin-mediated signaling because tenascin-C can also interact directly with heparan sulfate side chains of syndecans, which will compete with fibronectin (64). As such, syndecan-4 plays an important role throughout the tissue repair in communicating signals between the dynamic ECM and the cell.

In nervous system, neurons must often send their axons far across complex terrain. Successful navigation of this challenging extracellular wilderness and productive interactions with target cells require an array of local and long-range axon guidance cues that influence the cell motility machinery within the growing tip of the axon. Instructing axon guidance decisions to neuronal receptors requires interactions with ECM (65). Heparan sulfate PG is important for shaping and modulating the guidance factor landscape. The midline, a conserved feature of bilaterian organisms, functions as an organizing center and intermediate target that regulates the sides of the body (66). Slit, a secreted glycoprotein, was identified as a major repellent at the midline of the central nervous system. Binding of Slit to receptors of the Roundabout (Robo) family triggers cytoskeletal rearrangements within the axon growth cone, resulting in axon repulsion. Heparan sulfate PGs are critically involved in Slit–Robo signaling. Immunoprecipitation experiments using *Drosophila* cell extracts show that both Slit and Robo interact with syndecan (67). SPR experiments showed that Slit binds to heparin with a high affinity, $K_D \sim 0.33$ μM and the minimum heparin oligosaccharide size that binds to Slit is tetrasaccharide (68). A highly conserved basic sequence motif in the C-terminal domain of Slit is responsible for high-affinity heparan sulfate/heparin binding, but the leucine-rich

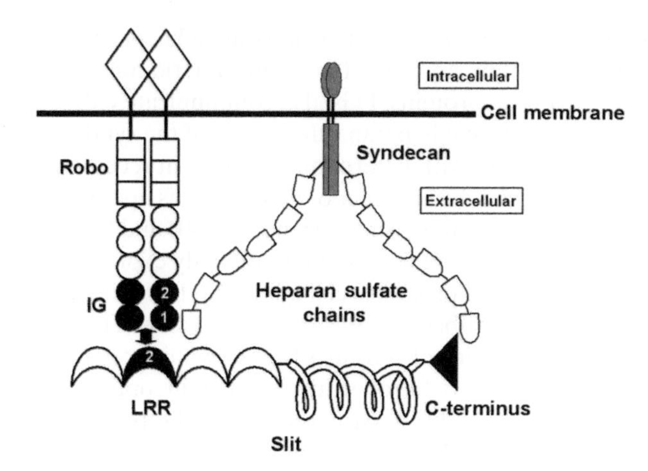

Figure 5 Schematic drawing of Slit, Robo, and syndecan (heparan sulfate PG).

repeat domains at the N-terminus of Slit also show appreciable heparan sulfate/heparin binding (69). This suggests that heparan sulfate chains may be required for capturing Slit at the cell surface of the Robo-expressing growth cone. Affinity chromatography showed the Slit-binding ectodomain of *Drosophila* Robo also displays a strong affinity to heparin within its IG1–2 region, suggesting heparan sulfate PGs may be required for formation of a specific ternary Slit–Robo–heparan sulfate signaling complex. Solid-phase experiment and gel filtration chromatography conformed the formation of the ternary Slit–Robo–heparan sulfate (70). Figure 5 demonstrates the interactions of heparan sulfate side chains with Slit and Robo. Based on these experimental data, it is conceivable that heparan sulfate has a dual role in modulating Slit–Robo signaling: the C-terminal high-affinity heparan sulfate-binding site may serve to concentrate Slit at the growth cone surface, while a weaker, but potentially more specific, site in the leucine-rich repeat region may be required for the formation of a ternary Slit–Robo–heparan sulfate signaling complex. Physiological proteolytic cleavage of Slit would separate these activities and could have a profound effect on Slit activity *in vivo* (69).

VIII. Lipid Transport and Clearance/Metabolism

Heparin and heparan sulfate are also involved in lipid transport and metabolic clearance.

Apolipoprotein E (apoE) is an important lipid transport protein in human plasma and brain. It mediates hepatic clearance of remnant lipoproteins as a high-affinity ligand for the low-density lipoprotein receptor (LDLR) family, including LDLR, LDLR-related protein (LRP), and cell surface heparan sulfate PGs (71). In the liver, heparan sulfate PGs facilitate the interaction of remnant particles with LRP, which is known as the heparan sulfate PG–LRP pathway, in which apoE initially interacts with heparan sulfate PG on the cell surface and is then transferred to the LRP for internalization (72). The ability of apoE to interact with members of the LDLR family and with heparan sulfate PG can also be significant for cell signaling events (73). ApoE binding to heparan sulfate PG inhibits the smooth muscle cell proliferation (74). In addition, the interaction of apoE with heparan sulfate PG has been implicated in neuronal growth and repair and also is involved in the progression of late onset familial Alzheimer's disease (75–77). ApoE is a 299-amino acid, single-chain protein and contains two independently folded functional domains, a 22-kDa N-terminal domain (residues 1–191) and a 10-kDa C-terminal domain (residues 222–299) (78). The N-terminal domain exists in the lipid-free state as a four-helix bundle of amphipathic α-helices and contains the LDL receptor-binding region (residues 136–150 in helix 4) (79). The amphipathic nature of the α-helix containing residues 136–150 is critical for normal binding to the LDL receptor (80). The C-terminal domain is also predicted to be a highly α-helical structure and contains the major lipid-binding region (81,82). The N- and C-terminal domains each contain a heparin-binding site (83,84). Heparin interacts with the N-terminal domain of apoE through forming salt bridges between sulfate groups of heparin and Arg-142 and Arg-145 residue of apoE (85). Site-directed mutagenesis study showed that Arg-142, Lys-143, Arg-145, Lys-146, and Arg-147 of the N-terminal domain are required for high-affinity binding to heparin, with Lys-146 participating in an ionic interaction with heparin and Lys-143 participating in a hydrogen bond (86). In the C-terminal domain of apoE, Lys-233 is crucial for the heparin interaction because of its high positive electrostatic potential. The C-terminal domain of apoE can form a stable tetramer in aqueous solution that results in the higher heparin-binding affinity to C-terminus in the lipid-free state than in the lipid-present state. However, only N-terminal site of apoE is involved in the heparin interaction of full-length apoE despite of the strong heparin binding to the C-terminus (87). A two-step mechanism of binding apoE to heparin was proposed on the basis of molecular modeling and SPR experiments. In first binding step, eight basic residues including Lys-146 of apoE directly contact with sulfate or carboxyl groups of the heparin chain through electrostatic interaction. Hydrogen bonding between Lys-143 residue of apoE and heparin, hydrophobic interaction between the shallow groove of the α-helix of apoE, and the saccharide chains of heparin are involved in the second apoE–heparin binding step. The first binding step contributes most of the overall free energy of binding (88). In the remnant lipoprotein metabolism process, because of the fast association of apoE in the first step to heparin through long range and nondirectional ionic interactions, apoE-enriched remnant particles can be captured rapidly by the abundant heparan sulfate PG on the cell surface (88) and then transfer to the LRP for internalization (72).

Heparan sulfate is involved in regulating the metabolism of lipoproteins and major neutral lipids such as cholesteryl esters, triglycerides, and fatty acids. In the metabolism of lipoproteins, the endothelial cell surface heparan sulfate PGs have been shown to bind very low density lipoproteins (VLDL) through electrostatic interaction between positively charged regions of lipoproteins and negatively charged carboxyl- and sulfo-groups of heparan sulfate chains. Lipoprotein lipase, binding to this cell surface heparan sulfate PGs also, digests VLDL in capillaries and releases free fatty acids to the circulation (89). Heparan sulfate also mediates the following metabolism of the remnant neutral lipids. The neutral lipolytic enzymes, such as human pancreatic cholesterol esterase and triglyceride lipase, have been shown to bind to membrane heparin *in vitro* (90). This interaction is saturable, concentration-dependent, and specific for the subfraction of intestinal heparin. The high concentration of heparin in the intestine can serve to concentrate and localize these neutral lipolytic enzymes on the intestinal membrane, thus facilitate the movement of neutral lipids in the circulation from the unfavorable aqueous milieu to the enterocyte membrane, where the lipolytic enzymes hydrolyze the fatty acid ester of cholesterol or glycerol and the products can be easily adsorbed by the intestine.

Heparan sulfate mediates the clearance of thrombin–AT III (TAT) complex from the circulation by hepatic receptors. As we discussed, AT III inhibits its cognate proteinase, thrombin, by forming covalent 1:1 stoichiometric TAT complex. Vitronectin (VN), a 78-kDa glycoprotein synthesized by the liver, plays an important role in the metabolism of TAT complex. TAT in serum or plasma has been found to exist in the form of a covalent ternary complex with VN (91,92). During this physiological process, hepatocyte-associated heparan sulfate PG has been identified to be a major binding site involved in the metabolism of VN–TAT both *in vivo* and *in vitro*. VN has been shown to neutralize the heparin catalysis of AT inhibition of thrombin and factor Xa. The dissociation constant for the interaction of VN to heparin is around 10–50 nM (93,94). TAT clearance in mice is inhibited by protamine that blocks TAT interaction with heparan sulfate PGs. In the presence of VN, TAT prefers to form a high molecular mass adduct VN–TAT before being cleared. *In vitro* cell culture experiments showed that VN–TAT bound to HepG2 cells and human hepatoma cell line was also degraded. Addition of heparin inhibited in this clearance process (95). These experiment data suggest TAT first interacts with VN to form ternary VN–TAT complexes and then this VN–TAT complex binds to liver heparan sulfate PGs, thus VN–TAT will be passed onto hepatic receptors and be removed from the circulation.

IX. Potential Therapeutic Roles of Heparin and Heparan Sulfate

For potential therapeutic applications, the strength of the interactions of heparin and heparan sulfate with different proteins are certainly important but so is the specificity of these interactions. Several questions on specificity need to be addressed: (a) Whether a specific oligosaccharide sequence within the polysaccharide chain is responsible for the binding to a given protein; (b) How different are the oligosaccharide sequences involved in binding different proteins, and (c) How specific their activities are? (96) As discussed in the previous sections, only a small portion of the polysaccharide often with a defined oligosaccharide structure is often critical for binding to a given protein.

For the well-studied anticoagulant property of heparin and heparan sulfate, a defined pentasaccharide sequence (Fig. 2) containing a 3-*O*-sulfo group in the central glucosamine residue is required to catalyze the AT III-mediated inhibition of factor Xa. Fondaparinux (Arixtra, Sanofi-Synthelabo), an antithrombin III-binding pentasacchride sequence with a methyl group stabilizing the anomeric end, has been approved for use in thromboprophylaxis following orthopedic surgery (97). Low molecular weight (LMW) heparins, prepared by periodate oxidation and borohydride reduction, lack a glucuronic acid residue found within the AT III pentasaccharide-binding site (Fig. 2) and thus do not have the anticoagulant activity measured by AT III-mediated anti-Xa assay (98). However, this LMW heparin retains its ability to regulate smooth muscle cell proliferation. The antiproliferation of pulmonary smooth muscle cell assay showed the *O*-hexanoyl derivative of this LMW heparin inhibits the growth of pulmonary smooth muscle cells better than heparin itself (99).

In the inflammatory process, heparin has been proposed to have a regulatory role in limiting inflammation through its interaction with proteins such as cytokines and adhesion molecules (100). Whole animal studies showed that exogenous heparin reduces the leukocyte rolling in postcapillary venules, which is an essential stage in cell adhesion and transport to sites of inflammation. Heparin has also the potential to treat the inflammatory diseases in humans, such as asthma, arthritis, and inflammatory bowel disease. Heparin delivered by inhalation prevented the bronchoconstriction response in asthmatic patients without altering the clotting parameters in blood (101–103). Heparin combined with sulfasalazine had been shown to treat rheumatoid arthritis and ulcerative colitis due to its ability to inhibit leukocyte rolling on the endothelium by binding to L- and P-selectins (104).

Heparan sulfate or heparan sulfate PGs are also involved in cancer treatment due to their roles in tumorigenesis, tumor progression, and tumor metastasis. Tumor cell surface heparan sulfate modulates angiogenesis through binding with growth factors, such as FGF2, and cytokines to promote growth factor signaling and tumor cell proliferation and progression. However, the endothelial cell surface heparan sulfate mediates FGF2 signaling to induce angiogenesis. Heparan sulfate also regulates tumor metastasis to sites, such as liver, lung, and spleen, by mediating interactions between tumor cells and platelets, endothelial cells, and host cells. Tumor cell surface heparan sulfate acts as ligands for P-selectin that mediates adhesion either to platelets or to the endothelial lining of the capillary system. This allows tumor cells to extravasate and enter blood stream, as well as to metastasize to other organs. Tumor cell surface heparan sulfate also mediates a local coagulation through interacting with coagulation serine proteases, such as thrombin, to promote the formation of a protective layer of fibrin around the tumor, thus prevent the tumor cell to be attacked by natural killer cells of the immune system. Exogenous heparin can compete with heparan sulfate during these processes and may represent a potent agent for cancer treatment. Clinical trials showed that unfractionated heparin and LMW heparin have the antitumor effects due to their anticoagulant properties, which can treat venous thromboembolism in cancer patients (6).

An increasing number of structurally defined fragments of heparin and heparan sulfate are confirmed to bind with different proteins and to regulate numerous biological processes. These heparin, heparan sulfate, and heparin-like molecules may offer a diverse group of therapeutic benefits beyond their anticoagulant activity. In fact, the major limitation in utilizing heparin as therapeutic agent remains its strong anticoagulant properties, which become a side effect leading to hemorrhagic complication (3). Synthesizing heparin-like analogues lacking the AT III-binding site such as LMW heparins may solve this problem to some extent. However, complicated synthesis strategies result in a second and possibly greater limitation. The original chemical synthesis of the AT III-binding pentasaccharide involved almost 50 steps and afforded the target compound in very low overall yields (105). The biosynthesis strategy, developed by Rosenberg and colleagues, uses less steps but was not scalable and afforded only minute quantity of product. Enzymatic synthesis starts with a polysaccharide isolated from *Escherichia coli* K5 and is followed by *N*-deacetylation, *N*-sulfation, enzymatic epimerization, and *O*-sulfation modifications using biosynthetic enzymes (106). Using recombinant *E. coli*-expressed enzymes and cofactor recycling AT III-binding heparin has been synthesized in multimilligram quantities (107).

In summary, the activities and specificities of heparin and heparan sulfate binding to numerous proteins regulate various physiological and pathophysiological processes. This has resulted in a number of novel therapeutic applications of heparin, heparan sulfate, and heparin-like analogues. Identification of specific oligosaccharide sequences that affect each particular biological process might someday enable the development of novel classes of therapeutics for the treatment on a wide range of diseases.

References

1. Sasisekharan R, Venkataraman G. Heparin and heparan sulfate: biosynthesis, structure and function. Curr Opin Chem Biol 2000; 4:626–631.
2. Tyrrell DJ, Kilfeather S, Page CP. Therapeutic uses of heparin beyond its traditional role as an anticoagulant. Trends Pharmacol Sci 1995; 16:198–204.
3. Capila I, Linhardt RJ. Heparin–protein interactions. Angew Chem Int Ed 2002; 41:390–412.
4. Jacques LB. Heparin: an old drug with a new paradigm. Science 1979; 206:528–533.
5. Linhardt RJ. Perspective: 2003 Claude S. Hudson award address in carbohydrate chemistry. Heparin: structure and activity. J Med Chem 2003; 46:2551–2554.
6. Sasisekharan R, Shriver Z, Venkataraman G, Narayanasami U. Roles of heparan-sulphate glycosaminoglycans in cancer. Nat Rev Cancer 2002; 2:521–528.
7. Hileman RE, Jennings RN, Linhardt RJ. Thermodynamic analysis of the heparin interaction with a basic cyclic peptide using isothermal titration calorimetry. Biochemistry 1998; 37:15231–15237.
8. Bae J, Desai UR, Pervin A, Caldwell EE, Weiler JM, Linhardt RJ. Interaction of heparin with synthetic antithrombin III peptide analogues. Biochem J 1994; 301(Pt. 1):121–129.
9. Olson ST, Chuang YJ. Heparin activates antithrombin anticoagulant function by generating new interaction sites (exosites) for blood clotting proteinases. Trends Cardiovasc Med 2002; 12:331–338.
10. Olson ST, Bjork I, Sheffer R, Craig PA, Shore JD, Choay J. Role of the antithrombin-binding pentasaccharide in heparin acceleration of antithrombin-proteinase reactions. Resolution of the antithrombin conformational

change contribution to heparin rate enhancement. J Biol Chem 1992; 267:12528–12538.
11. Jin L, Abrahams JP, Skinner R, Petitou M, Pike RN, Carrell RW. The anticoagulant activation of antithrombin by heparin. Proc Natl Acad Sci USA 1997; 94:14683–14688.
12. Johnson DJ, Li W, Adams TE, Huntington JA. Antithrombin-S195A factor Xa-heparin structure reveals the allosteric mechanism of antithrombin activation. Embo J 2006; 25:2029–2037.
13. Desai UR, Petitou M, Bjork I, Olson ST. Mechanism of heparin activation of antithrombin. Role of individual residues of the pentasaccharide activating sequence in the recognition of native and activated states of antithrombin. J Biol Chem 1998; 273:7478–7487.
14. Rusnati M, Tanghetti E, Dell'Era P, Gualandris A, Presta M. $\alpha v \beta 3$ Integrin mediates the cell-adhesive capacity and biological activity of basic fibroblast growth factor (FGF-2) in cultured endothelial cells. Mol Biol Cell 1997; 8:2449–2461.
15. Johnson DE, Williams LT, Gritli-Linde A, Lewis P, McMahon AP, Linde A. Structural and functional diversity in the FGF receptor multigene family. Adv Cancer Res 1993; 60:1–41.
16. Pellegrini L, Burke DF, Delft Fv, Mulloy B, Blundell TL. Crystal structure of fibroblast growth factor receptor ectodomain bound to ligand and heparin. Nature 2000; 407:1029–1034.
17. Robinson CJ, Harmer NJ, Goodger SJ, Blundell TL, Gallagher JT. Cooperative dimerization of fibroblast growth factor 1 (FGF1) upon a single heparin saccharide may drive the formation of 2:2:1 FGF1-FGFR2c-Heparin ternary complexes. J Biol Chem 2005; 280:42274–42282.
18. Schlessinger J, Plotnikov AN, Ibrahimi OA, Eliseenkova AV, Yeh BK, Yayon A, Linhardt RJ, Mohammadi M. Crystal structure of a ternary FGF-FGFR-Heparin complex reveals a dual role for heparin in FGFR binding and dimerization. Mol Cell 2000; 6:743–750.
19. Ibrahimi OA, Zhang F, Hrstka SC, Mohammadi M, Linhardt RJ. Kinetic model for FGF, FGFR, and proteoglycan signal transduction complex assembly. Biochemistry 2004; 43:4724–4730.
20. Kemp LE, Mulloy B, Gherardi E. Signalling by HGF/SF and Met: the role of heparan sulphate co-receptors. Biochem Soc Trans 2006; 34:414–417.
21. Lyon M, Deakin JA, Gallagher JT. The mode of action of heparan and dermatan sulfates in the regulation of hepatocyte growth factor/scatter factor. J Biol Chem 2002; 277:1040–1046.
22. Lietha D, Chirgadze DY, Mulloy B, Blundell TL, Gherardi E. Crystal structures of NK1-heparin complexes reveal the basis for NK1 activity and enable engineering of potent agonists of the MET receptor. Embo J 2001; 20:5543–5555.
23. Vander Kooi CW, Jusino MA, Perman B, Neau DB, Bellamy HD, Leahy DJ. Structural basis for ligand and heparin binding to neuropilin B domains. Proc Natl Acad Sci USA 2007; 104(15):6152–6157.
24. Parish CR. The role of heparan sulfate in inflammation. Nat Rev Immunol 2006; 6:633–643.
25. Shikata K, Suzuki Y, Wada J, Hirata K, Matsuda M, Kawashima H, Suzuki T, Iizuka M, Makino H, Miyasaka M. L-selectin and its ligands mediate

infiltration of mononuclear cells into kidney interstitium after ureteric obstruction. J Pathol 1999; 188:93–99.
26. Kawashima H, Watanabe N, Hirose M, Sun X, Atarashi K, Kimura T, Shikata K, Matsuda M, Ogawa D, Heljasvaara R, Rehn M, Pihlajaniemi T, Miyasaka M. Collagen XVIII, a basement membrane heparan sulfate proteoglycan, interacts with L-selectin and monocyte chemoattractant protein-1. J Biol Chem 2003; 278:13069–13076.
27. Celie JW AM, Keuning ED, Beelen RHJ, Drager AM, Zweegman S, Kessler FL, Soininen R, Born JVD. Identification of L-selectin binding heparan sulfates attached to collagen type XVIII. J Biol Chem 2005; 280:26965–26973.
28. McEver RP, Moore KL, Cummings RD. Leukocyte trafficking mediated by selectin-carbohydrate interactions. J Biol Chem 1995; 270:11025–11028.
29. Ma Y-Q, Geng J-G. Heparan sulfate-like proteoglycans mediate adhesion of human malignant melanoma A375 cells to P-selectin under flow. J Immunol 2000; 165:558–565.
30. Springer TA. Traffic signals for lymphocyte recirculation and leukocyte emigration: the multistep paradigm. Cell 1994; 76:301–314.
31. Middleton J, Neil S, Wintle J, Clark-Lewis I, Moore H, Lam C, Auer M, Hub E, Rot A. Transcytosis and surface presentation of IL-8 by venular endothelial cells. Cell 1997; 91:385–395.
32. Wang L, Fuster M, Sriramarao P, Esko JD. Endothelial heparan sulfate deficiency impairs L-selectin- and chemokine-mediated neutrophil trafficking during inflammatory responses. Nat Immunol 2005; 6:902–910.
33. Johnson Z, Kosco-Vilbois MH, Herren S, Cirillo R, Muzio V, Zaratin P, Carbonatto M, Mack M, Smailbegovic A, Rose M, Lever R, Page C, Wells TN, Proudfoot AE. Interference with heparin binding and oligomerization creates a novel anti-inflammatory strategy targeting the chemokine system. J Immunol 2004; 173:5776–5785.
34. Oberlin E, Amara A, Bachelerie F, Bessia C, Virelizier JL, Arenzana-Seisdedos F, Schwartz O, Heard JM, Clark-Lewis I, Legler DF, Loetscher M, Baggiolini M, Moser B. The CXC chemokine SDF-1 is the ligand for LESTR/fusin and prevents infection by T-cell-line-adapted HIV-1. Nature 1996; 382:833–835.
35. Mbemba E, Gluckman JC, Gattegno L. Glycan and glycosaminoglycan binding properties of stromal cell-derived factor (SDF)-1alpha. Glycobiology 2000; 10:21–29.
36. Sadir R, Baleux F, Grosdidier A, Imberty A, Lortat-Jacob H. Characterization of the stromal cell-derived factor-1alpha-heparin complex. J Biol Chem 2001; 276:8288–8296.
37. Shamri R, Grabovsky V, Gauguet JM, Feigelson S, Manevich E, Kolanus W, Robinson MK, Staunton DE, von Andrian UH, Alon R. Lymphocyte arrest requires instantaneous induction of an extended LFA-1 conformation mediated by endothelium-bound chemokines. Nat Immunol 2005; 6:497–506.
38. Mota MM, Pradel G, Vanderberg JP, Hafalla JC, Frevert U, Nussenzweig RS, Nussenzweig V, Rodriguez A. Migration of Plasmodium sporozoites through cells before infection. Science 2001; 291:141–144.
39. Dame JB, Williams JL, McCutchan TF, Weber JL, Wirtz RA, Hockmeyer WT, Maloy WL, Haynes JD, Schneider I, Roberts D, Sanders GS, Reddy EP, Diggs CL, Miller LH. Structure of the gene encoding the immunodominant surface

antigen on the sporozoite of the human malaria parasite Plasmodium falciparum. Science 1984; 225:593–599.
40. Cerami C, Frevert U, Sinnis P, Takacs B, Clavijo P, Santos MJ, Nussenzweig V. The basolateral domain of the hepatocyte plasma membrane bears receptors for the circumsporozoite protein of Plasmodium falciparum sporozoites. Cell 1992; 70:1021–1033.
41. Frevert U, Sinnis P, Cerami C, Shreffler W, Takacs B, Nussenzweig V. Malaria circumsporozoite protein binds to heparan sulfate proteoglycans associated with the surface membrane of hepatocytes. J Exp Med 1993; 177:1287–1298.
42. Rathore D, McCutchan TF, Garboczi DN, Toida T, Hernaiz MJ, LeBrun LA, Lang SC, Linhardt RJ. Direct measurement of the interactions of glycosaminoglycans and a heparin decasaccharide with the malaria circumsporozoite protein. Biochemistry 2001; 40:11518–11524.
43. Lindenbach BD, Rice CM. Unravelling hepatitis C virus replication from genome to function. Nature 2005; 436:933–938.
44. Rosa D, Campagnoli S, Moretto C, Guenzi E, Cousens L, Chin M, Dong C, Weiner AJ, Lau JY, Choo QL, Chien D, Pileri P, Houghton M, Abrignani S. A quantitative test to estimate neutralizing antibodies to the hepatitis C virus: cytofluorimetric assessment of envelope glycoprotein 2 binding to target cells. Proc Natl Acad Sci USA 1996; 93:1759–1763.
45. Flint M, McKeating JA. The role of the hepatitis C virus glycoproteins in infection. Rev Med Virol 2000; 10:101–117.
46. Lindenbach BD, Rice CM. Flaviviridae: the viruses and their replication. Baltimore: Lippincott Williams & Wilkins, 2001.
47. Penin F, Combet C, Germanidis G, Frainais PO, Deleage G, Pawlotsky JM. Conservation of the conformation and positive charges of hepatitis C virus E2 envelope glycoprotein hypervariable region 1 points to a role in cell attachment. J Virol 2001; 75:5703–5710.
48. Barth H, Schafer C, Adah MI, Zhang F, Linhardt RJ, Toyoda H, Kinoshita-Toyoda A, Toida T, Van Kuppevelt TH, Depla E, Von Weizsacker F, Blum HE, Baumert TF. Cellular binding of hepatitis C virus envelope glycoprotein E2 requires cell surface heparan sulfate. J Biol Chem 2003; 278:41003–41012.
49. Barth H, Schnober EK, Zhang F, Linhardt RJ, Depla E, Boson B, Cosset FL, Patel AH, Blum HE, Baumert TF. Viral and cellular determinants of the hepatitis C virus envelope-heparan sulfate interaction. J Virol 2006; 80:10579–10590.
50. Rider CC, Coombe DR, Harrop HA, Hounsell EF, Bauer C, Feeney J, Mulloy B, Mahmood N, Hay A, Parish CR. Anti-HIV-1 activity of chemically modified heparins: correlation between binding to the V3 loop of gp120 and inhibition of cellular HIV-1 infection in vitro. Biochemistry 1994; 33:6974–6980.
51. Vives RR, Imberty A, Sattentau QJ, Lortat-Jacob H. Heparan sulfate targets the HIV-1 envelope glycoprotein gp120 coreceptor binding site. J Biol Chem 2005; 280:21353–21357.
52. WuDunn D, Spear PG. Initial interaction of herpes simplex virus with cells is binding to heparan sulfate. J Virol 1989; 63:52–58.
53. Herold BC, Gerber SI, Polonsky T, Belval BJ, Shaklee PN, Holme K. Identification of structural features of heparin required for inhibition of herpes simplex virus type 1 binding. Virology 1995; 206:1108–1116.

54. Spear PG, Shieh MT, Herold BC, WuDunn D, Koshy TI. Heparan sulfate glycosaminoglycans as primary cell surface receptors for herpes simplex virus. Adv Exp Med Biol 1992; 313:341–353.
55. Lukashev ME, Werb Z. ECM signalling: orchestrating cell behaviour and misbehaviour. Trends Cell Biol 1998; 8:437–441.
56. Ricard-Blum S, Beraud M, Raynal N, Farndale RW, Ruggiero F. Structural requirements for heparin/heparan sulfate binding to type V collagen. J Biol Chem 2006; 281:25195–25204.
57. Delacoux F, Fichard A, Geourjon C, Garrone R, Ruggiero F. Molecular features of the collagen V heparin binding site. J Biol Chem 1998; 273:15069–15076.
58. Delacoux F, Fichard A, Cogne S, Garrone R, Ruggiero F. Unraveling the amino acid sequence crucial for heparin binding to collagen V. J Biol Chem 2000; 275:29377–29382.
59. Gallo R, Kim C, Kokenyesi R, Adzick NS, Bernfield M. Syndecans-1 and -4 are induced during wound repair of neonatal but not fetal skin. J Invest Dermatol 1996; 107:676–683.
60. Woods A, Longley RL, Tumova S, Couchman JR. Syndecan-4 binding to the high affinity heparin-binding domain of fibronectin drives focal adhesion formation in fibroblasts. Arch Biochem Biophys 2000; 374:66–72.
61. Salmivirta M, Elenius K, Vainio S, Hofer U, Chiquet-Ehrismann R, Thesleff I, Jalkanen M. Syndecan from embryonic tooth mesenchyme binds tenascin. J Biol Chem 1991; 266:7733–7739.
62. Midwood KS, Williams LV, Schwarzbauer JE. Tissue repair and the dynamics of the extracellular matrix. Int J Biochem Cell Biol 2004; 36:1031–1037.
63. Wenk MB, Midwood KS, Schwarzbauer JE. Tenascin-C suppresses Rho activation. J Cell Biol 2000; 150:913–920.
64. Midwood KS, Valenick LV, Hsia HC, Schwarzbauer JE. Coregulation of fibronectin signaling and matrix contraction by tenascin-C and syndecan-4. Mol Biol Cell 2004; 15:5670–5677.
65. Dickson BJ. Molecular mechanisms of axon guidance. Science 2002; 298:1959–1964.
66. Van Vactor D, Wall DP, Johnson KG. Heparan sulfate proteoglycans and the emergence of neuronal connectivity. Curr Opin Neurobiol 2006; 16:40–51.
67. Tanaka NK, Awasaki T, Shimada T, Ito K. Integration of chemosensory pathways in the Drosophila second-order olfactory centers. Curr Biol 2004; 14:449–457.
68. Zhang F, Ronca F, Linhardt RJ, Margolis RU. Structural determinants of heparan sulfate interactions with Slit proteins. Biochem Biophys Res Commun 2004; 317:352–357.
69. Hohenester E, Hussain S, Howitt JA. Interaction of the guidance molecule Slit with cellular receptors. Biochem Soc Trans 2006; 34:418–421.
70. Hussain SA, Piper M, Fukuhara N, Strochlic L, Cho G, Howitt JA, Ahmed Y, Powell AK, Turnbull JE, Holt CE, Hohenester E. A molecular mechanism for the heparan sulfate dependence of Slit-Robo signaling. J Biol Chem 2006; 281:39693–39698.
71. Cooper AD. Hepatic uptake of chylomicron remnants. J Lipid Res 1997; 38:2173–2192.
72. Mahley RW, Ji ZS. Remnant lipoprotein metabolism: key pathways involving cell-surface heparan sulfate proteoglycans and apolipoprotein E. J Lipid Res 1999; 40:1–16.

73. Swertfeger DK, Hui DY. Apolipoprotein E: a cholesterol transport protein with lipid transport-independent cell signaling properties. Front Biosci 2001; 6:D526–D535.
74. Swertfeger DK, Hui DY. Apolipoprotein E receptor binding versus heparan sulfate proteoglycan binding in its regulation of smooth muscle cell migration and proliferation. J Biol Chem 2001; 276:25043–25048.
75. Bazin HG, Marques MA, Owens AP, 3rd, Linhardt RJ, Crutcher KA. Inhibition of apolipoprotein E-related neurotoxicity by glycosaminoglycans and their oligosaccharides. Biochemistry 2002; 41:8203–8211.
76. Ji ZS, Pitas RE, Mahley RW. Differential cellular accumulation/retention of apolipoprotein E mediated by cell surface heparan sulfate proteoglycans. Apolipoproteins E3 and E2 greater than e4. J Biol Chem 1998; 273:13452–13460.
77. Mahley RW, Rall SC, Jr. Apolipoprotein E: far more than a lipid transport protein. Annu Rev Genomics Hum Genet 2000; 1:507–537.
78. Aggerbeck LP, Wetterau JR, Weisgraber KH, Wu CS, Lindgren FT. Human apolipoprotein E3 in aqueous solution. II. Properties of the amino- and carboxyl-terminal domains. J Biol Chem 1988; 263:6249–6258.
79. Wilson C, Wardell MR, Weisgraber KH, Mahley RW, Agard DA. Three-dimensional structure of the LDL receptor-binding domain of human apolipoprotein E. Science 1991; 252:1817–1822.
80. Zaiou M, Arnold KS, Newhouse YM, Innerarity TL, Weisgraber KH, Segall ML, Phillips MC, Lund-Katz S. Apolipoprotein E; -low density lipoprotein receptor interaction. Influences of basic residue and amphipathic alpha-helix organization in the ligand. J Lipid Res 2000; 41:1087–1095.
81. Segrest JP, Garber DW, Brouillette CG, Harvey SC, Anantharamaiah GM. The amphipathic alpha helix: a multifunctional structural motif in plasma apolipoproteins. Adv Protein Chem 1994; 45:303–369.
82. Westerlund JA, Weisgraber KH. Discrete carboxyl-terminal segments of apolipoprotein E mediate lipoprotein association and protein oligomerization. J Biol Chem 1993; 268:15745–15750.
83. Cardin AD, Hirose N, Blankenship DT, Jackson RL, Harmony JA, Sparrow DA, Sparrow JT. Binding of a high reactive heparin to human apolipoprotein E: identification of two heparin-binding domains. Biochem Biophys Res Commun 1986; 134:783–789.
84. Weisgraber KH, Rall SC, Jr, Mahley RW, Milne RW, Marcel YL, Sparrow JT. Human apolipoprotein E. Determination of the heparin binding sites of apolipoprotein E3. J Biol Chem 1986; 261:2068–2076.
85. Dong J, Peters-Libeu CA, Weisgraber KH, Segelke BW, Rupp B, Capila I, Hernaiz MJ, LeBrun LA, Linhardt RJ. Interaction of the N-terminal domain of apolipoprotein E4 with heparin. Biochemistry 2001; 40:2826–2834.
86. Libeu CP, Lund-Katz S, Phillips MC, Wehrli S, Hernaiz MJ, Capila I, Linhardt RJ, Raffai RL, Newhouse YM, Zhou F, Weisgraber KH. New insights into the heparan sulfate proteoglycan-binding activity of apolipoprotein E. J Biol Chem 2001; 276:39138–39144.
87. Saito H, Dhanasekaran P, Nguyen D, Baldwin F, Weisgraber KH, Wehrli S, Phillips MC, Lund-Katz S. Characterization of the heparin binding sites in human apolipoprotein E. J Biol Chem 2003; 278:14782–14787.
88. Futamura M, Dhanasekaran P, Handa T, Phillips MC, Lund-Katz S, Saito H. Two-step mechanism of binding of apolipoprotein E to heparin: implications

for the kinetics of apolipoprotein E-heparan sulfate proteoglycan complex formation on cell surfaces. J Biol Chem 2005; 280:5414–5422.
89. Olsson U, Ostergren-Lunden G, Moses J. Glycosaminoglycan-lipoprotein interaction. Glycoconj J 2001; 18:789–797.
90. Bosner MS, Gulick T, Riley DJ, Spilburg CA, Lange LG, 3rd. Receptor-like function of heparin in the binding and uptake of neutral lipids. Proc Natl Acad Sci USA 1988; 85:7438–7442.
91. de Boer HC, de Groot PG, Bouma BN, Preissner KT. Ternary vitronectin-thrombin-antithrombin III complexes in human plasma. Detection and mode of association. J Biol Chem 1993; 268:1279–1283.
92. Ill CR, Ruoslahti E. Association of thrombin-antithrombin III complex with vitronectin in serum. J Biol Chem 1985; 260:15610–15615.
93. Podack ER, Dahlback B, Griffin JH. Interaction of S-protein of complement with thrombin and antithrombin III during coagulation. Protection of thrombin by S-protein from antithrombin III inactivation. J Biol Chem 1986; 261:7387–7392.
94. Preissner KT, Muller-Berghaus G. Neutralization and binding of heparin by S protein/vitronectin in the inhibition of factor Xa by antithrombin III. Involvement of an inducible heparin-binding domain of S protein/vitronectin. J Biol Chem 1987; 262:12247–12253.
95. Wells MJ, Blajchman MA. In vivo clearance of ternary complexes of vitronectin-thrombin-antithrombin is mediated by hepatic heparan sulfate proteoglycans. J Biol Chem 1998; 273:23440–23447.
96. Fugedi P. The potential of the molecular diversity of heparin and heparan sulfate for drug development. Mini Rev Med Chem 2003; 3:659–667.
97. Bauer KA. New pentasaccharides for prophylaxis of deep vein thrombosis: pharmacology. Chest 2003; 124:364S–370S.
98. Islam T, Butler M, Sikkander SA, Toida T, Linhardt RJ. Further evidence that periodate cleavage of heparin occurs primarily through the antithrombin binding site. Carbohydr Res 2002; 337:2239–2243.
99. Garg HG, Hales CA, Yu L, Butler M, Islam T, Xie J, Linhardt RJ. Increase in the growth inhibition of bovine pulmonary artery smooth muscle cells by an O-hexanoyl low-molecular-weight heparin derivative. Carbohydr Res 2006; 341:2607–2612.
100. Page CP. One explanation of the asthma paradox: inhibition of natural anti-inflammatory mechanism by beta 2-agonists. Lancet 1991; 337:717–720.
101. Bowler SD, Smith SM, Lavercombe PS. Heparin inhibits the immediate response to antigen in the skin and lungs of allergic subjects. Am Rev Respir Dis 1993; 147:160–163.
102. O'Donnell WJ, Rosenberg M, Niven RW, Drazen JM, Israel E. Acetazolamide and furosemide attenuate asthma induced by hyperventilation of cold, dry air. Am Rev Respir Dis 1992; 146:1518–1523.
103. Ahmed T, Garrigo J, Danta I. Preventing bronchoconstriction in exercise-induced asthma with inhaled heparin. N Engl J Med 1993; 329:90–95.
104. Gaffney A, Gaffney P. Rheumatoid arthritis and heparin. Br J Rheumatol 1996; 35:808–809.
105. Petitou M, Duchaussoy P, Lederman I, Choay J, Sinay P, Jacquinet JC, Torri G. Synthesis of heparin fragments. A chemical synthesis of the pentasaccharide *O*-(2-deoxy-2-sulfamido-6-*O*-sulfo-alpha-D-glucopyranosyl)-

(1–4)-*O*-(beta-D-glucopyranosyluronic acid)-(1–4)-*O*-(2-deoxy-2-sulfamido-3,6-di-*O*-sulfo-alpha-D-glu copyranosyl)-(1–4)-*O*-(2-*O*-sulfo-alpha-L-idopyranosyluronic acid)-(1–4)-2-deoxy-2-sulfamido-6-*O*-sulfo-D-glucopyranose decasodium salt, a heparin fragment having high affinity for antithrombin III. Carbohydr Res 1986; 147:221–236.

106. Kuberan B, Beeler DL, Lech M, Wu ZL, Rosenberg RD. Chemoenzymatic synthesis of classical and non-classical anticoagulant heparan sulfate polysaccharides. J Biol Chem 2003; 278:52613–52621.
107. Chen J, Avci FY, Munoz EM, McDowell LM, Chen M, Pedersen LC, Zhang L, Linhardt RJ, Liu J. Enzymatic redesigning of biologically active heparan sulfate. J Biol Chem 2005; 280:42817–42825.

Carbohydrate Chemistry, Biology and Medical Applications
Hari G. Garg, Mary K. Cowman and Charles A. Hales

DOI: 10.1016/B978-0-08-054816-6.00011-2

Chapter 11

Carbohydrates and Cutaneous Wound Healing

ANDREW BURD AND LIN HUANG

Division of Plastic and Reconstructive Surgery, Department of Surgery, The Chinese University of Hong Kong, Prince of Wales Hospital, Hong Kong SAR, People's Republic of China

I. Introduction

Skin is the largest immunologically competent organ in the body (1,2). The skin has two layers, epidermis and dermis (Fig. 1). The epidermis is rich in cells that are specialized in the formation of keratin and are called keratinocytes. The epidermis consists of a stratified layer of cells stretching from the basement membrane below to the stratum corneum above. The area of the basement membrane far exceeds that of the surface of the stratum corneum because multiple dermal papillae project from the surface of the dermis and only about 12% of the basal cells are proliferating at any one time in normal skin. These papillae are formed by loose approximations of collagen bundles referred to as the papillary dermis, which lies on the relatively much thicker reticular dermis. Collagen bundles in the reticular dermis are thicker and more condensed. The dermis consists of an organized structure of elastin and collagen fibers between which is a thick, viscous fluid made up of glycosaminoglycans (GAGs) and hyaluronan (HA). This arrangement of the dermis gives the skin its major biomechanical properties, allowing stretching and recoil and deformation without destruction. The dermis contains a complex vascular arrangement of capillary and venous plexuses. Within the dermis are adnexal structures of ectodermal origin—hair follicles, sweat glands, and sebaceous glands—which are lined with keratinocytes. This arrangement becomes particularly important when considering the mechanisms of wound healing.

The skin performs a diverse range of vital body functions in addition to offering protection (3). When the skin is wounded, healing takes place by two principal processes: *regeneration* and *repair*. Regeneration is typified by the process of reepithelialization where the form and function of the previously injured tissue is completely replaced. Repair occurs when the dermis is breached and as a result of a macrophage-effected fibroproliferative process, a new dermal matrix is created, which does not have the same structure as the uninjured tissue and is apparent as scar tissue. Scarring is the cause for disability and deformity after injury to the skin and abnormal scars can exacerbate the problems (4). Both healing processes are extremely complex and involve cell proliferation and migration as well as extracellular matrix deposition, modeling, and remodeling. The aim of wound management is to achieve uncomplicated healing with minimal scarring.

Carbohydrates represent a ubiquitous range of compounds present in nature subserving a complex range of forms and functions. They have been used in a variety of forms with a range of functions to support and modulate the wound healing process. In this chapter, we focus on the topical application of exogenous carbohydrates in the modulation of cutaneous wound healing. There are three principal roles: carbohydrates as dressings, as wound healing modulators, and as tissue-engineering constructs to aid in the replacement or facilitate reconstruction of lost or damaged skin (Fig. 1).

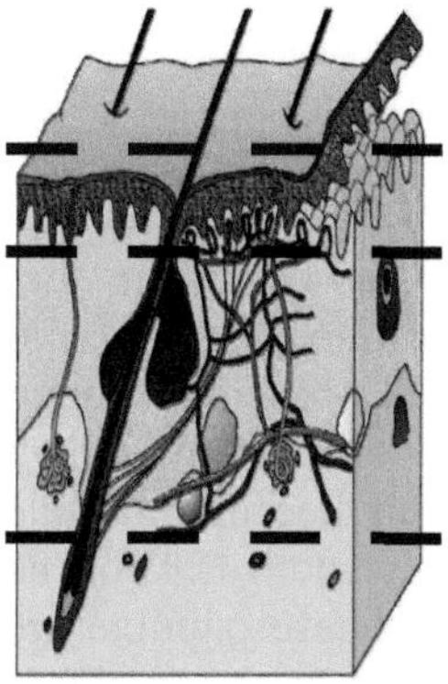

Layers	Nature of Healing	Process	Outcome	Role(s) of Carbohydrates
Epidermis	Regeneration	Cellular proliferation and migration	No scar	Superficial wounds **Dressings**
Dermis	Repair	1. Inflammation 2. Proliferation and matrix deposition 3. Remodeling	Scar	Partial thickness wounds **Dressings and/or modulation** Full-thickness wounds **Tissue engineered replacement**

Figure 1 An injury can be defined as a discontinuity in tissue integrity and healing as a process of restoring that integrity. Skin has two main forms of healing with very different outcomes in terms of appearance and function. Carbohydrates are used in a number of roles to enhance healing.

II. Carbohydrate Polymers as Wound Dressings

A. Wound Dressing

An ideal wound dressing would be protective, proteolytic, and capable of relieving pain and promoting healing.

1. Protection

The skin provides a number of protective functions. The stratum corneum acts as a barrier that prevents invasive infection, and sebum secreted by sebaceous glands has an antibacterial action. The burn wound is vulnerable to bacterial colonization and invasive bacterial contamination and sepsis. As the burn depth increases the potential reservoir for bacterial invasion increases before viable tissue and host defenses are reached. Burns dressing should have either an antibacterial barrier function or a bactericidal/bacteriastatic function or ideally both.

The dressing should also protect the regenerating epidermis from further mechanical injury.

2. Proteolytic

Apart from the most superficial injuries, the wound is going to result in a layer of dead tissue. The natural process of removing this tissue involves proteolysis and phagocytosis. The deeper the wound the more tissue has to be removed and the longer it will take for healing to occur. The ideal dressing would either support or augment the natural autolytic process or, even better, have an intrinsic enzymatic action.

3. Healing

The ideal dressing would promote healing through regeneration. Regeneration occurs when the epidermis is damaged but repair is the mechanism for dermal wounds healing. Repair is a macrophage-driven fibroproliferative process and collagen is rapidly but randomly deposited. The fibroblasts involved are recruited by the macrophages but their origin remains a subject of controversy. They are mesodermal cells and one possibility is that they are derived from a differentiated endothelial cell at the front end of sprouting capillary buds. Whatever their origin, they seem to have lost the blueprint for the original dermal architecture.

Apart from the selective promotion of specific cell behavior, the ideal wound dressing would also provide an environment which would be more conducive to healing. In this respect, the healing wound performs better in a moist healing environment but not in a macerated healing environment. As such, the dressing should have the physical ability to deal with the exudates produced by the wound but at the same time not allow the wound to dry out and the tissues to desiccate.

4. Pain Relieving

Partial-thickness burns can be extremely painful and one aspect of pain control is the nature of the dressing applied to the wound. In addition, when the dressing needs to be changed, it should be possible to remove it easily from the wound prior to reapplication.

B. Carbohydrate Polymer Dressings

Polysaccharides are of numerous variations in composition, structure, and function. Being naturally occurring biomolecules, polysaccharides have been an obvious choice for investigation as potential wound management aids (5). In recent years, it was recognized that not only can polysaccharides be produced with required physical characteristics for a wound dressing, but that the actual polysaccharide or polysaccharide derivative may itself actively participate in the process of wound healing. In this section, we will review a number of polysaccharides that are commonly used as wound dressings, the physical forms in which they are used, and also the biological properties which enable them to participate actively in the wound healing process.

1. *Cellulose*

Cellulose is a complex carbohydrate, $(C_6H_{10}O_5)_n$, that is composed of glucose units, forms the main constituent of the cell wall in most plants, and is important in the manufacture of numerous products, such as paper, textiles, pharmaceuticals, and explosives. In native form it is highly crystalline and rigid. It is insoluble in water and other ordinary solvents and exhibits marked properties of absorption. Because cellulose contains a large number of hydroxyl groups, it reacts with acids to form esters and with alcohols to form ethers. Cellulose derivatives include guncotton, fully nitrated cellulose, used for explosives; celluloid (the first plastic), the product of cellulose nitrates treated with camphor; collodion, a thickening agent; and cellulose acetate, used for plastics, lacquers, and fibers such as rayon.

Cellulose derivatives have been widely used as hydrophilic particles to distribute in the adhesive composite of a hydrocolloid dressing. Aquacel® (ConvaTec, United Kingdom) is a primary wound dressing made from sodium carboxymethylcellulose and produced as a textile fiber. The dressing absorbs and interacts with wound exudate to form a soft, hydrophilic, gas-permeable gel that traps bacteria and conforms to the contours of the wound while providing a microenvironment that is believed to facilitate healing. It has been tested in the treatment of partial-thickness burns (6) and deep diabetic foot ulcers (7), and for healing split-skin graft donor sites (8). The dressing has been shown to be safe, easy to apply, and also good in pain relief and promotion of epithelialization. Promogran (Johnson & Johnson, Cincinnati, Ohio), another wound dressing used for the management of exuding wounds, is composed of collagen and oxidized regenerated cellulose (ORC) matrix. In the presence of wound exudates, the matrix absorbs liquid and forms a soft, conformable, biodegradable gel that physically binds and inactivates matrix metalloproteases (MMPs) (9). The gel also binds naturally occurring growth factors within the wound and protects them from degradation by the proteases, and thereafter releases them back into the wound in an active form as the matrix slowly breaks down (10). Promogran has been indicated to accelerate healing in a number of randomized controlled trials (11,12).

2. *Dextran*

Dextran is a polyglucose biopolymer characterized by preponderance of α-1,6 linkage, and generally produced by enzymes from certain strains of *Leuconostoc* or *Streptococcus*. While formerly its principal utility was as a blood plasma substitute, dextran is also used in various fields such as pharmaceutical, photographic, agricultural, and food industries. Dextran fractions are characterized by their average molecular weights and molecular weight distributions. Dextran's chemical and physical properties depend on the strain of microorganism employed and the environmental conditions imposed on the bacterium during growth, or the reaction conditions where an enzymatic method of dextran production is employed. Dextran is neutral and water soluble even at high molecular weight. It is also biocompatible and biodegradable.

Dextran appeared to have beneficial activities in the treatment of wounds. To reduce its solubility, an emulsion polymerization of dextran using epichlorohydrin as the cross-linking agent was carried out, which produced insoluble beads that swell in water. Dextranomer (Pharmacia, Inc., Piscataway, NJ) is such a dressing with polysaccharide beads, 0.1–0.3 mm in diameter. It is highly hygroscopic and 1 g can absorb 4 ml of water and swell until it is saturated. It has been shown that bacteria and cellular debris present in the wound are taken up by capillary action and become trapped in the spaces between the beads. When the dressing is changed, this debris will be washed away (13). Dextran accelerates the polymerization of fibrin and also influences the structure of the fibrin clot with the diameter of the fibrin fibers being broader when dextran is present (14). Dextran derivatives that mimic the action of heparin have been shown to protect heparin-binding growth factors associated with wound healing, such as transforming growth factor (TGF)-β1 and fibroblast growth factor (FGF)-2 (15). The sulfated dextran derivatives were also shown to accelerate the collagen matrix organization and stimulate the human type-III collagen expression and induce apoptosis of myofibroblasts, a property which may be beneficial in the treatment of hypertrophic and keloid scar (16). These cross-linked dextran beads were found to be able to stimulate macrophages in the wound healing process (17). Positively charged cross-linked diethylaminoethyl dextran (CLDD) beads were also shown to enhance significantly the tensile properties of healing cutaneous wounds in both adult and geriatric rhesus (18).

3. *Alginates*

Alginate (alginic acid) is a viscous gum that is abundant in the cell walls of brown algae. Chemically, it is a linear copolymer with homopolymeric blocks of (1,4)-linked β-D-mannuronate (M) and its C-5 epimer α-L-guluronate (G) residues, respectively, covalently linked together in different sequences or blocks. Commercial varieties of alginate are extracted from seaweed, including the giant kelp *Macrocystis pyrifera*, *Ascophyllum Nodosum*, and various types of *Laminaria*.

Alginate absorbs water quickly. The high absorption is achieved via strong hydrophilic gel formation. Alginates partly dissolve on contact with wound fluid to form a gel as a result of the exchange of sodium ions in wound fluid for calcium ions in the dressing. Calcium alginate dressings such as Sorbsan (Pharma-Plast Ltd., ConvaTec, United Kingdom), Tegagen (3M Health Care Ltd., United Kingdom), and Kaltostat (ConvaTec) have been successfully used to cleanse a wide variety of secreting lesions, to control wound secretion levels, and to minimize bacterial contamination (19–21). There have been a few studies of the effect of alginate dressings on the processes of wound healing (19,22,23). The results indicated that calcium alginate may improve some cellular aspects of normal wound healing, that is, increasing the proliferation of fibroblast but decreasing fibroblast motility. On the contrary, the calcium alginate decreased the proliferation of microvascular endothelial cells and keratinocytes but had no effect on keratinocyte motility. In a number of clinical trials, alginates have been approved to be a significant improvement on conventional dressings in the treatment of diabetic and trophic foot ulcers (24), packing deep wounds (25), treatment of full-thickness pressure ulcers (26), and in healing split-skin graft donor sites (27,28). The dressings maintain a physiologically moist microenvironment that promotes healing and the formation of granulation tissue.

4. Chitin and Chitosan

Chitin, the second most abundant natural polysaccharide, is widely distributed in nature as the principal component of exoskeletons of crustaceans and insects as well as of cell walls of some bacteria and fungi. Like cellulose, it is a glucose-based unbranched polysaccharide. Partial deacetylation of chitin results in the production of chitosan, which is a polysaccharide comprising copolymers of glucosamine and *N*-acetyl glucosamine. It is also naturally present in some microorganisms and fungi.

Chitosan is a biocompatible and biodegradable polymer with its degradation products being known natural metabolites. Chitosan possesses the characteristics favorable for promoting rapid dermal regeneration and is being used as a wound healing accelerator in veterinary medicine. It is observed that both chitosan and chitin are chemoattractants for neutrophils, an early event essential in accelerated wound healing (29). Chitosan enhances the functions of inflammatory cells such as polymorphonuclear leukocytes (PMN) (phagocytosis, production of osteopontin and leukotriene B4) and macrophages [phagocytosis, production of interleukin (IL)-1, TGF-β1, and platelet-derived growth factor] (30). As a result, chitosan promotes granulation and organization. Chitin and chitosan also induce fibroblasts to release IL-8, which is involved in migration and proliferation of fibroblasts and vascular endothelial cells (30). Chitosan membrane as a wound dressing has been tested at the skin-graft donor site in patients, showing a positive effect on the reepithelialization and the regeneration of the granular layer (31). It was reported that chitosan membranes did not restrict normal human skin fibroblasts but impeded keloid fibroblasts by inhibiting type 1 collagen secretion and suggested a role for wound healing in keloid control (32,33).

Chitosan has also been applied as an antimicrobial agent (34). Its antimicrobial activity is well observed on a wide variety of microorganisms including fungi, algae, virus, and some bacteria (35). An engineered chitosan acetate dressing, HemCon® bandage (HemCon Medical Tech, Inc. Portland, Oregon), was reported to rapidly kill bacteria in highly contaminated wounds in mice before systemic invasion can take place, and is superior to alginate bandage and topical silver sulfadiazine treatment (36).

Chitosan has been combined with a variety of modified materials to improve the healing process. Photocrosslinked chitosan hydrogel containing FGF-2 accelerated the rate of healing in healing-impaired db/db mice (37). FGF-2 molecules were gradually released from chitosan hydrogelson their *in vivo* biodegradation. Another investigation on healing of second-degree burn wounds in rat model with chitosan gel formulation containing epidermal growth factor (EGF) demonstrated a better and faster epithelialization in the EGF-chitosan gel treatment group. Antibacterial agents chlorhexidine and silver sulfadiazine have also been introduced to the chitosan gel or membrane, and the controlled release of these antibiotics was found to be effective in controlling infection in wound healing (38). Chitosan in combination with alginate as polyelectrolyte complex (PEC) films displays greater stability to pH changes and is more effective in promoting accelerated healing of incisional wounds in a rat model (39).

Chitosan and its derivatives are also effective in regenerating the wounded skin tissue. A chitosan/collagen-based scaffold combining dermal stem cells and hair follicle epidermal stem cells was reported to provide a suitable substrate for the tridimensional growth of skin stem cells (40). An active laminin peptide-conjugated chitosan membrane has recently been demonstrated as a suitable scaffold material for delivering keratinocytes to the wound bed (41).

The physical and chemical properties of chitosan affect its biological performance. A study was done to correlate fibroblast responses with known chitosan material characteristics, including degree of deacetylation (DDA), molecular weight , and residual protein and ash contents (42). No relationship was found between DDA, wettability, molecular weight, and cell attachment or proliferation. However, a general trend was observed for increasing proliferation with increasing residual ash content and decreasing residual protein.

III. Carbohydrate in Wound Healing Modulation

Because of the vast number and range of carbohydrates, it is not surprising that there are well-established examples of carbohydrates that are applied to patients not as a dressing but as a means to change the biological processes of wound healing. In this section, we look at two such clinical applications, honey and heparin. These provide an interesting comparison that illustrates the complexities of "evidence" in clinical medicine. There is no doubt that honey has been used for many thousands of years and that there are many papers reporting clinical outcomes. Nevertheless, honey is a complex chemical compound. Being a natural product of such complexity, it is understandable that there will be variations in the nature and efficacy of the product, no matter how much care and attention is made to keep conditions of preparation, extraction, and production constant. The analogy would be in wine production and even the most inexperienced enologist will appreciate the wide variations in wines from different geographical regions even though derived from the same grape variety. Moreover, despite all attempts to achieve a reproducibility of product not every vintage from the same region will have the same characteristics. Heparin, on the other hand, is a precise and specific chemical with a particular formulation and will be constant in its chemical properties and reproducible in its biochemical interaction. In some respects, honey and heparin stand as paradigms of key carbohydrate strategies in wound healing modulation, exemplifying the paradoxes of the traditional approach and the Western scientific approach to medicine. There have been attempts to standardize honey for therapeutic application but this standardization is based on outcome and not on composition. The outcome measured is called the Unique Manuka Factor (UMF) and relates to the ability of honey to kill standardized bacteria.

A. Honey

Honey is a sweet syrupy substance produced by honeybees from the nectar of flowers. Its precise composition, and color, will depend on the source of the nectar. After collecting the nectar, bees return to their hive and begin to ingest and regurgitate the nectar a number of times until it is partially digested. This product is stored in honeycombs and then subjected to evaporative water loss. This causes a rise in sugar content that prevents fermentation of the honey by naturally occurring yeasts. Honey is a mixture of sugars and other compounds. With respect to carbohydrates, honey is mainly fructose (about 38.5%) and glucose (about 31.0%). The remaining carbohydrates include maltose, sucrose, and other complex carbohydrates. In addition, honey contains a wide array of vitamins, such as vitamin B_6, thiamin, niacin, riboflavin, and pantothenic acid. Essential minerals, including calcium, copper, iron, magnesium, manganese, phosphorus, potassium, sodium, and zinc, as well as several different amino acids have been identified in honey. Honey also contains several compounds that function as antioxidants. Known antioxidant compounds in honey are chrysin, pinobanksin, vitamin C, catalase, and pinocembrin.

1. Honey in Wound Healing

There is a long tradition of the use of honey in wound healing. Medical literature from ancient Egypt and Greece as well as the Ayurvedic traditions of India have described the use of honey alone or in combination with other substances as a wound healing agent (43). These traditional applications of honey have evolved into current evidence-based practices in which honey is used as a therapeutic agent to modulate wound healing in a multifactorial way. It promotes debridement, kills bacteria, creates a moist healing environment, and appears to actively promote healing (44).

A systematic review of the use of honey as a wound dressing was undertaken in 2001 (45). This review looked at seven randomized trials using burns of various depths as well as infected postoperative wounds. Comparators were polyurethane film, amniotic membrane, potatoes peel, and silver sulfadiazine. Honey compared favorably with all the comparators and showed a significantly shorter healing period, but the quality of the studies was considered to be low to give a high rating to the confidence of the usefulness of honey as a treatment. Nevertheless, the biological plausibility of such treatment was acknowledged.

Five years later, Molan in a review article revisited the evidence for the use of honey as a wound dressing (46). This was an interesting review written by a person with an established interest in undertaking research and clinical usage of honey. The review details positive findings of the therapeutic applications of honey in 12 randomized controlled trials involving 1965 participants and 5 clinical trials (not randomized) involving 97 participants. Further evidence of the effectiveness of honey on wound healing was detailed in 16 trials involving 533 wounds in animals. It was noted that there were also numerous case reports detailing the efficacy of honey. Of note, there are reports of the success of honey in the treatment of some particularly difficult organisms including Methicillin-resistant *Staphylococcus aureas* and Vancomycin-resistant enterococci (47). What is particularly interesting however are the comments concerning the abundant clinical evidence to support the use of honey compared with the relative lack of evidence supporting the use of other wound care products.

2. *Clinical Application*

While a significant amount of evidence exists concerning the potential applications of honey, there are a number of factors that perhaps contribute to the lack of clinical use. The first problem is the sourcing of *therapeutic honey*.

It has been recognized throughout the history of the medicinal use of honey that not all honeys are the same (48). Certainly all honeys have a high sugar content, which inhibits bacterial growth, and all are acidic in nature with a pH of 3.2–4.2, which again inhibits the growth of most pathogenic bacteria. There is, in addition, the production of hydrogen peroxide, which in the therapeutic honey is going to be a slow, low level production that will kill bacteria but not harm tissues. Specific, as yet unidentified plant-related factors have also been proposed by those promoting commercially available "therapeutic" honey; floral sources such as *Leptospermum scoparium* (Manuka) and *L. polygalifolium* (Medihoney) (49) are promoted as having additional antibacterial, antioxidant, or other effects. On the contrary, Subrahmanyam from India has published a number of studies confirming the positive effects of honey, including a prospective, randomized comparative study of superficial burn wound healing comparing honey and silver sulfadiazine (50). The honey-dressed wounds showed, on histological analysis, earlier subsidence of acute inflammatory changes together with better infection control and more rapid wound healing than did the silver sulfadiazine–treated wounds. What is significant about this study is that the honey used was "pure, unprocessed, undiluted, honey obtained from hives." The therapeutic action of honey, however, remains a matter of research and exploration. The role of hydrogen peroxide as the main antibacterial agent was looked at in a study reported by Henriques *et al.* (51). The background to this study was the observation that when honey is diluted, hydrogen peroxide is generated by glucose oxidase present in the honey. However, if catalase is added to the diluted honey, hydrogen peroxide is not produced and most raw honeys thus treated will lose their antibacterial activity. However, there are some "raw" honeys that maintain their antibacterial activity even when diluted in the presence of catalase. These honeys rely on components other than hydrogen peroxide for antibacterial effects. It has been proposed that phytochemicals may be responsible for these effects. The study looked at the free radical production and antioxidant potential of three groups of honey. The three honeys were: (a) antibacterial, nonperoxide producing honey; (b) antibacterial, peroxide producing honey; and (c) non-antibacterial honey. All three honeys were found to power antioxidant potential although the rate of "quenching" varied with (a) the Manuka honey, completely quenching added radicals in 5 min. The production and quenching of free radicals is however a complex balance in nature. Peroxide can lead to the formation of hydroxyl and superoxide which are both produced by mammalian cells to produce an antibacterial effect, but they are also produced by bacteria to cause tissue damage and permit bacterial invasion.

The wound healing effects of honey are of particular interest in the chronic wound. Topical applications of honey have been shown to change the wound healing profile from a static chronic phase to an active healing phase. Using a monocytic cell line, MonoMac-6 (MM6), Manuka, pasture, and jelly bush honey have all been shown to increase the secretion of TNF-α, IL-1β, and IL-6 (48). It is obvious that the mode of action of honey on wound healing is not a simple one and that a synergistic, multifactorial effect most probably occurs with a range of responses which will be determined by both host factors and "honey" factors. As such, this remains a very open-ended field of both clinical and laboratory research.

The other major problem is the practical application of honey. There is no consensus regarding the best method of application or the frequency of change. In most studies, it is applied daily or on alternate days directly onto the wound. Others use honey-soaked gauze and/or occlusive dressings. Molan has concisely listed the practical considerations when using honey clinically, and Table 1 is based on his published guidelines (44). Medical honey is certainly different from commercial consumption honey in that the latter is heated, which inactivates the enzyme responsible for the production of hydrogen peroxide. Sterilization of therapeutic honey used in "Western medicine" is achieved via gamma irradiation.

B. Heparin

Heparin is an example of GAGs. These are long, unbranched polysaccharides consisting of repeating disaccharide units. The unit consists of an *N*-acetyl hexosamine and a hexose or hexuronic group, either of which or both may be sulfated. These are physiologically reactive with a high density of negative charge and are acidic. Other GAGs include chondroitin sulfate, dermatan sulfate, keratin sulfate, heparin sulfate, and HA. Heparin is the most acidic and most sulfated of the GAGs and has been used in wound healing, particularly in burns. Indeed, it has been used parenterally, topically, by inhalation, by pellet, and also in tissue-engineered constructs in burns patients (52).

Research in heparin complexes and wound healing has indicated some potential modes of action. One *in vitro* study looked at the effect of a heparin–chitosan gel, which resulted in 90% reepithelialization of experimental wounds compared with 30% when covered with chitosan alone and 0% when just treated with

Table 1 Practical Considerations for the Clinical Use of Honey [Based on Molan (44)]

1. The amount of honey required on the wound relates to the amount of fluid exuding from the wound diluting it. The frequency of dressing changes required will depend on how rapidly the honey is being diluted by exudates. If there are no exudates, dressings need to be changed twice-weekly to maintain a "reservoir" of antibacterial components as they diffuse into the wound tissues.
2. To achieve best results the honey should be applied to an absorbent dressing prior to application. If applied directly to the wound, the honey tends to run off before a secondary dressing is applied to hold it in place.
3. In some situations, a "blister" of honey can be held on a wound using an adhesive film dressing. Honey can be used to treat cavity wounds in this way, although this approach is not suitable for heavily exuding wounds.
4. For moderately to heavily exuding wounds, a secondary dressing may be needed to contain seepage of diluted honey from the primary dressing.
5. Alginate dressings impregnated with honey are a good alternative to cotton/cellulose dressings, as the alginate converts into a honey-containing soft gel.
6. Any depressions or cavities in the wound bed need to be filled with honey in addition to using a honey-impregnated dressing. This is to ensure that the antibacterial components of the honey diffuse into the wound tissues.

heparin. The stimulatory effect of the heparin–chitosan complexes was related to the concentration of the heparin, and it was hypothesized that the effects were due to stabilization and activation of growth factors bound to the immobilized heparin (53). A recent report has described the enhanced healing of full-thickness wounds in genetically diabetic (db/db) mice using a synthetic extracellular matrix hydrogel film. The film comprises co-crosslinked thiolated derivatives of chondroitin-6-sulfate and heparin. The film alone accelerated wound repair but this effect could be enhanced in a dose-dependent way by adding basic fibroblast growth factor (bFGF). bFGF is recognized on a polypeptide that stimulates growth and differentiation of cells. It stimulates wound repair in experimental models but this effect cannot be translated into the clinical situation. It has been suggested that topical application fails as the bFGF is rapidly displaced from the wound bed by exudates and also, in soluble form, it is rapidly degraded. Within the dermis there is a natural reservoir for bFGF bound to heparin sulfate. The cross-linked GAGs hydrogel immobilized heparin, which mimics the naturally occurring heparin sulfate and binds bFGF. When the bFGF-loaded polymer was applied to the wounds, it provided a slow and sustained release of bFGF into the wound (54).

These two reports 10 years apart reflect the continuing interest in heparin combinations playing a multifactorial role in wound healing. The role of heparin alone has also received considerable attention. Considerable research has looked at the role of heparin in anticoagulation but in the context of cutaneous wound healing the focus is on topical application.

1. Topical Heparin

Saliba describes a protocol of topical applications using varying amounts of heparin at a concentration of 5000 IU/ml (55). The solution is sprayed onto the wound using a 5- or 10-cc syringe and a 30-gauge needle. Applications will be repeated every 5 min for 25–30 min. Following this initial treatment, the application is repeated two to three times a day using diminishing doses and duration of application until healing occurs. An alternative approach to the topical application when there are blisters present is to aspirate the blister fluid and inject heparin solution into the blister cavity. In Saliba's review, he reports that the topical application of heparin to burn wounds leads to a rapid relief from pain, rapid healing, and decreased scarring. The clinical results have been reported in meetings and meeting proceedings but there is a paucity of evidence in peer-reviewed journals. There are many published reports looking at the beneficial effects of systemic heparin administration in a wide range of conditions including burns. In addition, there are a large number of laboratory-based studies looking at the interaction between heparin matrices and cells. The angiogenic effects of heparin are well established (56) and laboratory studies have indicated that heparin can induce cell proliferation and collagen production in fibroblasts (57). The effects and mechanisms of topical heparin on wound healing are, however, neither well described nor understood and there is considerable scope for further research in this area.

IV. Carbohydrate in Skin Tissue Engineering

A. Tissue Engineering

When the term "tissue engineering" was officially coined at a National Science Foundation Workshop in the United States in 1988, it was understood to mean "the application of principles and methods of engineering and life sciences toward the fundamental understanding of structure–function relationships in normal and pathological mammalian tissues and the development of biological substitutes to restore, maintain, or improve tissue function." This concept has unfortunately led to some serious misconceptions that have resulted in the early promise of skin tissue engineering being slow to be realized in clinical practice. The misconception was that skin is a tissue, like cartilage, and would be relatively simple to address as a tissue-engineering challenge. Skin however is NOT a tissue but an extremely complex organ that brings into conjunction cells from three different embryological origins and serves multiple functions (3). The original futuristic claims of producing "off-the shelf" skin replacements have become far more restrained in their expectation and now tissue-engineering skin products are being described as skin substitutes to aid healing and repair, temporary skin replacements, and occasionally aids to regeneration (58).

1. Strategies of Skin Tissue Engineering

Strategies used to construct skin substitutes in tissue engineering are generally considered to be either *ex vivo* tissue manufacturing with guided generation or *in vivo* regeneration. The strategy of *ex vivo* tissue manufacturing is the technique initially most commonly associated with tissue engineering. In this approach, fibroblasts and/or keratinocyte are seeded into dermal matrix or scaffold and cocultured in a bioreactor or specialized culture system with some growth factors. The matrix provides a scaffold combining with the bioreactor providing cellular nutrients, allowing the cells to proliferate and differentiate in the *ex vivo* environment. When the procedure is completed, the skin substitute is implanted into the wound and further matures and integrates into the recipient tissues.

The other strategy is to guide the "bioengineering" of skin *in situ*. Such a strategy needs an understanding of the cellular and molecular interactions in tissue healing and development.

Skin substitutes are needed for wounds that arise from extensive tissue loss or damage. These may result from trauma, in particular burns, and in pathological conditions such as epidermolysis bullosa and acute exfoliative skin conditions. Such wounds may need either temporary or permanent closure with substitutes. The ideal skin substitute should

1. protect the wound, maintain a moist healing environment, and control protein and electrolyte loss;
2. prevent local infection and provide an environment for accelerated wound healing;
3. reduce pain and allow early mobilization;
4. be easy to handle and cost-effective;
5. be safe in terms of virus transmission and not provoke a strong immunological reaction; and
6. be readily available.

B. Carbohydrate and the Dermal Matrix

There are two basic roles for carbohydrates in the field of skin replacement. These relate to the bilaminar nature of the skin with the dermal matrix and the cellular epidermis. The matrix is a biomechanical construct and can be "made." As already described, there are collagenous and noncollagenous components in the human dermis. There is and has been a considerable amount of research to explore the roles of collagen and GAG polymers in matrix production. Indeed, this is a feature of research and reviews spanning the last two decades. The most successful commercial product is *Integra*, which was approved for clinical use in 1996 both in the United States and Europe.

1. Integra

Integra sets the standard in addressing the problem of *dermal repair* (59). Integra is a bilaminar material; the outer layer, silicone, is synthetic. The inner layer is derived from bovine type I collagen and shark-derived chondroitin sulfate combined in a tissue-engineering process to produce a biodegradable matrix. The design of this matrix is such that the process of degradation results in the formation of a highly organized autocollageous replacement. With concurrent vascularization the new "dermis" can be covered with autologous keratinocytes after the silicone outer layer is removed. The concept and practice come close to being realized in terms of permanent skin replacement. The process is, however, influenced by many factors and while the best results compare extremely favorably with autologous full-thickness and thick-split thickness grafts there is still a considerable range in outcomes that suggests more work needs to be done both in refining the product and defining the clinical procedures and applications.

Yannas, one of the inventors of Integra, was the coauthor of a paper published in 1996 looking at the, then, recent advances in biologically active analogous of the extracellular matrix (60). It was noted in this review that many GAGs had been grafted onto collagen, including chondroitin-6-sulfate, chondroitin-4-sulfate, heparan sulfate, heparin, dermatin sulfate, and keratin sulfate. It was noted that most research had focused on type I collagen and chondroitin-6-sulfate (the composition that had been successfully used in Integra). Yannas had also published work looking at decorin and aggrecan as substitutes for chondroitin-6-sulfate in a model of delayed wound contraction and enhanced regeneration (61). The outcome of these studies really confirmed the importance of the GAG rather than the protein core of a proteoglycan in the positive effects of the dermal analogue on wound healing. This observation has stood the test of time and a report of blended nanofibrous scaffolds containing collagen and a GAG had retained the collagen–chondroitin sulfate combination (62).

C. Carbohydrate in Cell Delivery

The second major role for carbohydrate in skin tissue engineering is in cell delivery. In this respect, the most widely used carbohydrate is *hyaluronan*.

1. Hyaluronan

This has a dual application both in matrix modification and in cell delivery. HA is attractive as a building block for new biocompatible and biodegradable polymers that have applications in tissue engineering. However, HA has poor biomechanical properties in its native form and a variety of chemical modifications have been devised to provide mechanically and chemically robust materials. The resulting HA derivatives have physicochemical properties that may significantly differ from those of the native polymer, but most derivatives retain the biocompatibility and biodegradability, and in some cases, the pharmacological properties, of native HA. The most commonly used modification of native HA is esterification of the carboxyl groups (63).

Esterified HA biomaterials have been prepared by alkylation of the tetra (*n*-butyl)ammonium salt of HA with an alkyl halide in dimethylformamide (DMF) solution. At higher percentages of esterification, the resulting HYAFF® materials (Fidia Advanced Biopolymers, Abano Terme, Italy) became insoluble in water. These HA esters can be extruded to produce membranes and fibers, lyophilized to obtain sponges, or processed by spray-drying, extraction, and evaporation to produce microspheres. These polymers show good mechanical strength when dry, but the hydrated materials are less robust. The degree of esterification influences the size of hydrophobic patches, which produces a polymer chain network that is more rigid and stable, and less susceptible to enzymatic degradation.

a. Structure and physical properties of HYAFF. The HYAFF series of semisynthetic HA derivatives consist of alcohol esters of HA, which vary in respect to the pendant alcohol group (benzyl and ethyl) and the level of esterification achieved. X-ray diffraction studies of the partial and total benzyl esters of HA (HYAFF 11p75 and HYAFF 11) indicate that the HYAFF derivatives exhibit a diffraction pattern similar to that of the unsubstituted HA. Molecular modeling based on the X-ray data demonstrates that the benzyl esters of HA have the same conformity and flexibility as unmodified HA. The ethyl ester of HA is known as HYAFF 7. Extruding HYAFF into organic solvents has allowed the production of fibers and membranes. The fibers can be woven and the membranes perforated to produce a wide range of biomaterials. These HYAFF-based biomaterials are biodegradable. The principal factor that determines the length of time it takes for the materials to degrade is the polymer used. The three most common polymers with increasing resistance to biodegradation are HYAFF 11p75<HYAFF 7<HYAFF 11. The length of time implanted products may remain can range from several weeks to several months and is also influenced by the physical conformation of the product and the site of the implantation (64). Degradation of the HYAFF occurs through an initial hydrolysis of the ester bond which releases free alcohol. Solubilization of the material then follows and the free alcohol and HA are broken down by their normal metabolic pathway. This means that a wide range of safe and biodegradable medical devices can be developed for specific clinical applications.

b. Animal studies. Studies on cutaneous wounds have been performed on pigs. In these studies, two biomaterials formulated from the partial benzyl ester of HA (HYAFF 11p75) were used. These were a fleece-like material HYALOFILL-F and a rope-like material HYALOFILL-R. Both forms of the material have shown good biocompatibility and were compared with Sorbsan, a calcium alginate preparation, in the healing of deep excisional punch wounds made in Yucatan micropigs.

The histological assessment of the wound healing process in this comparative study underlined some of the biological advantages of using HA and HA derivatives in wound healing. A principal advantage of HA over other biological materials is that it is characteristically inert or even inhibitory as far as inflammatory and immunological reactions are concerned. Some macrophages were observed transiently during the phase of degradation but these were presumed to be consuming the soluble biomaterial. These macrophages are therefore acting in a relatively unactivated form as far as expressing cytokines in the classical wound repair context. Thus in this animal model, the speed of healing and also the quality, in terms of reduced scarring, both appeared to be enhanced by using HYAFF formulations (65).

c. Clinical studies. Clinical application of HYAFF formulation has proved to demonstrate considerable promise. Laserskin™ is a membrane made entirely of 100% benzyl ester of low molecular weight HA (HYAFF 11). Transparent sheets were manufactured with rows of 40-μm diameter laser-drilled holes and 0.5-mm diameter mechanically drilled holes. The concept is that the membrane can act as a carrier for keratinocytes cultured in the laboratory, allowing easy handling, rapid preparation, and also the application of proliferating keratinocytes to the wound bed. A comparison of "clinical" take rates in a porcine kerato-dermal model demonstrated a significant reduction in take rates as a result of halving the keratinocyte seeding density on the membrane. The take rates of grafts grown on the membrane at a conventional seeding density and then transplanted to the dermal wound bed were comparable, and in some cases superior, to those of keratinocyte sheet grafts (66).

A further modification of this technique is to use a HYAFF scaffold in which to place fibroblasts. This is then applied to the wound and allowed to incorporate for several weeks before being overgrafted with the Laserskin carrying keratinocytes. In this manner, a bilaminar skin has been formed with both epidermal and dermal components (67–69). The use of Laserskin as a carrier for autologous keratinocytes has also been advocated in chronic wounds. In this situation, the keratinocytes appear to stimulate the underlying cells, leading to granulation tissue formation and progressive wound healing (70). HYALOFILL has been used in chronic wound situations and initial studies indicate that it has potential for converting a chronic wound to an acute wound. That is to say that it is effective in changing the cytokine profile and cellular dynamics in the wound and "kick-starting" the wound healing process toward repair (71,72). One possible explanation of this effect comes from a study of the comparative antioxidant effects of HYAFF 11p75, Aquacel (a carboxymethylcellulase material), and HA. Experimental data indicate that HYAFF 11p75 has a greater antioxidant capacity toward the superoxide radical due to the esterified benzyl groups providing alternative sites for superoxide radical attack other than the HA backbone of HYAFF 11p75 itself (73).

d. HA, wound healing, and scarring: A new perspective. In 1991, a paper was published suggesting that a new perspective should be taken on HA and wound healing (74). The background to this paper was the serendipitous discovery that HA extracted from human scar tissue and highly purified still had an identifiable collagen component using cyanogen bromide digests (75). This led to further studies of other HA preparations which indicated that all HA preparations had some protein "association." HA is a unique molecule which has a profound impact on the biological behavior of cells and tissues that are under the influence of multiple peptide factors. HA creates a permissive environment for cell proliferation and is able to suppress, to a degree, the inflammatory response of injured tissues. Nevertheless, it is essential to realize and establish that HA, alone, is not going to determine the outcome of the wound healing response. The reality is that HA can suppress the inflammatory response to wound healing through receptor-mediated interaction, it can promote cell movement by its physicochemical properties, and it can deliver the cytokine factors which are the major elements of intercellular communication via its structural conformation. HA facilitates wound healing and plays a role in scarless healing but it is not the primary agent. It may well be that as our understanding of the complexity of biological interactions develops, we will begin to unravel a number of different "languages." Thus we may find a "language" of growth, another of development, another of repair, and another of regeneration. HA has a role in all these processes but the defining differences most probably relate to the peptide factors that are active within the differing HA environments. To take this concept into the clinical situation it is likely that there will be a considerable increase in the number of "HA-plus" preparations, where the "plus" refers to cytokine combinations that can promote certain types of tissue behavior. One of the favored single-peptide factors popularized in wound healing is TGF-β (76,77). It has been proposed that application of a neutralizing antibody to TGF-β can control adult scarring (78). While this is too simplistic a concept, it is possible that combining HA and such an antibody may have more effect.

The proposal previously made was that HA acted like a "taxi" in the tissues taking the "passengers" cytokines from the cell of origin to their effector site. This concept may still well be valid in terms of considering a role of HA to be the facilitated distribution of cytokines in the remodeling matrix. It is evident, however, that HA has other potent contributions to play in the complex biological process of wound healing and scarring. Further developments in unraveling of this contribution are likely to focus on the electrochemical influences on molecular interactions and conformational change. The determination of the mechanisms concerning the control, direction, and potential extrinsic modulation of this energy is going to be a major challenge in biological research.

V. Conclusions

There is no doubt that carbohydrates play multiple and complex roles in the process of cutaneous wound healing. This chapter has been able to give only an overview of some of these roles and in many cases the mechanisms of action and effect are still in the process of being elucidated.

References

1. Burd A. New skin. Transplantation 2000; 70:1551–1552.
2. McGrath JA, Eady RAJ, Pope FM. Anatomy and organization of human skin. In: Burns T, Breathnach S, Cox N, Griffiths C, eds. Rook's Textbook of Dermatology, 7th edn. UK: Blackwell Publishing, 2004; vol. 1:3.1–3.84.
3. Archer CB. Functions of the skin. In: Burns T, Breathnach S, Cox N, Griffiths C, eds. Rook's Textbook of Dermatology, 7th edn. UK: Blackwell Publishing, 2004; vol. 1:4.1–4.12.
4. Burd A, Huang L. Hypertrophic response and keloid diathesis: two very different forms of scar. Plast Reconstr Surg 2005; 116:150e–157e.
5. Lloyd LL, Kennedy JF, Methacanon P, Paterson M, Knill CJ. Carbohydrate polymers as wound management aids. Carbohydr Polym 1998; 37:315–322.
6. Vloemans AF, Soesman AM, Kreis RW, Middelkoop E. A newly developed hydrofibre dressing, in the treatment of partial-thickness burns. Burns 2001; 27:167–173.

7. Piaggesi A, Baccetti F, Rizzo L, Romanelli M, Navalesi R, Benzi L. Sodium carboxyl-methyl-cellulose dressings in the management of deep ulcerations of diabetic foot. Diabet Med 2001; 18:320–324.
8. Barnea Y, Amir A, Leshem D, Zaretski A, Weiss J, Shafir R, Gur E. Clinical comparative study of aquacel and paraffin gauze dressing for split-skin donor site treatment. Ann Plast Surg 2004; 53:132–136.
9. Cullen B. The role of oxidized regenerated cellulose/collagen in chronic wound repair. Part 2. Ostomy Wound Manage 2002; 48(6 Suppl.):8–13. Review.
10. Lobmann R, Zemlin C, Motzkau M, Reschke K, Lehnert H. Expression of matrix metalloproteinases and growth factors in diabetic foot wounds treated with a protease absorbent dressing. J Diabetes Complicat 2006; 20:329–335.
11. Vin F, Teot L, Meaume S. The healing properties of Promogran in venous leg ulcers. J Wound Care 2002; 11:335–341.
12. Veves A, Sheehan P, Pham HT. A randomized, controlled trial of Promogran (a collagen/oxidized regenerated cellulose dressing) vs standard treatment in the management of diabetic foot ulcers. Arch Surg 2002; 137:822–827.
13. Heel RC, Morton P, Brogden RN, Speight TM, Avery GS. Dextranomer: a review of its general properties and therapeutic efficacy. Drugs 1979; 18:89–102.
14. Carlin G, Wik KO, Arfors KE, Saldeen T, Tangen O. Influences on the formation and structure of fibrin. Thromb Res 1976; 9:623–636.
15. Frank L, Lebreton-Decoster C, Godeau G, Coulomb B, Jozefonvicz J. Effect of a dextran derivative associated with TGF-beta 1 or FGF-2 on dermal fibroblast behaviour in dermal equivalents. J Biomater Sci Polym Ed 2004; 15:1463–1480.
16. Frank L, Lebreton-Decoster C, Godeau G, Coulomb B, Jozefonvicz J. Dextran derivatives modulate collagen matrix organization in dermal equivalent. J Biomater Sci Polym Ed 2006; 17:499–517.
17. Blanckmeister CA, Sussdorf DH. Macrophage activation by cross-linked dextran. J Leukoc Biol 1985; 37:209–219.
18. Burgess E, Hollinger J, Bennett S, Schmitt J, Buck D, Shannon R, Joh SP, Choi T, Mustoe T, Lin X, Skalla W, Connors D, Christoforou C, Gruskin E. Charged beads enhance cutaneous wound healing in rhesus nonhuman primates. Plast Reconstr Surg 1998; 102:2395–2403.
19. Gilchrist T, Martin AM. Wound treatment with Sorbsan—an alginate fibre dressing. Biomaterials 1983; 4:317–320.
20. Williams C. Tegagen alginate dressing for moderate to heavily exuding wounds. Br J Nurs 1998; 7:550–552.
21. Cihantimur B, Kahveci R, Özcan M. Comparing Kaltostat with Jelonet in the treatment of split-thickness skin graft donor sites. Eur J Plast Surg 1997; 20:260–263.
22. Motta GJ. Calcium alginate topical wound dressings: a new dimension in the cost-effective treatment for exudating dermal wounds and pressure sores. Ostomy Wound Manage 1989; 25:52–56.
23. Doyle JW, Roth TP, Smith RM, Li YQ, Dunn RM. Effects of calcium alginate on cellular wound healing processes modeled in vitro. J Biomed Mater Res 1996; 32:561–568.
24. Fraser R, Gilchrist T. Sorbsan calcium alginate fibre dressings in footcare. Biomaterials 1983; 4:222–224.

25. Dawson C, Armstrong MW, Fulford SC, Faruqi RM, Galland RB. Use of calcium alginate to pack abscess cavities: a controlled clinical trial. J R Coll Surg Edinb 1992; 37:177–179.
26. Sayag J, Meaume S, Bohbot S. Healing properties of calcium alginate dressings. J Wound Care 1996; 5:357–362.
27. O'Donoghue JM, O'Sullivan ST, Beausang ES, Panchal JI, O'Shaughnessy TP, O'Connor TP. Calcium alginate dressings promote healing of split skin graft donor sites. Acta Chir Plast 1997; 39:53–55.
28. Bettinger D, Gore D, Humphries Y. Evaluation of calcium alginate for skin graft donor sites. J Burn Care Rehabil 1995; 16:59–61.
29. Ueno H, Yamada H, Tanaka I, Kaba N, Matsuura M, Okumura M, Kadosawa T, Fujinaga T. Accelerating effects of chitosan for healing at early phase of experimental open wound in dogs. Biomaterials 1999; 20:1407–1414.
30. Ueno H, Mori T, Fujinaga T. Topical formulations and wound healing applications of chitosan. Adv Drug Deliv Rev 2001; 52:105–115.
31. Azad AK, Sermsintham N, Chandrkrachang S, Stevens WF. Chitosan membrane as a wound-healing dressing: characterization and clinical application. J Biomed Mater Res B Appl Biomater 2004; 69:216–222.
32. Zhu X, Chian KS, Chan-Park MB, Lee ST. Effect of argon-plasma treatment on proliferation of human-skin-derived fibroblast on chitosan membrane in vitro. J Biomed Mater Res A 2005; 73:264–274.
33. Chen XG, Wang Z, Liu WS, Park HJ. The effect of carboxymethyl-chitosan on proliferation and collagen secretion of normal and keloid skin fibroblasts. Biomaterials 2002; 23:4609–4614.
34. Kim KW, Thomas RL, Lee C, Park HJ. Antimicrobial activity of native chitosan, degraded chitosan, and O-carboxymethylated chitosan. J Food Prot 2003; 66:1495–1498.
35. Rabea EI, Badawy ME, Stevens CV, Smagghe G, Steurbaut W. Chitosan as antimicrobial agent: applications and mode of action. Biomacromolecules 2003; 4:1457–1465.
36. Burkatovskaya M, Tegos GP, Swietlik E, Demidova TN, Castano AP, Hamblin MR. Use of chitosan bandage to prevent fatal infections developing from highly contaminated wounds in mice. Biomaterials 2006; 27:4157–4164.
37. Ishihara M, Fujita M, Obara K, Hattori H, Nakamura S, Nambu M, Kiyosawa Y, Kanatani Y, Takase B, Kikuchi M, Maehara T. Controlled releases of FGF-2 and paclitaxel from chitosan hydrogels and their subsequent effects on wound repair, angiogenesis, and tumor growth. Curr Drug Deliv 2006; 3:351–358.
38. Alemdaroglu C, Degim Z, Celebi N, Zor F, Ozturk S, Erdogan D. An investigation on burn wound healing in rats with chitosan gel formulation containing epidermal growth factor. Burns 2006; 32:319–327.
39. Yan XL, Khor E, Lim LY. Chitosan-alginate films prepared with chitosans of different molecular weights. J Biomed Mater Res 2001; 58:358–365.
40. Shi C, Cheng T, Su Y, Mai Y, Qu J, Lou S, Ran X, Xu H, Luo C. Transplantation of dermal multipotent cells promotes survival and wound healing in rats with combined radiation and wound injury. Radiat Res 2004; 162:56–63.
41. Ikemoto S, Mochizuki M, Yamada M, Takeda A, Uchinuma E, Yamashina S, Nomizu Y, Kadoya Y. Laminin peptide-conjugated chitosan membrane:

application for keratinocyte delivery in wounded skin. J Biomed Mater Res 2006; 79A:716–722.
42. Hamilton V, Yuan Y, Rigney DA, Puckett AD, Ong JL, Yang Y, Elder SH, Bumgardner JD. Characterization of chitosan films and effects on fibroblast cell attachment and proliferation. J Mater Sci 2006; 17:1373–1381.
43. Lusby PE, Coombes A, Wilkinson JM. Honey: a potent agent for wound healing? J Wound Ostomy Continence Nurs 2002; 29:295–300.
44. Molan PC. Honey as a topical antibacterial agent for treatment of infected wounds. World Wide Wounds 2001, November; (http://www.worldwide-wounds.com/2001/november/Molan/honey-as-topical-agent.html).
45. Moore OA, Smith LA, Campbell F, Seers K, McQuay HJ, Moore RA. Systematic review of the use of honey as a wound dressing. BMC Complement Altern Med (2001); 1:2; (http://www.biomedcentral.com/1472–6882/1/2).
46. Molan PC. The evidence supporting the use of honey as a wound dressing. Low Extrem Wounds 2006; 5:40–54.
47. Natarajan S, Williamson D, Grey J, Harding KG, Cooper RA. Healing of an MRSA-colonized, hydroxyurea-induced leg ulcer with honey. J Dermatol Treat 2001; 12:33–36.
48. Tonks AJ, Cooper RA, Jones KP, Blair S, Parton J, Tonks A. Honey stimulates inflammatory cytokine production from monocytes. Cytokine 2003; 21:242–247.
49. Simon A, Sofka K, Wiszniewsky G, Blaser G, Bode U, Fleischhack G. Wound care with antibacterial honey (Medihoney) in pediatric hematology-oncology. Support Care Cancer 2006; 14:91–97.
50. Subrahmanyam M. A prospective randomized clinical and histological study of superficial burn wound healing with honey and silver sulfadiazine. Burns 1998; 24:157–161.
51. Henriques A, Jackson S, Cooper R, Burton N. Free radical production and quenching in honeys with wound healing potential. J Antimicrob Chemother 2006; 58:773–777.
52. Galvan L. Effects of heparin on wound healing. J Wound Ostomy Continence Nurs 1996; 23:224–226.
53. Kratz G, Arnander C, Swedenborg J, Back M, Falk C, Gouda I, Larm O. Heparin-chitosan complexes stimulate wound healing in human skin. Scand J Plast Reconstr Hand Surg 1997; 31:119–123.
54. Liu Y, Cai S, Shu XZ, Shelby J, Prestwich GD. Release of basic fibroblast growth factor from a crosslinked glycosaminoglycan hydrogel promotes wound healing. Wound Repair Regen 2007; 15:245–251.
55. Saliba MJ. Heparin in the treatment of burns: a review. Burns 2001; 27:349–358.
56. Folkman J, Shing Y. Control of angiogenesis by heparin and other sulfated polysaccharides. Adv Exp Med Biol 1992; 313:355–364.
57. Ferrao AV, Mason RM. The effect of heparin on cell proliferation and type-I collagen synthesis by adult human dermal fibroblasts. Biochem Biophys Acta 1993; 1180:225–230.
58. Metcalfe AD, Ferguson MWJ. Harnessing wound healing and regeneration for tissue engineering. Biochem Soc Trans 2005; 33:413–417.

59. Burke JF. Observations on the development and clinical use of artificial skin—an attempt to employ regeneration rather than scar formation in wound healing. Jpn J Surg 1987; 17:431–438.
60. Ellis DL, Yannas IV. Recent advances in tissue synthesis in vivo by use of collagen-glycosaminoglycan copolymers. Biomaterials 1996; 17:291–299.
61. Shafritz TA, Rosenberg LC, Yannas IV. Specific effects of glycosaminoglycans in an analog of extracellular matrix that delays wound contraction and induces regeneration. Wound Repair Regen 1994; 2:270–276.
62. Zhong S, Teo WE, Zhu X, Beuerman R, Ramakrishna S, Yung LYL. Formation of collagen-glycosaminoglycan blended nanofibrous scaffolds and their biological properties. Biomacromolecules 2005; 6:2998–3004.
63. Campoccia D, Doherty P, Radice M, Brun P, Giovanni A, Williams DF. Semisynthetic resorbable materials from hyaluronan esterification. Biomaterials 1998; 19:2101–2127.
64. Benedetti L, Cortivo R, Berti T, Berti A, Pea F, Mazzo M, Moras M, Abatangelo G. Biocompatibility and biodegradation of different hyaluronan derivatives (Hyaff) implanted in rats. Biomaterials 1993; 14:1154–1160.
65. Davidson JM, Nanney LB, Broadley KN, Whitsett JS, Aquino AM, Beccaro M, Rastrelli A. Hyaluronate derivatives and their application to wound healing: preliminary observations. Clin Mater 1991; 8:171–177.
66. Myers SR, Grady J, Soranzo C, Sanders R, Green C, Leigh IM, Navsaria HA. A hyaluronic acid membrane delivery system for cultured keratinocytes: clinical "take" rates in the porcine kerato-dermal model. J Burn Care Rehabil 1997; 18:214–222.
67. Harris PA, di Francesco F, Barisoni D, Leigh IM, Navsaria HA. Use of hyaluronic acid and cultured autologous keratinocytes and fibroblasts in extensive burns. Lancet 1999; 353:35–36.
68. Hollander DA, Soranzo C, Falk S, Windolf J. Extensive traumatic soft tissue loss: reconstruction in severely injured patients using cultured hyaluronan-based three-dimensional dermal and epidermal autografts. J Trauma 2001; 50:1125–1136.
69. Galassi G, Brun P, Radice M, Cortivo R, Zanon GF, Genovese P, Abatangelo G. In vitro reconstructed dermis implanted in human wounds: degradation studies of the HA-based supporting scaffold. Biomaterials 2000; 21:2183–2191.
70. Bernd A, Hollander D, Pannike A, Kippenberger S, Muller J, Stein M, Kaufmann R. Benzylester hyaluronic acid membranes as substrate for cultivation and transplantation of autologous keratinocytes for the treatment of non-healing wounds. J Invest Dermatol 1996; 107:450.
71. Edmonds M, Bates M, Doxford M, Gough A, Foster A. New treatments in ulcer healing and wound infection. Diabetes Metab Res Rev 2000; 16(Suppl. 1): S51–S54.
72. Georgina C. Wound repair: advanced dressing materials. Nurs Stand 2002; 17:49–53.
73. Moseley R, Leaver M, Walker M, Waddington RJ, Parsons D, Chen WY, Embery G. Comparison of the antioxidant properties of HYAFF-11p75, AQUACEL and hyaluronan towards reactive oxygen species in vitro. Biomaterials 2002; 23:2255–2264.

74. Burd DAR, Greco RM, Regauer S, Longaker MT, Siebert JW, Garg HG. Hyaluronan and wound healing: a new perspective. Br J Plast Surg 1991; 44:579–584.
75. Burd DAR, Siebert JW, Ehrlich HP, Garg HG. Human skin and post-scar hyaluronan: demonstration of association with collagen and other proteins. Matrix 1989; 9:322–327.
76. Sporn MB, Roberts AB. Transforming growth factor-B: recent progress and new challenges. J Cell Biol 1992; 119:1017–1021.
77. Ellis IR, Schor SL. Differential effects of TGF-β 1 on hyaluronan synthesis by fetal and adult skin fibroblasts: implications for cell migrationa and wound healing. Exp Cell Res 1996; 228:326–333.
78. Shah M, Foreman DM, Ferguson MWJ. Control of scarring in adult wounds by neutralizing antibody to transforming growth factor B. Lancet 1992; 339:213–214.

Carbohydrate Chemistry, Biology and Medical Applications
Hari G. Garg, Mary K. Cowman and Charles A. Hales

DOI: 10.1016/B978-0-08-054816-6.00012-4

Chapter 12

Carbohydrates in Human Milk and Infant Formulas

GÜNTHER BOEHM,[*,†] BERND STAHL,[*] JAN KNOL[‡] AND JOHANN GARSSEN[†,§]

**Numico Research, Friedrichsdorf, Germany*
[†]Sophia Children's Hospital, Erasmus University, Rotterdam, The Netherlands
[‡]*Numico Research, Wageningen, The Netherlands*
[§]*Utrecht Institute for Pharmaceutical Sciences, Utrecht University, Utrecht, The Netherlands*

I. Introduction

Human milk is widely accepted as the ideal nutrition for term infants because it provides all necessary nutrients for rapid growth in sufficient amounts, without overloading the functional capacity of the not yet fully developed gastrointestinal tract or metabolism. This is especially achieved by the high bioavailability of the nutrients, so that relatively low concentrations ensure a sufficient supply. In addition, human milk contains a lot of components that have a functional importance rather than a nutritive one (1,2). Breastfed infants develop differently compared to infants with artificial feeding (3). Apart from a reduced incidence of allergic or atopic diseases (4–6) as well as a reduced incidence of infections (7–9), the varying incidence of diabetes mellitus type I (10) and the better cognitive functions of breastfed infants in later life are also worth mentioning (11).

If, compared to breastfed infants, the same capacity for development is offered to formula-fed infants, functional components have to be found that are able to compensate for the differences still present.

This chapter is focused on the influence of breastfeeding on the postnatal development of intestinal microbiota, the possibilities to mimic this function with oligosaccharides of nonmilk origin, and the benefits regarding the postnatal development of the immune system.

II. Prebiotic Factors in Human Milk

Before birth, the infant is sterile. During vaginal delivery, the natural colonization of the infant starts with bacteria mainly from the vaginal and intestinal microbiota of the mother. For the further development of the intestinal microbiota of the infant, the diet plays an important role. During breastfeeding, the microbiota change within a short period to a flora dominated by bifidobacteria whereas the intestinal microbiota of infants fed formulas without prebiotics is characterized by a flora of a more adult type (12).

The prebiotic effect of human milk was intensively investigated over the last century. Several so-called "bifido-factors" have been identified. In particular, several milk proteins like lactoferrin or lactalbumin or urea as part of the nonprotein nitrogen fraction have been described as factors contributing to a beneficial composition of the intestinal microbiota. The low content of protein or the low phosphate concentration in breast milk also have been attributed to the bifidogenicity of breast milk but no proof is available yet. Although the effect of human milk on the postnatal development of the intestinal microbiota cannot be attributed to a single ingredient, there is evidence, however, that human milk oligosaccharides (HMOS) might play a key role in this matter (13,14).

Oligosaccharides occur in human milk at a concentration of up to 1 g/100 ml, while they are virtually absent in regular infant formulas. The structure of oligosaccharides in human milk is very complex (15) and the functional consequences of these very different structures remain to be elucidated. The molecules are synthesized in the breast starting with lactose at its reducing terminus. The core molecule is characterized by repetitive attachment of galactose and *N*-acetylglucosamine in β-glycosidic linkage to lactose. Although the structure of the core molecule results in a wide range of different molecules, the variety is even higher due to α-glycosidic linkages of fucose (neutral oligosaccharides) and fucose and/or sialic acid (acidic oligosaccharides) to the respective core molecules (16). Especially, the attachment of fucose is based on the secretor/Lewis blood group status of the individual mother. This results in at least four groups of individually composed patterns of milk oligosaccharides based on genetic factors (17).

There are many different functions attributed to HMOS (13,15,16) that might explain the great variety of the structures. With respect to the influence of intestinal microbiota, the neutral fraction of HMOS seems to be a key factor for the development of the intestinal microbiota typical for breastfed infants. Because the human intestine has no enzymes to cleave the α-glycosidic linkages of fucose and sialic acid as well as the β-glycosidic linkages in the core molecule, they are protected from digestion. The reduced or absent digestibility is a prerequisite for prebiotic activities of dietary compounds. On the other hand, many intestinal bacteria express glycosidases to cleave HMOS (18). This structural characteristic provides strong evidence that HMOS are preferentially synthesized to be metabolized by intestinal bacteria.

There is a wide range of molecular size distribution within the HMOS fraction. Since 1980, oligosaccharides are defined as carbohydrates with a degree of polymerization up to 10. However, as there is no physiological reason for this definition, oligosaccharides have been variously defined later, on ranging from a degree of polymerization of 2 up to 20 and more (13,19). Recently, the IUB-IUPAC Joint Commission on Biochemical Nomenclature stated that the borderline between oligo- and polysaccharides cannot be drawn strictly. However, the term oligosaccharide is commonly used to refer to defined structures as opposed to a polymer of unspecified length (20). Thus, even having molecules with a degree of polymerization significantly larger than 10, the HMOS are all described as oligosaccharides. The same approach is used for oligosaccharides of nonhuman milk origin as long as they have defined structures.

As HMOS are resistant to digestion, they reach the colon unchanged, where they can develop their prebiotic effect. Finally, they can be detected in the feces as well as in the urine of breastfed infants (21).

Apart from their prebiotic effects, there is also evidence that HMOS act as receptor analogues to inhibit the adhesion of pathogens on the epithelial surface (22) and interact directly with human immune cells (23) as well as with animal immune cells in preclinical experiments (24,25).

III. Oligosaccharides of Nonhuman Milk Origin

The development of new analytical techniques has significantly improved our knowledge about the structures of HMOS (26–28). Additionally, new preparation methods have been developed that allow purification of oligosaccharide structures, which is a prerequisite for identifying their biological effects (29,30).

However, there are still many questions remaining regarding the relationship between the structure of oligosaccharides and their biological function. For the future, it would be important to know which structural elements in HMOS play the key role for their function.

As the structure of HMOS is so complex, there is so far no possibility of producing identical structures for production of infant formulas [structural aspects of HMOS were intensively reviewed by Boehm and Stahl (15)].

For more than 50 years, HMOS have been identified as effective bifidogenic factors (14). In the search for alternatives to HMOS, oligosaccharides from milk of domestic animals as well as several oligosaccharides of nonmilk origin have been under investigation.

Milks of domestic animals are not an optimal source for prebiotic oligosaccharides. In comparison to human milk, the concentrations of oligosaccharides in these milks are much lower than in human milk and their structure is less complex (15,31). However, the β-glycosidic bond galactose is also characteristic for many of these oligosaccharides that protect them—like the HMOS—from digestion during passage through the human intestine. Linkages to fucose—in contrast to HMOS—are very rare whereas linkages of galactose and *N*-acetylglucosamine are dominant. The relation differs between the species. The oligosaccharides from domestic animals are extensively reviewed by Urashima et al. (32). Based on the structure of these oligosaccharides, it can be assumed that they are also effective as prebiotics in humans. However, the preparation of these compounds is difficult and therefore large-scale preparations have not been commercially available. Consequently, so far no clinical trial has been published using fractions of animal milk oligosaccharides as prebiotics.

Another alternative is the use of nonmilk oligosaccharides. In Table 1, those oligosaccharides are summarized for which a bifidogenic effect has been described in humans. Related to the use of prebiotics during infancy, the most experience exists for galacto-oligosaccharides (GOS) and fructans from the inulin type (fructo-oligosaccharides; FOS).

Fructans are linear or branched fructose polymers that are either β2–1-linked inulins or β2–6-linked levans. The inulin-type fructans can easily be extracted from plant sources and have been widely used as an ingredient for dietary products. In their natural sources, the molecule size is wide ranging (from polymerization degree 2 to more than 60). Because the biological activity depends on the molecule size, usually they are characterized as short-chain FOS (scFOS) representing a degree of polymerization up to 10 monomers and long-chain FOS (lcFOS) with a degree of polymerization of more than 10 monomers.

GOS are synthesized from lactose via an enzymatic transgalactosylation using a β-galactosidase mainly of bacterial origin (33). These OS consist of a chain of galactose monomers usually with a glucose monomer on the reducing terminus with a degree of polymerization significantly below 10 monomers.

Table 1 Most Important Oligosaccharides (OS) of Nonmilk Origin Already Used as Prebiotics in Human Nutrition

Trivial name	Structure[a]	Preparation
Galacto OS	[Gal(β1–3/4/6]$_n$Gal (β1–4)Glc	Enzymatic synthesis from lactose
Fructo OS/Inulin	[Fru(β2-]$_n$ 1)Glc	Extraction from natural sources and enzymatic synthesis
Palatinose/ Isomaltulose OS	[Glc(α1-]$_n$ 6)Fru	Enzymatic synthesis from sucrose
Soybean OS	[Gal(α1-]$_n$ 6)Glc (α1–6)Fru	Extraction from natural sources
Lactosucrose	[Gal(β1–4)]Glc(α1–2) Fru	Enzymatic synthesis from lactose
Xylo OS	[Xyl(β1]$_n$ - 4)Xyl	Enzymatic synthesis of, for example, corncob xylan
Galacturonic acid OS	[GalA(α1]$_n$ - 4)GalA	Enzymatic degradation of pectin

Source: Taken from Refs. (15,37).
[a]Structures presented according to the recommendations of Joint Commission on Biochemical Nomenclature (20).

IV. Prebiotic Effect of Nonmilk Oligosaccharides

Because there is a broad consensus that the intestinal microbiota plays an important physiological role for the host, many attempts have been made to influence the intestinal microbiota by dietary interventions.

In principle, there are two major strategies for influencing the flora. One is the use of living bacteria added to the food that must survive the gastrointestinal tract to be active in the colon ("bacteria for food"; probiotics) (34). The second strategy is the use of dietary ingredients that are nondigestible, reach the colon, and can be used by health-promoting colonic bacteria ("food for bacteria"; prebiotics) (35).

More recently, the latter prebiotic concept was revised. The authors come to the definition that prebiotics have to be resistant until they are fermented by the intestinal (i.e., not only colonic) flora. The balanced stimulation of growth and/or activity of the health-promoting bacteria in the gastrointestinal tract have to be demonstrated finally by performing studies in the target group to produce sound scientific data (36).

More recently, products combining both principles as "synbiotics" are under discussion (37).

The counts of fecal bifidobacteria or the percentage of fecal bifidobacteria of the total bacteria are generally accepted measurements to detect a prebiotic effect. Today, different types of nondigestible oligosaccharides (see Table 1) are on the world market (38). However, in the following text, the focus will be on oligosaccharides for which the prebiotic effect during infancy has clearly been documented. Prebiotic effects have been found with GOS (39,40), scFOS (41–44) galacturonic acid oligosaccharides (45), and a mixture of GOS and lcFOS (IMMUNOFORTIS) (46–54) (for details see Table 2) and for preterm infants only with the mixture of GOS/lcFOS (55–57) (for details see Table 3).

The most extensively studied prebiotic oligosaccharides are GOS and FOS, and together with the disaccharide lactulose, they are the only nonmilk oligosaccharides with an approved prebiotic status (36).

The low digestibility—an essential prerequisite to act as a prebiotic—has been demonstrated for GOS and lcFOS (58). A mixture of these two oligosaccharides has been introduced in the market in 2001. The bifidogenic effect of this mixture was shown in several studies in formula-fed preterm (Table 3) and term (Table 2) infants. As in the animal experiment, the effect depends on the dosage (46). At a concentration of 0.8 g/dl (this corresponds to the concentration of neutral oligosaccharides in human milk), the number of bifidobacteria was tantamount to that found in feces of breastfed infants. The bifidogenic effect was associated with a reduction of the stool pH (46,48) as well as a reduction of pathogenic bacteria (48,59).

Short-chain fatty acids (SCFAs) are of considerable importance to the physiological effect of the intestinal microbiota. They are the fermentation product of bacteria in the colon and are therefore an important characteristic feature of the intestinal microbiota (60). Compared to formula-fed infants, the profile of SCFAs differs considerably from that of breastfed infants. On supplementing an infant formula

Table 2 Clinical Trials with Prebiotic Oligosaccharides in Term Infants (Nutritional Intervention During the First Year of Life)

Prebiotic	Design	Study groups	Main outcome
GOS (2.0 g/dl)	r, p, p-c, d-b	Prebiotics (*n*=43) control (*n*=17) BM reference (*n*=20)	Increased counts of bifidobacteria and lactobacilli (39)
GOS (0.24 g/dl)	r, p, p-c, d-b	Prebiotics (*n*=69) control (*n*=52) BM reference (*n*=26)	Increased counts of bifidobacteria and lactobacilli, reduced fecal pH, increased total fecal short-chain fatty acids (40)
scFOS (1.5 and 3.0 g/dl)	r, p, d-b, cross-over single site	Prebiotics 1.5 g (*n*=28) Prebiotics 3.0 g (*n*=30) BM reference (*n*=14)	Increases counts of bifidobacteria but also counts of *Clostridium difficile* and concentration of their toxin in stool compared to BM (41)
scFOS (0.55 g/ 15 g cereals)	r, p, p-c, d-b	Prebiotics (*n*=63) control (*n*=60)	Decrease of severity of diarrheal disease(42)
scFOS (1.0, 2.0, 3.0 g/day)	r, p, p-c, d-b	Prebiotics 1.0 g (*n*=13) Prebiotics 2.0 g (*n*=11) Prebiotics 3.0 g (*n*=12)	No bifidogenicity, no side effects(43)
scFOS (1.5 and 3.0 g/dl)	r, p, p-c, d-b, multicenter	Prebiotics 1.5 g (*n*=72) Prebiotics 3.0 g (*n*=74) control (*n*=66)	Safe and supports normal growth (44)
AOS (0.2 g/dl and 0.2 g/ dl+GOS/ lcFOS)	r, p, p-c, d-b	Prebiotic AOS (*n*=16) Prebiotic AOS+GOS/ lcFOS/dl (*n*=15) control (*n*=15)	AOS alone has no effect on intestinal microbiota but reduces fecal pH (45)
GOS/lcFOS (0.4 and 0.8 g/dl)	r, p, p-c, d-b	Prebiotics 0.4 g (*n*=28) Prebiotics 0.8 g (*n*=28) Control (*n*=29) BM reference (*n*=15)	Increased counts of bifidobacteria, reduced fecal pH in a dose-dependent manner (46)
GOS/lcFOS (0.8 g/dl)	r, p, p-c, d-b	Prebiotic (*n*=28) Placebo (*n*=29) BM reference (*n*=15)	Increases counts of bifidobacteria and lactobacilli (47)
GOS/lcFOS (0.8 g/dl)	r, p, p-c, d-b	Prebiotic (*n*=21) Placebo (*n*=20)	Stimulation of entire flora including *Bifidobacterium* species (48)
GOS/lcFOS (0.8 g/dl)	r, p, p-c, d-b	Prebiotic (*n*=10) Control (*n*=10)	Bifidogenicity after introduction of solid food (49)
GOS/lcFOS (0.8 g/dl)	r, p, p-c, d-b	Prebiotic (*n*=19) Control (*n*=19)	No significant effects on bifidobacteria, increased fecal IgA (50)
GOS/lcFOS (0.8 g/dl)	r, p, p-c, d-b	Prebiotic (*n*=102) Control (*n*=104)	Increased counts of bifidobacteria, reduced incidence of atopic

Continued

Table 2 Cont'd

Prebiotic	Design	Study groups	Main outcome
			dermatitis at 6 and 24 months of life and antiallergic antibody profile at 6 months of age (51)
GOS/lcFOS (0.6 g/dl)	r, p, p-c, d-b	Prebiotic (n=86) Control (n=90)	Increased counts of bifidobacteria, increased fecal sIgA (52)
GOS/lcFOS (0.4 g/dl)	r, p, p-c, d-b	Prebiotic (n=14) Control (n=19)	Increase of bifidobacteria (53)
GOS/lcFOS (0.8 g/dl)	r, p, p-c, d-b	Prebiotic (n=8) Control (n=8) BM reference (n=8)	*Bifidobacterium* species similar to breastfed infants (54)

Abbreviations: GOS: galacto-oligosaccharides, FOS: fructo-oligosaccharides, AOS: acidic oligosaccharides deriving from pectin, sc: short chain, lc: long chain, BM: breast milk.

Table 3 Review Clinical Trials With Prebiotics in Preterm Infants (Nutritional Intervention During the First Year of Life)

Prebiotic	Design	Study groups	Main outcome
GOS/lcFOS 1.0 g/dl	r, p, p-c, d-b	+GOS/lcFOS (n=15) +placebo (n=15) HM reference (n=13)	Increasing counts of bifidobacteria, reduction of hardness of stools (55)
GOS/lcFOS 0.8 g/dl	r, p, p-c, d-b	+GOS/lcFOS (n=10) +placebo (n=10)	Reduction of gastrointestinal transit time, reduction of stool viscosity (56)
GOS/lcFOS 0.8 g/dl	r, p, p-c, d-b	+GOS/lcFOS (n=10) +placebo (n=10)	Statistically significant but small effect on reduction of gastric emptying time (57)

Abbreviations: GOS: galacto-oligosaccharides, FOS: fructo-oligosaccharides, sc: short chain, lc: long chain.

with a mixture of GOS/lcFOS, there was a pattern of SCFAs in feces that corresponded to the pattern found in the feces of breastfed infants (48). The SCFA pattern reflects the metabolic activity of the entire microbiota.

Thus, the similarity of fecal SCFAs between breastfed infants and infants fed a formula supplemented with the prebiotic mixture of GOS/lcFOS indicates that the given prebiotic mixture stimulates the entire flora toward the flora of breastfed infants (61).

There are several results available indicating that SCFA and pH influence the physiological role of intestinal cells.

In an *in vitro* model epithelial cells (T 84 cell line) were combined with myofibroblast cells (CDD-18 Co cell line) in a coculture. SCFAs as they appear in the feces of breastfed infants were used. These SCFAs stimulated mucin 2 production and improved the barrier integrity (62). The effect of SCFA on growth of pathogens like *Escherichia coli* (ETEC), *Enterococcus faecalis*, *Pseudomonas aeruginosa*, *Salmonella typhimurium*, and *Staphylococcus aureus* as well as of commensals has been studied *in vitro* at pH 7.5 (typical fecal pH in formula-fed infants) and at pH 5.5 (typical fecal pH in breastfed infants). SCFAs decreased the growth of pathogens in a dose-dependent manner but did not affect the growth of commensals. This effect was only seen at pH 5.5 but not at pH 7.5. Thus, these experiments indicated that achieving the same pH and SCFA pattern with prebiotics, as found in stools from breastfed infants, resulted in reduced growth of pathogens (63). This effect has clinical relevance during infancy. A reduction of fecal pathogens could be demonstrated in a study in preterm (59) as well as term infants (48).

There are two studies focusing on the analysis of the effect of prebiotics (both studies with the mixture of GOS/lcFOS) on the development of the different bifobacteria species (48,54). In these studies, it could be demonstrated that the prebiotics promoted *Bifidobacterium infantis* and depressed *B. adolescentis*. In one study (48), *B. adolestentis* dominated on the fifth day of life (70%) but the percentage was reduced to approximately 20% during a 6 weeks breastfeeding period. The same changes occur during feeding with the GOS/lcFOS mixture. The decrease of *B. adolescentis* was not seen in the group-fed formula without prebiotics. On the one hand, this was an indicator for the specific effect of the prebiotics. On the other hand, it was an important finding related to the immune system because there has been evidence that early colonization with specific microbiota might be associated with the development of allergic symptoms later in life. Björksten et al. (64) found that allergic infants in Estonia—a country with a low prevalence of allergy—were less often colonized by lactobacilli and bifidobacteria than allergic infants in Sweden—a country with a high prevalence of allergy. In contrast, the Estonian infants were more often colonized with aerobic pathogenic microorganisms, particularly coliforms and *S. aureus*, compared to the Swedish infants. Additionally, it was found that allergic infants had more adult-like species in their fecal flora, including *B. adolescentis*, compared with healthy infants. In the latter, *B. bifidum*, *B. infantis*, and *B. breve* predominated (65). Also, in Japanese infants suffering from atopic dermatitis, similar findings have been reported (66). This suggests that different bacterial species may have different functional effects on the immunological reaction of the host. Specific modulation of the composition of the intestinal microbiota through the use of prebiotics is therefore expected to have an impact on the functioning of the immune system.

In summary, the experimental data as well as the results of clinical trials have proven that substances with a structure different from the structure of HMOS are able to influence the intestinal microbiota. The most extensively studied prebiotic compound has been a mixture of GOS/lcFOS, which stimulated the development of intestinal microbiota comparable to those found in breastfed infants.

In summary, the studies performed in term infants during infancy indicate that a prebiotic effect achieved during breastfeeding is comparable with the effect that can be seen with ingredients with molecular structures different from HMOS.

Due to the very complex composition of the intestinal microbiota and considering the great variety of structures found in HMOS, it is now plausible that mixtures of different oligosaccharide types and chain lengths, which are composed to meet the different metabolic requirements of the different bacteria, will have a better chance of mimicking the prebiotic effect of breastfeeding than individual compounds.

V. Effect of Intestinal Microbiota on the Immune System: Preclinical Studies

There is accumulating evidence that the interaction between the intestinal microbiota and the gut plays an important role for the postnatal development of the immune system. However, the interactions between the intestinal epithelial and immune cells and the different species of the intestinal microbiota are very complex and not fully understood. The complexity of these interactions is based on the fact that on the one hand the human defense system consists of several layers, for example, of mechanical and chemical barriers (first line of defence) as well as innate and adaptive immunity (67) all of which can be influenced by microbiota (68).

Following the PASSCLAIM recommendation (69), studies in mice were recommended to substantiate conclusions related to immunological effects of dietary compounds. As for the prebiotic function, the most systematic studies were performed with a mixture of GOS/lcFOS. The experimental data have been intensively reviewed recently by Vos et al. (70).

In mice, it could be shown that GOS/lcFOS was bifidogenic in a dose-dependent manner, resulting in a reduction of the fecal pH and in a fecal SCFA pattern as found in human infants, thus, supporting the relevance of the animal data for the human situation (71).

In a mouse vaccination model adapted to study the effect of prebiotics, the animals were vaccinated twice with the Influvac (*Orthomyxovirus influenza*) vaccine (booster vaccination after 21 days). The response to the vaccination was measured at day 30 after the first vaccination. Parameters used to identify the response to vaccination were DTH response (skin response after local subcutaneous vaccine injection as an *in vivo* measure for Th1-mediated immunity), plasma titers of specific antibodies, *ex vivo* lymphocyte stimulation, T-cell proliferation, cytokine production, and natural killer cell activity. A specific prebiotic mixture (GOS/lcFOS) significantly stimulated the vaccination response in a dose-dependent manner and increased the DTH response indicating a modulation of the immune system toward a Th1-dominated immune response. This effect only occurred if the intervention with prebiotic nutrition started before the first vaccination. However, it was not observed if the prebiotics were fed after the first vaccination (71). This influence of treatment timing indicated that the prebiotic function was mainly mediated by the developing intestinal microbiota and probably not just due to direct interactions with the gut luminal and mesenterial immune cells. It might also indicate that the use of prebiotics for prevention was more relevant than any treatment approach.

In the same experiments, different classical fiber mixtures in a similar dose to the GOS/lcFOS mixture were tested. There was no effect of these fibers on the measured parameter of the immune system, indicating that different nondigestible carbohydrates react differently with respect to intestinal microbiota and immune function (71).

The same group studied the effects of a specific prebiotic mixture on the effect of allergic reaction in a mouse model using ovalbumin as antigen. The animals were sensitized by 10 g ovalbumin in alum and boosted 7 days later. The allergic reaction was measured before and after inhalation challenge (10 mg/ml ovalbumin; 20 min duration). The animals were challenged at 21, 24, and 27 days after first sensitization. Parameters used to identify the response to allergen exposure were airway responsiveness, bronchial lavage inflammatory cells, and antibody levels in plasma. Feeding the GOS/lcFOS mixture reduced significantly the allergic reaction against ovalbumin as demonstrated by reduction of bronchial constriction after metacholine application, reduction in inflammatory cells in the bronchial lavage fluid, and reduction in the IgE concentration in plasma (72,73).

In summary, the animal data allow the conclusion that prebiotics, like the mixture of GOS/lcFOS, modulate the immune system and provide a preventive effect with regard to the development of allergic diseases. This effect is mainly mediated by modulation of the intestinal microbiota.

VI. Effect of Intestinal Microbiota on the Immune System: Clinical Trials

There is increasing evidence that the interaction between the intestinal microbiota and the intestinal epithelial and immune cells plays a key role in the postnatal development of the immune system. First studies with probiotics (74) and synbiotics (40) demonstrate effects during infancy, and studies regarding the vaccination response in the elderly (75) indicate that the prebiotics might also influence the immune system. In particular, the animal experiments with prebiotics described above allow the hypothesis that prebiotics that are able to influence the composition of the entire intestinal microbiota toward microbiota found in breastfed infants might support the development of the immune system during infancy.

Clinical studies designed to prove this hypothesis should be focused on clinical outcome (incidence of infectious and allergic symptoms) and biomarkers representing the status of the immune system.

Based on the data derived from preclinical experiments with the mixture of GOS/lcFOS, a preventive study in term infants at risk for atopy was performed (51).

The study was designed to investigate the possible influence of this prebiotic mixture on the cumulative incidence of atopic dermatitis during the first 6 months of life in formula-fed infants at risk to develop allergy (paternal history of allergy). The study was performed as a prospective, double-blind, randomized, placebo (GOS/lcFOS was replaced by maltodextrin in the placebo formula) controlled study. Two hundred fifty nine infants with a family history of atopy were enrolled in the study. Fifty three infants left the trial before completing the study. The main reason for dropping out was the continuation or reestablishment of breastfeeding. One hundred two infants in the prebiotic group and 104 infants in the placebo group completed the study.

If the mother decided to start bottlefeeding, the infant was randomly assigned to one of two hydrolyzed protein formula groups (plus 0.8 g/100 ml prebiotics or maltodextrine as placebo).

The infants were seen on a monthly basis. The interview of the parents was based on a diary kept by the parents. The primary outcome of the study was the cumulative incidence of atopic dermatitis during the first 6 months of life. The skin was investigated for atopic dermatitis according to the diagnostic criteria described by Harrigan and Rabinowitz (76) and Muraro et al. (69). The severity of the skin alterations was scored by the SCORAD index based on extension, intensity of the skin symptoms, as well as on the subjective symptoms pruritus and sleep loss as recommended by the European Task Force on atopic dermatitis (77,78).

In a subgroup of 98 infants, the parents allowed the collection of stool for microbiological analysis. In a subgroup of 86 infants, the parents allowed blood samples to be drawn for analysis of antibodies.

Intestinal microbiota was measured using plating techniques. In the plasma samples, subsequently total immune globulins, cow's milk protein (CMP), and DTP-specific immune globulins were measured.

Feeding the prebiotic mixture resulted in a significant stimulation of fecal bifidobacteria compared to the placebo group. Ten infants (9.8%; 95 CI: 5.4–17.1%) in the GOS/lcFOS group and 24 infants (23.1%; 95 CI: 16.0–32.1%) in the placebo group developed atopic dermatitis. The severity of the dermatitis was not influenced by diet. Supplementation of GOS/lcFOS resulted in a significant reduction of the plasma level of total IgE, IgG1, IgG2, and IgG3-Igs, whereas no effect on IgG4 was observed. CMP-specific IgG1 was significantly decreased. Other CMP-specific immune globulins and DTP-specific immune globulins were not affected at all by GOS/lcFOS supplementation indicating that the prebiotics induced an antiallergic immune globulins profile in this cohort of infants at risk (79).

More recently, the authors of this study reported results from a 2 years follow-up, which confirmed the findings at 6 months (80).

The effect of a prebiotic diet on the incidence of infectious symptoms was studied in a healthy infant population fed either a standard formula (n=162) or a formula supplemented with the prebiotic mixture of GOS/lcFOS (n=164). After 9 months, the prebiotic diet resulted in a reduction of the cumulative incidence of recurrent respiratory tract infections ($\leq$3 episodes) and of diarrhea (67).

These observations are in line with the findings that breastfeeding results in reduced incidence of atopic and allergic diseases (4–6) and reduced incidence of infections (7–9). Although the different effects of breastfeeding are of multifactorial origin, the prebiotic oligosaccharides of breast milk might play a key role.

VII. Conclusion

The data demonstrate a significant and biological relevant effect of dietetic prebiotics on the postnatal development of the immune system. The most conclusive data exists for a mixture of GOS/lcFOS. The mechanism behind the immune modulatory effects of the studied prebiotic oligosaccharides is not fully understood yet. However, the finding in the human trial is in accordance with the results obtained from animal models demonstrating an active strengthening of the immune system. This indicates that these prebiotics would serve as an effective and safe tool for prevention of infection and allergies.

References

1. Oddy WH. The impact of breast milk on infant and child health. Breastfeed Rev 2002; 10:5–18.
2. Hamosh M. Breastfeeding: unravelling the mysteries of mother's milk. Medscape Womens Health 1996; 16:4–9.
3. Davis MK. Breastfeeding in chronic disease in childhood and adolescence. Padiatr Clin North Am 2001; 48:125–141.
4. Garofalo RP, Goldman AS. Expression of functional immunmodulatory and anti-inflammatory factors in human milk. Clin Perinatol 1999; 26:361–378.
5. Halken S, Host A. Prevention. Curr Opin Allergy Clin Immunol 2001; 1:229–236.
6. Kelly D, Coutts AG. Early nutrition and the development of immune function in the neonate. Proc Nutr Soc 2000; 59:177–185.
7. Howie PW, Forsyth JS, Ogston SA, Clark A, du Florey VC. Protective effect of breast feeding against infection. Br Med J 1990; 300:11–18.
8. Hanson LA, Korotkova M. The role of breastfeeding in prevention of neonatal infection. Semin Neonatol 2002; 7:275–281.
9. Chien PF, Howie PW. Breast milk and the risk of opportunistic infection in infancy in industrialized and non-industrialized settings. Adv Nutr Res 2001; 10:69–104.
10. Wasmuth HE, Kolb H. Cow's milk and immune-mediated diabetes. Proc. Nutr Soc 2000; 59:573–579.
11. Morley R, Lucas A. Nutrition and cognitive development. Br Med Bull 1997; 53:123–134.
12. Harmsen HJ, Wildeboer-Veloo AC, Raangs GC, Wagendorp AA, Klijn N, Bindels J, Welling GW. Analysis of intestinal flora development in breast fed and formula fed infants by using molecular identification and detection methods. J Pediatr Gastroenterol Nutr 2000; 30:61–67.
13. Kunz C, Rudloff S, Baier W, Klein N, Strobel S. Oligosaccharides in human milk: structural, functional and metabolic aspects. Ann Rev Nutr 2000; 20:699–722.

14. György P, Norris RF, Rose CS. A variant of Lactobacillus bifidus requiring a special growth factor. Arch Biochem Biophys 1954; 48:193–201.
15. Boehm G, Stahl B. Oligosaccharides. In: Mattila-Sandholm T, ed. Functional Dairy Products, Cambridge: Woodhead Publication, 2003; 203–243.
16. Newburg DS, Neubauer SH. Carbohydrates in milk. In: Jensen RG, Thompson MP, eds. Handbook of Milk Composition, San Diego: Academic Press, 1995; 273–349.
17. Thurl S, Henker J, Siegel M, Tovar K, Sawatzki G. Detection of four human milk groups with respect to Lewis blood group dependent oligosaccharides. Glycoconj J 1997; 14:795–799.
18. Hill MJ. Bacterial fermentation of complex carbohydrate in the human colon. Eur J Can Prev 1995; 4:353–358.
19. British Nutrition Foundation. Complex Carbohydrates in Foods: Report of the British Nutrition's Task Force. London: Chapmann and Hall, 1990.
20. McNaught AD. IUB-IUPAC Joint Commission on Biochemical Nomenclature (JCBN). Nomenclature of carbohydrates, recommendations 1996. Carbohydr Res 1997; 297:1–92.
21. Coppa GV, Pierani P, Zampini L, Bruni S, Carloni I, Gabriello O. Characterization of oligosaccharides in milk and feces of breast fed infants by high-performance anion-exchange chromatography. Adv Exp Med Biol 2001; 501:307–314.
22. Barthelson R, Mobasseri A, Zopf D, Simon P. Adherence of *Streptococcus pneumoniae* to respiratory epithelial cells is inhibited by sialylated oligosaccharides. Infect Immunol 1998; 66:83–89.
23. Eiwegger T, Stahl B, Schmitt JJ, Boehm G, Gerstmayr M, Pichler J, Dehlinek E, Urbanek R, Szépfalusi Z. Human milk derived oligosaccharides and plant derived oligosaccharides stimulate cytokine production of cord blood T-cells in vitro. Pediatr Res 2004; 56:536–540.
24. Velupillai P, Harn DA. Oligosaccharide-specific induction of interleukin 10 production by B220+ cells from schistosome-infected mice: a mechanism for regulation of CD4+ T-cell subsets. Proc Natl Acad Sci USA 1994; 91:18–22.
25. Terrazas LI, Walsh K, Piskorska D, McGuire E, Harn DA. The schistosome oligosaccharide lacto-N-neotetraose expands Gr1(+) cells that secrete anti-inflammatory cytokines and inhibit proliferation of naive CD4(+) cells: a potential mechanism for immune polarization in helminth infections. J Immunol 2001; 167: 5294–5303.
26. Finke B, Stahl B, Pfenninger A, Karas M, Daniel H, Sawatzki G. Analysis of high-molecular-weight oligosaccharides from human milk by liquid chromatography and MALDI-MS. Anal Chem 1999; 71:3755–3762.
27. Coppa GV, Pierani P, Zampini L, Bruni S, Carloni I, Gabrielli O. Characterization of oligosaccharides in milk and feces of breast-fed infants by high performance anion exchange chromatography. Adv Exp Med Biol 2001; 501:307–314.
28. Mank M, Stahl B, Boehm G. 2,5 Dihydroxybenzoic acid butylamine (DHBB) and other ionic liquid matrices for enhanced MALDI-MS analysis for biomolecules. Anal Chem 2004; 76:2938–2950.
29. Finke B, Stahl B, Pritschet M, Facius D, Wolfgang J, Boehm G. Preparative continuous annular chromatography (P-CAC) enables the large-scale fractionation of fructans. J Agric Food Chem 2002; 50:4743–4748.

30. Geisser A, Hendrich T, Boehm G, Stahl B. Separation of lactose from human milk oligosaccharides with simulated moving bed chromatography. J Chromatogr A 2005; 1092:17–23.
31. Boehm G, Stahl B. Oligosaccharides from milk. J Nutr 2007; 137:1S–3S.
32. Urashima T, Nakamura T, Saito T. Biological significance of milk oligosaccharides—homology and heterogeneity of milk oligosaccharides among mammalian species. Milk Sci 1997; 46:211–220.
33. Kinsella JE, Taylor SL, eds. Commercial β-galactosidase. Food processing. Adv Food Nutr Res. San Diego: Academic Press, 1995.
34. Fuller R. Probiotics in man and animals. J Appl Bacteriol 1989; 66:365–378.
35. Gibson GR, Roberfroid MB. Dietary modulation of the human colonic microbiota: introducing the concept of prebiotics. J Nutr 1955; 125:1401–1412.
36. Gibson GR, Probert HM, Van Loo JAE, Rastall RA, Roberfroid MB. Dietary modulation of the human colonic microbiota: updating the concept of prebiotics. Nutr Res Rev 2004; 17:259–275.
37. Kukkonen K, Savilathi E, Haathela T, Juntunen-Backman K, Korpela R, Poussa T, Tuure T, Kuitunen M. Probiotics and prebiotic galacto-oligosaccharides in the prevention of allergic diseases: a randomized, double-blind, placebo-controlled trial. J Allergy Clin Immunol 2007; 119:192–198.
38. Roberfroid M. Prebiotics: the concept revisited. J Nutr 2007; 137:830S–837S.
39. Yahiro M, Nishikawa I, Murakami Y, Yoshida H, Ahiko K. tudies on application of galactosyl lactose for infant formula. II. Changes of fecal characteristics on infant fed galactosyl lactose. Reports of Research Laboratory, Snow Brand Milk Products 1998; 78:27–32.
40. Ben X, Zhou X, Zhao W, Yu W, Pan W, Zhang W, Wu S, Van Beusekom M, Schaafsma A. Supplementation of milk formula with galacto-oligosaccharides improves intestinal micro-flora and fermentation in term infants. Chin Med J 2004; 117:927–931.
41. Saarveda J, Tscherina A, Moore N, Abi-Hanna A, Coletta F, Emenhiser C, Yolken R. Gastrointestinal function in infants consuming a weaning food supplemented with oligofructose, a prebiotic. J Pediatr Gastroenterol Nutr 1999; 29:95.
42. Guesry PR, Bodanski H, Tomsit E, Aeschlimann JM. Effect of 3 doses of fructo-oligosaccharides in infants. J Pediatr Gastroenterol Nutr 2000; 31:S252.
43. Euler AR, Mitchell DK, Kline R, Pickering LK. Prebiotic effect of fructo-oligosaccharide supplemented term infant formula at two concentrations compared with unsupplemented formula and human milk. J Pediatr Gastroenterol Nutr 2005; 40:157–164.
44. Bettler J, Euler AR. An evaluation of the growth of term infants fed formula supplemented with fructo-oligosaccharides. Int J Probiotics Prebiotics 2006; 1:19–26.
45. Fanaro S, Jelinek J, Stahl B, Boehm G, Kock R, Vigi V. Acidic oligosaccharides from pectin hydrosylate as new component for infant formulae: effect on intestinal flora, stool characteristics, and pH. J Pediatr Gastroenterol Nutr 2005; 41:186–190.
46. Moro G, Minoli I, Mosca M, Jelinek J, Stahl B, Boehm G. Dosage related bifidogenic effects of galacto- and fructo-oligosaccharides in formula fed term infants. J Pediatr Gastroenterol Nutr 2002; 34:291–295.

47. Schmelze H, Wirth S, Skopnik H, Radke M, Knol J, Böckler HM, Brönstrup A, Wells J, Fush C. Randomized double-blind study on the nutritional efficiacy and bifidogenicity of a new infant formula containing partially hydrolysed protein, a high β-palmitic acid level, and nondigestible oligosaccharides. J Pediatr Gastroenteriol Nutr 2003; 36:343–351.
48. Knol J, Scholtens P, Kafka C, Steenbakkers J, Groß S, Helm K, Klarczyk M, Schöpfer H, Böckler HM, Wells J. Colon microflora in infants fed formula with galacto- and fructo-oligosaccharides: more like breast fed infants. J Pediatr Gastroenterol Nutr 2005; 40:36–42.
49. Scholtens P, Alles M, Bindels J, Linde van der E, Toolbom JJM Knol J. Bifidogenic effect of solid weaning foods with added prebiotic oligosaccharides: a randomized controlled clinical trial. J Pediatr Gastroenterol Nutr 2006; 42:553–559.
50. Bakker-Zierikzee AM, Tol EA, Kroes H, Alles MS, Kok FJ, Bindels JG. Faecal sIgA secretion in infants fed on pre- or probiotic infant formula. Pediatr Allergy Immunol 2006; 17: 134–140.
51. Moro G, Arslanoglu S, Stahl B, Jelinek J, Wahn U, Boehm G. A mixture of prebiotic oligosaccharides reduces the incidence of atopic dermatitis during the first six months of age. Arch Dis Child 2006; 91:814–819.
52. Alliet P, Scholtens P, Raes M, Vandenplas Y, Kroes H, Knol J. An infant formula containing a specific prebiotic mixture of GOS/lc FOS leads to higher faecal secretory IgA in infants. J Pediatr Gatsroenterol Nutr 2007; 44(Suppl 1):e179.
53. Desci T, Arato A, Balogh M, Dolinary T, Kanjo AH, Szabo E, Varkonyi A. Randomized placebo controlled double blind study on the effect of prebiotic oligosaccharides on intestinal flora in healthy term infants (translation from Hungarian language). Orvosi Heliap 2005; 146:2445–2450.
54. Rinne MM, Gueimonde M, Kalliomäki M, Hoppu U, Salminen SJ, Isolauri E. Similar bifidogenic effects of prebiotic-supplemented partially hydrolyzed infant formula and breastfeeding on infant gut microbiota. FEMS Immunol Med Microbiol 2005; 43:59–65.
55. Boehm G, Lidestri M, Casetta P, Jelinek J, Negretti F, Stahl B, Marini A. Supplementation of an oligosaccharide mixture to a bovine milk formula increases counts of faecal bifidobacteria in preterm infants. Arch Dis Childh 2002; 86:F178–F181.
56. Mihatsch WA, Hoegel J, Pohlandt F. Prebioitc oligosaccharides reduce stool viscosity and accelerate gastrointestinal transport in preterm infants. Acta Paediatr 2006; 95:843–848.
57. Indrio F, Riezzo G, Montagna O, Valenzano E, Mautone A, Boehm G. Effect of a prebiotic mixture of short chain galacto-oligosaccharides and long chain fructo-on gastric motility in preterm infants. J Pediatr Gastroenterol Nutr 2007; 44(Suppl 1):e217.
58. Moro G, Stahl B, Jelinek J, Boehm G, Coppa GV. Dietary prebiotic oligosaccharides are detectable in faeces of formula fed infants. Acta Paediatr 2005; 94(Suppl 449):27–30.
59. Knol J, Boehm G, Lidestri L, Negretti F, Jelinek J, Agosti M, Stahl B, Marini A, Mosca F. Increase of fecal bifidobacteria due to dietary oligosaccharides induces a reduction of clinically relevant pathogen germs in the

faeces of formula-fed preterm infants. Acta Paediatr 2005; 94(Suppl 449):31–33.
60. Siigur U, Ormisson A, Tamm A. Faecal short-chain fatty acids in breast-fed and bottle-fed infants. Acta Paediatr 1993; 82:536–538.
61. Boehm G, Stahl B, van Laere K, Knol J, Fanaro S, Moro G, Vigi V. Prebiotics in infant formulas. J Clin Gastroenterol 2004; 38(Suppl 2):S76–S79.
62. Willemsen LE, Koetsier MA, van Deventer SJ, van Tol EA. Short chain fatty acids stimulate epithelial mucin 2 expression through differential effects on prostaglandin E(1) and E(2) production by intestinal myofibroblasts. Gut 2003; 52:1442–1447.
63. Van Limpt C, Crienen A, Vriesema A, Knol J. Effect of colonic short chain fatty acids, lactate and pH on the growth of common gut pathogens. Pediatr Res 2004; 56:487.
64. Björksten B, Sepp E, Julge K, Voor T, Mikelsaar M. Allergy development and intestinal flora during the first year of life. J Allergy Clin Immunol 2001; 108:516–520.
65. Ouwehand AC, Isiolauri E, He F, Hashimoto H, Benoo Y, Salimine S. Differences in *Bifidobacterium* flora composition in allergic and healthy infants. J Allergy Clin Immunol 2001; 108:144–145.
66. Watanabe S, Narisawa Y, Arase S, Okamatsu H, Ikenaga T, Tajiri Y, Kumemura M. Differences in fecal microflora between patients with atopic dermatitis and healthy control subjects. J Allergy Clin Immunol 2003; 111: 587–591.
67. Boehm G, Stahl B, Garssen J, Bruzzese E, Moro G, Arslanoglu S. Prebiotics in infant formulas: immune modulators during infancy. Nutrafoods 2005; 4:51–57.
68. Favier CF, de Vos WM, Akkermans AD. Development of bacterial and bifidobacterial communities in faeces of newborn babies. Anaerobe 2003; 9:219–229.
69. Muraro A, Dreborg S, Halken S, Host A, Niggemann B, Aalberse R, Arshad SH, von Berg AV, Carlsen K-H, Duschen K, Eigenmann P, Hill D, Jones C, Mellon M, Oldeus G, Oranje A, Pascual C, Prescott S, Sampson H, Svartengren M, Vandenplas Y, Wahn U, Warner JA, Warner JO, Wickman M, Zeiger RS. Dietary prevention of allergic diseases in infants and small children. Part II: evaluation of methods in allergy prevention studies and sensitization markers. Definitions and diagnostic criteria for allergic diseases. Pediatr Allergy Immunol 2004; 14:196–205.
70. Vos AP, M'rabet L, Stahl B, Boehm G, Garssen J. Immune modulatory effects and potential working mechanisms of orally applied non-digestible carbohydrates. Critical reviews. Immunology 2007; 207:97–190.
71. Vos AP, Haarman M, van Ginkel JWH, Knol J, Garssen J, Stahl B, Boehm G, M'Rabet L. Dietary supplementation of neutral and acidic oligosaccharides enhances Th1 dependent vaccination responses in mice. Pediatric Allergy and Immunol 2007; 10:309–312.
72. Boehm G, Jelinek J, Knol J, M'Rabet L, Stahl B, Vos P, Garssen J. Prebiotic and immune response. J Pediatr Gastroenterol Nutr 2004; 39:772–773.
73. Vos AP, van Esch B, M'Rabet L, Folkerts G, Garssen J. Dietary supplementation with specific oligosaccharide mixtures decreases parameters of allergic asthma in mice. Intern Immunopharmacol 2007; 7:1502–1507.

74. Kalliomäki M, Salminen S, Arvilommi H, Kero P, Koskinen P, Isolauri E. Probiotics in primary prevention of atopic diseases: a randomized placebo-controlled trial. Lancet 2001; 357:1076–1079.
75. Bornet FR, Brous F. Immune-stimulating and gut health-promoting properties of short-chain fructo-oligosaccharides. Nutr Rev 2002; 45:326–334.
76. Harrigan E, Rabinowitz LG. Atopic dermatitis. Pediatr Allergy Immunol 1999; 19:383–389.
77. Kunz B, Oranje AP, Labreze L, Stalder JF, Ring J, Taieb A. Clinical validation and guidelines for the SCORAD index: consensus report of the European Task Force on atopic dermatitis. Dermatology 1997; 195:10–19.
78. Severity scoring of atopic dermatitis: the SCORAD index. Consensus report of the European Task Force on atopic dermatitis. Dermatology 1993; 186:23–31.
79. Garssen J, Arslanoglu S, Boehm G, Faber J, Knol J, Ruiter B, Guido M. A mixture of short chain galacto-oligosaccharides and long chain fructo-oligosaccharides induces an anti-allergic immunoglobulin profile in infants at risk. J Pediatr Gastroenterol Nutr 2007; 40(Suppl 1):e122.
80. Arslanoglu S, Moro G, Schmitt J, Boehm G. Early dietary intervention with a mixture of prebiotic oligosaccharides reduces the allergy associated symptoms and infections during the first 2 years of life. J Pediatr Gastroenterol Nutr 2007; 40(Suppl 1):e129.

Carbohydrate Chemistry, Biology and Medical Applications
Hari G. Garg, Mary K. Cowman and Charles A. Hales

DOI: 10.1016/B978-0-08-054816-6.00013-6

Chapter 13

Role of Cell Surface Carbohydrates in Development and Disease

MICHIKO N. FUKUDA,* TOMOYA O. AKAMA* AND KAZUHIRO SUGIHARA†

**Glycobiology Program, Cancer Research Center, Burnham Institute for Medical Research, La Jolla, CA 92037, USA*
†Department of Obstetrics and Gynecology, Hamamatsu University School of Medicine, Hamamatsu 431-3192, Japan

I. Introduction

The outer surface of mammalian cells is covered by glycoproteins and glycolipids. Substantial biochemical and immunochemical evidence suggests that cell surface carbohydrates play significant roles in development and health (1,2). However, functional studies of cell surface carbohydrates still leave many questions unanswered. In the last decade, genetic approaches and sophisticated chemical analyses have enabled us to reveal the function of specific carbohydrate structures *in vivo*, and as a result we are just starting to understand the role of carbohydrates in development and disease. In this chapter, we discuss the roles of cell surface carbohydrates in development, while focusing on embryo implantation, spermatogenesis, and tissue maturation. We also describe the role of carbohydrates in cancer and discuss therapeutic applications of carbohydrate-based drug discovery.

II. The Role of Cell Surface Carbohydrates in Reproduction and Development

In the field of reproductive biology and embryology, it has been assumed that cell surface carbohydrates play important roles (3–6). However, these hypotheses are difficult to test because embryonic development is dynamic and the material of interest is often too limited to allow chemical analysis. Analyzing early stage embryos requires well-trained hands and skills that many biochemists and molecular biologists have only recently developed. Nonetheless, many attractive hypotheses await testing by new technologies.

A. Embryo Implantation

Implantation and subsequent placentation is a unique mammalian form of reproduction. A fertilized mammalian egg autonomously develops into a blastocyst, which must be successfully implanted in the uterus to develop further into a fetus. Initial adhesion of the embryo to the uterus occurs via the apical cell membrane of two polarized epithelial cells, the trophoblast of blastocysts, and surface epithelial cells of the endometrium. This adhesion is unique because it is apical–apical adhesion between two epithelial cells, whereas generally apical cell surfaces of epithelia are nonadhesive.

Although embryo implantation occurs in all mammals, many specifics of the process of embryo implantation differ significantly among mammalian species (7). Thus, some mechanisms underlying human embryo implantation are likely unique to humans. For example, ectopic pregnancy occurs at a rate of 0.25–1.4% in all pregnancies in humans (8,9), but extrauterine implantation does not occur in rodents (10–12) and is extremely rare in nonhuman primates (13).

1. Embryo Implantation and Cell Surface Carbohydrates

Many reports suggested a role of cell surface carbohydrates in embryo implantation in the mouse. For example, type 1H and Le^y antigens, both containing Fucα1→2Gal terminal structure, were implicated in mouse blastocyst implantation (4,14–16). Two fucosyltransferase genes, *Fut1* and *Fut2*, responsible for Fucα1→2Gal terminal structure of these antigens were cloned and evaluated for their contribution to embryo implantation (17,18). However, neither *Fut1* nor *Fut2* nulls showed reproductive failure (18). This study excluded an essential role of Fucα1→2Gal terminated glycans in fertility in the mouse.

Recently, a carbohydrate-binding protein, L-selectin, has been proposed as an adhesion molecule for human embryo implantation (19). L-Selectin is expressed on the surface of lymphocytes and interacts with sulfated and fucosylated carbohydrates expressed on lymph node endothelial cells (20–26). Carbohydrate ligand for L-selectin in the lymph node is closely related to MECA-79 antigen (Fig. 1). In human endometrium, the expression of MECA-79 antigen is hormonal cycle dependent but the presence of an embryo is not required (27). In the mouse, L-selectin nulls reproduce normally (28). Mutant mice lacking sulfotransferases (22,23,29–32) and glycosyltransferases (21,31,32) required for synthesis of L-selectin ligand exhibited no defects in reproduction. Therefore, it is clear that L-selectin is not required for the mouse embryo implantation. It thus appears that L-selectin is uniquely involved in human embryo implantation.

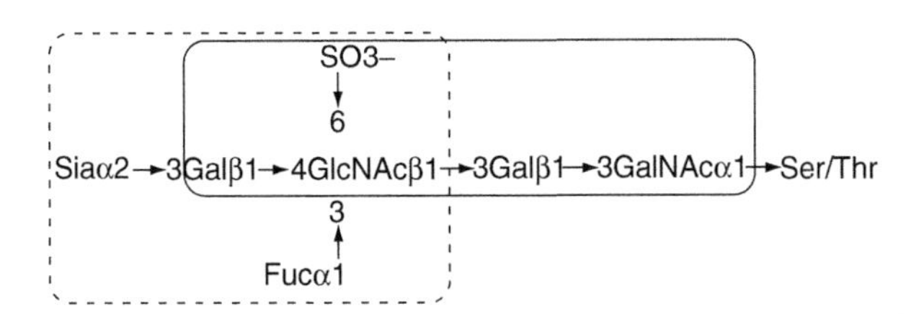

Figure 1 Structure of L-selectin ligand and MECA-79 epitope. L-Selectin recognizes sulfated sialyl Lewis X structure (27) presented by *O*-glycans and *N*-glycans, marked by dotted-lines. MECA-79 antibody is specific to sulfated extended core 1 *O*-glycan marked by solid-lines, but MECA-79 antibody also recognizes this oligosaccharide with sialic acid and fucose (24,28,29).

2. *Involvement of Heparin-Binding Epidermal Growth Factor-Like Growth Factor in Embryo Implantation*

Many studies have established the mechanism underlying the initial adhesion of mouse embryo to the uterus. Thus, ErbB4 (one of ErbB family receptor tyrosine kinases) expressed on the blastocyst surface binds to the membrane-bound form heparin-binding epidermal growth factor (HB-EGF) expressed by uterine epithelia (33–35). These studies indicated that the implanting blastocyst induces HB-EGF in the maternal epithelia in a spatially and temporally dependent manner, demonstrating interaction between maternal cells and implanting embryo prior to the adhesion (35).

A unique apical cell adhesion molecule between human trophoblastic cells and endometrial epithelial cells has been identified and designated trophinin (36). Trophinin is not part of a known gene family in mammals. The trophinin gene is mapped to the X-chromosome locus, which is linked to the evolution of mammals (37). Trophinin protein is expressed strongly in trophectoderm cells at the embryonic pole in the monkey blastocyst where it adheres to the maternal epithelium (36). Trophinin is expressed in cells in human embryo implantation sites in normal and ectopic pregnancies (38,39). Significant differences have been found in the expression of trophinin between mouse and human. Trophinin expression in the mouse uterus is dependent on estrogen secreted from ovary regardless of the presence or absence of blastocysts (40). Furthermore, trophinin gene knockout mouse did not show a defect in embryo implantation (41). These findings indicate that trophinin does not play an essential role in mouse embryo implantation. Nonetheless, the hypothesis that trophinin mediates the initial adhesion of human embryo implantation was strengthened by cell-type-specific apical cell adhesion activity (36,37), embryo-dependent trophinin expression by maternal epithelia in humans (Fig. 2) (39), and regulation of ErbB4 by trophinin (42,43).

A recent study showed that trophinin not only functions as an adhesion molecule but also plays a role in signal transduction required for ErbB4 activation in human trophoblastic cells (42,43). Thus, trophinin has an activity responsible for its ability to transform silent trophectoderm cells into an aggressive trophoblast for

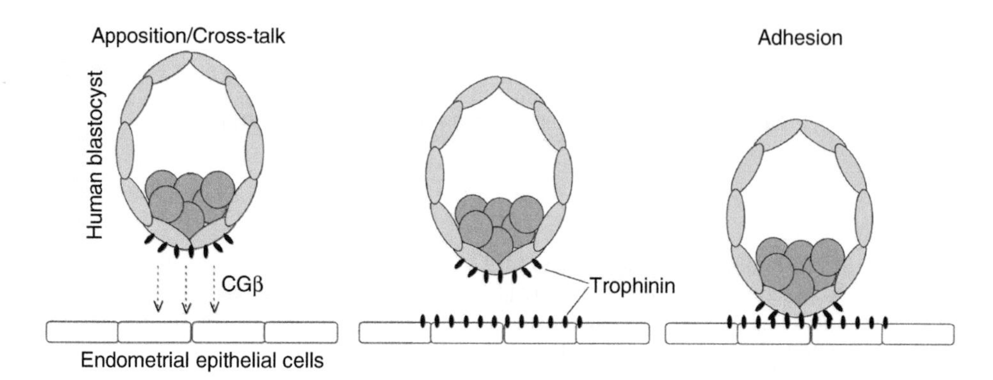

Figure 2 Initial adhesion of a human blastocyst to endometrial surface epithelia mediated by trophinin. A human blastocyst secretes glycoprotein hormone, hCG. It is likely that a human blastocyst expresses homophilic adhesion molecule trophinin on the trophectoderm surface at embryonic pole in a similar manner as a monkey blastocyst does (36). When the blastocyst comes close to the endometrial epithelia cells, high concentration of hCG derived from blastocyst affects locally the maternal epithelia (left) and induces trophinin expression in narrowly limited region of endometrial epithelia (center) (38). The blastocyst and endometrial epithelia then adhere to each other through trophinin–trophinin binding (right).

invasion upon apical adhesion (44–46). Thus, the emerging figure is that trophinin functions upstream of ErbB4/HB-EGF as part of a pathway conserved in mouse and human embryo implantation (42,43).

Expression of human trophinin by uterine epithelial cells was narrowly limited to the implantation site (36,38,39). Human chorionic gonadotrophin (hCG) secreted from the blastocyst induced trophinin expression by the maternal epithelia (39) for adhesion (Fig. 2). The beta chain of hCG, or CGβ, is unique to primates and is not highly homologous between primates and humans (47,48), providing a molecular basis for mechanisms unique to human embryo implantation (39).

B. Spermatogenesis

Many reports showed remarkable staining patterns of testes with lectins and monoclonal anti-carbohydrate antibodies, suggesting that specific developmental stages of spermatogenic cells are associated with unique cell surface carbohydrate structures (49–55). A mouse gene knockout of UDP-galactose:ceramide galactosyltransferase that is responsible for seminolipid (a sulfated glycolipid) resulted in male infertility because of massive apoptosis of spermatogenic cells (56,57). A mutant mouse lacking a sialyltransferase required for GM2/GD2 glycolipid synthesis resulted in a spermatogenesis defect probably because of impaired secretion of testosterone from Leydig cells (58). These phenotypes indicated that specific carbohydrates play critical roles in spermatogenesis.

Mutant mice with defective spermatogenesis were found in α-mannosidase IIx (MX) gene knockout mice (59). MX is an enzyme closely related to the Golgi *N*-glycan processing α-mannosidase II (MII) (60,61). Both MX and MII are expressed in a wide variety of tissues and cell types. MII homozygous null mice exhibit phenotypes indicative of dyserythropoiesis (62), a condition similar to the human genetic disease, congenital dyserythropoietic anemia type II (63,64). The mild phenotype shown by MII nulls sharply contrasted with embryonic lethality resulting from *N*-acetylglucosaminyltransferase-I (GnT-I) null mutations (65,66) and the early neonatal lethality seen in GnT-II null mice (67,68) (Fig. 3). Structural analysis of *N*-glycans synthesized by MII null mice showed accumulation of five-mannosyl hybrid-type structures, consistent with the substrate specificity of MII. However, *N*-glycan processing in MII null mice was not completely blocked, since MII null mouse tissues synthesize complex-type oligosaccharides downstream (69). This observation suggested the existence of an alternative pathway, independent of MII. MX was proposed as one of candidate enzymes for that alternate pathway.

MX gene-deficient mice were born and grew to adulthood without illness, but MX null males were infertile (59). The testes of MX null males were smaller than those of wild-type or heterozygous littermates. Electron microscopy showed prominent intercellular spaces surrounding MX null spermatocytes, suggesting a failure of germ cell adhesion to Sertoli cells within the seminiferous tubules.

Quantitative structural analyses of *N*-glycans showed that wild-type testes contained significant amounts of GlcNAc-terminated complex type *N*-glycans (59). However, in MX null testes, levels of GlcNAc-terminated oligosaccharides were reduced. An *in vitro* assay for adhesion of spermatogenic cells to Sertoli cells showed that a GlcNAc-terminated triantennary and fucosylated *N*-glycan structure (Fig. 4) plays a key role in germ cell/Sertoli cell adhesion. Collectively, MX null spermatogenic cells failed to adhere to Sertoli cells in seminiferous tubules and were prematurely released from the testis to the epididymis.

MX was considered to be playing a subsidiary role to MII. MX gene knockout mouse revealed an unexpected important function of MX in spermatogenesis. Therefore, the genetic approach elucidated the specific role of cell surface carbohydrates in specific cell types.

C. Embryonic Development

1. *Embryonic and Postnatal-Lethality in MX/MII Double Deficient Mutant*

Although it is widely assumed that cell surface carbohydrates play a significant role in mammalian embryonic development, the first evidence for this hypothesis came from the GnT-I gene knockout mouse, which dies at early embryonic stages (65,66). We generated MII/MX double-null mutant mice by crossing MII- and MX-null mice. All double-null embryos survived until E15, but some double nulls died between E15 and the day of birth (E18). The majority of double nulls died soon after birth, with many double-null neonates dying after gasping for air at birth, suggesting that neonatal lethality is due to respiratory failure (69).

Lung tissue of double-null neonates had less alveolar air space in comparison to lung tissue from wild-type mice or mice mutant in either MII or MX. The pulmonary epithelial cell layer of double nulls appeared thicker than that of either

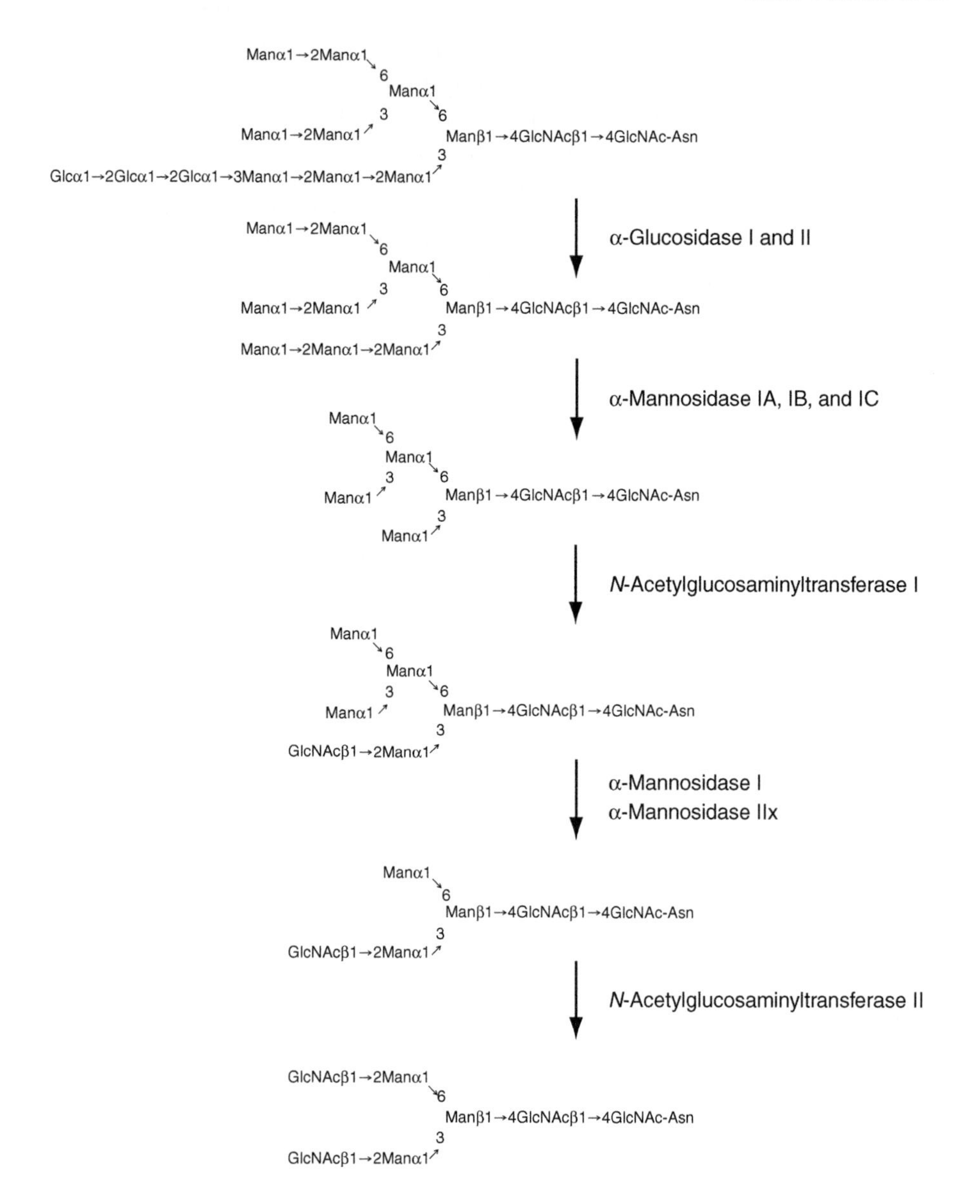

Figure 3 Biosynthetic pathway of *N*-glycans. *N*-glycan processing pathway. The pathway is based on the previously established pathway (69–71) and recent finding on MX (69). In the endoplasmic reticulum, $Glc_3Man_9GlcNAc_2$ linked to dolichol is transferred from dolichol to nascent polypeptides. Immediately after this transfer, α-glucosidase I and α-glucosidase II in the endoplasmic reticulum remove α1,2- and α-1,3-glucose residues, respectively. Removal of mannose residues continues in the Golgi by α1,2-mannosidase Is (IA, IB, and IC), which produces $Man_5GlcNAc_2$. To this substrate, *N*-acetylglucosaminyltransferase I (GnT-I) adds GlcNAc to form $GlcNAc_1Man_5GlcNAc_2$. Then α-mannosidase II (MII) and α-mannosidase IIx (MX) remove the α1,3- and α1,6-linked mannose residues to form $GlcNAc_1Man_3GlcNAc_2$. *N*-acetylglucosaminyltransferase II (GnT-II) then produces $GlcNAc_2Man_3GlcNAc_2$ for further modification. *N*-glycan structures before *N*-acetylglucosaminyltransferase I are called high mannose type, those between *N*-acetylglucosaminyltransferase I and II are called hybrid type, and those after *N*-acetylglucosaminyltransferase II are called complex type.

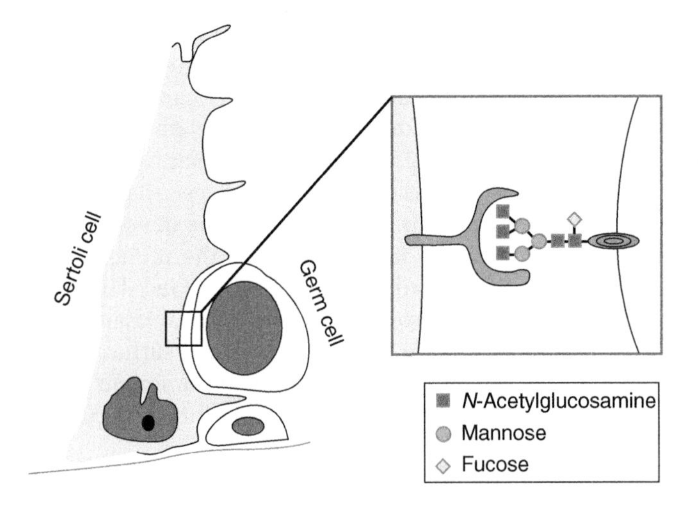

Figure 4 Triantennary *N*-glycans involved in germ cell and Sertoli cell interaction in the mouse testis. Spermatogenesis takes place in the tubal structure called seminiferous tube in the testis. Inside of seminiferous tube is lined by Sertoli cells, the tall epithelial cells. When germ stem cells at the base of Sertoli cells differentiate to spermatocytes, the cells move upward interacting with Sertoli cells. The MX gene knockout mouse revealed that fucosylated GlcNAc-terminated triantennary *N*-glycan (insert) plays an important role in germ cell adhesion to Sertoli cells (59).

wild type or single nulls, suggesting a difficulty in alveolar expansion in double nulls.

Livers of double nulls were also frequently damaged in both neonates and E15 embryos. Apoptotic signals indicative of cell death were increased in damaged areas of the double-null liver. Interestingly, in double-null mice, abnormal morphologies were not apparent in other tissues, such as heart, small intestine, or pancreas.

Carbohydrate analysis showed the complete absence of the complex type *N*-glycans in double nulls, indicating that lack of both MII and MX leads to arrest of *N*-glycan processing at hybrid-type structures (Fig. 3).

Analysis of enzymatic activity showed that mouse MX specifically removes two mannosyl residues from $GlcNAc_1Man_5GlcNAc_2$ to produce $GlcNAc_1$-$Man_3GlcNAc_2$. Therefore, MX specifically hydrolyzes the same oligosaccharide substrate that MII does, meaning MX is an MII isozyme (69). Results obtained by analyzing MII/MX double mutants indicate that MX is responsible for the previously proposed "alternate pathway" (62).

2. *Glycosylation Enzymes Essential for Embryonic Development*

GnT-I nulls die *in utero* around embryonic day 10 (65,66). GnT-II nulls show genetic background-dependent postnatal lethality 1–4 weeks after birth (67,68). For example, GnT-II nulls back-crossed to C57/BL6 strain died in 4 weeks, while the mutant mice back-crossed to ICR strain survived for more than 6 weeks. It is striking that the phenotypes of MII and MX double nulls fall between these two. This evidence clearly indicates that disruption of *N*-glycan processing at an earlier step causes more severe effects on embryonic and postnatal development. Because GnT-III, -IV, and -V add additional branches to the biantennary structure shown at the bottom of Fig. 3, these findings further suggest that terminal structures and their activities are important for normal embryonic development and maintenance of health in the adult stage. Indeed, none of the mutant mice deficient of GnT-III, -IV, and -V are embryonic lethal (70–74). GnT-III nulls are healthy except that they showed reduced progression of diethylnitrosamine-induced hepatoma (69). GnT-IV nulls are diabetic due to reduced cell surface expression of glucose transporter (75). GnT-V nulls showed kidney autoimmune disease, enhanced delayed-type hypersensitivity, and increased susceptibility to experimental autoimmune encephalomyelitis (76,77).

In chimeric embryos containing inactivated GnT-I cells, GnT-I null cells in bronchi failed to produce the epithelial layer (72). Therefore, lung defects are common features in mouse mutant for genes catalyzing *N*-glycan biosynthesis. Mouse lines defective in *N*-glycan synthesis may therefore serve as good models for studying respiratory failure in humans.

Three *N*-glycan processing α-1,2-mannosidases, IA, IB, and IC, have been identified (78–81). They have similar activities with only slight differences in the order of mannose removal from $Man_{8-9}GlcNAc_2$. In the mouse, IB is strongly expressed in testis but weakly expressed or absent in many other tissues. IB null fetuses survive until term, and null neonates are indistinguishable from littermates. However, within several hours of birth they progressively display signs of respiratory distress. Histology of neonatal lungs demonstrates thickening of alveolar septae and significant hemorrhage (82). Gene knockout mice for IA and IC have not been generated yet.

One fucosyltransferase, Fut8, transfers fucose to innermost GlcNAc of *N*-glycans. Fut8-null mice die shortly after birth because of abnormal lung development (83). Lungs from Fut8-null mice show enlarged alveoli and much empty space, due to excessive degradation of extracellular matrix by overexpressed metalloproteases.

In summary, gene knockout mouse technology has allowed us to determine the function of cell surface carbohydrates in mouse embryos in recent years. Some glycosyltransferases-deficient mutant mouse showed severe phenotypes, including embryonic lethality. However, we have not yet identified the mechanisms underlying these phenotypes. Identification of critical carbohydrate structure(s), specific proteins carrying specific carbohydrate structure(s), and the counter-receptors for those carbohydrates should be identified in future studies.

III. The Role of Cell Surface Carbohydrates in Cancer

Cancer is the number one cause of death in developed countries. The cancer mortality rate has remained high in the last 30 years (84) in part because there is no effective treatment for adenocarcinoma, the most aggressive form of cancer including prostate, breast, lung, colon, stomach, and endometrial cancers (85). Normal cellular counterparts of adenocarcinomas are epithelial cells, whose apical surfaces are covered by a thick layer known as the glycocalyx. When epithelial cells are transformed, the structure of cell surface carbohydrates changes (86–88). Monoclonal antibodies raised against adenocarcinoma are often specific to the cell surface carbohydrates expressed by cancer (89).

Many studies have been carried out using anti-carbohydrate antibodies for cancer diagnosis. These studies reveal a correlation between particular carbohydrate antigens and poor clinical prognosis, including metastasis (88,90–92). Therefore, specific carbohydrate structures expressed on the cancer cell surface may play a role in metastasis. The section below describes two hypotheses on carbohydrate-dependent cancer metastasis by selectin-dependent and selectin-independent mechanisms.

A. Selectin-Dependent Metastasis

Several studies consistently indicate that colon cancer patients with sialyl Lewis a (sLe^a-) and/or sialyl Lewis x (sLe^x-)positive tumors show poorer survival than do patients with sLe^a/sLe^x negative tumors (93–98). Because both sLe^a and sLe^x antigens are ligands for E-selectin (Fig. 5), sLe^a- and/or sLe^x-expressing cancer cells should adhere to E-selectin expressing endothelial cells, providing the mechanism for E-selectin-dependent hematogenous metastasis (90). *In vitro* assays of sLe^a- or sLe^x-expressing cancer cells indeed show adherence to cultured endothelial cells expressing E-selectin (99,100). Although normal healthy endothelial cells do not express E-selectin, primary tumors in cancer patients secrete inflammatory cytokines such as TNFα, which induce E-selectin (90,101,102). This hypothesis of E-selectin induction in cancer patients by cytokines derived from the primary

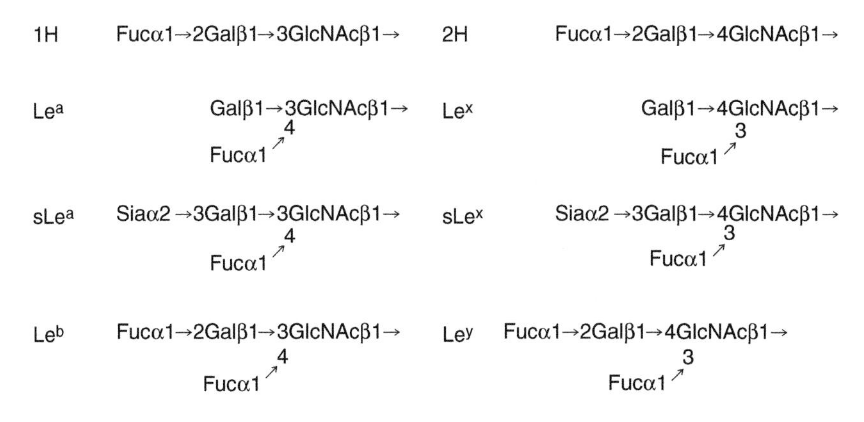

Figure 5 Structures of cancer-associated fucosylated and/or sialylated carbohydrate antigens. Blood group type 1-based (left) and type 2-based (right) structures are shown. E-selectin binds both sLe^a and sLe^x, but E-selectin does not bind to H, Le^a, Le^x, Le^b, and Le^y antigens.

tumor is supported by observation of elevated levels of soluble E-selectin in cancer patients' sera (103). The evidence collectively suggests that sLe^{a}- and/or sLe^{x}-antigen positive cancer cells metastasize through an E-selectin-mediated mechanism.

B. Selectin-Independent Metastasis

Many reports indicate that malignant cancer cells express carbohydrate antigens other than sLe^{a} and sLe^{x} antigens. These reports show a correlation between expression of particular cell surface carbohydrate antigens and poor prognosis in cancer patients (104,105). Examples of these are seen in Lewis B (Le^{b}) and Lewis Y (Le^{y}) antigens overexpressed by breast, lung, ovarian, and endometrial cancers (104,106–110).

We investigated carbohydrate-dependent lung colonization by cancer cells in mice (111–113). While mouse B16 melanoma cells do not express sLe^{x} antigen, they become sLe^{x} positive following transfection with FUT-3 (111). Upon intravascular injection, transfected B16 cells colonize the lung. Colonization is inhibited by anti-sLe^{x} antibody and also by selectin ligand carbohydrate mimicry peptide. Together these results indicated that lung colonization of transfected B16 is carbohydrate dependent (111,112). However, lung colonization occurred in E-selectin/P-selectin double-null mice, excluding involvement of these selectins (113). These data suggested strongly that an as yet unidentified carbohydrate-binding receptor is expressed in lung vasculature, and that this receptor is responsible for carbohydrate-dependent metastasis. Further studies should identify the carbohydrate-binding receptor(s) responsible for metastasis, which is an important step for development of reliable cancer diagnosis and therapy. A chemically synthesized peptide that binds to novel endothelial receptor (112) may be used as a therapeutic agent against the potential metastatic sites in the vasculature.

IV. Perspective

Gene knockout experiments in mice clearly indicate that cell surface carbohydrates play many vital roles in development and health. Evidence suggests that cell surface carbohydrates function in many cellular activities by interacting with their microenvironment *in vivo*. Future studies should identify mechanisms underlying activation of cells by cell surface carbohydrates, which are vital for understanding the activities of malignant cancer cells.

References

1. Varki A. Biological roles of oligosaccharides: all of the theories are correct. Glycobiology 1993; 3:97–130.
2. Hakomori S. Tumor malignancy defined by aberrant glycosylation and sphingo (glyco)lipid metabolism (review). Cancer Res 1996; 56:5309–5318.
3. Dell A, Morris HR, Easton RL, Patankar M, Clark GF. The glycobiology of gametes and fertilization. Biochim Biophys Acta 1999; 1473:196–205.

4. Kimber SJ. Molecular interactions at the maternal-embryonic interface during the early phase of implantation. Semin Reprod Med 2000; 18:237–253.
5. Aplin JD, Kimber SJ. Trophoblast-uterine interactions at implantation. Reprod Biol Endocrinol 2004; 2:48.
6. Carson DD, DeSouza MM, Kardon R, Zhou X, Lagow E, Julian J. Mucin expression and function in the female reproductive tract. Human Reprod update 1998; 4:459–464.
7. Carson DD, Bagchi I, Dey SK, Enders AC, Fazleabas AT, Lessey BA, Yoshinaga K. Embryo implantation. Dev Biol 2000; 223:217–237.
8. Brenner PF, Roy S, Mishell DR, Jr. Ectopic pregnancy. A study of 300 consecutive surgically treated cases. Jama 1980; 243:673–676.
9. Graczykowski JW, Mishell DRJ. Ectopic pregnancy. In: Lobo RA, Mishell DRJ, Paulson RJ, Shoupe D, eds. Mishell's Textbook of Infertility, Contraception, and Reproductive Endocrinology. Masachusetts, USA: Blackwell Science, 1997; 623–637.
10. Orsini MW, McLaren A. Loss of the zona pellucida in mice, and the effect of tubal ligation and ovariectomy. J Reprod Fertil 1967; 13:485–499.
11. Tutton DA, Curr DH. The fate of trophoblast within the oviduct in the mouse. Gynecol Obstet Invest 1984; 17:18–24.
12. Pauerstein CJ, Eddy CA, Koong MK, Moore GD. Rabbit endosalpinx suppress ectopic implantation. Fertil Steril 1990; 54:522–526.
13. Schlabritz-Loutsevitch NE, Hubbard GB, Frost PA, Cummins LB, Dick EJ Jr, Nathanielsz PW, McDonald TJ. Abdominal pregnancy in a baboon: a first case report. J Med Primatol 2004; 33:55–59.
14. Lindenberg S, Sundberg K, Kimber SJ, Lundblad A. The milk oligosaccharide, lacto-N-fucopentaose I, inhibits attachment of mouse blastocysts on endometrial monolayers. J Reprod Fertil 1988; 83:149–158.
15. Kimber SJ, Lindenberg S, Lundblad A. Distribution of some Galb1–3(4) GlcNAc related carbohydrate antigens on the mouse uterine epithelium in relation to the peri-implantation period. J Reprod Imm 1988; 12:297–313.
16. Zhu ZM, Kojima N, Stroud MR, Hakomori S-I, Fenderson BA. Monoclonal antibody directed to Ley oligosaccharide inhibits implantation in the mouse. Biol Reprod 1995; 52:903–912.
17. Domino SE, Zhang L, Lowe JB. Molecular cloning, genomic mapping, and expression of two secretor blood group alpha(1,2)fucosyltransferase genes differentially regulated in mouse uterine epithelium and gastrointestinal tract. J Biol Chem 2001; 276:23748–23756.
18. Domino SE, Zhang L, Gillespie PJ, Saunders TL, Lowe JB. Deficiency of reproductive tract alpha(1,2)fucosylated glycans and normal fertility in mice with targeted deletions of the FUT1 or FUT2 alpha(1,2)fucosyltransferase locus. Mol Cell Biol 2001; 21:8336–8345.
19. Genbacev OD, Prakobphol A, Foulk RA, Krtolica AR, Ilic D, Singer MS, Yang ZQ, Kiessling LL, Rosen SD, Fisher SJ. Trophoblast L-selectin-mediated adhesion at the maternal-fetal interface. Science 2003; 299:405–408.
20. Lowe JB. Selectin ligands, leukocyte trafficking, and fucosyltransferase genes. Kidney Int 1997; 51:1418–1426.
21. Yeh JC, Hiraoka N, Petryniak B, Nakayama J, Ellies LG, Rabuka D, Hindsgaul O, Marth JD, Lowe JB, Fukuda M. Novel sulfated lymphocyte

homing receptors and their control by a Core1 extension beta 1,3-N-acetylglucosaminyltransferase. Cell 2001; 105:957–969.
22. Hemmerich S, Bistrup A, Singer MS, van Zante A, Lee JK, Tsay D, Peters M, Carminati JL, Brennan TJ, Carver-Moore K, Leviten M, Fuentes ME, Ruddle NH, Rosen SD. Sulfation of L-selectin ligands by an HEV-restricted sulfotransferase regulates lymphocyte homing to lymph nodes. Immunity 2001; 15:237–247.
23. Hemmerich S, Rosen SD. Carbohydrate sulfotransferases in lymphocyte homing. Glycobiology 2000; 10:849–856.
24. Mitsuoka C, Sawada-Kasugai M, Ando-Furui K, Izawa M, Nakanistri H, Nakamura S, Ishida H, Kiso M, Kannagi R. Identification of a major carbohydrate capping group of the L-selectin ligand on high endothelial venules in human lymph nodes as 6-sulfo sialyl Lewis X. J Biol Chem 1998; 273:11225–11233.
25. Mitoma J, Petryniak B, Hiraoka N, Yeh JC, Lowe JB, Fukuda M. Extended core 1 and core 2 branched *O*-glycans differentially modulate sialyl Lewis X-type L-selectin ligand activity. J Biol Chem 2003; 278:9953–9961.
26. Mitoma J, Bao X, Petryanik B, Schaerli P, Gauguet JM, Yu SY, Kawashima H, Saito H, Ohtsubo K, Marth JD, Khoo KH, von Andrian UH, Lowe JB, Fukuda M. Critical functions of *N*-glycans in L-selectin-mediated lymphocyte homing and recruitment. Nat Immunol 2007; 8:409–418.
27. Lai TH, Shih Ie M, Vlahos N, Ho CL, Wallach E, Zhao Y. Differential expression of L-selectin ligand in the endometrium during the menstrual cycle. Fertil Steril 2005; 83(Suppl 1):1297–1302.
28. Frenette PS, Wagner DD. Insights into selectin function from knockout mice. Thromb Haemost 1997; 78:60–64.
29. Kawashima H, Petryniak B, Hiraoka N, Mitoma J, Huckaby V, Nakayama J, Uchimura K, Kadomatsu K, Muramatsu T, Lowe JB, Fukuda M. N-acetylglucosamine-6-*O*-sulfotransferases 1 and 2 cooperatively control lymphocyte homing through L-selectin ligand biosynthesis in high endothelial venules. Nat Immunol 2005; 6:1096–1104.
30. Hiraoka N, Kawashima H, Petryniak B, Nakayama J, Mitoma J, Marth JD, Lowe JB, Fukuda M. Core 2 branching beta1,6-*N*-acetylglucosaminyltransferase and high endothelial venule-restricted sulfotransferase collaboratively control lymphocyte homing. J Biol Chem 2004; 279:3058–3067.
31. Mal P, Thall AD, Petryniak B, Rogers CE, Smith PL, Marks RM, Kelley RJ, Gersten KM, Cheng G, Saunders TL, Camper SA, Camphausen RT, Sullivan FX, Isogai Y, Hindsgaul O, Von Andrian VH, Lowe JB. The alpha(1,3) fucosyltransferase FucTVII controls leukocyte trafficking through an essential role in L-, E-, and P-selectin ligand biosynthesis. Cell 1996; 86:643–653.
32. Homeister JW, Daugherty A, Lowe JB. Alpha(1,3)fucosyltransferases FucT-IV and FucT-VII control susceptibility to atherosclerosis in apolipoprotein E-/- mice. Arterioscler Thromb Vasc Biol 2004; 24:1897–1903.
33. Paria BC, Elenius K, Klagsbrun M, Dey SK. Heparin-binding EGF-like growth factor interacts with mouse blastocysts independently of ErbB1: a possible role for heparan sulfate proteoglycans and ErbB4 in blastocyst implantation. Development 1999; 126:1997–2005.
34. Raab G, Kover K, Paria BC, Dey SK, Ezzell RM, Klagsbrun M. Mouse preimplantation blastocysts adhere to cells expressing the transmembrane form of heparin-binding EGF-like growth factor. Development 1996; 122:637–645.

35. Paria BC, Reese J, Das SK, Dey SK. Deciphering the cross-talk of implantation: advances and challenges. Science 2002; 296:2185–2188.
36. Fukuda MN, Sato T, Nakayama J, Klier G, Mikami M, Aoki D, Nozawa S. Trophinin and tastin, a novel cell adhesion molecule complex with potential involvement in embryo implantation. Genes Dev 1995; 9:1199–1210.
37. Pack S, Tanigami A, Ledbetter DH, Sato T, Fukuda MN. Assignment of trophoblast/endometrial epithelium cell adhesion molecule trophinin gene TRO to human chromosome bands Xp11.22→p11.21 by in situ hybridization. Cytogenet Cell Genet 1997; 79:123–124.
38. Suzuki N, Nakayama J, Shih IM, Aoki D, Nozawa S, Fukuda MN. Expression of trophinin, tastin, and bystin by trophoblast and endometrial cells in human placenta. Biol Reprod 1999; 60:621–627.
39. Nakayama J, Aoki D, Suga T, Akama TO, Ishizone S, Yamaguchi H, Imakawa K, Nadano D, Fazleabas AT, Katsuyama T, Nozawa S, Fukuda MN. Implantation-dependent expression of trophinin by maternal fallopian tube epithelia during Tubal pregnancies: possible role of human chorionic gonadotrophin on Ectopic pregnancy. Am J Pathol 2003; 163:2211–2219.
40. Suzuki N, Nadano D, Paria BC, Kupriyanov S, Sugihara K, Fukuda MN. Trophinin expression in the mouse uterus coincides with implantation and is hormonally regulated but not induced by implanting blastocysts. Endocrinology 2000; 141:4247–4254.
41. Nadano D, Sugihara K, Paria BC, Saburi S, Copeland NG, Gilbert DJ, Jenkins NA, Nakayama J, Fukuda MN. Significant differences between mouse and human trophinins are revealed by their expression patterns and targeted disruption of mouse trophinin gene. Biol Reprod 2002; 66:313–321.
42. Sugihara K, Sugiyama D, Byrne J, Wolf DP, Lowitz KP, Kobayashi Y, Kabir-Salmani M, Nadano D, Aoki D, Nozawa S, Nakayama J, Mustelin T, Ruoslahti E, Yamaguchi N, Fukuda MN. Trophoblast cell activation by trophinin ligation is implicated in human embryo implantation. Proc Natl Acad Sci USA 2007; 104:3799–3804.
43. Fukuda MN, Sugihara K. Signal transduction in human embryo implantation. Cell Cycle 2007; 6:1153–1156.
44. Knoth M, Larsen JF. Ultrastructure of a human implantation site. Acta Obstet Gynecol Scand 1972; 51:385–393.
45. Lindenberg S. Ultrastructure in human implantation: transmission and scanning electron-microscopy. Baillieres Clin Obstet Gynecol 1991; 5:1–14.
46. Enders AC. Cytology of human early implantation. Res Reprod 1976; 8:1–2.
47. Pierce J, Parsons TF. Glycoprotein hormones: structure and Function. Ann Rev Biochem 1981; 50:465–495.
48. Maston GA, Ruvolo M. Chorionic gonadotropin has a recent origin within primates and an evolutionary history of selection. Mol Biol Evol 2002; 19:320–335.
49. Lee MC, Damjanov I. Anatomical distribution of lectin-binding sites in mouse testis and epididymis. Differentiation 1984; 27:74–81.
50. Arenas MJ, Madrid JF, Bethencourt FR, Fraile B, Paniagua R. lectin histochemistry of the human testis. Int J Androl 1998; 21:332–342.
51. Ertl C, Wrobel KH. Distribution of sugar residues in the bovine testis during postnatal ontogenesis demonstrated with lectin-horseradish peroxidase conjugates. Histochemistry 1992; 97:161–171.

52. Kurohmaru M, Kobayashi H, Kanai Y, Hattori S, Nishida T, Hayashi Y. Distribution of lectin in the testis of the musk shew. J Anat 1995; 187:323–329.
53. Eddy EM, Muller CH, Lingwood CA. Preparation of monoclonal antibody to sulfatoxygalactosylglycerolipid by in vitro immunization with a glycolipid-glass conjugate. J Immunol Methods 1985; 81:137–146.
54. Symington FW, Fenderson BA, Hakomori S. Fine specificity of a monoclonal anti-testicular cell antibody for glycolipids with terminal *N*-acetyl-D-glucosamine structure. Mol Immunol 1984; 21:877–882.
55. Fenderson BA, O'Brien DA, Millette CF, Eddy EM. Stage-specific expression of three cell surface carbohydrate antigens during murine spermatogenesis detected with monoclonal antibodies. Dev Biol 1984; 103:117–128.
56. Honke K, Hirahara Y, Dupree J, Suzuki K, Popko B, Fukushima K, Fukushima J, Nagasawa T, Yoshida N, Wada Y, Taniguchi N. Paranodal junction formation and spermatogenesis require sulfoglycolipids. Proc Natl Acad Sci USA 2002; 99:4227–4232.
57. Fujimoto H, Tadano-Aritomi K, Tokumasu A, Ito K, Hikita T, Suzuki K, Ishizuka I. Requirement of seminolipid in spermatogenesis revealed by UDP-galactose: ceramide galactosyltransferase-deficient mice. J Biol Chem 2000; 275:22623–22626.
58. Takamiya K, Yamamoto A, Furukawa K, Zhao J, Fukumoto S, Yamashiro S, Okada M, Haraguchi M, Shin M, Kishikawa M, Shiku H, Aizawa S. Complex gangliosides are essential in spermatogenesis of mice: possible roles in the transport of testosterone. Proc Natl Acad Sci USA 1998; 95:12147–12152.
59. Akama TO, Nakagawa H, Sugihara K, Narisawa S, Ohyama C, Nishimura S, O'Brien AD, Moremen KW, Millan JL, Fukuda MN. Germ cell survival through carbohydrate-mediated interaction with Sertoli cells. Science 2002; 295:124–127.
60. Misago M, Liao Y-F, Kudo S, Eto S, Mattei MG, Moremen KW, Fukuda MN. Molecular cloning and expression of cDNAs encoding human α-mannosidase II and a novel α-mannosidase IIx isozyme. Proc Natl Acad Sci USA 1995; 92:11766–11770.
61. Moremen KW, Robbins PW. Isolation, characterization, and expression of cDNAs encoding murine α-mannosidase II, a Golgi enzyme that controls conversion of high mannose to complex *N*-glycans. J Cell Biol 1991; 115:1521–1534.
62. Chui D, Oh-Eda M, Liao YF, Panneerselvam K, Lal A, Marek KW, Freeze HH, Moremen KW, Fukuda MN, Marth JD. Alpha-mannosidase-II deficiency results in dyserythropoiesis and unveils an alternate pathway in oligosaccharide biosynthesis. Cell 1997; 90:157–167.
63. Fukuda MN. Congenital dyserythropoietic anaemia type II (HEMPAS) and its molecular basis. Baillieres Clin Haematol 1993; 6:493–511.
64. Fukuda MN. HEMPAS. Hereditary erythroblastic multinuclearity with positive acidified serum lysis test. Biochim Biophys Acta 1999; 1455:231–239.
65. Metzler M, Gertz A, Sarker M, Schachter H, Scrader JM, Marth JD. Complex asparagine-linked oligosaccharides required for morphogenic events during post-implantation development. EMBO J 1994; 13:2056–2065.
66. Ioffe E, Stanley P. Mice lacking *N*-acetylglucosaminyltransferase I activity die at midgestation, revealing an essential role for complex or hybrid N-linked carbohydrates. Proc Natl Acad Sci USA 1994; 91:728–732.

67. Wang Y, Tan J, Sutton-Smith M, Ditto D, Panico M, Campbell RM, Varki NM, Long JM, Jaeken J, Levinson SR, Wynshaw-Boris A, Morris HR, Le D, Dell A, Schachter H, Marth JD. Modeling human congenital disorder of glycosylation type IIa in the mouse: conservation of asparagine-linked glycan-dependent functions in mammalian physiology and insights into disease pathogenesis. Glycobiology 2001; 11:1051–1070.
68. Wang Y, Schachter H, Marth JD. Mice with a homozygous deletion of the Mgat2 gene encoding UDP-*N*-acetylglucosamine:alpha-6-D-mannoside beta1,2-*N*-acetylglucosaminyltransferase II: a model for congenital disorder of glycosylation type IIa. Biochim Biophys Acta 2002; 1573:301–311.
69. Akama TO, Nakagawa H, Wong NK, Sutton-Smith M, Dell A, Morris HR, Nakayama J, Nishimura S, Pai A, Moremen KW, Marth JD, Fukuda MN. Essential and mutually compensatory roles of α-mannosidase II and α-mannosidase IIx in *N*-glycan processing in vivo in mice. Proc Natl Acad Sci USA 2006; 103:8983–8988.
70. Yang X, Tang J, Rogler CE, Stanley P. Reduced hepatocyte proliferation is the basis of retarded liver tumor progression and liver regeneration in mice lacking *N*-acetylglucosaminyltransferase III. Cancer Res 2003; 63:7753–7759.
71. Priatel JJ, Sarkar M, Schachter H, Marth JD. Isolation, characterization and inactivation of the mouse Mgat3 gene: the bisecting *N*-acetylglucosamine in asparagine-linked oligosaccharides appears dispensable for viability and reproduction. Glycobiology 1997; 7:45–56.
72. Granovsky M, Fata J, Pawling J, Muller WJ, Khokha R, Dennis JW. Suppression of tumor growth and metastasis in Mgat5-deficient mice. Nat Med 2000; 6:306–312.
73. Demetriou M, Granovsky M, Quaggin S, Dennis JW. Negative regulation of T-cell activation and autoimmunity by Mgat5 *N*-glycosylation. Nature 2001; 409:733–739.
74. Ohtsubo K, Takamatsu S, Minowa MT, Yoshida A, Takeuchi M, Marth JD. Dietary and genetic control of glucose transporter 2 glycosylation promotes insulin secretion in suppressing diabetes. Cell 2005; 123:1307–1321.
75. Kornfeld R, Kornfeld S. Assembly of asparagine-linked oligosaccharides. Annu Rev Biochem 1985; 54:631–664.
76. Schachter H. The 'yellow brick road' to branched complex *N*-glycans. Glycobiology 1991; 1:453–461.
77. Ioffe E, Liu Y, Stanley P. Essential role for complex *N*-glycans in forming an organized layer of bronchial epithelium. Proc Natl Acad Sci USA 1996; 93:11041–11046.
78. Herscovics A. Importance of glycosidases in mammalian glycoprotein biosynthesis. Biochim Biophys Acta 1999; 1473:96–107.
79. Tremblay LO, Herscovics A. Characterization of a cDNA encoding a novel human Golgi alpha 1, 2-mannosidase (IC) involved in *N*-glycan biosynthesis. J Biol Chem 2000; 275:31655–31660.
80. Tremblay LO, Herscovics A. Cloning and expression of a specific human alpha 1,2-mannosidase that trims Man9GlcNAc2 to Man8GlcNAc2 isomer B during *N*-glycan biosynthesis. Glycobiology 1999; 9:1073–1078.
81. Tremblay LO, Campbell Dyke N, Herscovics A. Molecular cloning, chromosomal mapping and tissue-specific expression of a novel human alpha1,2-mannosidase gene involved in *N*-glycan maturation. Glycobiology 1998; 8:585–595.

82. Tremblay LO, Nagy Kovacs E, Daniels E, Wong NK, Sutton-Smith M, Morris HR, Dell A, Marcinkiewicz E, Seidah NG, McKerlie C, Herscovics A. Respiratory distress and neonatal lethality in mice lacking Golgi alpha1,2-mannosidase IB involved in *N*-glycan maturation. J Biol Chem 2007; 282:2558–2566.
83. Wang X, Inoue S, Gu J, Miyoshi E, Noda K, Li W, Mizuno-Horikawa Y, Nakano M, Asahi M, Takahashi M, Uozumi N, Ihara S, Lee SH, Ikeda Y, Yamaguchi Y, Aze Y, Tomiyama Y, Fujii J, Suzuki K, Kondo A, Shapiro SD, Lopez-Otin C, Kuwaki T, Okabe M, Honke K, Taniguchi N. Dysregulation of TGF-beta1 receptor activation leads to abnormal lung development and emphysema-like phenotype in core fucose-deficient mice. Proc Natl Acad Sci USA 2005; 102:15791–15796.
84. Jemal A, Siegel R, Ward E, Murray T, Xu J, Thun MJ. Cancer statistics, 2007. CA Cancer J Clin 2007; 57:43–66.
85. Sporn MB. The war on cancer: a review. The Lancet 1996; 347:1377–1381.
86. Hakomori S. Glycosylation defining cancer malignancy: new wine in an old bottle. Proc Natl Acad Sci USA 2002; 99:10231–10233.
87. Kobata A, Amano J. Altered glycosylation of proteins produced by malignant cells, and application for the diagnosis and immunotherapy of tumours. Immunol Cell Biol 2005; 83:429–439.
88. Fukuda M. Possible roles of tumor-associated carbohydrate antigens. Cancer Res 1996; 56:2237–2244.
89. Kannagi R. Monoclonal anti-glycolipid antibodies. Method Enzymol 2000; 312:160–179.
90. Kannagi R, Izawa M, Koike T, Miyazaki K, Kimura N. Carbohydrate-mediated cell adhesion in cancer metastasis and angiogenesis. Cancer Sci 2004; 95:377–384.
91. Irimura T. Cancer metastasis determined by carbohydrate-mediated cell adhesion. Adv Exp Med Biol 1994; 353:27–34.
92. Kim YS, Varki A. Perspectives on the significance of altered glycosylation of glycoproteins in cancer. Glycoconj J 1997; 14:569–576.
93. Nakayama T, Watanabe M, Katsumata T, Teramoto T, Kitajima M. Expression of sialyl Lewis(a) as a new prognostic factor for patients with advanced colorectal carcinoma. Cancer 1995; 75:2051–2056.
94. Nakamori S, Kameyama M, Imaoka S, Furukawa H, Ishikawa O, Sasaki Y, Izumi Y, Irimura T. Involvement of carbohydrate antigen sialyl Lewis(x) in colorectal cancer metastasis. Dis Colon Rectum 1997; 40:420–431.
95. Sato T, Nishimura G, Nonomura A, Miwa K, Miyazaki I. Serological studies on CEA, CA 19–9, STn and SLX in colorectal cancer. Hepatogastroenterology 1999; 46:914–919.
96. Sasaki A, Kawano K, Inomata M, Shibata K, Matsumoto T, Kitano S. Value of serum carbohydrate antigen 19–9 for predicting extrahepatic metastasis in patients with liver metastasis from colorectal carcinoma. Hepatogastroenterology 2005; 52:1814–1819.
97. Kouri M, Pyrhonen S, Kuusela P. Elevated CA19–9 as the most significant prognostic factor in advanced colorectal carcinoma. J Surg Oncol 1992; 49:78–85.
98. Matsui T, Kojima H, Suzuki H, Hamajima H, Nakazato H, Ito K, Nakao A, Sakamoto J. Sialyl Lewisa expression as a predictor of the prognosis of colon

carcinoma patients in a prospective randomized clinical trial. Jpn J Clin Oncol 2004; 34:588–593.

99. Takada A, Ohmori K, Yoneda T, Tsuyuoka K, Hasegawa A, Kiso M, Kannagi R. Contribution of carbohydrate antigens sialyl Lewis A and sialyl Lewis X to adhesion of human cancer cells to vascular endothelium. Cancer Res 1993; 53:354–361.
100. Sawada R, Lowe JB, Fukuda M. E-selectin-dependent adhesion efficiency of colonic carcinoma cells is increased by genetic manipulation of their cell surface lysosomal membrane glycoprotein-1 expression levels. J Biol Chem 1993; 268:12675–12681.
101. Khatib AM, Kontogiannea M, Fallavollita L, Jamison B, Meterissian S, Brodt P. Rapid induction of cytokine and E-selectin expression in the liver in response to metastatic tumor cells. Cancer Res 1999; 59:1356–13561.
102. Zetter BR. The cellular basis of site-specific tumor metastasis. N Engl J Med 1990; 322:605–612.
103. Hebbar M, Krzewinski-Recchi MA, Hornez L, Verdiere A, Harduin-Lepers A, Bonneterre J, Delannoy P, Peyrat JP. Prognostic value of tumoral sialyltransferase expression and circulating E-selectin concentrations in node-negative breast cancer patients. Int J Biol Markers 2003; 18:116–122.
104. Miyake M, Taki T, Hitomi S, Hakomori S. Correlation of expression of H/Le(y)/Le(b) antigens with survival in patients with carcinoma of the lung. N Engl J Med 1992; 327:14–18.
105. Poropatich C, Nozawa S, Rojas M, Chapman WB, Silverberg SG. MSN-1 antibody in the evaluation of female genital tract adenocarcinomas. Int J Gynecol Pathol 1990; 9:73–79.
106. Madjd Z, Parsons T, Watson NF, Spendlove I, Ellis I, Durrant LG. High expression of Lewis y/b antigens is associated with decreased survival in lymph node negative breast carcinomas. Breast Cancer Res 2005; 7: R780–787.
107. Iwamori M, Sakayori M, Nozawa S, Yamamoto T, Yago M, Noguchi M, Nagai Y. Monoclonal antibody-defined antigen of human uterine endometrial carcinomas is Leb. J Biochem (Tokyo) 1989; 105:718–722.
108. Kim YS, Itzkowitz SH, Yuan M, Chung Y, Satake K, Umeyama K, Hakomori S. Lex and Ley antigen expression in human pancreatic cancer. Cancer Res 1988; 48:475–482.
109. Abe K, Hakomori S, Ohshiba S. Differential expression of difucosyl type 2 chain (LeY) defined by monoclonal antibody AH6 in different locations of colonic epithelia, various histological types of colonic polyps, and adenocarcinomas. Cancer Res 1986; 46:2639–2644.
110. Inufusa H, Adachi T, Kiyokawa T, Nakatani Y, Wakano T, Nakamura M, Okuno K, Shiozaki H, Yamamoto S, Suzuki M, Ando O, Kurimoto M, Miyake M, Yasutomi M. Ley glycolipid-recognizing monoclonal antibody inhibits procoagulant activity and metastasis of human adenocarcinoma. Int J Oncol 2001; 19:941–946.
111. Ohyama C, Tsuboi S, Fukuda M. Dual roles of sialyl Lewis X oligosaccharides in tumor metastasis and rejection by natural killer cells. Embo J 1999; 18:1516–1525.

112. Fukuda MN, Ohyama C, Lowitz K, Matsuo O, Pasqualini R, Ruoslahti E, Fukuda M. A peptide mimic of E-selectin ligand inhibits sialyl Lewis X-dependent lung colonization of tumor cells. Cancer Res 2000; 60:450–456.
113. Zhang J, Nakayama J, Ohyama C, Suzuki M, Suzuki A, Fukuda M, Fukuda MN. Sialyl Lewis X-dependent lung colonization of B16 melanoma cells through a selectin-like endothelial receptor distinct from E- or P-selectin. Cancer Res 2002; 62:4194–4198.

Carbohydrate Chemistry, Biology and Medical Applications
Hari G. Garg, Mary K. Cowman and Charles A. Hales

DOI: 10.1016/B978-0-08-054816-6.00014-8

Chapter 14

Therapeutic Use of Hyaluronan-Based Products

ENDRE A. BALAZS* AND PHILIP A. BAND†

**Matrix Biology Institute, Edgewater, NJ 07020, USA*
†New York University School of Medicine, Departments of Pharmacology and Orthopaedic Surgery, New York, NY 10016, USA

I. Introduction

The development of hyaluronan preparations suitable for parenteral application has had a major impact on clinical practice in multiple medical specialties, often enabling the introduction of entirely new therapeutic procedures. Hyaluronan-based products are firmly established for ophthalmic surgery (viscosurgery), intra-articular treatment of arthritis (viscosupplementation), and tissue augmentation (viscoaugmentation). Hyaluronan is also used to control postsurgical adhesions and scar formation, to promote wound healing, and for cell-based tissue repair and drug delivery, though these applications have had less impact on the medical needs they address, to date. In addition, numerous published reports describe off-label clinical uses, and many preclinical studies have described new hyaluronan formulations and indications. We will here review the molecular basis for hyaluronan's current medical indications, describe how product formulations have evolved to address specific needs, and provide an overview of the ongoing work to expand hyaluronan's therapeutic applications.

II. Ophthalmic Viscosurgery

A. Development

In 1970 at an international meeting of ophthalmic surgeons, one of the authors (E.A.B.) of this chapter reported his work on the use of purified hyaluronan solutions in eye surgery. The contents of this report, published in the following years (1–7), established that hyaluronan solutions, after being purified from proteins, nucleic acids, endotoxin, and other inflammatory agents, can be used in various surgical procedures, provided the solution has certain viscous and elastic properties. The report was the result of a decade of laboratory research (Boston Biomedical Research Institute), industrial development (Biotrics, Inc., Arlington, MA), and clinical studies in academic ophthalmology departments (Boston, Stockholm, Essen, Paris, Zurich). The biocompatibility of this product produced from human umbilical cord and rooster comb was tested in nearly 500 owl monkey eyes. Owl monkeys were used because their viscous vitreous is liquid, and half of it (1 ml) can be replaced by test solutions without causing traumatic injury to the eye. The clinical studies established the usefulness of the highly elastoviscous hyaluronan solutions as vitreous replacements in retinal detachment surgery and as viscoelastic tools for surgical manipulations. The same solutions were injected into the anterior chamber of human patients in keratoplasty, cataract surgery, and iridectomy. The rationale for using these highly elastoviscous hyaluronan solutions was that they should serve as a "mechanical buffer" between the retina and the gel vitreous, the cornea and lens, and the cornea and the vitreous in aphakic eyes. The same mechanical properties of the elastoviscous fluid made it a "soft tool" to manipulate sensitive tissues during surgery. An equally important reason for using hyaluronan in eye surgery was its effect on wound healing. Animal studies with this hyaluronan preparation showed that it prevented excessive fibrous tissue formation during wound healing (8). This effect is important after retinal detachment surgery, cataract surgery, and keratoplasty. Healon®, the product that established ophthalmic viscosurgery, was developed by scientists who had long experience with the rheological properties, as well as the biological activity, of hyaluronan. Most importantly, they worked from the very beginning in close cooperation with ophthalmic surgeons, who expressed a need for such products to be used for replacement of the vitreous after retinal surgery and protection of the corneal endothelium during keratoplasty (corneal transplantation). In 1976, Biotrics, the developer of the noninflammatory fraction of hyaluronan, licensed the manufacturing and marketing rights worldwide (under the trade name Healon) to the Swedish drug company, Pharmacia AB (Uppsala) (9).

B. Hyaluronan for Use in Intraocular Lens Implantation

The increased use of plastic intraocular lens after the extraction of cataractous lenses dramatically increased the need for the use of viscoelastic solutions of hyaluronan (10–13). The well-known protective effect of Healon on the corneal endothelium provided safer use of more aggressive surgical procedures, which in turn permitted the use of more sophisticated artificial lens designs. The use of this procedure, called viscosurgery, was extended to trauma surgery of the eye. Healon was launched first in Europe and then in 1980 in the United States. In the early 1980s, in rapid succession, several books were published on viscosurgery of the eye (14–16). A few years later, the first competitive products appeared containing hyaluronan and another polysaccharide, chondroitin sulfate, or made of water-soluble cellulose derivatives (hydroxypropyl methylcellulose).

One decade later, Pharmacia introduced Healon 5® that was formulated with a higher concentration (2.3%), but with hyaluronan of the same molecular weight (4 million) as in Healon (1%). At this time, the ophthalmic surgeons worldwide could select from more than a dozen available ophthalmic viscosurgical devices (OVDs). Many of these products are made from cellulose derivatives and had extremely low viscosity and no elasticity, and others are made entirely or partially from hyaluronan but with significantly lower viscosity than the original Healon. Only a few competitors copied Healon, Healon GV®, and Healon 5. The original reason for introducing a low-viscosity product was the advantages they offered: less expensive, and when left in the eye after some surgical procedures, they did not cause postoperative rise of intraocular pressure. The highly viscous and elastic products had to be washed out before the wound was closed at the conclusion of the surgery. While the postoperative washout did not cause any medical problems, the issue remained a point of contention in evaluating the commercial worth of various products.

The availability of so many OVDs with substantial rheological differences created some confusion, mostly because the utility of the devices itself changed rapidly as new surgical techniques were developed. The diversity of the available OVDs itself stimulated the development of new surgical techniques. The classification of OVDs based on their rheological properties was the work of Steve Arshinoff (17). Healon 5, with the highest viscosity (7 mPa s), was called a "viscoadaptive"; products with very high viscosity (2–4.8 mPa s) were called "super viscous-cohesive"; products with high viscosity (0.1–1.0 mPa s) were called "viscous-cohesive"; and products with low viscosity (<0.1 mPa s) were called "dispersive" OVDs. The viscoadaptive product maintained the space during every step of the surgery, resisting breakup during the turbulence caused by the instruments, and was easy to wash out after surgery—thus the name "viscoadaptive." The viscous-cohesive products were similar but less effective. The viscous-dispersive products tend to be retained on the surface of the corneal endothelium longer, protecting it during surgery, which can be useful in complicated cases. This classification system helped surgeons to select and identify the best product for certain procedures. The availability of the viscoadaptive product permitted the development of the so-called "soft-shell" technique. With this technique, the dispersive (low-viscosity) OVD was injected first into the anterior chamber on the surface of the lens. Then, the adaptive OVD was introduced over it, without mixing the two layers. The combination of the two types of OVDs produced the best surgical results (18).

Most ophthalmic surgeons believe that the ideal viscosurgical tool for cataract surgery that includes plastic lens implantation has not yet been marketed. Such an ideal product must be able to create space, separate and stabilize tissue, prevent leakage of aqueous, stop bleeding, protect endothelial and other cells from mechanical damage caused by surgical tools, and be easily removed from the eye after surgery. Such an ideal OVD must be built from hyaluronan because this molecule is present in the human vitreous and in other eye tissues. Hyaluronan has unique elastic properties that other molecules in the vertebrate body do not have. Its rubber-like elasticity is greater than that of the natural rubber. It is likely that more than one OVD product with very different rheological properties is needed to satisfy all viscosurgical needs in various surgical situations.

C. Other Viscosurgical Uses

The most extensive use of viscosurgery is in cataract surgery and intraocular lens implantation; however, it is used for many other surgical treatments of diseases of the eye. The usefulness of highly elastoviscous hyaluronan solutions was discovered to make and maintain space, protect and manipulate tissues, and prevent excessive scar formation and adhesions. This very same functionality of hyaluronan-based OVDs led to other applications. The usefulness of Healon was tested early in various glaucoma surgeries. It was shown that "viscoelastics" enable surgeons to make safer and more aggressive procedures. In recent developments of glaucoma surgery, viscoelastics played an important role in the creation of a fistula between the anterior chamber and the subconjuntival space to restore normal outflow paths for the aqueous. They have been used to deepen and maintain the anterior chamber in trabeculectomy and to facilitate the outflow of aqueous into the superchoroidal space (cyclodialysis) (19). It was found that viscoelastics could be used to control bleeding, which in turn prevents postoperative recurrence of adhesions (synechiae). This confirmed early findings about the role of viscoelastic hyaluronan fluids in preventing postsurgical adhesions in various connective tissues. Recently, a new procedure was developed for glaucoma therapy based on the use of hyaluronan of the greatest elastoviscous properties (Healon 5), called viscocanalostomy, meaning the use of viscoelastic solutions to open a new canal for outflow of aqueous (20). Viscoelastics are also used in strabismus surgery, lachrymal duct, and ptosis surgeries. For more details of current uses of viscoelastic hyaluronan solutions in ophthalmic surgery, see Ref. (21).

III. Intra-Articular Hyaluronan: Viscosupplementation

Intra-articular hyaluronan (IA-HA) injections are widely used to treat osteoarthritis (OA). This procedure is often referred to as viscosupplementation (22) because it involves the replacement of pathologic synovial fluid with viscoelastic hyaluronan-based solutions or gels. In the United States, IA-HA is specifically labeled as an intra-articular analgesic and is indicated to treat pain associated with knee OA when conservative measures and simple analgesics fail (e.g., acetaminophen). In other parts of the world, IA-HA is also approved for treatment of joints other than the knee and in some countries for arthritic conditions other than OA. The molecular basis for this application of hyaluronan, and the history of its development, has been recently reviewed (23). In this chapter, we will update the clinical evidence and describe the different types of hyaluronan formulations available in the United States.

A. Clinical Evidence

The use of IA-HA is now recommended in the guidelines for treating knee OA published by the American College of Rheumatology (24) and the European Union of Leagues Against Rheumatic Disease (25). The most comprehensive systematic review of randomized clinical trials (RCTs) evaluating IA-HA for the treatment of knee OA was recently published in the Cochrane Library of evidence-based medicine (26). The Cochrane review identified 63 RCTs of IA-HA, of which 37 compared IA-HA to intra-articular saline or another control (placebo) intra-articular procedure. The review concluded that the benefit of IA-HA was statistically significant and clinically important. Several other systematic reviews of IA-HA published in medical journals drew similar conclusions (27,28); though in one, the clinical importance of the benefit was considered no better than the clinical benefit of nonsteroidal anti-inflammatory drugs (NSAIDs) over acetaminophen (28). Several randomized pragmatic trials have evaluated the incremental clinical benefit of adding IA-HA into the treatment paradigm for knee OA and its cost-effectiveness (29–31), reporting that the availability of IA-HA provides clinically important benefits and good health economic value. IA-HA treatment is reimbursed as part of standard health insurance in many countries, including the United States (32–34).

B. IA-HA Products and Treatment Regimens

Five IA-HA products are currently available in the United States (35–39). All of them are also available worldwide (see Table 1). Two of these products (Supartz® and Hyalgan®) have an average molecular weight (avg MW) of 0.6–1.0 million. Two other products (Orthovisc®, Euflexxa®) have higher avg MW (1–3.6 million) and the fifth product (Synvisc®) has an even higher avg MW (6 million) and contains both fluid and gel components. Synvisc is made of two modified hyaluronans

Table 1 Products Available in the United States

Trade names	Hyaluronan concentration (%)	Hyaluronan source	Average MW[a]	Treatment schedule on label
Supartz (Smith & Nephew)	1	Avian	0.6–1 million	3–5 weekly injections
Hyalgan (Sanofi-Aventis)	1	Avian	0.6–0.7 million	3–5 weekly injections
Euflexxa (Ferring)	1	Bacterial capsule	2.4–3.6 million	3 weekly injections
Synvisc[b] (Genzyme)	0.8 (hylan)	Avian	6 million+gel particles	3 weekly injections
Orthovisc (J&J)	1.5	Avian	1.0–2.9 million	3 or 4 weekly injections

[a]Reported by the manufacturer.
[b]Hylan GF-20 (generic name).

called hylan A and hylan B. Hylan A is a fluid and hylan B is a water-insoluble gel that is present in the mixture as small, irregular particles. These gel particles (hylan B) are also used in tissue augmentation (see Section IV). Synvisc is formulated by mixing eight parts (volumes) of hylan A fluid with two parts hylan B gel; therefore, the generic name hylan GF-20.

The difference in avg MW of the different products greatly influences their rheological properties, namely their viscosity, elasticity, and pseudoplasticity. Synvisc was designed to imitate the rheology of the hyaluronan in healthy human synovial fluid, and adding hylan B gel to it increases the intra-articular residence time of the product.

Differences have been reported in the "chemical purity" of the five hyaluronan products (40). The clinical importance of their differences is not clear, but it continues to be evaluated in RCTs.

Primarily, local reactions were reported after viscosupplementation with all the products available in the United States. These reactions are described as transient, suggesting local, mild inflammatory reactions, manifested in pain, often not influencing the longer beneficial effect of the treatment. Several publications reported a higher rate of these types of local reactions after viscosupplementation with Synvisc. Other publications reported the occurrence of granulation as well. The only head-to-head trials that allowed multiple treatment courses found that the increased rate of local reactions to Synvisc was only statistically significant during repeat treatment. Since the major difference between Synvisc and the other products available worldwide is the presence of hylan B gel particles, it is tempting to speculate that the particles are responsible for the differences in local reactions. The fact that animal studies established that the gel particles can penetrate some areas of the synovial tissue and reside substantially longer in tissues of the joint than the fluid component (hylan A) further confirms this speculation. Proper injection technique is important to the clinical success of viscosupplementation, particularly the complete removal of all synovial fluid exudate during arthrocentesis and accurate placement of the IA-HA product into the synovial cavity. An RCT comparing Synvisc to appropriate care reported an overall improvement in safety because of a decrease in NSAID-related gastrointestinal side effects (41), illustrating that the occurrence of local, self-resolving reactions may be a reasonable trade-off for the systemic side effects associated with other therapeutics for OA (30).

Many RCTs have recently been published evaluating IA-HA in joints other than the knee, generally with either saline injections or steroid injections as the comparator. Of particular interest was a large RCT of 660 patients with chronic shoulder pain, of which 60% had radiological signs of glenohumeral OA. The trial found that IA-HA was significantly more effective than saline in patients with glenohumeral OA and that concurrent adhesive capsulitis or rotator cuff tear did not impede efficacy (42).

Little work has been done to determine the effect of the volume of hyaluronan injected on the treatment regimen for IA-HA, particularly as these relate to the needs of individual patients with greatly varying volumes of synovial fluid exudate. Only two products that are approved in the Unites States specify a fixed number of injections (three weekly injections; see Table 1). The label of the remaining three products allows a treatment course consisting of three, four, or five injections, but does not provide guidance as to how clinicians should determine the appropriate number of injections for an individual patient. Several open studies conducted in patients with hip or ankle OA have evaluated flexible injection regimens in which the number of injections administered is adjusted based on the patient's response to treatment (43–45). A recent RCT reported that a single 6-ml injection of Synvisc was significantly better than intra-articular saline for 6 months, suggesting that it may be unnecessary for patients to receive a course of three weekly injections (46).

The potential clinical and economic advantage of a single injection treatment regimen has generated increased interest in the intra-articular use of chemically modified hyaluronan. Presumably, these efforts are based on the rationale that increasing the intra-articular residence time of hyaluronan will improve its effectiveness and enable a single injection treatment regimen. Some manufacturers believe that this can be achieved with products made of gel particles produced by chemical cross-linking procedures, such as the hyaluronan gel (hylan B) present in Synvisc (20% per volume). An epoxide cross-linked hyaluronan marketed in Europe and currently under review by Food and Drug Administration (FDA) (Durolane®, Q-Med, Uppsala Sweden) was found to provide significant improvements over saline at 6 weeks, but not at the trial's primary endpoint (6 months) (47). It should be noted that the latter product is composed of only gel particles with no added hyaluronan fluid, representing an important departure from prior IA-HA products developed for viscosupplementation. The clinical safety and effectiveness of products that contain only gel particles and dissolve slowly or not at all during their residence in the joint remains to be established.

C. Analgesia

The only clinically proven effect of viscosupplementation is local analgesia. This analgesia was originally discovered by injecting elastoviscous hyaluronan solutions to replace pathological synovial fluid in horses and humans, and demonstrating pain relief that lasts longer than the residence of the injected hyaluronan in the-joint. This analgesic effect was first objectively proven in horses with traumatic arthritis using "force plate" studies that measured the difference between the length of time the horse willingly placed its body weight on the leg that had an arthritic (painful) joint before and after treatment. These "objective" studies on pain were the basis of approval of hyaluronan for the first time in the United States as a drug for the treatment of arthritic pain in veterinary medicine (48). These studies also established that nonelastoviscous solutions containing the same concentration of hyaluronan but with lower avg MW did not have an analgesic effect. Nearly a decade later Japanese authors, using behavioral pain models in rat knee joints and in the mouse peritoneal cavity, confirmed that the analgesic effect of hyaluronan injected into these tissues depended on both molecular weight and concentration. Below a certain avg MW even high concentration solutions were ineffective. The pain-causing agents injected into the joint and abdomen were urate crystals, bradykinin, and prostaglandin (49–51).

Table 2 Hyaluronan Dermal Fillers Approved by FDA

Trade name, manufacturer	Cross-linked by	Source of hyaluronan	Hyaluronan conc (mg/ml)	Average particle size (μm)
Hylaform[a], Genzyme	DVS[b]	Avian	5.5	500
Hylaform[a]-Plus, Genzyme	DVS	Avian	5.5	700
Captique[c] Genzyme	DVS	Bacterial capsule	5.5	500
Restylane, Q-Med	BDDE[b]	Bacterial capsule	20	250
Perlane, Q-Med	BDDE	Bacterial capsule	20	940–1090
Juvederm Ultra (24HV), Allergan	BDDE	Bacterial capsule	24	NA
Juvederm Ultra-Plus (30HV); Allergan	BDDE	Bacterial capsule	24	NA
CTA (connective tissue augmentation	NA	Bacterial capsule	28	NA

FDA as a supplement to the Hylaform approval.
NA: Data not available.
[a]The distribution of Hylaform products was temporarily suspended as of July 2007.
[b]DVS, divinyl sulfone; BDDE, butanediol diglycidyl ether (1,2,3,4-diepoxybutane).
[c]Captique was never tested in randomized clinical trials, but was approved by FDA as a supplement to the Hylaform approval.

products are the reagents used to cross-link hyaluronan, the hyaluronan concentration in the particles, and particle size in the final injectable product.

The difference between DVS and butanediol diglycidyl ether (BDDE) cross-linking is illustrated by the lower hyaluronan concentration in the hydrated gels produced using DVS. Although the equilibrium hydration and hyaluronan concentration of cross-linked hyaluronan depend on the conditions used during cross-linking (e.g., hyaluronan and cross-linker concentration, and their ratio), it is notable that DVS cross-linking produces a solid gel (Hylaform) that is 99.5% solvent (physiologic saline), and yet extends the residence time of the implanted hyaluronan from days to months. The BDDE cross-linked gels have a higher hyaluronan concentration at equilibrium hydration and contain ~98% water. This difference between BDDE and DVS cross-linking probably reflects the difference in their selectivity for carboxyl groups over hydroxyl groups. Under the cross-linking conditions used, DVS is selective for the hydroxyl groups of hyaluronan. BDDE can react with both hydroxyl and carboxylate groups of hyaluronan.

Comparing Restylane/Perlane to the Juvederm family of products illustrates that even when using the same cross-linking agent, the performance characteristics can be varied in the finished products. The US labeling for the Juvederm products was recently revised to extend the duration of effectiveness claim out to 1 year, the longest for any hyaluronan-based dermal augmentation product.

The difference between avian-derived (chicken comb tissue) and fermentation-derived sources of hyaluronan is illustrated by comparison of Hylaform and Captique, which differ only in their hyaluronan source. There are no differences between the two products in concentration or particle size, and no difference is claimed or reported in any physical property of the two.

Cross-linked hyaluronan is injected as gel particles, which are irregular in shape, greatly variable in size, and broadly distributed in a given product. The size and solidity (elasticity) of the gel particles depend on the cross-linking method, processing conditions, and the concentration of hyaluronan in the finished product. Particle size in particular is widely accepted as influencing clinical performance characteristics (67), though this has never been demonstrated in clinical trials. Nonetheless, product extensions were developed containing gel particles with both larger and smaller average size distributions (Table 2). The larger particle variations are believed to be more suitable for correction of deep dermal and subdermal layers, and to persist longer at the injection site. The smaller particle variations are believed to be more suitable for subtle correction of the superficial dermis. Note that Perlane, the product with the largest particle size, has the only label indication specifying its suitability for injection into the deep dermis and superficial cutis, whereas Hylaform and Restylane are indicated for injection into the mid and deep dermis.

The effect of the cross-linking process on clinical performance is still poorly understood. Very few RCTs have directly evaluated the consequences of these variations on safety or effectiveness outcomes, and there has never been a trial in which a single variable was individually tested (e.g., particle size or source). A recently reported RCT directly compared Perlane to Hylaform for the treatment of nasolabial folds in 150 patients (68), providing a clinically relevant comparison of DVS and BDDE cross-linking. Perlane was found to have a significantly longer duration of correction, with 75% of Perlane-treated sites maintaining correction at 6 months, compared with 38% of Hylaform-treated sites. Conversely, the frequency of injection site reactions was higher for Perlane than for Hylaform; 22.6% versus 7.3% for swelling, 20.0% versus 5.3% for pain, and 14.0% versus 8.0% for redness. It is unclear whether these differences in clinical outcome result from differences in particle size distribution or cross-linking methods.

Cross-linked hyaluronan has also been suggested for the augmentation of tissues other than skin, such as the vocal chord folds and the urinary sphincter. Animal studies demonstrated that DVS cross-linked hyaluronan (hylan B) was safe and effective for augmentation of vocal folds (69,70). Clinical trials have confirmed these early findings (71), but no cross-linked hyaluronan product has been approved for this indication in the United States. During the 1990s, DVS cross-linked hyaluronan (hylan B) was also evaluated for augmentation of the urinary sphincter to treat incontinence (Biomatrix, unpublished data), but was never tested clinically. As of this writing, no cross-linked hyaluronan product is approved in the United States for the treatment of urinary incontinence, and no clinical trials have been reported testing any type of cross-linked hyaluronan for this indication. Hyaluronan is used in the treatment of urinary incontinence, as the suspension vehicle for dextranomer particles (cross-linked dextran) in a product called Deflux® (Q-Med, Uppsala, Sweden). Deflux is endoscopically injected into the submucosa of the urinary bladder, close to the ureteral opening. The dextranomer particles stimulate a connective tissue reaction that provides the ultimate tissue bulking effect.

V. Hyaluronan for Antiadhesion, Wound Healing, and Matrix Engineering

Early studies demonstrated that elastoviscous hyaluronan solutions could influence granulation tissue formation, and scarring (8). These studies established a firm foundation for developing hyaluronan-based products for the control of postsurgical adhesions, wound healing, and tissue engineering. The term matrix engineering was coined by one of the authors (E.A.B.) to describe the use of hyaluronan-based matrices to control, direct, or augment tissue regenerative processes (72,73). Ongoing efforts to bring this concept into clinical practice will be described in this section.

A. Antiadhesion

"The implantation of hyaluronic acid jellies, sheets, and membranes is a type of matrix engineering. Hyaluronic acid influences the invasion and activities of cells participating in the acute and chronic inflammatory process. It is proposed that implantation of hyaluronic acid should be used to prevent fibrous tissue formation and consequent development of adhesion and scars." With these words, viscoseparation with hyaluronan was introduced into therapeutics (72). Highly elastoviscous hyaluronan solutions (called jellies at the time) regulated the mitogen-induced stimulation of peripheral blood lymphocytes (74). In dogs and owl monkeys, hyaluronan jellies significantly improved the healing of intra-articular wounds. In rabbits, hyaluronan reduced adhesion formation around tendons after injuries. Granulation tissue formation and development of scar tissues and adhesion were prevented when hyaluronan was introduced into the wound after injury or around implanted foreign bodies (8). Further studies using Healon demonstrated that elastoviscous hyaluronan solutions can prevent adhesion and excessive scar formation in various surgical models including repair of profundus tendon in the fingers of monkeys, adhesions of the hallucis longus tendon in rabbits, abrasion-caused adhesions in rabbit cervical horn [for review, see Ref. (75)].

The application of hyaluronan solutions to control postsurgical adhesion is limited by their short residence time and the difficulty of keeping fluid barriers at the tissue sites where adhesion control is necessary. To improve their applicability for adhesion control, methods were developed to cross-link hyaluronan to produce solid materials that could maintain a longer lasting antiadhesive barrier after surgery (60). The advantage of hylan B gel is its solidity and longer residence time at the site of implantation, making it preferable because adhesions may develop some time after surgery. In animal models, hylan B gel was found to reduce posttraumatic and postsurgical adhesions, and to minimize excessive scar formation (76–78).

Non-cross-linked hyaluronan derivatives with reduced water solubility have also been tested as antiadhesion barriers in animal models and human clinical trials. Seprafilm® (Genzyme, Inc.) is an antiadhesion barrier dressing containing a mixture of hyaluronan and carboxymethyl cellulose derivatized with carbodiimide. Seprafilm was found to reduce adhesions in animal models and human trials (79,80) and more recently to provide significant improvement in surgical outcomes (81). Seprafilm is approved in the United States to reduce postsurgical adhesions after pelvic and abdominal surgery (Summary of Safety and Effectiveness [SSE]).

For surgical indications where adhesion control devices are considered as Class II medical devices, essentially like wound dressings, FDA has cleared several hyaluronan-based products for marketing without clinical trial data based on their substantial equivalence to preexisting devices. Examples of products approved by this so-called 510(K) process include Sepragel® Sinus (hylan B, Genzyme), Seprapak® (hyaluronan and carboxymethyl cellulose derivatized with carbodiimide, Genzyme), and Merogel® (HYAFF hyaluronan benzyl ester, Fidia). Sepramesh® is a polypropylene mesh use for tissue support during hernia repair. It is coated on its viscera-contacting surface with hyaluronan and carboxymethyl cellulose derivatized with carbodiimide (Genzyme), and is intended to reduce the incidence of adhesions between the mesh and the bowel.

B. Viscoprotection and Wound Healing

The potential of hyaluronan to protect tissue surfaces and facilitate wound healing was recognized early in its medical development. Hyaluronan solutions in a purified form have been used from the early 1970s to protect the surface of the cornea (82,83). The cornea, the transparent "skin" of the eye, is highly innervated and consequently more sensitive to dryness and pain than the skin. Hyaluronan is used to protect the cornea from dryness as a substitute for tears when the production of tears is impaired. Hyaluronan cannot be called artificial tears because tears do not contain this molecule and they are not viscous. Elastoviscous hyaluronan solutions have been used for decades as eye drops for three reasons. First, because of their viscosity, they flow out slowly from the conjunctival sacs; second, because of their elasticity, the fast movements of blinking do not remove them easily from the surface of the cornea, extending their residence time; and third, they retain water at body temperature and do not lose water between blinking. Elastoviscous hyaluronan eye drops have been available worldwide for the protection of corneal epithelium not only in pathological conditions when the tear formation is impaired (dry-eye syndrome), but also as a wetting solution for hard contact lenses and to protect the cornea from dryness when the frequency of blinking decreases, such as in healthy eyes as they focus on television and computer screens for extended periods of time (84). The same hyaluronan solutions were also used as natural wetting agents on the surface of the cornea during cataract extractions and lens implantations, as well as in vitreoretinal surgery (85).

The utilization of hyaluronan-based materials to treat skin wounds is based on principles similar to those described for the control of postsurgical adhesions. The glycosaminoglycan composition of skin undergoes an orderly series of changes during wound healing, and is particularly enriched in hyaluronan during the first several days after wounding (86–88). Early studies demonstrated that hyaluronan (Healon) at concentrations between 5 and 9 mg/ml decreased subcutaneous scar formation after surgical incisions (89). Exogenously applied 1% hyaluronan solutions were found to facilitate the healing of partial thickness skin wounds in rats with alloxan-induced diabetes (90). Early studies have demonstrated that hyaluronan controls multiple activities of lymphomyeloid cells relevant to wound healing and that the effect depends on its molecular weight and concentration (74). Fetal wounds are known to exhibit scarless healing, a fact that has been attributed to the higher concentration of hyaluronan in fetal compared to adult wounds (91–93).

As described above for adhesion control, several hyaluronan-containing products have been cleared by FDA as wound dressings based on their substantial equivalence to preexisting dressings. These include Hyalofil®, Hyalomatrix® fibrous nonwoven pad, and Bionect® products (all produced by Fidia Advanced Biomaterials, Terme, Italy and marketed by Convatec); Hycoat® (marketed by Hymed); Spinco Second Skin®; and LAM® Wound gel (LAM).

C. Matrix Engineering

The concept of matrix engineering is now being applied to the development of products containing live cells incorporated into hyaluronan-containing matrices, an approach commonly referred to as tissue engineering. This application is based on the observation by many investigators during the past three decades that hyaluronan participates in wound healing and tissue regeneration. The utility of hyaluronan for matrix engineering is based on its biological participation in tissue regenerative processes, as described above. The biological role of hyaluronan, coupled with the universal biocompatibility of purified high molecular weight hyaluronan, make it an obvious candidate as a carrier matrix for implantation of cells and for tissue regeneration.

The use of matrices based on hyaluronan to support cell attachment and multiplication has been studied in animal models and clinical trials. The behavior of different cell types on cross-linked hyaluronan materials has been characterized, including the way proteins influence cell–hyaluronan interactions (94,95). Recent studies have demonstrated that the chemical structure and viscoelastic properties of hyaluronan-based matrices strongly influence the cellular response to these materials, with more rigid materials inducing a more stretched and organized cytoskeleton (96). Several types of derivatized hyaluronan have been used as a matrix for the *ex vivo* expansion of keratinocytes, fibroblasts, and chondrocytes; and subsequently shown in clinical trials to be safe and effective for the repair of skin and cartilage defects, respectively (97–101). Although the clinical data reported thus far appear promising, these approaches to tissue repair have yet to enter US medical practice.

New hyaluronan derivatives are being continually developed because of their potential commercial and medical value. Consequently, our understanding continues to expand with respect to the ways in which different chemical modifications of hyaluronan can affect its physical and biological properties (102). Thiol-modified derivatives of hyaluronan-containing chondroitin sulfate, heparin, peptides, and protein have recently been tested for scar-free healing and adhesion prevention. Copolymers of hyaluronan and other glycosaminoglycans have been produced using multiple derivatization methods, and recently reported data suggest profound effects on the healing of skin wounds, abdominal and tendon injuries, and defects of soft tissues and cartilage (103). Both the chemical composition and the physical structure of hyaluronan-based materials can influence the signaling activity and expression of cells embedded in them (104). Promising animal data has been reported (105), but remains to be confirmed in clinical trials.

Understanding cellular responses to hyaluronan and its derivatives so that these responses can be integrated with tissue repair processes represents the next horizon. Ongoing work continues to characterize how different chemical modifications of hyaluronan influence cellular responses and performance characteristics (106–108). It is well established that biopolymers, such as hyaluronan and other glycosaminoglycans of the intercellular matrix, play a significant role in tissue development and regeneration. Artificial intercellular matrices made from these molecules and their derivatives can, therefore, provide significant advantages over synthetic polymers and can be expected to become important tools for tissue engineering. We propose that semisolid and solid structures built for the augmentation, repair, reconstruction, and development of tissues should be called "biomatrices" when they are built from molecules derived from the intercellular matrix.

References

1. Balazs EA, Freeman MI, Klöti R, Meyer-Schwickerath G, Regnault F, Sweeney DB. The utilization of hyaluronic acid for intravitreal injection. In: Solanes MP, ed. ACTA XXI Concilium Ophthalmologicum Mexico, Part I. Amsterdam: Excerpta Medica, 1971; 555–558.
2. Balazs EA, Freeman MI, Klöti R, Meyer-Schwickerath G, Regnault F, Sweeney DB. Hyaluronic acid and the replacement of vitreous and aqueous humor. In: Streiff EB, ed. Modern Problems in Ophthalmology (Secondary Detachment of the Retina, Lausanne, 1970). Basel: S. Karger, 1972; 3–21.
3. Klöti R. Hyaluronsäure als Glaskörpersubstituent. Schweiz Ophthal Ges 1972; 165:351–359.
4. Regnault F. Acide hyaluronique intravitéen et cryocoagulatin dans le traitement des formes graves de décollement de la rétine. Bull Soc Ophthalmol Fr 1971; 84:106–112.
5. Meyer-Schwickerath G. Further experience with Healon® (hyaluronic acid) in retinal detachment surgery. Mod Probl Ophthal 1974; 12:384.
6. Algvere P. Intravitreal implantation of a high-molecular hyaluronic acid in surgery for retinal detachment. Acta Ophthalmol 1971; 49:975–976.
7. Edmund J. Comments on the clinical use of Healon and a short survey of the use of intraocular injection of hyaluronic acid. In: Irvine AR, O'Malley C, eds. Advances in Vitreous Surgery. Springfield, IL: C. Thomas, 1976; 624–625.

8. Rydell N, Balazs EA. Effect of intra-articular injection of hyaluronic acid on the clinical symptoms of osteoarthritis and on granulation tissue formation. Clin Orthop 1971; 80:25–32.
9. Balazs EA. Ultrapure hyaluronic acid and the use thereof. US Patent 4,141,973 (1979).
10. Balazs EA, Miller D, Stegmann R. Viscosurgery and the use of Na-hyaluronate in intraocular lens implantation. In: International Congress and First Film Festival on Intraocular Implantation, Cannes, France, 1979;1–6.
11. Graue EL, Polack FM, Balazs EA. The protective effect of Na-hyaluronate to corneal endothelium. Exp Eye Res 1980; 31:119–127.
12. Pape LG, Balazs EA. The use of sodium hyaluronate (Healon®) in human anterior segment surgery. Ophthalmol 1980; 87:699–705.
13. Miller D. Use of Na-hyaluronate in anterior segment eye surgery. Am Intraocular Implant Soc J 1980; 6:13–15.
14. Meyer-Schwickerath G, ed. Viskochirurgie des Auges: Beiträge des ersten nationalen Healon®-Symposiums, October 15 & 16, 1982.
15. Miller D, Stegmann R, eds. Healon (Sodium Hyaluronate): A Guide to Its Use in Ophthalmic Surgery. New York, NY: John Wiley & Sons, 1983.
16. Eisner G, ed. Ophthalmic Viscosurgery: A Review of Standards, Techniques and Applications. Montreal, Canada: Medicopea, 1986.
17. Arshinoff SA. Dispersive-cohesive viscoelastic soft shell technique. J Cataract Refract Surg 1999; 25:167–173.
18. Arshinoff S. Ultimate soft-shell technique and AcrySof Monarch injector cartridges. J Cataract Refract Surg 2004; 30:1809–1810.
19. Alpar JJ. Sodium hyaluronate (Healon) in glaucoma filtering procedures. Ophthalmic Surg 1986; 17:724–730.
20. Carassa RG, Bettin P, Fiori M, Brancato R. Viscocanalostomy versus trabeculectomy in white adults affected by open-angle glaucoma: a 2-year randomized, controlled trial. Ophthalmology 2003; 110:882–887.
21. Buratto L, Giardini P, Bellucci R, eds. Viscoelastics in Ophthalmic Surgery. Thorofare, NJ: Slack, Inc., 2000.
22. Balazs EA, Denlinger JL. Viscosupplementation: a new concept in the treatment of osteoarthritis. J Rheumatol 1993; 20(Suppl 39):3–9.
23. Balazs EA. Viscosupplementation for treatment of osteoarthritis: from initial discovery to current status and results. Surg Technol Int 2004; 12: 278–289.
24. American College of Rheumatology Subcommittee on Osteoarthritis. Recommendations for the medical management of osteoarthritis of the hip and knee. Arthr Rheum 2000; 43:1905–1915.
25. Jordan KM, Arden NK, Doherty M, Bannwarth B, Bijlsma JW, Dieppe P, Günther K, Hauselmann H, Herrero-Beaumont G, Kaklamanis P, Lohmander S, Leeb B, Lequesne M, Mazieres B, Martin-Mola E, Pavelka K, Pendleton A, Punzi L, Serni U, Swoboda B, Verbruggen G, Zimmerman-Gorska I, Dougados M. EULAR Recommendations 2003: an evidence based approach to the management of knee osteoarthritis. Report of a Task Force of the Standing Committee for International Clinical Studies Including Therapeutic Trials (ESCISIT). Ann Rheum Dis 2003; 62:1145–1155.
26. Bellamy N, Campbell J, Robinson V, Gee T, Bourne R, Wells G. Viscosupplementation for the treatment of osteoarthritis of the knee (Cochrane

Review) (comparing hyaluronan and hylan derivatives). Art. No. CD005321. DOI:10.1002/14651858.CD005321. Cochrane Database Syst Rev 2005; 1–429.
27. Wang CT, Lin J, Chang CJ, Lin YT, Hou SM. Therapeutic effects of hyaluronic acid on osteoarthritis of the knee: A meta-analysis of randomized controlled trials. J Bone Joint Surg Am 2004; 86-A:538–545.
28. Lo GH, LaValley M, McAlindon T, Felson DT. Intra-articular hyaluronic acid in treatment of knee osteoarthritis: a meta-analysis. JAMA 2003; 290:3115–3121.
29. Kahan A, Lieu PL, Salin L. Prospective randomized study comparing the medicoeconomic benefits of Hylan GF-20 vs. conventional treatment in knee osteoarthritis. Joint Bone Spine 2003; 70:276–281.
30. Raynauld J-P, Torrance GW, Band PA, Goldsmith CH, Tugwell P, Walker V, Schultz M, Bellamy N. A prospective, randomized, pragmatic, health outcomes trial evaluating the incorporation of hylan GF-20 into the treatment paradigm for patients with knee osteoarthritis (Part 1 of 2): clinical results. Osteoarth Cartilage 2002; 10:506–517.
31. Torrance GW, Raynauld JP, Walker V, Goldsmith CH, Bellamy N, Band PA, Schultz M, Tugwell P. A prospective, randomized, health outcomes trial evaluating the incorporation of hylan G-F 20 into the treatment paradigm for patients with knee osteoarthritis (Part 2 of 2): economic results. Osteoarth Cartilage 2002; 10:518–527.
32. Intra-articular injections for osteoarthritis of the knee. Med Lett Drugs Ther 2006; 48:25–27.
33. Special report: intra-articular hyaluronan for osteoarthritis of the knee. Technol Eval Cent Asess Program Exec Summ 2005; 19:1–2.
34. Intra-articular hyaluronan injections for treatment of osteoarthritis of the knee. Tecnologica MAP Suppl 1998; 6–10.
35. United States Food and Drug Administration. Center for Devices and Radiological Health. New Device Approval Letter: Synvisc®—PMA P940015. Summary of Safety and Effectiveness Data. August 8. 1997. In: www.accessdata.fda.gov/scripts/cdrh/cfdocs/cfPMA/pma.cfm?ID=5689.
36. United States Food and Drug Administration. Center for Devices and Radiological Health. New Device Approval Letter: Supartz™—PMA P980044. Summary of Safety and Effectiveness Data. January 24, 2001. In: www.fda.gov/cdrh/pdf/P010029.html.
37. United States Food and Drug Administration. Center of Devices and Radiological Health. New Device Approval Letter: Nuflexxa—PMA P010029. Summary of Safety and Effectiveness Data. December 3, 2004. In: www.fda.gov/cdrh/pdf/P010029.html.
38. United States Food and Drug Administration Center for Devices and Radiological Health. New Device Approval Letter: Orthovisc® High Molecular Weight Hyaluronan—PMA P030019. Summary of Safety and Effectiveness Data. February 4, 2004. In: www.fda.gov/cdrh/pdf3/p030019.html.
39. United States Food and Drug Administration. Center for Devices and Radiological Health. New Device Approval. Hyalgan®—PMA P950027. May 28, 1997. In: www.fda.gov.cdrh/pdf/P950027.pdf.
40. Ohshima Y, Yokota S, Kasama K, Ono H. Comparative studies on levels of proteins, bacterial endotoxins and nucleic acids in hyaluronan preparations used to treat osteoarthritis of the knee—could residual proteins and bacterial endotoxins relate to complications. Jpn Pharmacol Ther 2004; 32:655–662.

41. Wolfe MN, Lichtenstein DR, Singh G. Gastrointestinal toxicity of nonsteroidal anti-inflammatory drugs. N Eng J Med 1999; 340:1888–1899.
42. Altman RD, Moskowitz R, Jacobs S, Daley M, Udell J, Levin R. A double-blind, randomized trial of intra-articular sodium hyaluronate for the treatment of chronic shoulder pain. In: 69th Annual Meeting of the American College of Rheumatology, November 12–17, 2005, San Diego, CA.
43. Conrozier T, Vignon E. Is there evidence to support the inclusion of viscosupplementation in the treatment paradigm for patients with hip osteoarthritis? Clin Exp Rheumatol 2005; 23:711–716.
44. Conrozier T, Bertin P, Mathieu P, Charlot J, Bailleul F, Treves R, Vignon E, Chevalier X. Intra-articular injections of hylan G-F 20 in patients with symptomatic hip osteoarthritis: an open-label, multicentre, pilot study. Clin Exp Rheumatol 2003; 21:605–610.
45. Witteveen AGH, Giannini S, Guido G, Jerosch J, Lohrer H, van Dijk CN. Prospective study of the safety and efficacy of hylan G-F 20 (Synvisc®) in patients with symptomatic ankle osteoarthritis. In: American College of Rheumatology 70th Annual Meeting, Washington, D.C., November 10–15, 2006.
46. Chevalier X, Jerosch J, Luyten FP, van Dijk CN, Goupille P, Scott D, Bailleul F, Pavelka K. A double-blind, randomized placebo-controlled evaluation of the efficacy and safety of a single dose of 6 mL of hylan G-F 20 in patients with symptomatic osteoarthritis of the knee. In: EULAR Barcelona, Spain, June 13–16, 2007.
47. Altman RD, Akermark C, Beaulieu AD, Schnitzer T. Efficacy and safety of a single intra-articular injection of non-animal stabilized hyaluronic acid (NASHA) in patients with osteoarthritis of the knee. Osteoarthritis Cartilage 2004; 12:642–649.
48. Balazs EA, Denlinger JL. Sodium hyaluronateand joint function. Equine Vet Sci 1985; 5:217–228.
49. Gotoh S, Onaya JI, Abe M, Miyazaki K, Hamai A, Horie K, Tokuyasu K. Effects of the molecular weight of hyaluronic acid and its action mechanisms on experimental joint pain in rats. Ann Rheum Dis 1993; 52:817–822.
50. Miyazaki K, Gotoh S, Ohkawara H, Yamaguchi T. Studies on analgesic and anti-inflammatory effects of sodium hyaluronate (SPH). Pharmacometrics 1984; 28:1123–1135.
51. Aihara S, Murakami N, Ishii R, Kariya K, Azuma Y, Hamada K, Umemoto J, Maeda S. Effects of sodium hyaluronate on the nociceptive response of rats with experimentally induced arthritis. Folia Pharmacol Jpn 1992; 100: 359–365.
52. Pozo MA, Balazs EA, Belmonte C. Reduction of sensory responses to passive movements of inflamed knee joints by hylan, a hyaluronan derivative. Exp Brain Res 1997; 116:3–9.
53. Gomis A, Pawlak M, Balazs EA, Schmidt RF, Belmonte C. Effects of different molecular weight elastoviscous hyaluornan solutions on articular nociceptive afferents. Arthr Rheum 2004; 50:314–326.
54. Balazs EA. Analgesic effect of elastoviscous hyaluronan solutions and the treatment of arthritic pain. Cells, Tissues, Organs 2003; 174:49–62.
55. Gomis A, Miralles A, Schmidt RF, Belmonte C. Nociceptive nerve activity in an experimental model of knee joint osteoarthritis of the guinea pig: effect of intra-articular hyaluronan application. Pain 2007; 130:126–136.

56. Balazs EA. Viscoelastic properties of hyaluronan and its therapeutic use. In: Garg HG, Hales CA, eds. Chemistry and Biology of Hyaluronan. Amsterdam: Elsevier, Ltd., 2004; 415–455.
57. Piacquadio D, Jarcho M, Goltz R. Evaluation of hylan b gel as a soft-tissue augmentation implant material. J Am Acad Dermatol 1997; 36: 544–549.
58. Klein AW. Soft tissue augmentation 2006: filler fantasy. Dermatol Ther 2006; 19:129–133.
59. Lindqvist C, Tveten S, Bondevik BE, Fagrell D. A randomized, evaluator-blind, multicenter comparison of the efficacy and tolerability of Perlane versus Zyplast in the correction of nasolabial folds. Plast Reconstr Surg 2005; 115:282–289.
60. Balazs EA, Leshchiner E, Larsen N, Band P. Hyaluronan biomaterials: medical applications. In: Wise DL, ed. Handbook of Biomaterials and Applications. New York: Marcel Dekker Inc., 1995; 1693–1715.
61. United States Food and Drug Administration Center for Devices and Radiological Health. New Device Approval Letter: Hylaform® (hylan B gel)—PMA P030032. Summary of Safety and Effectiveness Data. April 22, 2004. In: www.fda.gov/cdrh/pdf3/p030032.html.
62. United States Food and Drug Administration Center for Devices and Radiological Health. New Device Approval Letter: Restylane® Injectable Gel—PMA P020023. Summary of Safety and Effectiveness Data. December 12, 2003. In: www.fda.gov/cdrh/pdf2/P020023.html.
63. United States Food and Drug Administration Center for Devices and Radiological Health. New Device Approval Letter: PERLANE® Injectable Gel—PMA P040024/S6. Summary of Safety and Effectiveness Data. May 2, 2007. In: www.fda.gov/cdrh/pdf4/P040024s006.html.
64. United States Food and Drug Administration. Center for Devices and Radiological Health. New Device Approval Letter: JUVÉDERM™—PMA P050047. Summary of Safety and Effectiveness Data. June 2, 2006. In: www.fda.gov/cdrh/pdf5/P050047b.pdf.
65. United States Food and Drug Administration. Center for Devices and Radiological Health. New Device Approval Letter: Hylaform® Plus (hylan B gel)—PMA P030032/S001. Summary of Safety and Effectiveness Data. October 13, 2004. In: www.fda.gov/cdrh/pdf3/p030032s001b.pdf.
66. United States Food and Drug Administration. Center for Devices and Radiological Health. New Device Approval Letter: Cosmetic Tissue Augmentation product (CTA)—PMA P050033. Summary of Safety and Effectiveness Data. December 20, 2006. In: www.fda.gov/cdrh/pdf5/p050033b.pdf.
67. Monheit GD, Coleman KM. Hyaluronic acid fillers. Dermatol Ther 2006; 19:141–150.
68. Carruthers A, Carey W, De Lorenzi C, Remington K, Schachter D, Sapra S. Randomized, double-blind comparison of the efficacy of two hyaluronic acid derivatives, restylane perlane and hylaform, in the treatment of nasolabial folds. Dermatol Surg 2005; 31:1591–1598; discussion 1598.
69. Hertegård S, Dahlqvist Å, Laurent C, Borzacchiello A, Ambrosio L. Viscoelastic properties of rabbit vocal folds after augmentation. Otolaryngol Head Neck Surg 2003; 128:401–406.
70. Hallén L, Johannson C, Laurent C. Cross-linked hyaluronan (Hylan B gel): a new injectable remedy for treatment of vocal fold insufficiency—an animal study. Acta Otolaryngol (Stockh) 1999; 119:107–111.

71. Hertegård S, Hallen L, Laurent C, Lindstrom E, Olofsson K, Testad P. Cross-linked hyaluronan versus collagen for injection treatment of glottal insufficiency: 2-year follow-up. Acta Oto-Larynologica 2004; 124:1208–1214.
72. Balazs EA. Hyaluronic acid and matrix implantation. 2nd edn., Arlington, MA: Biotrics, Inc., 1971.
73. Balazs EA, Band PA, Denlinger JL, Goldman AI, Larsen NE, Leshchiner EA, Leshchiner A, Morales B. Matrix engineering. Blood Coagul Fibrinolysis 1991; 2:173–178.
74. Balazs EA, Darzynkiewicz Z. The effect of hyaluronic acid on fibroblasts, mononuclear phagocytes and lymphocytes. In: Kulonen E, Pikkarainen J, eds. Biology of the Fibroblast (Papers of the Symposim held in Turku, Finland, 1972), London: Academic Press, 1973; 237–252.
75. Weiss C. The control of adhesions with hylan polymers. In: Kennedy JF, Phillips GO, Williams PA, Hascall VC, eds. Hyaluronan Volume 2 Biomedical, Medical and Clinical Aspects, Cambridge, UK: Woodhead, 2002; 483–490.
76. Weiss C, Levy HJ, Denlinger J, Suros J, Weiss H. The role of Na-hylan in reducing postsurgical tendon adhesions. Bull Hosp Jt Dis 1986; 46:9–15.
77. Weiss C, Suros JM, Michalow A, Denlinger J, Moore M, Tejeiro W. The role of Na-hylan in reducing postsurgical tendon adhesions: Part 2. Bull Hosp Jt Dis 1987; 47:31–39.
78. St. Onge R, Weiss C, Denlinger JL, Balazs EA. A preliminary assessment of Na-hyaluronate injection into "no man's land" for primary flexor tendon repair. Clin Orthop 1980; 146:269–275.
79. Diamond M. Reduction of adhesions after uterine myomectomy by Seprafilm membrane (HAL-F): a blinded, prospective, randomized, multicenter clinical study. Fertil Steril 1996; 66:904–910.
80. Becker J, Dayton M, Fazio V, Beck D, Stryker S, Wexner S, Wolff B, Roberts P, Smith L, Sweeney S, Moore M. Prevention of postoperative abdominal adhesions by a sodium hyaluronate-based bioresorbable membrane: a prospective, randomized, double-blind multicenter study. J Am Coll Surg 1996; 183:297–306.
81. Fazio VW, Cohen Z, Fleshman JW, van Goor H, Bauer JJ, Wolff BG, Corman M, Beart RW, Jr., Wexner SD, Becker JM, Monson JR, Kaufman HS, Beck DE, Bailey HR, Ludwig KA, Stamos MJ, Darzi A, Bleday R, Dorazio R, Madoff RD, Smith LE, Gearhart S, Lillemoe K, Gohl J. Reduction in adhesive small-bowel obstruction by Seprafilm adhesion barrier after intestinal resection. Dis Colon Rectum 2006; 49:1–11.
82. Denlinger JL. Biotrics, Inc., Report; 1976.
83. Polack FM, McNiece MT. The treatment of dry eyes with Na hyaluronate (Healon®). Cornea 1982; 1:133–136.
84. Acosta MC, Gallar J, Belmonte C. The influence of eye solutions on blinking and ocular comfort at rest and during work at video display terminals. Exp Eye Res 1999; 68:663–669.
85. Arshinoff SA, Khoury E. HsS versus a balanced salt solution as a corneal wetting agent during routine cataract extraction and lens implantation. J Cataract Refract Surg 1997; 23:1221–1225.
86. Alexander SA, Donoff RB. The glycosaminoglycans of open wounds. J Surg Res 1980; 29:422–429.

87. Balazs EA, Holmgren HJ. Effect of sulfomucopolysaccharides on growth of tumor tissue. Proc Soc Exp Biol Med 1949; 72:142–145.
88. Holmgren HJ, Balazs EA. Experimental studies on wound healing. Nord Med 1950; 43:471–474.
89. Rydell N, Freeman MI, Balazs EA. Decreased granulation tissue reaction after instillation of hyaluronic acid. Boston, MA: Retina Foundation, Dept. of Connective Tissue 1969.
90. Abatangelo G, Martelli M, Vecchia P. Healing of hyaluronic acid-enriched wounds: histological observations. J Surg Res 1983; 35:410–416.
91. Dillon PW, Keefer K, Blackburn JH, Houghton PE, Krummel TM. The extracellular matrix of the fetal wound: hyaluronic acid controls lymphocyte adhesion. J Surg Res 1994; 57:170–173.
92. Mast BA, Diegelmann RF, Krummel TM, Cohen IK. Scarless wound healing in the mammalian fetus. Surgery, Gynecol Obstetrics 1992; 174:441–451.
93. Mast BA, Flood LC, Haynes JH, Depalma RL, Cohen IK, Diegelmann RF, Krummel TM. Hyaluronic acid is a major component of the matrix of fetal rabbit skin and wounds: implications for healing by regeneration. Matrix 1991; 11:63–68.
94. Balazs EA, Eliezer IK, Dennebaum RA, Larsen NE, Whetstone JL. Cell attachment and growth on solid hyaluronan (hylan B gel). In: Kennedy JF, Phillips GO, Williams PA, Hascall VC, eds. Hyaluronan Volume 2 Biomedical, Medical and Clinical Aspects, Cambridge, UK: Woodhead, 2002; 33–38.
95. Ghosh K, Ren X-D, Shu XZ, Prestwich GD, Clark RAF. Fibronectin functional domains coupled to hyaluronan stimulate adult human dermal fibroblast responses critical for wound healing. Tissue Eng 2006; 12:601–613.
96. Ghosh K, Pan Z, Guan E, Ge S, Liu Y, Nakamura T, Ren X-D, Rafailovich M, Clark RAF. Cell adaptation to a physiologically relevant ECM mimic with different viscoelastic properties. Biomaterials 2007; 28:671–679.
97. Fujimori Y, Ueda K, Fumimoto H, Kubo K, Kuroyanagi Y. Skin regeneration for children with burn scar contracture using autologous cultured dermal substitutes and superthin auto-skin grafts: preliminary clinical study. Ann Plast Surg 2006; 57:408–414.
98. Caravaggi C, De Giglio R, Pritelli C, Sommaria M, Dalla Noce S, Faglia E, Mantero M, Clerici G, Fratino P, Dalla Paola L, Mariani G, Mingardi R, Morabito A. HYAFF 11-based autologous dermal and epidermal grafts in the treatment of noninfected diabetic plantar and dorsal foot ulcers: a prospective, multicenter, controlled, randomized clinical trial. Diabetes Care 2003; 26:2853–2859.
99. Gobbi A, Kon E, Berruto M, Francisco R, Filardo G, Marcacci M. Patellofemoral full-thickness chondral defects treated with hyalograft-C: a clinical, arthroscopic, and histologic review. Am J Sports Med 2006; 34:1763–1773.
100. Nehrer S, Domayer S, Dorotka R, Schatz K, Bindreiter U, Kotz R. Three-year clinical outcome after chondrocyte transplantation using a hyaluronan matrix for cartilage repair. Eur J Radiol 2006; 57:3–8.
101. Marcacci M, Berruto M, Brocchetta D, Delcogliano A, Ghinelli D, Gobbi A, Kon S, Pederzini L, Rosa D, Sacchetti GL, Stefani G, Zanasi S. Articular

cartilage engineering with hyalograft C: 3-year clinical results. Clin Orthop Relat Res 2005; 435:96–105.
102. Shu XZ, Prestwich G. Therapeutic biomaterials from chemically-modified hyaluronan. In: Garg HG, Hales CA, eds. Chemistry and Biology of Hyaluronan, Amsterdam: Elsevier, Ltd., 2004; 475–504.
103. Prestwich GD, Shu XZ, Liu Y, Cai S, Walsh JF, Hughes CW, Ahmad S, Kirker KR, Yu B, Orlandi RR, Park AH, Thibeault SL, Duflo S, Smith ME. Injectable synthetic extracellular matrices for tissue engineering and repair. Adv Exp Med Biol 2006; 585:125–133.
104. Shu XZ, Ahmad S, Liu Y, Prestwich GD. Synthesis and evaluation of injectable, in situ cross-linkable synthetic extracellular matrices for tissue engineering. J Biomed Mater Res A 2006; 79:902–912.
105. Prestwich G. Synthetic extracellular matrices for 3-D cell growth: applications in reparative medicine and cell biology. In: Seventh (7th) International Conference on Hyaluronan (Book of Abstracts). Charleston, SC, April 22–27, 2007; 8.
106. Orlandi RR, Shu XZ, McGill L, Petersen E, Prestwich GD. Structural variations in a single hyaluronan derivative significantly alter wound-healing effects in the rabbit maxillary sinus. Laryngoscope 2007; 117:1288–1295.
107. Proctor M, Proctor K, Shu XZ, McGill LD, Prestwich GD, Orlandi RR. Composition of hyaluronan affects wound healing in the rabbit maxillary sinus. Am J Rhinol 2006; 20:206–211.
108. Price RD, Das-Gupta V, Leigh IM, Navsaria HA. A comparison of tissue-engineered hyaluronic acid dermal matrices in a human wound model. Tissue Eng 2006; 12:2985–2995.

Carbohydrate Chemistry, Biology and Medical Applications
Hari G. Garg, Mary K. Cowman and Charles A. Hales

DOI: 10.1016/B978-0-08-054816-6.00015-X

Chapter 15

Drug Delivery and Medical Applications of Chemically Modified Hyaluronan

LUIS Z. AVILA,* DIEGO A. GIANOLIO,* PAUL A. KONOWICZ,* MICHAEL PHILBROOK,† MICHAEL R. SANTOS† AND ROBERT J. MILLER*

**Genzyme Corporation, Drug and Biomaterial R&D, 152 Second Ave; USA Waltham, MA 02451*
†Genzyme Corporation, Drug and Biomaterial R&D, 49 New York Avenue, Framingham, MA 01701-8805, USA

I. Introduction

Hyaluronan (HA) is a linear naturally occurring polyanionic polysaccharide that is ubiquitous in nature and is produced virtually by every tissue in higher organisms and some bacteria (1,2). HA has excellent biocompatibility and is readily catabolized and cleared *in vivo* (3–7). For these reasons, there has been a significant commercial focus on the development of products either from HA or from chemically modified derivatives of HA.

HA and HA plus chemically cross-linked HA preparations are the principle components in several viscosupplements for patients with early-stage osteoarthritis of the knee (8,9). Products such as Artz®, Hyalgan®, Euflexxa™, and Synvisc® have been approved as devices for the relief of pain associated with early-stage osteoarthritis (10). These products were initially designed to restore the rheology of the synovial fluid in diseased joints to more normal levels (11). The short residence time of this exogenous HA relative to the indicated duration of pain relief in the joint suggests that the exact molecular mechanism for this pain relief is complex and not well understood. Recent data from the clinic has shown that patients' knees that have been injected with Synvisc have synovial fluid that has both higher concentration and higher molecular weight HA (12–15). This result suggests that HA-based viscosupplements might stimulate HA synthases (HAS 1 and 2) leading to increased high molecular weight HA. Thus the pain relief might be due, in part, to a restoration of the normal HA in the joint. Several clinical studies with intra-articular injections of viscosupplements have shown significant pain relief in the carpometacarpal joint (16), temporomandibular joint (17–19), facet joint (20), and the hip (21,22). These studies indicate that pain relief from intra-articular injections of HA-based viscosupplements is not specific to the knee.

HA has been used in implantable medical devices for the reduction of post-surgical adhesions. One such device, Seprafilm®, has been on the market for ten years and is still the market leader for adhesion control in general surgical applications. Solutions of HA have been shown to act as tissue protectants and have been shown to affect adhesion formation in animal models (23–26).

Recently, cross-linked gels of HA have transformed the esthetics and dermatology areas. These HA-based products are remarkably safe and serve as tissue-bulking agents to correct wrinkles and other skin defects. Restylane® (27–29), Hylaform® (30–33), Captique™ (34), and Juvederm™ (35) are a few representative examples of HA-based dermal fillers that have been approved for skin depression in nasolabial folds.

HA clearly has demonstrated its chemical utility and biocompatibility for use in medical devices and for drug formulations. The wealth of approved products using HA affords a diverse platform from which drugs can be either admixed or conjugated covalently for controlled release applications. This chapter reviews recent uses of HA for drug delivery and controlled release applications.

II. Conjugation of Active Agents to Hyaluronan

We previously described chemical modification approaches that have been used to derivatize HA (36). This section focuses on recent literature reports on conjugation of active moieties to HA. Several earlier articles dealing with this topic have been published elsewhere (37–42).

Most reported chemical conjugations of active agents to HA involve the attachment of these agents through the carboxylic functionality of HA. The final HA conjugate is obtained either after appending the active agent to the molecule through an amide or ester linkage or by sequentially building the active agent on the molecule by a series of synthetic steps. Modification of the naturally occurring primary and secondary hydroxyl groups of HA and of the aldehyde at the reducing end has also been reported.

Another method to introduce a reactive group on HA involves the use of *Streptomyces* hyaluronidase (β-eliminase) to generate a conjugated α,β-unsaturated carboxylic acid at the nonreducing end (43,44) to afford an electrophilic site that can react with appropriate nucleophiles such as thiols. HA oligosaccharides with a free aldehyde group at the nonreducing end can be synthesized by ozonolysis of the double bond followed by reduction of the resulting ozonide (45). *N*-Deacetylation of HA allows access to the amine group for subsequent conjugation to a chemical or biological agent. However, deacetylation requires harsh chemical treatment with aqueous hydrazine under carefully controlled conditions to minimize HA depolymerization (46).

Conjugation of active agents to cross-linked HA and composite polymer gels is also synthetically achievable. In a majority of the reported strategies, a significant amount of the reactive sites remain available after the cross-linking step and as a consequence are suitable for further conjugation using diffusible active agents (47).

A. Conjugation of Analgesics

Local anesthetics are useful for reducing acute pain, but the short duration of this effect precludes them from use in solely managing postoperative pain. To prolong the duration of local anesthesia, bupivacaine was conjugated to native HA and divinyl sulfone (DVS) cross-linked hylan A (hylan B particles) using a hydrolyzable imide linkage (Fig. 1). Bupivacaine was released from the conjugated gel *in vitro* at a slower rate ($t_{1/2}=16.9\pm0.2$ h) when compared to a simple admixture of bupivacaine in hylan B ($t_{1/2}=0.4\pm0.1$ h). The selective release of bupivacaine from the gel was achieved by conjugating this agent through a β-thioether in the linker. Liquid chromatography–mass spectral analysis confirmed that bupivacaine was released from the gel unaltered (48).

Similarly, modified HA bearing hydrazide and aldehyde functionalities that cross-link upon mixing was used as a vehicle for bupivacaine (49). A 2% (w/v) cross-linked HA doubled the duration of block in a rat sciatic nerve blockade model, without a statistically significant increase in myotoxicity.

The opioid drugs morphine and codeine as well as the analog naloxone were conjugated to HA via two types of hydrolytically labile ester bonds and drug release kinetics from these conjugates were systematically investigated (50).

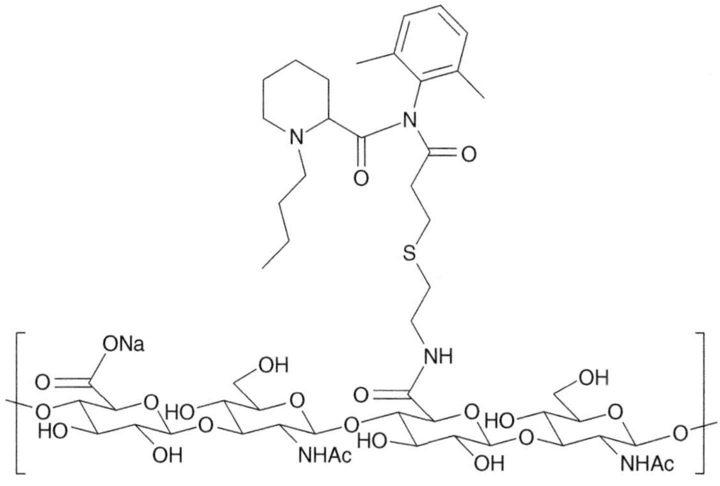

Figure 1 Structure of bupivacaine conjugated to the Hyaluronan (HA) backbone by a hydrolyzable imide bond.

B. Conjugation of Chemotherapy Agents

Although paclitaxel (PTX) has shown tremendous potential as an anticancer drug, its utility has been compromised by its poor aqueous solubility. PTX conjugation to HA was carried out to increase water solubility and potentially target it toward cells that overexpress HA cell surface receptors (e.g., CD44).

One conjugation approach exploited the higher nucleophilicity of acyl hydrazides compared to amines during EDC-mediated couplings (39,51,52). This use of difunctional dihydrazides has also been reported to allow facile access to other HA conjugates (Fig. 2).

HA-PTX conjugates generated using this approach showed selective toxicity toward human cancer cell lines that are known to overexpress HA receptors while no toxicity was observed toward a mouse fibroblast cell line at the same concentration. Release of PTX from the conjugate occurred after cleavage of the labile 2′ ester linkage.

A related fluorescent-HA-PTX conjugate showed cell-specific binding and uptake using flow cytometry and fluorescent confocal microscopy, indicating direct correlation of uptake with selective toxicity (53).

The use of *tert*-butylammonium salts of HA allowed dissolution of HA into organic solvents, and hence the synthesis of HA-PTX conjugate by esterification of the carboxylate anion of HA with PTX (Fig. 3). PTX loading was optimized at 20% (w/w), increasing by 500-fold the PTX concentration that could be achieved using an aqueous method. Histologic examination revealed that HA-PTX conjugate was well tolerated in vivo (54).

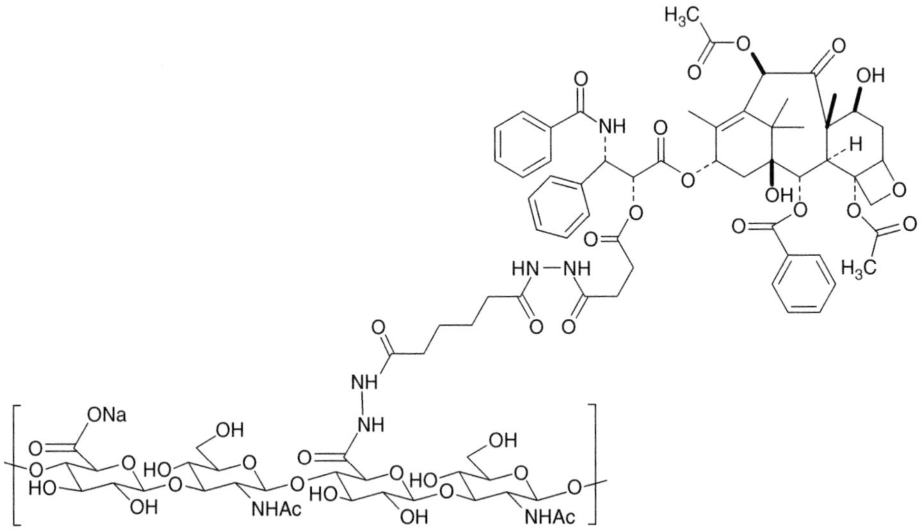

Figure 2 Structure of paclitaxel conjugated via an acyl hydrazide bond to the Hyaluronan (HA) backbone.

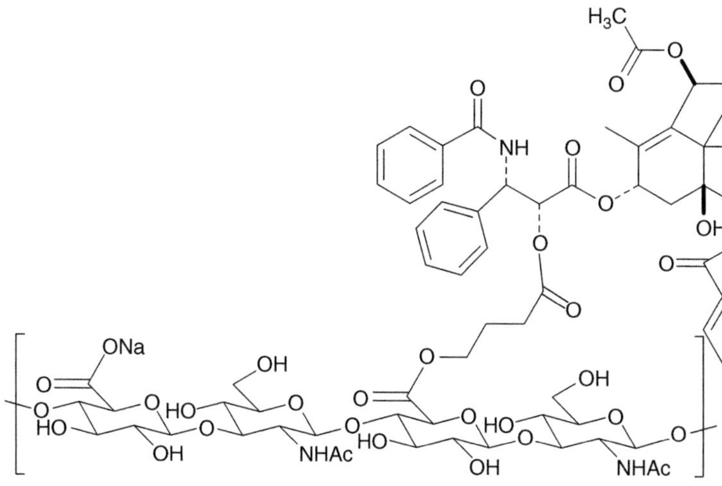

Figure 3 Structure of paclitaxel conjugated via an ester bond to the Hyaluronan (HA) backbone.

A third approach to prepare HA-taxol conjugate involves a labile succinate ester (Fig. 4). These hydrophobic PTX conjugates were amenable to the preparation of multilayer macromolecular assemblies of bioactive polyelectrolytes (38).

PTX conjugated to HA using a 4-hydroxybutanoic acid linker to increase the solubility has been reported (55). The HA-PTX conjugate was significantly more

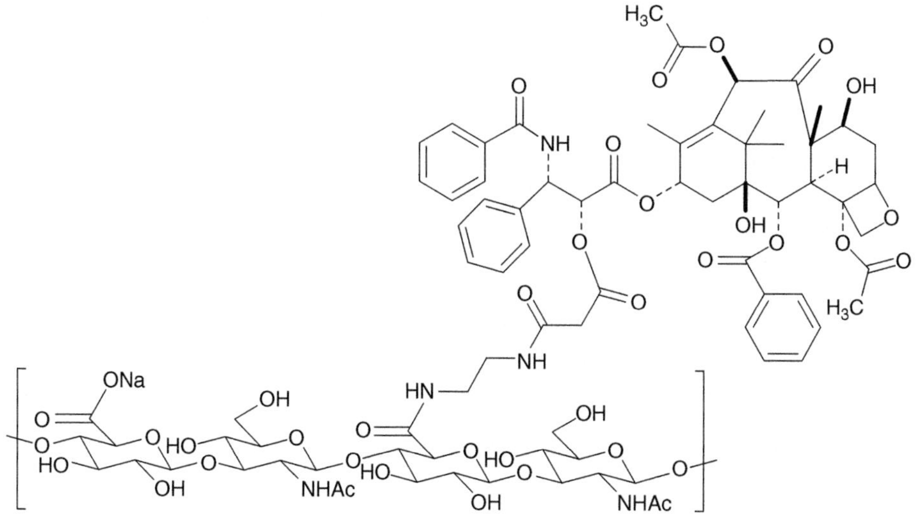

Figure 4 Structure of paclitaxel conjugated via an amide bond to the Hyaluronan (HA) backbone.

water soluble than the native PTX molecule. The potential therapeutic applications of HA-PTX conjugates were further expanded to a dual therapy by combining a radionuclide prepared by the addition of 99mTc-pertechnetate to HA-PTX bioconjugate (ONCOFID-P) with 100% radiochemical purity and stability in a phosphate buffer. Intraperitoneal, intravesical, and oral administrations showed that all the 99mTc-ONCOFID-P remained at the administration site, suggesting potential utility of ONCOFID-P as a local therapeutic agent for the treatment of superficial cancers (56).

Other anti-proliferatives have been similarly conjugated to HA. A cross-linked hydrogel that contained a covalently linked derivative of the drug mitomycin C (MMC) was prepared in two steps by initially coupling MMC-aziridinyl-*N*-acrylate with thiol-modified HA followed by cross-linking with poly(ethylene glycol) diacrylate (PEGDA). When implanted into a rat peritoneal cavity, these HA-MMC films reduced the thickness of fibrous tissue formed surrounding the implanted films, suggesting potential as anti-fibrotic barriers for the reduction of postsurgical adhesions (57,58).

Other cytotoxic drug conjugates, HA doxorubicin (HA-DOX), and hydroxypropylmethacrylamide (HPMA) copolymer-DOX containing HA as a side chain (HPMA-HA-DOX), were studied *in vitro* for their anti-proliferative property against a number of human tumor cell lines. The cytotoxicity of HPMA-HA-DOX bioconjugate was found to be higher than that of the nontargeted HPMA-DOX conjugate against human breast cancer (HBL-100), ovarian cancer (SKOV-3), and colon cancer (HCT-116) cells. Fluorescence confocal microscopy revealed that the targeted HPMA-HA-DOX conjugates were internalized more efficiently by cancer cells and that internalization of the polymer conjugates correlated with their cytotoxicity (59). Both conjugates showed minimal cytotoxicity toward mouse fibroblast NIH 3T3 cells. Finally, MMC conjugated to cross-linked HA hydrogels was shown to reduce postoperative adhesions (58).

C. Conjugation of Nitric Oxide Donors

Nitric oxide (NO) and hyaluronic acid are both known to play an important role in the wound-healing process. Novel HA-NO donors were prepared by first reacting spermidine with HA using EDC-coupling chemistry. Gaseous NO was then covalently bound to the secondary amine groups of the spermidine under pressure (5 atm). These structures, named NONO-ates, exhibited controlled NO release under physiological *in vitro* conditions. NONO-ates were also synthesized with concomitant cross-linking using Ugi's four-component reaction with formaldehyde, cyclohexylisocyanide, and spermidine (60).

D. Conjugation of Radionuclides and Metals

The conjugation of radionuclides to HA has been proposed as an anti-proliferative therapy. The combination of the hemocompatibility of HA-coated surfaces and the anti-proliferative effects of an appropriate radiotherapy inhibited hyperplasia following revascularization procedures. Radioactive devices were obtained by coating plasma-functionalized surfaces like stents or catheters with a HA-diethylenetriamine pentaacetic acid (DTPA) conjugate (HA-DTPA); the conjugate in turn was complexed with a γ- or β-emitting radionuclide. Therapeutic doses of the emitters, yttrium and indium, were conveniently loaded onto the surfaces and remained stable over 2 weeks with a radionuclide loss of about 6% (61).

An HA-boron conjugate has been prepared for use as an anti-proliferative therapy using boron neutron capture (BNC). To prepare this conjugate, a carborane was linked onto HA via an ester bond. A degree of substitution of ~30% was reported (62). This conjugate was designed to deliver boron atoms to target cells via HA-mediated uptake. *In vitro* experiments showed that this conjugate was non-toxic toward a variety of human tumor cells of different histotypes. It specifically interacted with CD44 as the native unconjugated HA, and the concentration of boron taken up by the tumor cells was sufficient for successful BNC therapy (62).

Other metal-containing conjugates include HA gadolinium DTPA derivatives. These have potential as tumor-specific contrast agents for magnetic resonance imaging (MRI). These conjugates were synthesized by initially reacting ethylenediamine with carboxylic acid groups followed by covalent linkage of DTPA to aminated HA. The level of conjugation of DTPA was estimated by a colorimetric assay, isothermal titration calorimetry (ITC), and nuclear magnetic resonance (^{1}H-NMR) spectroscopy (63). MRI signal enhancement was reported using a coating grown by a layer-by-layer assembly. A conjugate of HA with gadolinium-coordinated DTPA is used with chitosan to coat polymeric interventional tools. The visibility under MRI can be easily tailored by controlling the number of layers in these multilayer coatings (64).

E. Conjugation of Nucleic Acid

Disulfide cross-linked HA has also been reported for the preparation of novel nanogels that physically encapsulated small interfering RNA (siRNA) (65). These nanogel particles were fabricated by an inverse water-in-oil emulsion method. Thiol-conjugated HA was ultrasonically cross-linked in the presence of green fluorescence protein (GFP) and siRNA to afford nanogels. The nanogels were readily internalized by HA receptor positive cells (HCT-116) having HA-specific CD44 receptors on the surface via receptor-mediated endocytosis. The *in vitro* release rates of siRNA from the HA nanogels were modulated by changing the concentration of glutathione in the buffer solution, indicating that the degradation of the nanogels modulated the release pattern of siRNA.

Synthesis of poly(L-lysine)-grafted hyaluronic acid (PLL-g-HA) comb-type copolymers has recently been described. HA chains were covalently coupled at the ε-amino groups of polylysine by reductive amination reaction (66). These PLL-g-HA derivatives were shown to form complexes with plasmid DNA. The complex showed stability against nuclease degradation and was internalized by liver cells (67–70).

Controlled release of human vascular endothelial growth factor (VEGF) or basic fibroblast growth factor (bFGF) from hydrogels composed of thiol-modified

HA, gelatin, and heparin (15 kDa) was evaluated both *in vitro* and *in vivo*. The hypothesis was that inclusion of small quantities of heparin in these gels would regulate growth factor (GF) release over an extended period, while still maintaining the *in vivo* bioactivity of released GFs (71).

III. Conjugation of Peptides and Proteins to Hyaluronan

A. Peptides

HA has also been conjugated with peptides for local or targeted drug and antibody delivery systems as well as with matrices for immobilization of GF and cells. Such examples include the synthesis and characterization of bioconjugates of natural polymers such as HA, alginate, and dextran with peptides specific for *Bacillus subtilis* and *B. anthracis* (72). These peptide conjugates were synthesized by creating thioether bonds by the reaction between HA maleimide derivatives and thiol groups on the peptides.

A similar approach was used to prepare a thiol-modified HA (3,3′-dithiobis-propanoic dihydrazide) (HA-DTPH) that was subsequently reacted with an Arg-Gly-Asp (RGD) sequence and then cross-linked with PEGDA to create a biomaterial that supported cell attachment, spreading, and proliferation (73).

HA hydrogels and mixed polymer hydrogels have been synthesized using a photo-cross-linking strategy. Glycidyl methacrylate-HA was conjugated with acrylated PEG and PEG peptides to give composite hydrogels. These hydrogels were cyto-compatible, biologically active, and had a decreased rate of hyaluronidase degradation compared with native HA (74).

B. Proteins and Antibodies

Hydrogels were synthesized by cross-linking HA with DVS and poly(ethylene glycol)-functionalized divinyl sulfone (VS-PEG-VS). These gels were loaded with vitamin E succinate (VES) and bovine serum albumin (BSA), as models of anti-inflammatory proteins and drugs, and their release kinetics were measured *in vitro*. The rate of release from HA-VS-PEG-VS-HA hydrogels was faster than that from HA-DVS-HA hydrogels, presumably because of the lower cross-linked density in the former (75).

An antibody [immunoglobulin G (IgG)] releasing system was prepared as antagonists of the receptors Nogo-66 and NgR, receptors that have been shown to inhibit neuronal regeneration and are potential therapeutic agents for Central nervous system (CNS) axonal injuries, such as spinal cord and brain trauma. The antibody was covalently attached to an HA hydrogel using a hydrolytically labile hydrazone linkage. The bioactivity of antibody released from hydrogel was retained as demonstrated by indirect immunofluorescence technique (76). These modified hydrogels delivered antibodies and could potentially serve as scaffolds for neural regeneration following their implantation into injured tissue (77).

HA derivatives that could be candidates for immunosuppressive drugs, not only in organ transplantation but also in diseases where interferon-γ (IFN-γ) is overexpressed, are conjugates composed of a phospholipase A2 inhibitor linked to HA. This HA conjugate inhibited the induction of major histocompatibility class II after IFN-γ stimulation in a dose-dependent manner (78).

C. Growth Factors

A cross-linked hydrogel matrix with an immobilized GF, fibronectin (FN), has been reported. Cysteine-tagged fibronectin functional domains (FNfds) were coupled to a homobifunctional PEG derivative and these PEG-modified FNfd solutions containing a bifunctional PEG-based cross-linker were coupled to thiol-modified HA (HA-DTPH) to obtain a cross-linked hydrogel matrix. FN facilitated dermal fibroblast migration during normal wound healing, and the FN functional domains coupled to an HA backbone stimulated wound repair. When implanted in porcine cutaneous wounds, these biocompatible acellular matrices induced rapid and *en masse* recruitment of stromal fibroblasts (79).

The stability and activity of recombinant GFs administered locally for the repair of damaged tissue can be directly influenced by the physical structure and chemical composition of the delivery matrix (80). A fibroblast growth factor-2 (FGF-2) delivery system was synthesized by the conjugation of a structure-stabilizing HA with heparin (HP) that has inherent specific binding sites for members of the FGF family. The hypothesis was that the inclusion of small amounts of HP in these gels would regulate GF release over an extended period, while still maintaining the *in vivo* bioactivity of released GFs. These matrices were formed either by stable amine or by labile imine bonds by coupling amine-modified HA with oxidized HP. Recombinant human FGF-2 rapidly bound to the heparin segment of the HA-heparin (HAHP) conjugate. The FGF-2 was released *in vitro* from the imine-bonded (HAHPi) gels in the form of FGF-2-HP complexes through the hydrolysis of the imine bonds. In contrast, release from the more stable amine-bonded (HAHPa) gels required treatment with free HP or enzymatic digestion of the HA segment. Analysis of the released FGF-2 showed that the HAHP conjugate gels increased both the stability and the activity of the GF (81).

D. Lipids

Lipids have been conjugated to HA using an aldehyde group introduced by oxidation of vinyl groups that were introduced at the nonreducing terminal glucuronic acid of HA by treatment with bacterial hyaluronidase. The resulting aldehyde-functionalized HA was coupled to dipalmitoyl phosphatidylethanolamine (DPPE) in a reductive amination step. This methodology can conceivably be extended to link molecules such as biotin, polymers, or proteins to HA (82).

E. Miscellaneous Cross-linked Biomaterials

Thermoresponsive HAs were prepared by graft polymerization of *N*-isopropylacrylamide (NIPAM) on HA using dithiocarbamate. The dithiocarbamate acted as an initiator, transfer agent, and terminator in the reaction. These poly-NIPAM-grafted HAs were water soluble at room temperature and precipitated at temperatures above 34°C (83).

Glycidyl methacrylate derivatized dextran and/or HA were cross-linked into hydrogels in the presence of photoinitiators and UV radiation in multi-well inserts. The materials showed good cytocompatibility *in vitro* using smooth muscle cell migration/proliferation assay (84).

Ugi chemistry presents an interesting approach to manipulate HA to form novel types of polymer networks (i.e., hydrogels) and/or multiple emulsions. Studies focused on using HA as the carboxylic acid component in the multicomponent Ugi chemistry to generate new biomaterials of possible interest to the field of ophthalmology and for potential controlled delivery systems have been reported (85).

Similarly, other gels were prepared with different degrees of cross-linking by reacting *N*-deacetylated HA with formaldehyde and cyclohexylisocyanide under aqueous conditions for the Ugi reaction. In this example, HA provided both the amine and the carboxylic acid components. The gels were mechanically stable and exhibited good water uptake, which was strongly dependent on the extent of cross-linking. *N*-deacetylated HA samples were also selectively *N*-sulfated or *O*-sulfated; the former exhibits anticoagulant properties well exceeding those of the latter (46). HA was used in other aqueous Ugi reactions to give polymeric networks. HA supplied the carboxyl component and the amine component was supplied by lysine, which also served as a cross-linking agent. The structural and physicochemical properties of the hydrogels were studied using solid-state NMR spectroscopy and measurements of swelling in water and in aqueous NaCl solution (86).

HA can also be "self" cross-linked to give an auto-cross-linked polysaccharide (ACP) gel that is fully biocompatible, has prolonged *in vivo* residence time, and improved mechanical properties with respect to native HA for use in various surgical applications (87). Mixed polymer gels have also been prepared using carbodiimide chemistry. A mixed polymer hydrogel was prepared by cross-linking HA with glycol chitosan in aqueous solution using water-soluble carbodiimide at nearly neutral pH and room temperature (88).

Other novel mixed polymer systems have included glycosaminoglycans (GAGs) modified with multiblock copolymers to form self-assembling, cross-linker free hydrogels. For example, aminooxy pluronic derivatives were coupled with HA to give amphiphilic hydrogels with the ability to bind to hydrophobic surfaces (89).

IV. Drug Formulations Using Hyaluronan

A. Hyaluronan/Drug Admixtures in Oncology

Particular interest has also been devoted to HA as a potential delivery vehicle for chemotherapeutics. Research has suggested that HA may enhance chemotherapeutic activity, help to target drug delivery to CD44 and RHAMM receptors, promote transdermal delivery, and provide a solid support for delivery of chemotherapeutics to tissue surfaces.

B. Hyaluronan-Mediated Enhanced Chemotherapeutic Activity

Multidrug resistance is an intrinsic problem in cancer chemotherapy. Unfortunately, as therapy has become more and more effective, acquired resistance has also become common (90). The most common reason for acquisition of resistance to a broad range of anticancer drugs is expression of one or more Adenosine triphosphate (ATP)-dependent transporters that detect and eject anticancer drugs from cells, but other mechanisms of resistance, including insensitivity to drug-induced apoptosis and induction of drug-detoxifying mechanisms, probably play an important role in drug resistance (91–97).

HA has been shown to produce synergistic therapeutic effects when coadministered with 5-fluorouracil (5-FU) and with PTX. HA formulated with 5-FU enhanced the cellular uptake and cytotoxicity of the drug compared to unformulated 5-FU (98). A similar effect has been observed when PTX was coadministered with HA (99). The admixture of PTX and HA significantly reduced the migration of Lewis lung carcinoma cells in a synergistic fashion and markedly improved the life span of mice seeded with tumor cells compared with PTX alone or HA alone.

C. Hyaluronan-Mediated Targeted Delivery of Liposomes

Aggressive tumor cell growth and migration are dependent on cell–cell and cell–extracellular matrix interactions (100). HA plays a pivotal role in tumor invasion and metastasis by binding to overexpressed CD44 and RHAMM receptors on tumor cell surfaces (101). This interaction has prompted many investigators to evaluate HA-conjugated drugs for targeted delivery to tumor cells.

One of the approaches to couple drugs to HA is to encapsulate the drug in liposomes that have HA as a constituent of the liposome (LIP) membrane. The HA acts as a hydrophilic coat to promote long-term circulation of encapsulated drug as well as to target the overexpressed CD44 and RHAMM receptors on tumors (102,103). This targeting approach was demonstrated with MMC-loaded HA-LIP in which the cytotoxic effect of the targeted delivery system increased drug potency by 100-fold, in cells overexpressing, but not in cells underexpressing, HA receptors. HA-LIP formulations loaded with DOX and with MMC were also evaluated *in vivo* for their ability to target tumors in mice. The HA-LIP formulations showed longer circulation half-lives and higher drug accumulations in tumor-bearing lungs. Both HA-LIP formulations showed that tumor progression, metastatic burden, and survival were all superior in animals receiving drug-loaded HA-LIP compared with controls.

Similarly, phosphatidylethanolamine lipid derivatives containing HA oligosaccharides (HA-PE) have been incorporated into liposomes to target tumor cells that express CD44 (104). HA-targeted liposomes (HALs) avidly bound to the CD44-high-expressing B16F10 murine melanoma cell line but not to the CV-1 African green monkey kidney cells, which express CD44 at low levels. HALs binding to B16F10 were inhibited by an anti-CD44 monoclonal antibody. HALs delivering DOX were significantly more potent than the nonencapsulated DOX in cells expressing high levels of CD44, which suggested that HALs may be useful as targeted drug carrier to treat CD44-expressing tumors.

D. Hyaluronan-Mediated Enhanced Transdermal Delivery of Chemotherapeutics

Hydration is the most widely used and safest method to increase skin penetration of both hydrophilic and lipophilic permeants (105). Additional water within the stratum corneum could alter permeant solubility and thereby modify partitioning from the vehicle into the membrane. Also, increased skin hydration may swell and open the structure of the stratum corneum, leading to an increased penetration.

HA has been used as a hydrophilic carrier to deliver diclofenac topically for the treatment of premalignant skin lesions such as actinic keratoses (AK) (106–112) and for colon-26 adenocarcinoma (113–115). HA is the preferred carrier for transdermal delivery of diclofenac, as it has been shown to enhance the partitioning of the drug into the skin compared to other vehicles (116). Furthermore, in a clinical trial, the safety and efficacy of 3% diclofenac in 2.5% HA gel have been evaluated as a topical treatment for actinic keratosis (108). Patients treated with HA-diclofenac showed significantly lower target and cumulative lesion number scores and lesion total thickness scores compared to the placebo group. The treatment with 3.0% diclofenac in 2.5% HA gel was effective when used for 60 days and was well tolerated in patients with AK.

Topical delivery of diclofenac in 2.5% HA has also been shown to inhibit the development and angiogenesis of colon-26 adenocarcinoma (112). In this work, HA-diclofenac reduced proliferation and viability of colon-26 adenocarcinoma cells *in vitro*, inhibited tumor prostaglandin synthesis, and retarded angiogenesis in mice (114). Analysis of the tumor vasculature showed that vascular development was retarded by 12 days. The effects of HA-diclofenac were likely related to enhanced transdermal delivery and binding properties of HA.

E. Hyaluronan/Drug Admixtures for Wound Healing

HA plays an important role in tissue repair. It is known to influence a number of events critical to successful wound healing, including inflammation, cell migration, angiogenesis, re-epithelialization, and scar formation (117). Naturally occurring HA has a short residence time in tissues because of cell metabolism and hyaluronidase activity. HA is hydrophilic and can absorb large quantities of water to maintain tissue hydration levels necessary for healing. It has been used in wound healing for the treatment of venous, mixed etiology leg ulcers, diabetic foot ulceration (118–121), postsurgical wounds (122,123), and burns (124–127).

Increasing evidence suggests that fibrin deposition is an important pathogenic component of venous ulceration and that fibrin removal could accelerate ulcer healing. In a clinical study, 1% HA was formulated with recombinant tissue plasminogen activator (tPA) for acceleration of ulcer healing (121). Daily topical application of tPA/HA was applied to patients at escalating doses of 0.25, 0.5, and 1.0 mg/ml of tPA for 4 weeks. The trial showed a direct correlation between mean ulcer re-epithelialization, fibrin removal, and the dose of topically applied tPA. Furthermore, topically applied tPA with HA appeared to be a safe and promising agent for treating venous ulcers.

HA has also been formulated with 1% silver sulfadiazine (HA-SSD) for evaluation as a topical treatment of second-degree burns. In a comparative, multicenter, randomized trial, the topical formulation of both HA-SSD and SSD alone were effective and well tolerated (127). However, the HA-SSD showed a significantly more rapid re-epithelialization of burns resulting in a 4.5-day shorter healing time compared to SSD alone. The observed shorter time to healing caused by the HA-SSD formulation may be attributed to the enhanced wound-healing activity of HA.

Hyiodine (high molecular weight HA combined with KI_3 complex) is a nonadhesive wound dressing that was studied to determine the effects on functional properties of human keratinocytes and leukocytes, and of U937 and HL60 cell lines (128). KI_3 complex has been shown to inhibit the viability and proliferation of the cells tested. However, hyiodine did not have any significant effect on these cells. The expression of CD11b, CD62L, and CD69 on PMNL, monocytes, and lymphocytes, as well as the oxidative burst of blood neutrophils, was not changed. Hyiodine inhibited the Phorbol myristate factor (PMA)-activated oxidative burst and significantly increased the production of IL-6 and Tumor necrosis factor (TNF)-α by lymphocytes. The HA content of hyiodine reduced the toxic effect of KI_3 complex and accelerated the wound-healing process by increasing the production of inflammatory cytokines.

F. Hyaluronan/Drug Admixtures for Orthopedic Applications

High molecular weight HA is a major component of normal synovial fluid attributing to its viscoelastic nature while providing shock-absorption and lubrication to joints (129,130). Injection of HA into osteoarthritic joints has become common practice to restabilize joint homeostasis while reducing inflammation and providing temporary pain relief. HA has shown both biocompatibility and anti-inflammatory properties in the joint, thereby making it an ideal candidate for use as a carrier for local drug delivery in many orthopedic indications.

G. Hyaluronan as a Bioresorbable Drug Carrier for Osteomyelitis

Osteomyelitis can occur following joint surgeries leading to complications (131). Prophylactic administration of a local antibiotic after orthopedic surgery could be beneficial in preventing serious postoperative complications. HA has been investigated as a biodegradable carrier for the local administration of antibiotics. Demineralized bone (DBM) in a high molecular weight HA has proven to be a viable delivery vehicle. When loaded with gentomycin and vancomycin, it has maintained the activity of antibiotics *in vitro* while sustaining their release over 48 hours (132).

Similarly, HA films or HA gels (133,134) containing vancomycin, gentomycin, tobramycin, or minomycin have shown potential use for the treatment of osteomyelitis.

Delivery of antibiotics from such formulations resulted in suppressed bacterial growth, and thus antibacterial activity, in an *in vivo* rabbit model. Furthermore, these formulations have proven to be effective in an agar diffusion test, direct contact test, mouse model, and rabbit osteomyelitis model without disturbing normal bone ingrowth.

H. Hyaluronan as a Delivery Vehicle for Cartilage and Bone Repair Applications

HA has become a common choice as a carrier for local administration of GFs for use in cartilage repair. Evaluation of biocompatible HA scaffolds as carriers for GFs to induce chondrogenesis has been carried out with promising results (135–137). When scaffolds loaded with gelatin microspheres containing transforming growth factor-beta1 (TGF-β1) were administered subcutaneously in nude mice, they were able to increase mesenchymal stem cell (MSC) proliferation and glycosaminoglycan synthesis *in vitro* as well as induce ectopic cartilage formation at 3 weeks. Similar HA scaffolds loaded with gelatin microspheres containing bFGF have improved osteochondral defects in a rabbit model. Furthermore, cross-linked collagen-HA matrices loaded with recombinant human growth and development factor-5 showed evidence of both osteogenic and chondrogenic events when implanted in an ectopic site.

Bone repair is also an area where HA has been used as a delivery device. DBM has been shown to induce osteogenesis as well as promote bone repair in numerous *in vivo* settings (138–143). The physical properties and superior biocompatibility of HA have improved the handling properties while maintaining the biological activity of DBM (144–146). In an athymic mouse model, DBM-HA admixtures retained their activity when compared to saline controls. Also, in a mouse cortical bone defect model, the use of DBM-HA was attributed to expedited bone formation. HA/DBM putty also showed enhanced iliac crest bone grafts in a rabbit posterolateral spinal fusion model by improving the fusion rates as compared to bone grafts alone. It is believed that HA provides a matrix for attachment of appropriate osteoprogenitor cells and allows for degradation of this matrix for the ingrowth of new bone.

HA as a carrier for GF delivery has also been investigated for orthopedic bone repair. In general, members of the TGF superfamily of GFs, including bone morphogenic protein-2 (BMP-2) and bone morphogenic GF-7, are used as bioactive agents. Cross-linked HA gels have shown the ability to deliver GFs in combination with MSC (147). In a rat calvarial defect model, BMP-2 and MSC-loaded acrylated HA hydrogels showed the highest levels of osteocalcin and mature bone compared to controls at 4 weeks, as observed from histological analysis.

I. Hyaluronan/Drug Admixtures for Adhesion Prevention

As normal wound healing proceeds, adhesions occur in a majority of abdominal surgical patients (148,149) and can lead to significant complications (150–152). HA/carboxymethylcellulose (HA/CMC) films have the ability to provide a biocompatible, mucoadhesive, physical barrier that separates tissues long enough to reduce such adhesions (123,153,154).

However, HA/CMC films do not prevent all postsurgical adhesions. This led to the idea of incorporating therapeutic agents into these devices to improve their overall efficacy. HA has been shown to localize and control delivery of PTX (155,156) to surgical sites (157). One percent and 5% PTX-loaded HA films containing 10% glycerol, to improve the handling properties, were used in a rat cecal abrasion model. Both drug-loaded films resulted in a statistically significant increase in the number of animals with no adhesions as well as a decrease in the mean incidence of adhesions as compared to no treatment or control HA films. No toxicity was observed with the treatments; however, excess fluid present in the abdominal viscera was observed in the 5% PTX-loaded HA films at the time of necropsy.

Similarly, HA films loaded with camptothecin (CPT) have shown the ability to prevent postsurgical adhesions (158). Carbodiimide cross-linked HA films containing 20% (w/w) glycerol were used as the carrier. These films were loaded with CPT at concentrations of 0, 0.6, 2.5, and 7.5% (w/w) and evaluated in a rat cecal sidewall abrasion model. Compared to no treatment, all of the HA films, including controls, significantly reduced the mean strength and area of adhesions. The addition of 0.6, 2.5, and 7.5% (w/w) CPT gave improved adhesion scores compared to control HA films [0% (w/w)]. HA films containing CPT exhibited the desired physical, biocompatible as well as controlled release properties that are important in preventing adhesion formation.

V. Conclusion/Summary

Conjugation of drugs to HA holds great promise for the generation of a new class of polymer-based therapeutics. These polymer-based systems can serve not only a biomaterials-based function, such as the separation or bulking of tissue, but also as concomitant drug delivery systems for the local delivery of therapeutics. Drugs conjugated to HA could also serve to target the drug to cells or tissues in the body that are rich in HA-binding receptors, such as CD44, RHAMM, TLR2, and TLR4.

Lastly, one of the most promising uses of chemically modified HA is in the area of tissue engineering. HA is found in the extracellular matrix of virtually all tissues so its use as a cell delivery vehicle is obvious. The combination of cell delivery and drug attachment to HA offers a very versatile material for the development of sophisticated products to address complex and unmet medical needs.

Acknowledgment

The authors kindly thank Pradeep Dhal, Ph.D., for his thoughtful scientific and editorial comments.

References

1. Meyer K, Palmer JW. The polysaccharide of the vitreous humor. J Biol Chem 1934; 107:629–634.
2. Weissmann B, Meyer K. The structure of hyalobiuronic acid and of hyaluronic acid from umbilical cord. J Am Chem Soc 1954; 76:1753–1757.

3. Page-Thomas DP, Bard D, King B, Dingle JT. Clearance of proteoglycan from joint cavities. Ann Rheum Dis 1987; 46:934–937.
4. Lebel L. Clearance of hyaluronan from the circulation. Adv Drug Deliv Rev 1991; 7:221–235.
5. Owen SG, Francis HW, Roberts MS. Disappearance kinetics of solutes from synovial fluid after intra-articular injection. Br J Clin Pharm 1994; 38:349–355.
6. Lindenhayn K, Heilmann H-H, Niederhausen T, Walther H-U, Pohlenz K. Elimination of tritium-labeled hyaluronic acid from normal and osteoarthritic rabbit knee joints. Eur J Clin Chem Clin Biochem 1997; 35:355–363.
7. Antonas KN, Fraser JRE, Muirden KD. Distribution of biologically labeled radioactive hyaluronic acid injected into joints. Ann Rheum Dis 1973; 32:103–111.
8. Gomis A, Pawlak M, Balazs EA, Schmidt RF, Belmonte C. Effects of different molecular weight elastoviscous hyaluronan solutions on articular nociceptive afferents. Arthritis Rheum 2004; 50:314–326.
9. Mori S, Naito M, Moriyama S. Highly viscous sodium hyaluronate and joint lubrication. Int Orthop 2002; 26:116–121.
10. Bellamy N, Campbell J, Robinson V, Gee T, Bourne R, Wells G. Viscosupplementation for the treatment of osteoarthritis of the knee. Cochrane Database Syst Rev 2006; (2):CD005321.
11. Weiss C. Hyaluronan and hylan in the treatment of osteoarthritis. In: Kennedy JF, Phillips GO, Williams PA, Hascall VC, eds. Hyaluronan. Cambridge: Woodhead Publishing, 2002; 467–481.
12. Wang CT, Lin YT, Chiang BL, Lin YH, Hou SM. High molecular weight hyaluronic acid down-regulates the gene expression of osteoarthritis-associated cytokines and enzymes in fibroblast-like synoviocytes from patients with early osteoarthritis. Osteoarthritis Cartilage 2006; 14:1237–1247.
13. Homandberg GA, Ummadi V, Kang H. High molecular weight hyaluronan promotes repair of IL-1 beta-damaged cartilage explants from both young and old bovines. Osteoarthritis Cartilage 2003; 11:177–186.
14. Recklies AD, White C, Melching L, Roughley PJ. Differential regulation and expression of hyaluronan synthases in human articular chondrocytes, synovial cells and osteosarcoma cells. Biochem J 2001; 354:17–24.
15. Creamer P. Intra-articular corticosteroid treatment in osteoarthritis. Curr Opin Rheumatol 1999; 11:417–421.
16. Coaccioli S, Pinoca F, Puxeddu A. Short-term efficacy of intra-articular injection of hyaluronic acid in osteoarthritis of the first carpometacarpal joint in a preliminary open pilot study. Clin Ter 2006; 157:321–325.
17. List T, Tegelberg A, Haraldson T, Isacsson G. Intra-articular morphine as analgesic in temporomandibular joint arthralgia/osteoarthritis. Pain 2001; 94:275–282.
18. Kim C-H, Lee B-J, Yoon J, Seo K-M, Park J-H, Lee J-W, Choi E-S, Hong J-J, Lee Y-S, Park J-H. Therapeutic effect of hyaluronic acid on experimental osteoarthrosis of ovine temporomandibular joint. J Vet Med Sci 2001; 63:1083–1089.
19. Kubota E, Kubota T, Matsumoto J, Shibata T, Murakami KI. Synovial fluid cytokines and proteinases as markers of temporomandibular joint disease. J Oral Maxillofac Surg 1998; 56:192–198.

20. Fuchs S, Erbe T, Fischer H-L, Tibesku CO. Intraarticular hyaluronic acid versus glucocorticoid injections for nonradicular pain in the lumbar spine. J Vasc Interv Radiol 2005; 16:1493–1498.
21. Van Den Bekerom MPJ, Mylle G, Rys B, Mulier M. Viscosupplementation in symptomatic severe hip osteoarthritis: a review of the literature and report on 60 patients. Acta Orthop Belg 2006; 72:560–568.
22. Qvistgaard E, Christensen R, Torp-Pedersen S, Bliddal H. Intra-articular treatment of hip osteoarthritis: a randomized trial of hyaluronic acid, corticosteroid, and isotonic saline. Osteoarthritis Cartilage 2006; 14:163–170.
23. Diamond MP. Reduction of de novo postsurgical adhesions by intraoperative precoating with Sepracoat (HAL-C) solution: a prospective, randomized, blinded, placebo-controlled multicenter study. The Sepracoat Adhesion Study Group. Fertil Steril 1998; 69:1067–1074.
24. Ballore L, Orru F, Nicolini F, Contini SA, Galletti G, Gherli T. Experimental results of the use of hyaluronic acid-based materials (CV Seprafilm and CV Sepracoat) in postoperative pericardial adhesions. Acta Biomed Ateneo Parmense 2000; 71:159–166.
25. Ozmen MM, Aslar AK, Terzi MC, Albayrak L, Berberoglu M. Prevention of adhesions by bioresorbable tissue barrier following laparoscopic intraabdominal mesh insertion. Surg Laparosc Endosc Percutan Tech 2002; 12:342–346.
26. Van't Riet M, de Vos Van Steenwijk PJ, Bonthuis F, Marquet RL, Steyerberg EW, Jeekel J, Bonjer HJ. Prevention of adhesion to prosthetic mesh: comparison of different barriers using an incisional hernia model. Ann Surg 2003; 237:123–128.
27. Delorenzi C, Weinberg M, Solish N, Swift A. A multicenter study of the efficacy and safety of subcutaneous non-animal stabilized hyaluronic acid in aesthetic facial contouring: interim report. Dermatol Surg 2006; 32: 205–211.
28. Lowe NJ, Grover R. Injectable hyaluronic acid implant for malar and mental enhancement. Dermatol Surg 2006; 32:881–885.
29. Verpaele A, Strand A. Restylane SubQ, a non-animal stabilized hyaluronic acid gel for soft tissue augmentation of the mid- and lower face. Aesthet Surg 2006; 26:S10–S17.
30. Pollack S. Some new injectable dermal filler materials: hylaform, restylane, and artecoll. J Cutan Med Surg 1999; 3(Suppl. 4):S27–S35.
31. Hallen L, Johansson C, Laurent C. Cross-linked hyaluronan (Hylan B gel): a new injectable remedy for treatment of vocal fold insufficiency—an animal study. Acta Otolaryngol 1999; 119:107–111.
32. Carruthers A, Carey W, de Lorenzi C, Remington K, Schachter D, Sapra S. Randomized, double-blind comparison of the efficacy of two hyaluronic acid derivatives, restylane perlane and hylaform, in the treatment of nasolabial folds. Dermatol Surg 2005; 31:1591–1598.
33. Manna F, Dentini M, Desideri P, de Pita O, Mortilla E, Maras B. Comparative chemical evaluation of two commercially available derivatives of hyaluronic acid (hylaform from rooster combs and restylane from streptococcus) used for soft tissue augmentation. J Eur Acad Dermatol Venereol 1999; 13:183–192.
34. Monheit GD. Hyaluronic acid fillers: hylaform and captique. Facial Plast Surg Clin North Am 2007; 15:77–84.

65. Shu XZ, Liu Y, Luo Y, Roberts MC, Prestwich GD. Disulfide cross-linked hyaluronan hydrogels. Biomacromolecules 2002; 3:1304–1311.
66. Maruyama A, Takei Y. Synthesis of polyampholyte comb-type copolymers consisting of poly(L-lysine) backbone and hyaluronic acid side chains for DNA carrier. Methods Mol Med 2001; 65:1–9.
67. Takei Y, Ikejima K, Enomoto N, Maruyama A, Sato N. Genetic manipulation of liver sinusoidal endothelial cells. Keio University International Symposia for Life Sciences and Medicine 2005; 13:129–134.
68. Takei Y, Maruyama A, Ferdous A, Nishimura Y, Kawano S, Ikejima K, Okumura S, Asayama S, Nogawa M, Hashimoto M, Makino Y, Kinoshita M, Watanabe S, Akaike T, Lemasters JJ, Sato N. Targeted gene delivery to sinusoidal endothelial cells: DNA nanoassociate bearing hyaluronan-glycocalyx. FASEB J 2004; 18:699–701.
69. Takei Y, Maruyama A, Ferdous A, Ikejima K, Enomoto N, Kitamura T, Nishimura Y, Asayama S, Nogawa M, Akaike T, Lemasters JJ, Sato N. Genetic manipulation of sinusoidal endothelial cells. Cells Hepatic Sinusoid 2001; 8:142–143.
70. Nogawa M, Asayama S, Akaike T, Takei Y, Maruyama A. Comb-type copolymer consisting of a poly(L-lysine) backbone and hyaluronic acid side chains for a cell specific gene carrier. Proc Int Symp Controlled Release Bioactive Mater 1999; 26:821–822.
71. Pike DB, Cai S, Pomraning KR, Firpo MA, Fisher RJ, Shu XZ, Prestwich GD, Peattie RA. Heparin-regulated release of growth factors in vitro and angiogenic response in vivo to implanted hyaluronan hydrogels containing VEGF and bFGF. Biomaterials 2006; 27:5242–5251.
72. Sharma NK, Levon K. Synthesis and characterization of bioconjugates of natural polymers and peptides for the detection of bacterial spores. Polym Prepr (Am Chem Soc, Div Polym Chem) 2004; 45:430–431.
73. Shu XZ, Ghosh K, Liu Y, Palumbo FS, Luo Y, Clark RA, Prestwich GD. Attachment and spreading of fibroblasts on an RGD peptide-modified injectable hyaluronan hydrogel. J Biomed Mater Res Part A 2004; 68A:365–375.
74. Leach JB, Schmidt CE. Characterization of protein release from photocrosslinkable hyaluronic acid-polyethylene glycol hydrogel tissue engineering scaffolds. Biomaterials 2004; 26:125–135.
75. Hahn SK, Jelacic S, Maier RV, Stayton PS, Hoffman AS. Anti-inflammatory drug delivery from hyaluronic acid hydrogels. J Biomater Sci, Polym Ed 2004; 15:1111–1119.
76. Tian WM, Zhang CL, Hou SP, Yu X, Cui FZ, Xu QY, Sheng SL, Cui H, Li HD. Hyaluronic acid hydrogel as Nogo-66 receptor antibody delivery system for the repairing of injured rat brain: in vitro. J Control Release 2005; 102:13–22.
77. Hou S, Tian W, Xu Q, Cui F, Zhang J, Lu Q, Zhao C. The enhancement of cell adherence and inducement of neurite outgrowth of dorsal root ganglia co-cultured with hyaluronic acid hydrogels modified with Nogo-66 receptor antagonist in vitro. Neuroscience (San Diego, CA, United States) 2006; 137:519–529.
78. Yard BA, Yedgar S, Scheele M, van Der Woude D, Beck G, Heidrich B, Krimsky S, van Der Woude FJ, Post S. Modulation of IFN-g-induced

immunogenicity by phosphatidylethanolamine-linked hyaluronic acid. Transplantation 2002; 73:984–992.
79. Ghosh K, Ren X-D, Shu XZ, Prestwich GD, Clark RAF. Fibronectin functional domains coupled to hyaluronan stimulate adult human dermal fibroblast responses critical for wound healing. Tissue Eng 2006; 12:601–613.
80. Kuhn W, Schinz H. Lavandulic acid and its transformation products. Helv Chim Acta 1952; 35:2008–2015.
81. Liu L-S, Ng C-K, Thompson AY, Poser JW, Spiro RC. Hyaluronate-heparin conjugate gels for the delivery of basic fibroblast growth factor (FGF-2). J Biomed Mater Res 2002; 62:128–135.
82. Oohira A, Kushima Y, Tokita Y, Sugiura N, Sakurai K, Suzuki S, Kimata K. Effects of lipid-derivatized glycosaminoglycans (GAGs), a novel probe for functional analyses of GAGs, on cell-to-substratum adhesion and neurite elongation in primary cultures of fetal rat hippocampal neurons. Arch Biochem Biophys 2000; 378:78–83.
83. Ohya S, Nakayama Y, Matsuda T. Thermoresponsive artificial extracellular matrix for tissue engineering: hyaluronic acid bioconjugated with poly(N-isopropylacrylamide) grafts. Biomacromolecules 2001; 2:856–863.
84. Trudel J, Massia SP. Assessment of the cytotoxicity of photocrosslinked dextran and hyaluronan-based hydrogels to vascular smooth muscle cells. Biomaterials 2002; 23:3299–3307.
85. Crescenzi V, Tomasi M, Francescangeli A. New routes to hyaluronan-based networks and supramolecular assemblies. In: Abatangelo G, Weigel PH, eds. New Frontiers in Medical Sciences: Redefining Hyaluronan. Amsterdam: Elsevier, 2000; 173–180.
86. Crescenzi V, Francescangeli A, Capitani D, Mannina L, Renier D, Bellini D. Hyaluronan networking via Ugi's condensation using lysine as cross-linker diamine. Carbohydr Polym 2003; 53:311–316.
87. Renier D, Bellato P, Bellini D, Pavesio A, Pressato D, Borrione A. Pharmacokinetic behaviour of ACP gel, an autocrosslinked hyaluronan derivative, after intraperitoneal administration. Biomaterials 2005; 26:5368–5374.
88. Wang W. A novel hydrogel crosslinked hyaluronan with glycol chitosan. J Mater Sci Mater Med 2006; 17:1259–1265.
89. Gajewiak J, Cai S, Shu XZ, Prestwich GD. Aminooxy pluronics: synthesis and preparation of glycosaminoglycan adducts. Biomacromolecules 2006; 7:1781–1789.
90. Monceviciūte-Eringiene E. Neoplastic growth: the consequence of evolutionary malignant resistance to chronic damage for survival of cells (review of a new theory of the origin of cancer). Med Hypotheses 2005; 65:595–604.
91. Fojo AT, Menefee M. Microtubule targeting agents: basic mechanisms of multidrug resistance (MDR). Semin Oncol 2005; 32:S3–S8.
92. Fojo T, Bates S. Strategies for reversing drug resistance. Oncogene 2003; 22:7512–7523.
93. Tan B, Piwnica-Worms D, Ratner L. Multidrug resistance transporters and modulation. Curr Opin Oncol 2000; 12:450–458.
94. Lautier D, Canitrot Y, Deeley RG, Cole SP. Multidrug resistance mediated by the multidrug resistance protein (MRP) gene. Biochem Pharmacol 1996; 52:967–977.

95. Rafferty JA, Hickson I, Chinnasamy N, Lashford LS, Margison GP, Dexter TM, Fairbairn LJ. Chemoprotection of normal tissues by transfer of drug resistance genes. Cancer Metastasis Rev 1996; 15:365–383.
96. Stewart J, Gorman NT. Multi-drug resistance genes in the management of neoplastic disease. J Vet Intern Med 1991; 5:239–247.
97. Tsuruo T. Multidrug resistance: a transport system of antitumor agents and xenobiotics. Princess Takamatsu Symp 1990; 21:241–251.
98. Brown T, Fox R. Patent WO 0205852.
99. Yin DS, Ge ZQ, Yang WY, Liu CX, Yuan YJ. Inhibition of tumor metastasis in vivo by combination of paclitaxel and hyaluronic acid. Cancer Lett 2006; 243:71–79.
100. Erickson AC, Barcellos-Hoff MH. The not-so innocent bystander: the microenvironment as a therapeutic target in cancer. Expert Opin Ther Targets 2003; 7:71–88.
101. Slevin M, Krupinski J, Gaffney J, Matou S, West D, Delisser H, Savani RC, Kumar S. Hyaluronan-mediated angiogenesis in vascular disease: uncovering RHAMM and CD44 receptor signaling pathways. Matrix Biol 2007; 26:58–68.
102. Peer D, Margalit R. Tumor-targeted hyaluronan nanoliposomes increase the antitumor activity of liposomal doxorubicin in syngeneic and human xenograft mouse tumor models. Neoplasia 2004; 6:343–353.
103. Peer D, Margalit R. Loading mitomycin C inside long circulating hyaluronan targeted nano-liposomes increases its antitumor activity in three mice tumor models. Int J Cancer 2004; 108:780–789.
104. Eliaz RE, Szoka FC, Jr.. Liposome-encapsulated doxorubicin targeted to CD44: a strategy to kill CD44-overexpressing tumor cells. Cancer Res 2001; 61:2592–2601.
105. Cevc G. Drug delivery across the skin. Expert Opin Investig Drugs 1997; 6:1887–1937.
106. Pirard D, Vereecken P, Melot C, Heenen M. Three percent diclofenac in 2.5% hyaluronan gel in the treatment of actinic keratoses: a meta-analysis of the recent studies. Arch Dermatol Res 2005; 297:185–189.
107. Rivers JK. Topical 3% diclofenac in 2.5% hyaluronan gel for the treatment of actinic keratoses. Skin Therapy Lett 2004; 9:1–3.
108. Rivers JK, Mclean DI. An open study to assess the efficacy and safety of topical 3% diclofenac in a 2.5% hyaluronic acid gel for the treatment of actinic keratoses. Arch Dermatol 1997; 133:1239–1242.
109. Jarvis B, Figgitt DP. Topical 3% diclofenac in 2.5% hyaluronic acid gel: a review of its use in patients with actinic keratoses. Am J Clin Dermatol 2003; 4:203–213.
110. Wolf JE, Taylor JR, Tschen E, Kang S. Topical 3.0% diclofenac in 2.5% hyaluronan gel in the treatment of actinic keratoses. Int J Dermatol 2001; 40:709–713.
111. Peters DC, Foster RH. Diclofenac/hyaluronic acid. Drugs Aging 1999; 14:313–319.
112. Moore AR, Willoughby DA. Hyaluronan as a drug delivery system for diclofenac: a hypothesis for mode of action. Int J Tissue React 1995; 17:153–156.
113. Seed MP, Brown JR, Freemantle CN, Papworth JL, Colville-Nash PR, Willis D, Somerville KW, Asculai S, Willoughby DA. The inhibition of

colon-26 adenocarcinoma development and angiogenesis by topical diclofenac in 2.5% hyaluronan. Cancer Res 1997; 57:1625–1629.
114. Freemantle C, Alam CA, Brown JR, Seed MP, Willoughby DA. The modulation of granulomatous tissue and tumour angiogenesis by diclofenac in combination with hyaluronan (HYAL EX-0001). Int J Tissue React 1995; 17:157–166.
115. Seed MP, Freemantle CN, Alam CAS, Colville-Nash PR, Brown JR, Papworth JL, Somerville KW, Willoughby DA. Apoptosis induction and inhibition of colon-26 tumor growth and angiogenesis: findings on COX-1 and COX-2 inhibitors in vitro & in vivo and topical diclofenac in hyaluronan. Adv Exp Med Biol 1997; 433:339–342.
116. Brown MB, Hanpanitcharoen M, Martin GP. An in vitro investigation into the effect of glycosaminoglycans on the skin partitioning and deposition of NSAIDs. Int J Pharm 2001; 225:113–121.
117. Mccarty MF. Glucosamine for wound healing. Med Hypotheses 1996; 47:273–275.
118. Taddeucci P, Pianigiani E, Colletta V, Torasso F, Andreassi L, Andreassi A. An evaluation of Hyalofill-F plus compression bandaging in the treatment of chronic venous ulcers. J Wound Care 2004; 13:202–204.
119. Vazquez JR, Short B, Findlow AH, Nixon BP, Boulton AJ, Armstrong DG. Outcomes of hyaluronan therapy in diabetic foot wounds. Diabetes Res Clin Pract 2003; 59:123–127.
120. Caravaggi C, de Giglio R, Pritelli C, Sommaria M, Dalla Noce S, Faglia E, Mantero M, Clerici G, Fratino P, Dalla Paola L, Mariani G, Mingardi R, Morabito A. HYAFF 11-based autologous dermal and epidermal grafts in the treatment of noninfected diabetic plantar and dorsal foot ulcers: a prospective, multicenter, controlled, randomized clinical trial. Diabetes Care 2003; 26:2853–2859.
121. Falanga V, Carson P, Greenberg A, Hasan A, Nichols E, McPherson J. Topically applied recombinant tissue plasminogen activator for the treatment of venous ulcers. Dermatol Surg 1996; 22:643–644.
122. Burns JW, Skinner K, Colt J, Sheidlin A, Bronson R, Yaacobi Y, Goldberg EP. Prevention of tissue injury and postsurgical adhesions by precoating tissues with hyaluronic acid solutions. J Surg Res 1995; 59:644–652.
123. Burns JW, Skinner K, Colt MJ, Burgess L, Rose R, Diamond MP. A hyaluronate-based gel for the prevention of postsurgical adhesions: evaluation in two animal species. Fertil Steril 1996; 66:814–821.
124. Esposito G, Gravante G, Filingeri V, Delogu D, Montone A. Use of hyaluronan dressings following dermabrasion avoids escharectomy and facilitates healing in pediatric burn patients. Plast Reconstr Surg 2007; 119:2346–2347.
125. Cass DL, Meuli M, Adzick NS. Scar wars: implications of fetal wound healing for the pediatric burn patient. Pediatr Surg Int 1997; 12:484–489.
126. Chung JH, Kim HJ, Fagerholmb P, Cho BC. Effect of topically applied Na-hyaluronan on experimental corneal alkali wound healing. Korean J Ophthalmol 1996; 10:68–75.
127. Costagliola M, Agrosì M. Second-degree burns: a comparative, multicenter, randomized trial of hyaluronic acid plus silver sulfadiazine vs. silver sulfadiazine alone. Curr Med Res Opin 2005; 21:1235–1240.

128. Frankova J, Kubala L, Velebny V, Ciz M, Lojek A. The effect of hyaluronan combined with KI3 complex (Hyiodine wound dressing) on keratinocytes and immune cells. J Mater Sci Mater Med 2006; 17:891–898.
129. Fam H, Bryant J, Kontopoulou M. Rheological properties of synovial fluids. Biorheology 2007; 44:59–74.
130. Ghosh P, Guidolin D. Potential mechanism of action of intra-articular hyaluronan in osteoarthritis: are its effects molecular weight dependant? Semin Arthritis Rheum 2002; 32:10–37.
131. Sierra R, Trousdale R, Pagnano M. Above-the-knee amputation after a total knee replacement: prevalence, etiology, and functional outcome. J Bone Joint Surg 2003; 85:1000–1004.
132. Heijink A, Yaszemski M, Patel R, Rouse M, Lewallen D, Hanssen A. Local antibiotic delivery with OsteoSet, DBX, and Collagraft. Clin Orthop Rel Res 2006; 451.
133. Matsuno H, Yudoh K, Hashimoto M, Himeda Y, Miyoshi T, Yoshida K, Kano S. Antibiotic-containing hyaluronic acid gel as an antibacterial carrier: usefulness of sponge and film-formed HA gel in deep infection. J Orthop Res 2006; 24:321–326.
134. Matsuno H, Yudoh K, Hashimoto M, Himeda Y, Miyoshi T, Yoshida K, Kano S. A new antibacterial carrier of hyaluronic acid gel. J Orthop Sci 2006; 11:497–504.
135. Fan H, Hu Y, Li X, Wu H, Lv R, Bai J, Wang J, Qin L, Fan H, Hu Y, Li X, Wu H, Lv R, Bai J, Wang J, Qin L. Ectopic cartilage formation induced by mesenchymal stem cells on porous gelatin-chondroitin-hyaluronate scaffold containing microspheres loaded with TGF-beta1. Int J Artif Organs 2006; 29:602–611.
136. Yamazaki T, Tamura J, Nakamura T, Tabata Y, Matsusue Y. Repair of full-thickness defects in rabbit articular cartilage using bFGF and hyaluronan sponge. Key Eng Mater 2004; 254–256:1099–1102.
137. Ng C, Daverman R, Merck A, Thompson A, Liu L, Pohl J, Heidaran M, Spiro R. Induction of chondro- and osteo-genesis by implanted collagen-hyaluronan biomatrix with GDF-5. In: Kennedy JF, Phillips GO, Williams PA, Hascall VC, eds. Hyaluronan. Cambridge: Woodhead Publishing, 2002; 109–112.
138. Berven S, Tay BK, Kleinstueck FS, Bradford DS. Clinical applications of bone substitutes in spine surgery: consideration of mineralized and demineralized preparations and growth factor supplementation. Eur Spine J 2001; 10:S169–S177.
139. Bolander M, Balain G. The use of demineralized bone matrix in the repair of segmental defects: augmentation with extracted matrix proteins and a comparison with autologous grafts. J Bone Joint Surg 1986; 68:1264–1274.
140. Einhorn T, Lane J, Burstein A. The healing of segmental bone defects induced by demineralized bone matrix: a radiographic and biomechanical study. J Bone Joint Surg Am 1984; 66:274–279.
141. Oakes D, Lee C, Lieberman J. An evaluation of human demineralized bone matrices in a rat femoral defect model. Clin Orthop 2003; 413:281–290.
142. Rabie A, Deng Y, Samman N, Hagg U. The effect of demineralized bone matrix on the healing of intramembranous bone grafts in rabbit skull defects. J Dent Res 1996; 75:1045–1051.

143. Shen W, Chung K, Wang G. Demineralized bone matrix in the stabilization of porous-coated implants in bone defects in rabbits. Clin Orthop 1993; 293:346–352.
144. Gertzman AA, Sunwoo MH. A pilot study evaluating sodium hyaluronate as a carrier for freeze-dried demineralized bone powder. Cell Tiss Bank 2001; 2:87–94.
145. Colnot C, Romero D, Huang S, Helms J. Mechanisms of action of demineralized bone matrix in the repair of cortical bone defects. Clin Orthop Relat Res 2005; 435:69–78.
146. Yee A, Bae H, Friess D. Augmentation of rabbit posterolateral spondylodesis using a novel demineralized bone matrix-hyaluronan putty. Spine 2003; 28:2435–2440.
147. Kim J, Kim I, Cho T, Lee K, Hwang S, Tae G, Noh I, Lee S, Park Y, Sun K. Bone regeneration using hyaluronic acid-based hydrogel with bone morphogenic protein-2 and human mesenchymal stem cells. Biomaterials 2007; 28:1830–1837.
148. Ellis H. The cause and prevention of post operative intraperitoneal adhesions. Surg Gynecol Obstet 1971; 133:497–511.
149. Wiebel M-A, Majno G. Peritoneal adhesions and their relation to abdominal surgery. Am J Surg 1973; 126:345–353.
150. Di Zerega G. The cause and prevention of postsurgical adhesions: a contemporary update. Prog Clin Biol Res 1993; 381:1–18.
151. Di Zerega G. Contemporary adhesion prevention. Fertil Steril 1994; 61:219–235.
152. Dobell A, Jain A. Catastrophic hemorrhage during redo sternotomy. Ann Thorac Surg 1984; 37:273–278.
153. Burns J, Colt M, Burgess L, Skinner K. Preclinical evaluation of Seprafilm™ Bioresorbable membrane. Eur J Surg Suppl 1997; 577:40–48.
154. Becker J, Dayton M, Fazio V, Beck D, Stryker S, Wexner S, Wolff B, Roberts P, Smith L, Sweeney S, Moore M. Prevention of postoperative abdominal adhesions by a sodium hyaluronate-based bioresorbable membrane: a prospective randomized, double-blind multicentre study. J Am Coll Surg 1996; 183:297–306.
155. Burt H, Jackson J, Bains S, Liggins R, Oktaba A, Arsenault A, Hunter W. Controlled delivery of taxol from microspheres composed of a blend of ethylenevinyl acetate copolymer and poly(d,l-lactic acid). Cancer Lett 1995; 88:73–79.
156. Rowinsky E, Cazenave L, Donehower R. Review: taxol: a novel investigational antimicrotubule agent. J Natl Cancer Inst 1990; 82:1247–1259.
157. Jackson JK, Skinner KC, Burgess L, Sun T, Hunter WL, Burt HM. Paclitaxel-loaded crosslinked hyaluronic acid films for the prevention of postsurgical adhesions. Pharm Res 2002; 19:411–417.
158. Cashman J, Burt H, Springate C, Gleave J, Jackson J. Camptothecin-loaded films for the prevention of postsurgical adhesions. Inflamm Res 2004; 53:355–362.

Carbohydrate Chemistry, Biology and Medical Applications
Hari G. Garg, Mary K. Cowman and Charles A. Hales

DOI: 10.1016/B978-0-08-054816-6.00016-1

Chapter 16

Carbohydrate Microarrays as Essential Tools of Postgenomic Medicine

XICHUN ZHOU,* GREGORY T. CARROLL,† CRAIG TURCHI* AND DENONG WANG‡

**ADA Technologies, Inc., 8100 Shaffer Parkway, Littleton, CO 80127, USA*
†Department of Chemistry, Columbia University, 3000 Broadway, MC 3157, New York, NY 10027, USA
‡Department of Genetics, Stanford Tumor Glycome Laboratory, Stanford University School of Medicine, Beckman Center B007, Stanford, CA 94305, USA

I. Introduction

Carbohydrates, like nucleic acids and proteins, are essential biological molecules carrying important biological information. Carbohydrates are prominently displayed on the surface of cell membranes and expressed by virtually all secretory proteins in bodily fluids. This is achieved by the events of posttranslational protein modification called glycosylation. Importantly, expression of cellular glycans, in the form of either glycoproteins or glycolipids, is differentially regulated. Cell display of precise complex carbohydrates is characteristically associated with the stages or steps of embryonic development, cell differentiation, as well as transformation of normal cells to abnormally differentiated tumor or cancer cells (12,22,23,28). Sugar moieties are also abundantly expressed on the outer surfaces of the majority of viral, bacterial, protozoan, and fungal pathogens. Many sugar structures are pathogen specific, which makes them important molecular targets for pathogen recognition, diagnosis of infectious diseases, and vaccine development (16,20,29,40,48,65).

Exploring the biological information content in carbohydrates is one of the current focuses of postgenomic research and technology development. Biophysical, biochemical, and immunological methods have proven very valuable in studying carbohydrate–carbohydrate and carbohydrate–protein interactions. For example, X-ray crystallographic and NMR spectroscopic techniques have been employed to determine binding modes between carbohydrates and proteins. Surface plasmon resonance spectroscopy and isothermal titration calorimetry (ITC) can provide information on the binding affinities of carbohydrates to proteins. Many well-established immunochemical methods have been applied to determine the specificity and cross-reactivity of carbohydrate–antibody and carbohydrate–lectin interactions. These classical approaches were, however, designed to monitor carbohydrate-based molecular recognition on a one-by-one basis and have limited analytical power or throughput in practical applications.

A pressing need is, thus, the establishment of high-throughput technologies to enable the large-scale, multiplex analysis of carbohydrates and their cellular receptors. These include especially the characterization of immunological properties of carbohydrates that are important for medical applications of carbohydrate antigens and interactions of carbohydrates with other biomolecules or intact cells that play key roles in establishing comprehensive biological functions of essentially all existing living organisms. In parallel with developing microarray-based high-throughput technologies for nucleic acids (5,15,49) and proteins (37,38,54), significant progress has been made in developing carbohydrate microarrays (1,21,25,32,46,64,67,68,70).

In this chapter, we attempt to illustrate a few examples, with a focus on infectious diseases, to discuss the medical application of carbohydrate microarrays. We also discuss the principles for construction of various platforms of carbohydrate microarrays. This information may be helpful in selecting the proper technologies to address biomedical questions related to carbohydrates.

II. Carbohydrate Microarrays as Essential Tools in the Postgenomics Era

In the past few years, a number of experimental approaches have been applied to construct carbohydrate microarrays (1,21,25,32,46,64,67,68,70). In spite of their technological differences, these carbohydrate microarrays are all solid-phase binding assays for carbohydrates and their interactions with other biological molecules. They share a number of common characteristics and technical advantages. First, they contain the capacity to display a large panel of carbohydrates in a limited chip space. Second, each carbohydrate is spotted in an amount that is drastically smaller than that required for a conventional molecular or immunological assay. Thus, the bioarray platform makes an effective use of carbohydrate substances. Third, they have high detection sensitivity. The microarray-based assays have higher detection sensitivity than most conventional molecular and immunological assays. This was attributed to the fact that the binding of a molecule in solution phase to an immobilized microspot of ligand in the solid phase has minimal reduction of the molar concentration of the molecule in solution (19). Therefore, in a microarray assay, it is much easier to have a binding equilibrium take place and result in a high sensitivity.

Carbohydrate microarrays constructed by various methods may differ in their technical features and suitability for a given practical application. Some platforms may be applied complementarily to solve biological questions. The method of nitrocellulose-based immobilization of carbohydrate-containing macromolecules, including polysaccharides, glycoproteins, and glycolipids, is suitable for the high-throughput construction of carbohydrate antigen microarrays (25,62,63,67,68). This platform of carbohydrate microarrays is readily applicable for the large-scale immunological characterization of carbohydrate antigens and anti-carbohydrate antibodies. It is also useful for the initial screening of carbohydrate-binding proteins, such as those newly identified by the human genome project with preserved carbohydrate-binding domains and are predicted to have carbohydrate-binding properties. However, the detection specificity of this carbohydrate microarray would be at the level of a carbohydrate antigen, not a glycoepitope, if the native carbohydrate antigens were spotted. This is owing to the fact that many carbohydrate antigens display multiple antigenic determinants or glycoepitopes. Examining the finer details of the binding properties would require the use of microarrays of defined oligosaccharide sequences. Oligosaccharide array-based binding assays can be applied, in combination with saccharide competition assays, to decipher precise saccharide components of a specific antigenic determinant or glycoepitope (25,64).

We present here a few examples of medical applications of carbohydrate microarray technologies. The studies summarized below involve the use of carbohydrate microarrays to study emerging infectious agents, including SARS-CoV, *Influenza virus A*, and *Bacillus anthracis* (*B. anthracis*). This progress highlights the potential of the relatively nascent carbohydrate microarray technologies in exploring the mysteries of life shrouded in the structure of carbohydrates. The areas that require carbohydrate microarrays are far beyond infectious diseases in medicine.

A. Recognition of Autoimmunogenic Reactivity of SARS-CoV

Severe acute respiratory syndrome (SARS) is an emerging infectious disease that became a global fear in 2003. A previously unrecognized coronavirus, SARS-CoV, is responsible for the epidemic spread of SARS (24,35). In an effort to understand the immunogenic properties of carbohydrate structures expressed by the virus, a carbohydrate microarray printed on nitrocellulose-coated microglass chips was applied (68). This study involves three steps of experimental investigation. The first step is to perform a carbohydrate microarray characterization of the antibody responses to the virus. The second step focuses on identification of lectins and/or antibodies that are specific for the glycoepitopes that are recognized by the pathogen-elicited antibodies. This provides specific structural probes to enable the third step of investigation, that is, to probe the glycoepitopes of the pathogens using specific lectins or antibodies identified by steps 1 and 2.

The first step, microarray analysis, revealed that immunization of horses with a preparation of inactivated SARS-CoV induced antibodies specific for an abundant human glycoprotein asialo-orosomucoid (ASOR). Since the horse antisera has no reactivity toward agalacto-orosomucoid (AGOR), which lacks galactose in the upper stream nonreducing ends, the glycoepitopes with terminal galactose may contribute significantly to the antigenic reactivity of SARS-CoV. This glycan array finding gave an important lead in terms of identifying appropriate immunological probes to further characterize the glycoepitopes expressed by SARS-CoV. A microarray containing a panel of galactose-containing complex carbohydrates was created to scan for immunological probes specific for the ASOR-glycans. The lectin PHA-L was shown to be highly specific for the spotted ASOR preparation. The latter is known to be specific for the glycoepitopes that are composed of tri-antennary Galβ1–4GlcNAc (Tri-II) or multiantennary Galβ1–4GlcNAc (m-II) (see Fig. 1 for an asialo-Tri-II structure of *N*-glycans) (72). With this specific probe, the authors characterized SARS-CoV-infected and -uninfected cells. PHA-L was found to stain cells infected with SARS-CoV and this reactivity could be inhibited by ASOR but not AGOR. The authors concluded, therefore, that the glycoepitopes Tri-II or m-II of ASOR are highly expressed by SARS-CoV-infected cells.

These observations raise important questions about whether autoimmune responses are in fact elicited by SARS-CoV infection in human and other animal species and whether such autoimmunity contributes to SARS pathogenesis. ASOR is an abundant human serum glycoprotein and the ASOR-type complex carbohydrates are also expressed by other host glycoproteins (13,43). Thus, the human immune system is generally nonresponsive to these "self" carbohydrate structures. However, when similar sugar moieties were expressed by a viral glycoprotein, their cluster configuration could differ significantly from those displayed by a cellular glycan, and in this manner, it generates a novel "nonself" antigenic structure. Much remains to be learned regarding the specificity and cross-reactivity of the carbohydrate-mediated molecular recognition and its role in the "self/nonself" immune discrimination.

B. Deciphering the Sugar Codes for Selective Viral Entry of Host Cells (Influenza A)

Glycan arrays have also been applied to study the interaction between the *Influenza virus A* and its cellular receptors (51–53). This virus recognizes specific saccharides on the host's epithelial cells and utilizes these cellular glycans as receptors to initiate an infection. An antigenic protein on the virus' coat, hemagglutinin (HA), recognizes sialic acid-terminated glycans (see Fig. 1 for sialic acid-terminated glycans in the Tri-II sugar chain configuration). In addition, HA can distinguish between different kinds of sialic acid–galactose linkages. For example, HA variants adapted to humans recognize an α2–6 linkage whereas strains specific for

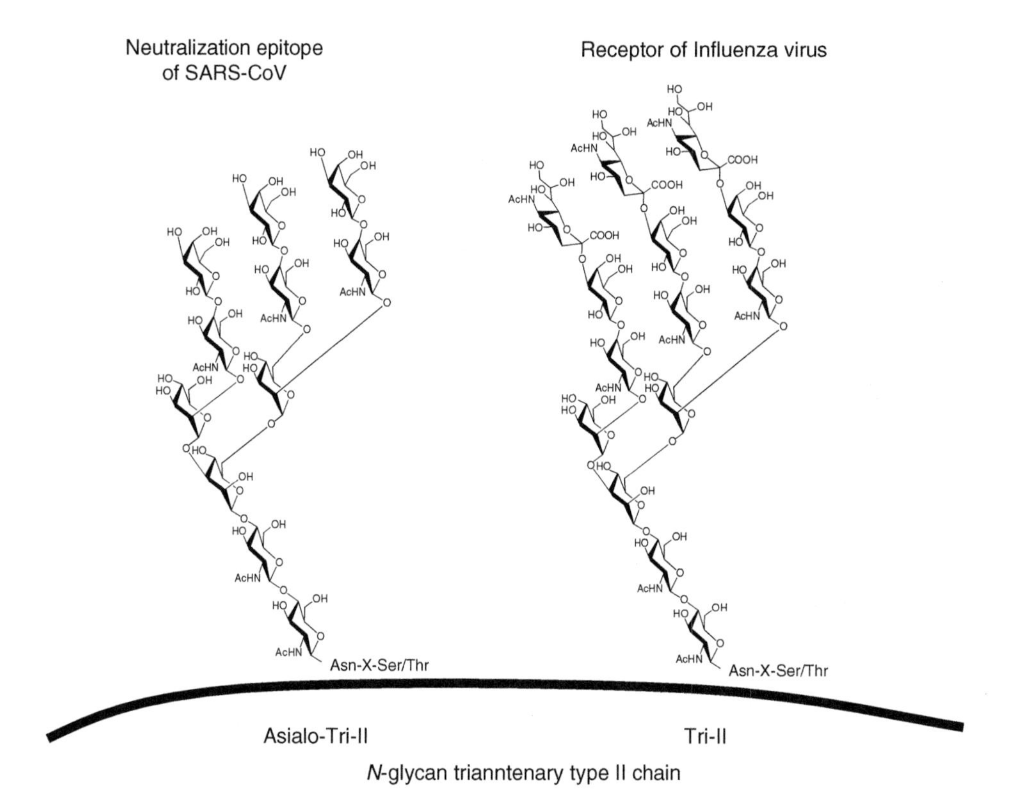

Figure 1 Characteristic carbohydrate moieties serve as markers for biological recognition. *N*-Glycan type II chains (Galβ1→4GlcNAc) in the tri-antennary cluster configurations with (Tri-II) and without sialic acid terminal residues (asialo-Tri-II). The asialo-Tri-II sugar moieties but not the Tri-II structures are specifically targeted by a horse-neutralization antibody of SARS-CoV (68).

birds recognize an α2–3 linkage. The specificity of a variety of HAs toward carbohydrates containing sialic acid residues attached via α2–3, α2–6, and α2–8 linkages was screened using a glycan array. The assay showed that specific mutations control the specificity of a given HA that selectively binds to a given linkage. The microarray could also be used to probe the effect of charge, size, sulfation, fucosylation, and sialylation on the binding specificity. The binding specificity between two previous pandemic strains of HA, 18NY and 18SC, could be distinguished by the microarray since the 18NY recognizes both α2–6 and α2–3 linkages whereas 18SC recognizes only α2–6 linkages. Only two mutations are required to make the avian strain sufficiently virulent toward humans. This study illustrates an example that glycan array analysis, in conjunction with mutation studies, helps in understanding and predicting how pathogenic strains can become virulent toward humans.

C. Identification of Immunogenic Sugar Moieties of *B. anthracis* Exosporium

Recent effort in hunting for the highly specific immunogenic sugar moiety of *B. anthracis* further demonstrates the potential of glycan array technologies in biomarker identification (64). *B. anthracis*, the etiological agent of anthrax infection, is a gram-positive, rod-shaped bacterium. The most lethal form of human anthrax is the pulmonary infection caused by inhaled spores. In view of the risk of *B. anthracis* spores as a biological weapon of mass destruction (WMD) (69), it is necessary to achieve the capacity for the rapid and specific detection of *B. anthracis* spores in various conditions (41,71). It is also important to develop new vaccines to block the anthrax infection at its initial stage before spore germination takes place (11,60). In this context, identification of highly specific immunogenic targets that are displayed on the outermost surfaces of *B. anthracis* spores is of utmost importance.

A photogenerated oligosaccharide array was introduced to facilitate identification and characterization of anthrax spore-specific immunogenic carbohydrate moieties (64). In essence, the authors utilized a photoactive surface on glass substrates for covalent immobilization and micropatterning of carbohydrates (7). They then applied this glycan array to probe specific antibodies that were elicited by immunizations using anthrax spores. The authors assumed that if *B. anthracis* spores express potent immunogenic carbohydrate moieties, immunization with the spores would elicit antibodies specific for these sugar structures. Such antibody reactivities would then be detected by glycan arrays that display the corresponding sugar structures. A schematic overview of this biomarker identification strategy is shown in Fig. 2.

This investigation demonstrates that IgG antibodies elicited by the native antigens of the *B. anthracis* spore recognize synthetic anthrose-containing sugar moieties. The saccharide-binding reactivities correlate directly with the sizes of the saccharides displayed by the glycan arrays. The terminal anthrose monosaccharide is marginally reactive and the anthrose-containing tetrasaccharides highly reactive, regardless of their anomeric configuration. However, the smaller saccharide units, including the anthrose mono-, di-, and trisaccharides, are potent inhibitors of the specific antibody reactivities to the tetrasaccharides displayed either by the photogenerated glycan arrays or by a bovine serum albumin (BSA) conjugate on an enzyme-linked immunosorbent assay (ELISA) microtiter plate. It was, thus, concluded that the anthrose-containing tetrasaccharide is immunogenic in its native configuration as displayed by the exosporium BclA glycoprotein and its terminal trisaccharide unit is essential for the constitution of a highly specific antigenic determinant.

Since the anthrose-containing carbohydrate moiety is displayed on the outermost surfaces of *B. anthracis* spores (14,55) and its expression is highly specific for the spore of *B. anthracis* (14), this unique tetrasaccharide is likely an important immunological target. Its applications may include identification of the presence of *B. anthracis* spores, surveillance and diagnosis of anthrax infection, and development of novel vaccines targeting the *B. anthracis* spore. In general, this glycan

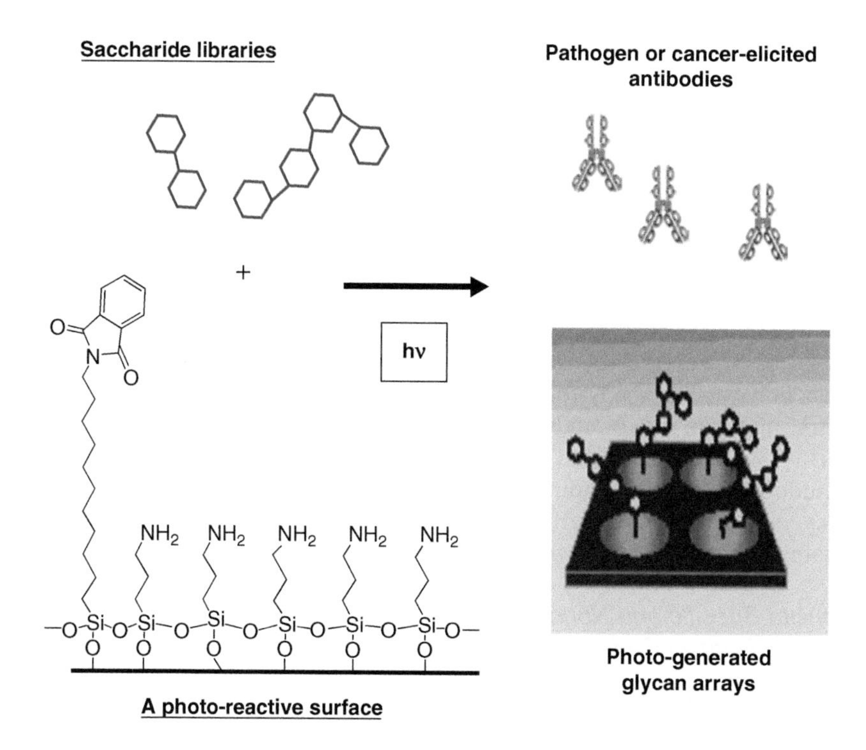

Figure 2 Photogenerated glycan arrays for rapid identification of pathogen-specific immunogenic sugar moieties. Saccharide preparations were dissolved in saline (0.9% NaCl) at a given concentration and spotted using a high-precision robot (PIXSYS 5500C, *Cartesian Technologies Irvine, CA*) onto the phthalimide-amine-coated slides. The printed slides were subjected to UV irradiation (300 nm) for 1 h to activate the photocoupling of carbohydrates to the surface. Pathogen-specific antisera were then applied on the glycan arrays to identify the potential immunogenic sugar moieties of a given pathogen (64).

array-based biomarker hunting strategy is likely applicable for exploring the immunogenic carbohydrate moieties of other microbial pathogens.

III. Progress in Developing Complementary Platforms of Carbohydrate Microarrays

In order to make the best use of available technologies of carbohydrate microarrays, it is important to conceptually understand the design and chemical principles of different carbohydrate microarray platforms. Different platforms may be technically complementary and can be applied in combination in addressing biomedical questions. For this purpose, we outline below a number of carbohydrate array platforms based on the chemical principle of array construction. These include technologies that directly utilize underivatized carbohydrates in microarray construction, technologies that require chemical modification of carbohydrates before microarray fabrication, methods of noncovalent immobilization of carbohydrates, and methods of covalent coupling of saccharides on array substrates. There are also technologies that are designed to display saccharides in defined orientations or specific cluster configurations in order to resemble the native configuration of functional carbohydrate ligands.

A. Carbohydrate Microarrays Fabricated by Using Underivatized Carbohydrates

The use of underivatized saccharides for microarray construction has the unique advantage of preserving the native structures of the carbohydrate molecules. It requires, however, a ready-to-use microarray surface with appropriate surface chemistry that can be directly used to fabricate comprehensive carbohydrate microarrays with underivatized carbohydrates from a wide range of sources. Methods include noncovalent binding of underivatized carbohydrate probes on a chip by passive adsorption and methods for covalently immobilizing underivatized carbohydrates on a slide surface by appropriate chemical-linking techniques.

1. Nonsite-Specific and Noncovalent Immobilization of Underivatized Carbohydrates in Microarrays

Noncovalent adsorption of native carbohydrate probes on a substrate surface is the simplest way to prepare carbohydrate microarrays. This method relies on the formation of a variety of noncovalent interactions between the surface and the arrayed carbohydrates. In addition to its simplicity and high-throughput characteristics in array construction, these approaches may be favorable in supporting the preservation of the native structure of spotted carbohydrate antigens since there is no need to modify the carbohydrates before microarray application. However, given that the saccharides are noncovalently immobilized on an array substrate, the efficiency of immobilization must be verified for each spotted carbohydrate.

Wang *et al.* (67) described a method of noncovalent immobilization of unmodified carbohydrates for construction of carbohydrate microarrays. They applied robotic microarray spotters to array the carbohydrates onto nitrocellulose-coated glass slides without any chemical modification. After air-drying at room temperature, the spotted arrays are ready for application. A wide range of carbohydrate antigens, including polysaccharides, glycoproteins, proteoglycans, and semisynthetic glycoconjugates, were tested by spotting them on the substrate and then probed with specific antibodies and lectins to verify the epitopes preserved in the carbohydrate microarrays.

The investigators showed that the nitrocellulose-coated glass chip is a ready-to-use substrate for carbohydrate microarrays by proving the fact that (i) carbohydrate-containing macromolecules of various structural configuration can be immobilized on a nitrocellulose-coated glass slide without chemical conjugation; (ii) the immobilized carbohydrate antigens are able to preserve their immunological properties and solvent accessibility; (iii) the system reaches the sensitivity, specificity, and capacity to detect a broad range of antibody specificities in clinical specimens; and (iv) this technology allows highly sensitive detection, as compared with other existing technologies, of the broad range of carbohydrate–lectin/antibody interactions. This strategy takes advantage of the existing cDNA microarray system, utilizing a spotter and a scanner for an efficient production and application of carbohydrate microarrays. In addition to the carbohydrate microarray application, this bioarray platform has been extended to spot microarrays of proteins (ProtoArray®, Invitrogen, CA) and cell lysates (8), as well as lectins for glycan profiling analysis (Procognia, United Kingdom).

However, this substrate is unlikely applicable for the immobilization of unmodified mono- and oligosaccharides. Using fluorescein-labeled preparations of α(1,6)dextrans, ranging from 20 to 2000 kDa, and inulin of 3.3 kDa, the authors investigated whether the size and molecular weight (MW) of saccharides influence their surface immobilization. They found that the efficiency of immobilization was dependent on the molecular mass of the spotted carbohydrates; the larger molecules were better retained than the smaller molecules. The reduced capacity of surface immobilization of smaller saccharide chains is likely owing to the fact that saccharide immobilization is based on passive interactions between spotted saccharides and the nitrocellulose-coated glass slides.

A practical way to compensate this weakness was described, which involves the use of glycoconjugates, either oligosaccharide–protein conjugates (67,68) or neoglycolipid (NGL) conjugates (25), for construction of the epitope-defined microarrays. In order to examine whether desired glycoepitopes or antigenic determinants are preserved after immobilization, the authors stained the microarrays using well-characterized monoclonal antidextran antibodies. These include antibodies bearing either the groove-type or the cavity-type antibody-combining sites (10,65). The former recognizes the internal linear chain of α(1,6)dextrans; while the latter is specific for the terminal nonreducing end structure of the polysaccharide. Results of this analysis confirmed that the desired glycoepitopes were well preserved by the spotted polysaccharide α(1,6)dextrans and by oligosaccharide–protein conjugates, that is, isomaltotriose (IM3) and isomaltoheptaose (IM7) coupled to BSA (67). Recently, Feizi's group further demonstrated the use of this platform for the construction of NGL-based oligosaccharide microarrays (25).

The nitrocellulose polymer substrate was a fully nitrated derivative of cellulose, in which the free hydroxyl groups are substituted by nitro groups, and is thus hydrophobic in nature. Researchers have shown that the immobilization of proteins on nitrocellulose surfaces relies on hydrophobic interactions. However, polysaccharides, being rich in hydroxyl groups, are hydrophilic in nature (42,61). The molecular forces for the carbohydrate–nitrocellulose interaction remain to be characterized, but it has been suggested that the three-dimensional (3D) microporous configuration of the nitrocellulose on the slides and the macropolymer characteristics of polysaccharides play important roles for the stable immobilization of many polysaccharides on the nitrocellulose surface. The polysaccharide molecules immobilized onto the nitrocellulose film are in a nonsite-specific format (Fig. 3).

A surface-modified polystyrene substrate provides another type of polymer surface that can be directly used to prepare carbohydrate microarrays through

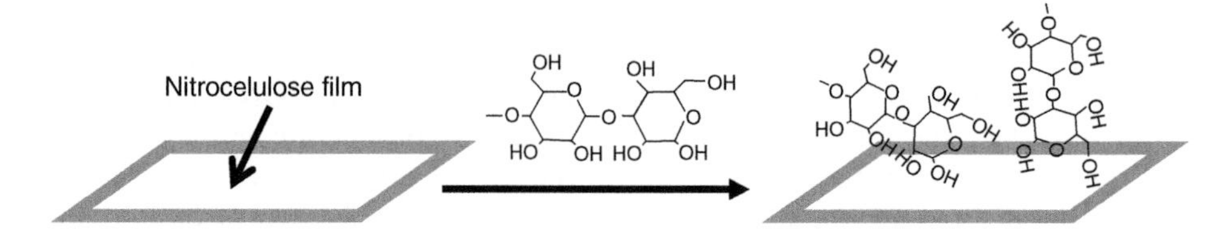

Figure 3 Noncovalent and nonspecific immobilization of underivatized carbohydrates on a nitrocellulose-coated slide.

the noncovalent immobilization of underivatized polysaccharides. The polystyrene substrate is produced by injection moulding of black polystyrene and the surface is modified by oxidation. This type of slide is commercially available from Nunc Roskilde, Denmark (http://www.nuncbrand.com/). Willats *et al.* of University of Leeds, United Kingdom, showed the applicability of this slide surface to produce comprehensive microarrays of polysaccharides, glycoproteins, proteoglycans, and cell extracts (70). These carbohydrate microarrays were directly fabricated by applying 50 pL of polysaccharide solution per spot on the black styrene substrate with a pitch of 375 μM. Probing these arrays using previously characterized monoclonal antibodies and a phage-derived antibody, the predicted patterns of antibody binding were observed.

2. *Covalent Immobilization of Underivatized Carbohydrates in Microarrays*

Covalent attachment is often preferred over other types of immobilization modules, such as those based on noncovalent bonds, including *van der Waals* forces, hydrogen bonds, hydrophobic forces, and ionic bonds in aqueous solutions, and various affinity-based binding reactions. Covalent bond formation provides a more stable linkage between the carbohydrate and the array substrate. Since the coupling efficiencies of the carbohydrate moieties are more readily controlled, the immobilization reproducibility is likely independent of the differences in the structures of carbohydrate probes.

A number of investigators have put effort in developing general, simple, and efficient array substrates that can be applied to a range of unprotected and unmodified oligosaccharides and polysaccharides. Four types of surface-functionalized substrates and related chemical-linking techniques have been reported to date for fabrication of carbohydrate microarrays through covalent immobilization of underivatized carbohydrates irrespective of the carbohydrate size.

Zhou *et al.* reported the slide surface containing aminooxy– groups as a platform for immobilizing an array of oligosaccharides through the formation of an oxime bond with the carbonyl group at the reducing end of a given carbohydrate (73). The use of aminooxyacetyl-terminated self-assembled monolayers for the immobilization of carbohydrates takes advantage of the oxime formation reaction between a highly reactive amine group of the nucleophilic aminooxyacetyl group and the carbonyl group at the reducing end of suitable carbohydrates via irreversible condensation.

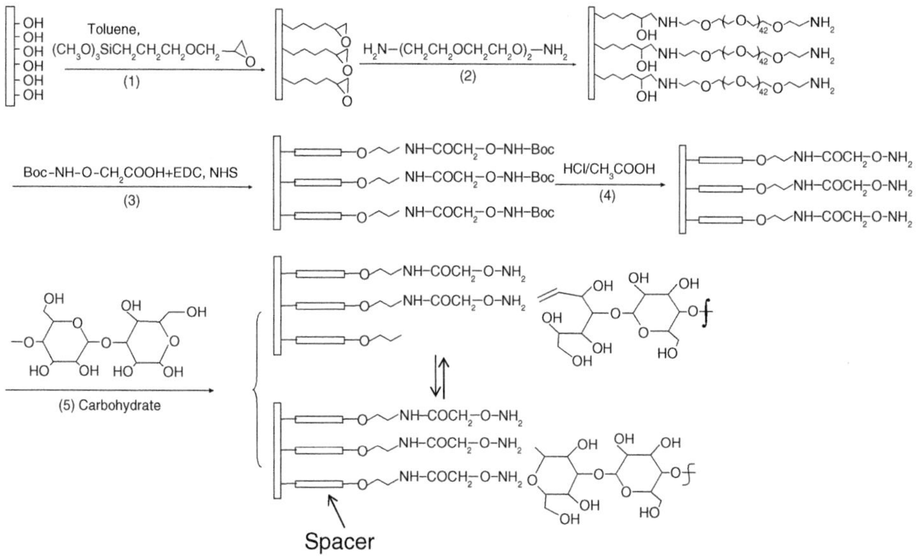

Figure 4 Chemical procedure for the preparation of aminooxyacetyl-functionalized glass slides and the immobilization of underivatized oligosaccharides. The aminooxyacetyl groups react selectively with the carbonyl group at the reducing end of carbohydrates via an irreversible condensation while the penta(ethylene glycol) groups serve as spacer arms and prevent the nonspecific adsorption of protein to the monolayer.

Aminooxyacetyl-functionalized glass slides were prepared in four steps starting with a (3-glycidyloxypropyl)trimethoxysilane (GPTS)-functionalized glass slide. The synthesis of the functionalized glass slide is presented in Fig. 4. The glycidyl group of the GPTS monolayer was treated with diamino-poly(ethylene glycol) (PEG), resulting in a PEG monolayer end-functionalized with an amine. The amine groups were then coupled to the carboxyl groups of an *N*-Boc-Aoa-OH that was activated with a hydroxyl succimide group. Free aminooxyacetyl groups were then obtained upon treatment of the glass slide with HCl/acetic acid in order to remove the Boc– group. The aminooxyacetyl groups on the slide surface reacted with formyl groups at the reducing ends of the oligosaccharides to form oxime bonds. In contrast to reductive amination, the sugar structure was preserved after coupling; equilibrium between the closed-ring and the open-ring forms might occur at the surface of the support. This chemical-linking technique reported requires only a few modification steps on the surface, allowing for the functional chips to be created in a timely manner making it an attractive method for preparing carbohydrate microarrays in individual laboratories.

The poly(ethylene glycol) layer on the glass slides provides essentially complete resistance to unwanted protein adsorption and other nonspecific interactions at the surface and ensures that only specific interactions between soluble proteins and immobilized ligands occur. The poly(ethylene glycol) containing carbohydrate microarrays showed lower background signal even without commonly used blocking procedures such as treating the substrate with bovine serum albumin or other blocking proteins to passivate the surface. This excellent control over unwanted adsorption has also been reported with monolayers presenting oligo(ethylene glycol) groups (9,30). The poly(ethylene glycol) also functions as a spacer between the carbohydrate and the substrate, which is expected to increase the accessibility of proteins to the binding site of the carbohydrates.

Since the carbohydrates are chemically linked to the aminooxy-functionalized substrate surface through the reducing end, the carbohydrate moieties are immobilized in a well-defined orientation (i.e., site-specific immobilization). The main advantages of an aminooxy-functionalized substrate for the fabrication of carbohydrate microarrays include (i) the ease of formation of oxime bonding under mild conditions between the underivatized carbohydrates and the slide surface; (ii) good stability of the oxime bonding under a wide range of pH; (iii) the monosaccharides are in a ring-closed format which will not affect protein binding in an irrelevant manner, allowing for a more accurate evaluation of the protein-binding function of the carbohydrate.

To demonstrate the utility of this chemistry for the immobilization of carbohydrates and the use of the arrayed carbohydrates for parallel determination of protein–carbohydrate interactions, Zhou *et al.* (73) printed 10 oligosaccharides on the aminooxyacetyl-functionalized glass slide. The arrayed substrates were kept in a humidified chamber at room temperature overnight, washed with water, and dried. These conditions permitted near quantitative immobilization using minimal quantities of carbohydrate conjugates. After incubation and washing away the unbound oligosaccharide, the remaining aminooxyacetyl groups on the substrate were inactivated by treatment of the glass slides with succinic anhydride [10 mM in dimethylformamide (DMF)] overnight followed by rinsing with DMF to remove physically adsorbed succinic anhydride.

To investigate the carbohydrate–protein-binding properties of the fabricated carbohydrate microarrays, identical arrays were treated separately with three biotin-labeled lectins [with a concentration of 2 μM in phosphate-buffered saline/tween (PBST)] for 2 h, and then washed with PBST. Detection of the bound analyte was subsequently achieved by incubating the microarray with Cy3-streptavidin at a final concentration of 5 μg/ml, and then imaged with a confocal array scanner. Figure 5 shows the results of the analyte characterization on the carbohydrate microarray. As expected, the oligosaccharides were found to bind to their specific lectin proteins. For example, the carbohydrate array probed with Con A showed significant fluorescence intensity in the spots arrayed with mannose, glucose, and *N*-acetylglucosamine (GlcNAc) (Fig. 5A). Analysis of the fluorescence intensity further reveals that the binding of Con A to the oligosaccharides is in the order of mannose $>$ glucose $\geq$ GlcNAc. The affinity binding difference of the arrayed oligosaccharides is consistent with solution-phase assays. A weak signal was obtained in the spots that were arrayed with maltooligosaccharide which has 4–10 units of α-glucose. This could be the result of an inefficient immobilization of the sugar on the substrate due to the reducing activity of the formyl groups of the

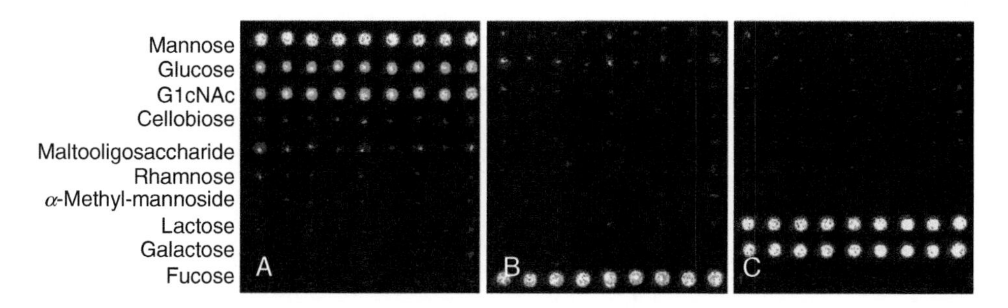

Figure 5 Fluorescence image of oligosaccharide microarrays spotted with 10 oligosaccharide probes with identical carbohydrate chips that were separately incubated with each of 3 biotin-labeled lectins (0.1 mg/ml in PBST) for 1 h, washed with PBST 3 times for 5 min each, and stained with 5 μg/ml of Cy3-strptavidin and evaluated by confocal fluorescence microscopy. Fluorescence images of oligosaccharide microarrays probed with (A) concanavali A (Con A), (B) LoTus Tetragonolobus (LT), and (C) Erythrina Gristagalli (EC).

maltooligosaccharide. In addition, no signal was observed in the spots arrayed with methyl-α-mannoside in which the C1 position was substituted with a methyl group. This result suggests that methyl-α-mannoside could not be immobilized on the glass surface. Probing the microarrays with the two other lectins also gave the expected results: *Lotus tetragonolobus* bound to the spots presenting α-fucose (Fig. 5B), whereas *Erythrina cristagalli* bound only to spots presenting lactose and galactose (Fig. 5C). Nonspecific adsorption was not observed on the spots arrayed with cellobiose and rhamnose and essentially no fluorescence was obtained from surfaces without carbohydrates. The weak signal obtained from the glucose spots when the microarray was probed with *L. tetragonolobus* may be caused by the weak cross-reaction of the lectin *L. tetragonolobus.* However, this weak cross-interaction that gave less than 8% of signal compared with the specific interaction would not affect the rapid determination of the presence or absence of specific carbohydrate epitopes.

Overall, these results demonstrated that the binding of lectins with the prepared carbohydrate microarrays are specific and multiple-analytic characterization can be achieved on the aminooxyacetyl-functionalized slide with good selectivity. Furthermore, periodate oxidation of the immobilized oligosaccharides with $NaIO_4$ resulted in the loss of lectin binding. These experiments verify that the fabricated carbohydrate microarray is well suited for the selective identification of carbohydrate-binding proteins.

Figure 6 shows the dose–response curves of lectins applied to the fabricated oligosaccharide microarrays. It was apparent that an increase in lectin concentration resulted in a corresponding increase in the fluorescence emitted from the arrayed spots, and saturation of affinities was obtained at high concentration. The calculated limit of detection (LOD, the concentration which gives a fluorescent signal higher than the background + three standard deviation units) was determined to be approximately 0.008 μg/ml for Con A, which is lower than the microtiter plate assay developed by Hatakeyama.

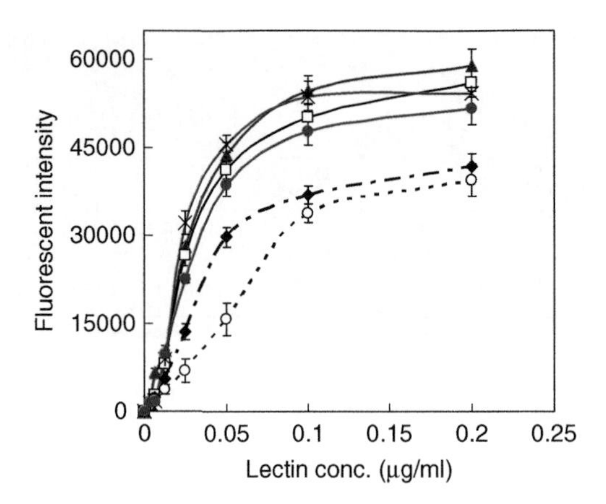

Figure 6 Dose–response binding curve of lectins to the arrayed oligosaccharides. Con A binding on mannose (▲), glucose (♦), and G1cNAc (O); EC on lactose (×) and galactose (□); LT on fucose (•). Each data point represents the average value of the mean signal±SD (standard deviation) of 18 replicate spots from 2 slides.

Carbohydrate arrays have the characteristics required for quantitative assays of multiple protein–carbohydrate interactions with minimal quantities of reagents. To assess the utilization of carbohydrate microarrays fabricated on aminooxyacetyl-functionalized slides for quantitative assays, α-methyl mannose was applied to inhibit Con A binding to the immobilized mannose spots. A series of mixtures containing biotin-labeled Con A (2 μM in PBST) and α-methyl mannose (0–4 mM in PBST) was applied on the microarray surface and incubated for 1 h at 25°C. The substrates were rinsed with PBST, stained with 5 μg/ml of Cy3-streptavidin, and analyzed with a fluorescence scanner to quantify the amount of bound Con A on the spots of mannose and glucose. The amount of lectin that bound to the chips for each concentration of soluble ligand (i.e., α-methyl mannose) is shown in Fig. 7. The IC_{50} (concentration of inhibitor required to prevent 50% of lectin binding to array spot) was determined using α-methyl mannose as an inhibitor of Con A binding to glucose and mannose. The results verified that the microarray spots of mannose (IC_{50}=60 μM) competed more effectively with the soluble carbohydrate for Con A than that of glucose (IC_{50}=23 μM). The relative binding affinities of these carbohydrates for Con A is consistent with those obtained in previous studies (31).

In a similar method, Lee and Shin (36) have developed hydrazide-coated glass slides to immobilize a wide range of carbohydrates including mono-, di-, and oligosaccharides in a simple, efficient, and chemoselective fashion. Preliminary protein-binding experiments show that carbohydrate microarrays prepared by this method are suitable for the high-throughput analysis of carbohydrate–protein interactions.

The advantages of the above two methods for covalent immobilization of underivatized carbohydrates rely on the ease of formation and on the good

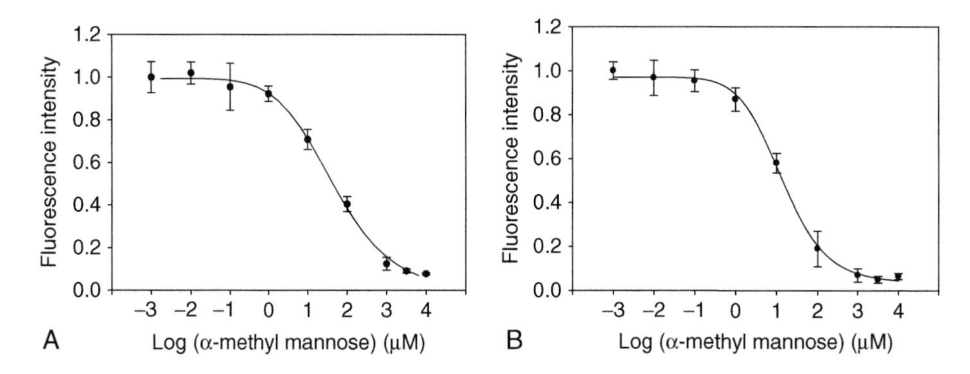

Figure 7 Quantitative inhibition assays in oligosaccharide microarray. (A) Determination of the concentration of soluble α-methyl mannose to inhibit 50% of Con A binding to spotted mannose. (B) Determination of the concentration of α-methyl mannose to inhibit 50% of Con A binding to spotted glucose. Each data point represents the mean±SD for 10 spots from 2 independent experiments.

stability of the oxime linkage and hydrazide linkage for oligosaccharides up to pH 9. However, reactions of aminooxy or hydrazide groups with free carbohydrates are slow when the carbohydrate MW increases because of the reducing activity of the aldehyde group of the carbohydrates. To improve the immobilization of larger carbohydrates, such as polysaccharides, on the aminooxy- and hydrazide-functionalized slide substrates, Zhou *et al.* have been investigating the utilization of microwave radiation energy to facilitate the fabrication of carbohydrate microarrays on a 3D polymer film bearing aminooxy- and hydrazide-functional groups (Zhou X., personal communication). Using microwave radiation to facilitate the reaction, oligosaccharides and polysaccharides can be covalently linked on the aminooxy- and hydrazide-functionalized surface within minutes.

Carroll *et al.* developed a method for covalent immobilization of underivatized mono-, oligo-, and polysaccharides onto glass substrates functionalized with self-assembled photoactive phthalimide chromophores (7). Upon exposure to UV light, the photoactive aromatic carbonyls presumably react with the C–H groups of the sugars by hydrogen abstraction followed by radical recombination to form a covalent bond (Fig. 8). Immobilization of unmodified carbohydrates by this approach was demonstrated to be much less dependent on the MWs of the spotted carbohydrates compared to a nitrocellulose-coated slide. Furthermore, for oligosaccharides the grafting efficiency was shown to be much higher than nitrocellulose. However, the method of photocoupling, which is expected to target any CH– group on the sugar rings with varying specificity depending on the structure of the ring, may interfere with the protein-binding specificity of monosaccharides. Wang *et al.* recently used this approach to generate a glycan array containing a large panel of synthetic carbohydrates and characterized their antigenic reactivities with pathogen-specific antibodies (64). As described in Section I, this investigation led to the discovery of a highly specific carbohydrate moiety of *B. anthracis* spores.

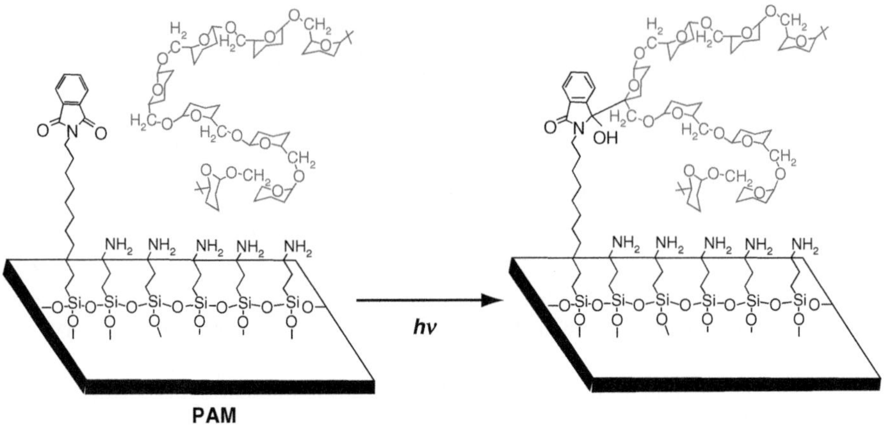

Figure 8 Covalent immobilization of underivatized carbohydrates on a phthalimide-functionalized slide surface by photo-immobilization.

Another type of photoactive microarray platform, based on dextran-coated glass slides (PhotoChips from CSEM, Switzerland) was reported by Sprenger's group (2). The dextran-based polymer OptoDex is functionalized with aryl-trifluoromethyl-diazirine groups. On illumination, aryl-trifluoromethyl-diazirine groups form reactive carbenes, which can undergo a variety of reactions with a vicinal molecule that result in covalent bond formation including insertion into σ and π bonds, addition of a nucleophile or electrophile and hydrogen abstraction. The authors have demonstrated that this substrate immobilizes polysaccharides and glycoligands. However, since the aryl-trifluoromethyl-diazirine-functionalized surface can react with any type of biomolecule, this type of array substrate is not suitable for preparing carbohydrate microarrays with unpurified carbohydrate extractions from cells or plants. Precaution must be made when applying this platform for serological studies since antidextran natural antibodies are frequently detected in human circulation (33,67).

The above slide surfaces and linking techniques provide the feasibility to fabricate microarrays of carbohydrates by using underivatized carbohydrate moieties. These methods are especially useful when working with complex oligosaccharides isolated from natural sources and when derivatized carbohydrates are not available. In many cases, the glycoepitopes contained in these microarrays are reactive toward appropriate antibodies, lectins, or other carbohydrate-binding partners of defined carbohydrate-binding specificities. However, these methods of saccharide immobilization are not expected to be site specific. Instead, the carbohydrates are attached without control of the orientation of saccharide display. A given saccharide spotted may, thus, present a glycoepitope in a spectrum of different configurations, although one or a few might be predominant. Thus, in order to further characterize the fine specificity of carbohydrate binding, especially the orientation effect of epitope display, one may want to explore the technologies described in the subsequent sections.

B. Carbohydrate Microarrays Fabricated by Using Derivatized Carbohydrates

Derivatized carbohydrates, termed glycoligands, are carbohydrate moieties with functional tags prepared by chemical modification. Glycoligands provide more flexibility in the selection of array substrates and chemical-linking techniques for carbohydrate microarrays. Most importantly, the use of glycoligands in combination with properly functionalized surfaces allows for the site-specific immobilization of carbohydrates onto the substrates. With these technical features, it is possible to construct carbohydrate microarrays with control over the ways of presentation of carbohydrate moieties for molecular recognition. These characteristics are important for achieving the specificity or selectivity of carbohydrate–protein interactions that play importance roles in cell–cell communication, signaling, and modulation of immune responses (12,39,65). Microarray presentation of the native configurations of glycoepitopes is likely a challenging issue that requires substantial and relatively long-term collaborative efforts by carbohydrate researchers and microarray experts.

Specific technical considerations in exploring this approach may include (i) the feasibility of preparing carbohydrate derivatives; (ii) the spacer between the glycoligands and the slide surface should provide optimal presentation of glycans and prevent nonspecific binding of proteins; (iii) the suitability of materials comprising the chip, for example, a functionalized glass slide versus a metallic surface; and (iv) the availability of tagged carbohydrate ligands for desired chip substrates.

The Consortium for Functional Glycomics (www.functionalglycomics.org) (3) has provided remarkable support to this field, building a library of about 200 synthetic glycoligands, which represent the most typical terminations and core fragments of mammal glycoproteins and glycolipids. A similar set of biotinylated oligosaccharides (~180 in total) are also available in the Consortium (www.functionalglycomics.org) (4). The number of described and well-characterized 2-aminopyridine derivatives of *N*-glycans (56–58) reaches several hundred; many of these derivatives are commercially available. With the advances being made in chemical and chemoenzymatic syntheses, increasing numbers of carbohydrate derivatives of known oligosaccharide sequences will become available for fabrication of carbohydrate microarrays.

1. Noncovalent Immobilization of Derivatized Carbohydrates in Microarrays

Because of the small molecular size and hydrophilic nature, most oligosaccharides cannot be directly immobilized onto nitrocellulose or black polystyrene surfaces for microarray applications. The oligosaccharide probe can be modified with a tag or coupled to a larger carrier molecule for noncovalent immobilization. A research group led by Feizi has developed oligosaccharide microarrays by noncovalently immobilizing NGLs on nitrocellulose (25,44). The oligosaccharides were obtained by chemical or enzymatic methods by using glycoproteins, glycolipids, proteoglycans, polysaccharides, or whole organs, or from chemical synthesis. The chemical derivatives of the oligosaccharides were synthesized by reductive amination of the oligosaccharides to the amino phospholipid 1,2-dihexadecyl-sn-glycero-3-phosphoethanolamine or its anthracene-containing fluorescent analogue. The immobilization efficiency of the NGLs on nitrocellulose was found to be high irrespective of the size of carbohydrates. The carbohydrate-binding proteins were investigated with known monoclonal antibodies, the E- and L-selectins, a chemokine (RANTES), and a cytokine. Binding was detected by colorimetric ELISA-type methods. It was shown that carbohydrate-binding proteins could single out their ligands, not only in arrays of homogeneous, structurally defined oligosaccharides but also in an array of heterogeneous *O*-glycan fractions derived from brain glycoproteins. The unique feature of this carbohydrate microarray technology is that deconvolution strategies are included with mass spectrometry for further determining the sequences of ligand-positive components within mixtures.

Wong's group developed a method for fabricating oligosaccharide arrays, which is a noncovalent but site-specific immobilization. In essence, they applied aliphatic derivatives of monosaccharides and oligosaccharides onto a polystyrene 96-well microtiter plate (6). They found that the carbohydrates were efficiently immobilized when the saturated hydrocarbon chain was between 13 and 15 carbons in length. Several di- to hexasaccharides containing terminal galactose, glucose, and/or fucose residues were chemically modified with a C_{14}-saturated hydrocarbon chain. Figure 9 illustrates the attachment of the modified carbohydrates to microtiter plate surfaces. All the sugars were stable after repeated washings and elicited the predicted binding signals with the lectins Ricinus B chain, Con A, and *Tetragonolobus purpurea*.

In addition, Wong and colleagues (6,21) reported that azide-derivatized forms of galactose and several azide-derivatized neutral and sialic acid containing di- to tetrasaccharides were immobilized onto aliphatic alkyne-coated plastic microtiter plate surfaces. These saccharides were immobilized on the surfaces by a 1,3-dipolar cycloaddition reaction between the azide and alkyne groups (Fig. 10). The noncovalent attachment also allowed convenient characterization of the lipid-linked products by mass spectrometry, as well as the detection of lectin binding. Using Guanosine diphosphate (GDP)-fucose and α-1,3-fucosyltransferase, fucosylation of sialyl-*N*-acetyllactosamine was carried out within the wells, showing that the surface is well suited for the high-throughput identification of enzyme inhibitors.

2. *Covalent Immobilization of Derivatized Carbohydrates in Microarrays*

Several other types of carbohydrate derivatives have been used for the fabrication of carbohydrate microarrays. Thiolated carbohydrate derivatives were immobilized on heterogeneous self-assembled monolayers that present maleimide end-functionalized and OH end-functionalized penta(ethylene glycol) chains on glass slides (31,47). The maleimide group provides an appropriate functionality that reacts efficiently with thiol-terminated glycoligands, whereas the penta(ethylene glycol) chains prevent the nonspecific adsorption of protein to the substrate. The penta(ethylene glycol) chain also works as a spacer arm to reduce the steric hindrance during protein binding to carbohydrates on the surface.

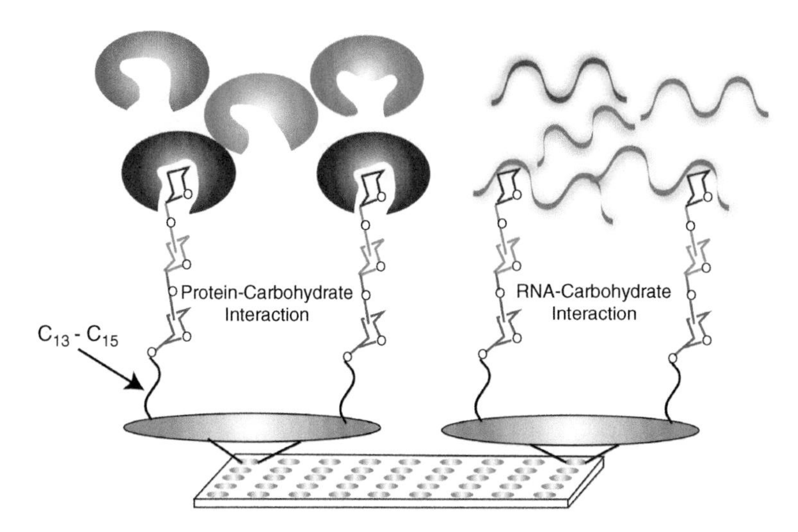

Figure 9 Carbohydrate microplate arrays prepared by the noncovalent immobilization of aliphatic alkyne-derivatized carbohydrates to microtiter plate surfaces. Carbohydrates can then be screened against a variety of biologically important substrates such as lectins and RNA.

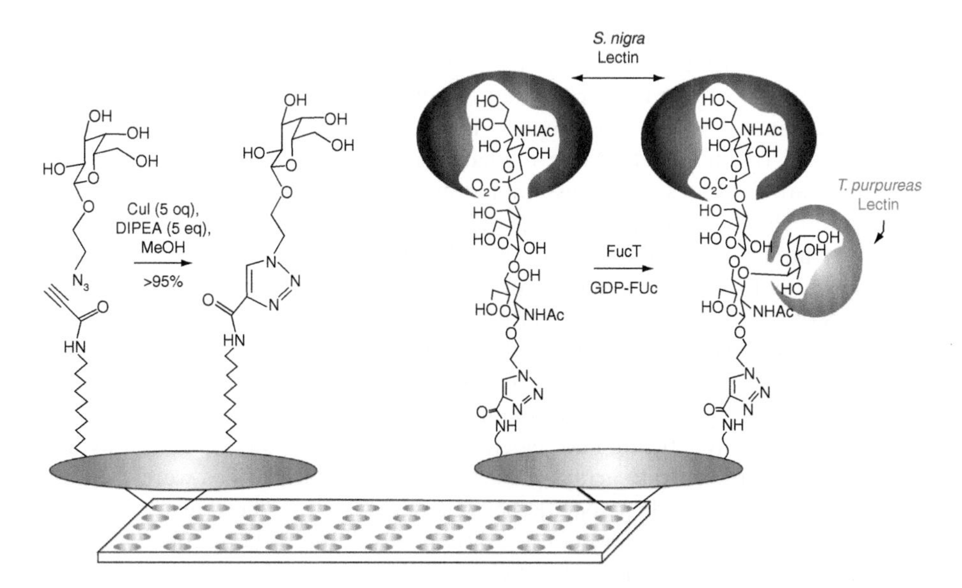

Figure 10 Carbohydrate microplate arrays prepared by the noncovalent immobilization of azide-derivatized carbohydrates to microtiter plates via a 1,3-dipolar cycloaddition reaction between alkynes and azides. Carbohydrates displaying terminal azides can be captured on microtiter plate surfaces through a terminal alkyne attached to a long, aliphatic tether and screened directly on the microtiter plate surface.

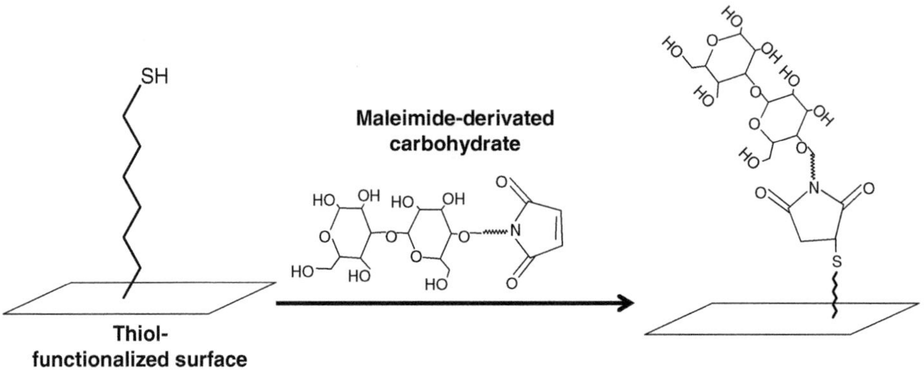

Maleimide-derivated carbohydrate

Thiol-functionalized surface

Figure 11 Carbohydrate microarrays prepared by covalent immobilization of maleimide-derivatized carbohydrates onto a thiol-functionalized substrate surface.

In contrast to the above approach, Shin's group prepared carbohydrate microarrays by covalent immobilization of maleimide-derivatized carbohydrates to thiol-functionalized glass slides (45,46) (see Fig. 11). Lectin-binding experiments showed that carbohydrates with different structural features selectively bound to the corresponding lectins with relative binding affinities that correlated with those obtained from solution-based assays. The author also demonstrated the fabrication of carbohydrate microarrays that contained more diverse carbohydrate probes. Enzymatic glycosylations on glass slides were consecutively performed to generate carbohydrate microarrays that contained the complex oligosaccharide, sialyl Le^x.

Mrksich and coworkers (32) reported a chemical strategy for preparing carbohydrate arrays by the Diels–Alder-mediated immobilization of cyclopentadiene-derivatized carbohydrates to self-assembled monolayers that present benzoquinone and penta(ethylene glycol) groups (Fig. 12). Modification of the gold surface was initiated by immersing gold-coated glass slides into a mixture of alkanethiols with (1%) and without (99%) appended hydroquinone groups to produce self-assembled monolayers of hydroquinone and penta(ethylene glycol) groups. Chemical or electrochemical oxidation was then performed to convert hydroquinone to benzoquinone groups. Finally, the cyclopentadiene-derivatized monosaccharides were covalently immobilized on the gold surface through the Diels–Alder reaction. This reaction was found to be highly efficient and selective for the immobilization of carbohydrates on the surface. Carbohydrate arrays presenting 10 monosaccharides were then evaluated by profiling the binding specificities of several lectins. These arrays were also used to determine the inhibitory concentrations of soluble carbohydrates for lectins and to characterize the substrate specificity of β-1,4-galactosyltransferase.

Blixt *et al.* (3) constructed a diverse glycan microarray by using standard robotic microarray printing technology to couple amine-derivatized glycoligands to an *N*-hydroxysuccinimide (NHS)-functionalized slide. The array comprises 200 synthetic and natural glycan sequences representing major glycan structures of glycoproteins and glycolipids. This array uses commercially available amine-reactive

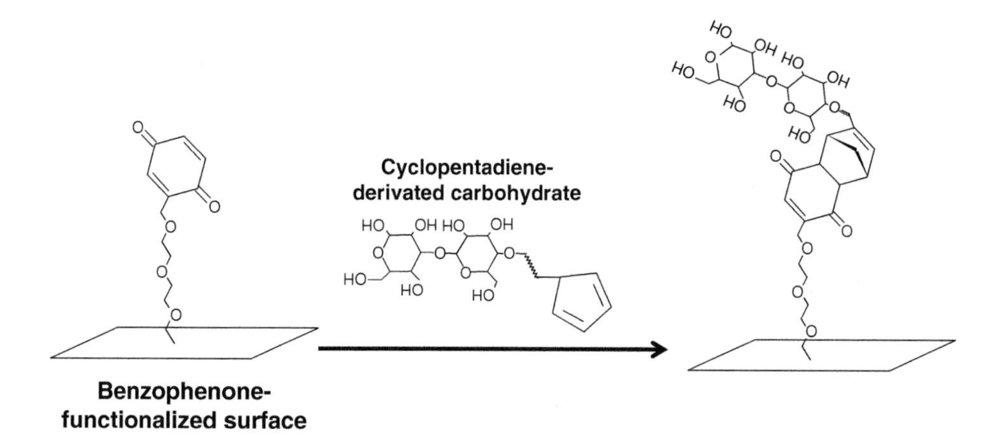

Figure 12 Carbohydrate microarrays prepared by covalent immobilization of cyclopentadiene-derivatized carbohydrates to a benzoquinone-functionalized substrate.

NHS-functionalized glass slides, which allow rapid covalent coupling of amine-functionalized glycans or glycoconjugates. The fabricated glycan microarray has shown utility for profiling the specificity of a diverse range of glycan-binding proteins, including C-type lectins, siglecs, galectins, anti-carbohydrate antibodies, lectins from plants and microbes, and intact viruses.

A microarray substrate for covalent immobilization of aminophenyl-derivatized carbohydrates is commercially available (GlycoChip, Glycominds, Lod, Israel). This substrate is functionalized with an oligomer of 1,8-diamino-3,6-dioxaoctan. Schwarz *et al.* reported the application of this substrate to fabricate oligosaccharide microarrays by using *p*-aminophenyl-derivatized carbohydrates via a cyanurchloride-activated linker (50). This approach allows the covalent attachment of glycans containing a terminal aliphatic amine by forming an amide bond under aqueous conditions at room temperature. The fabricated oligosaccharide microarray was used to analyze the glycan-binding antibody repertoire in a pool of affinity-purified IgG collected from a healthy human population. In addition, a novel anti-cellulose antibody was detected that binds specifically to β4-linked saccharides with a preference for glucopyranose over galactopyranose residues with the oligosaccharide microarray.

The group led by Waldmann has prepared carbohydrate microarrays by using Staudinger reactions between phosphane-functionalized glass slides and azide-derivatized carbohydrate moieties (34). The glass slide surface was first functionalized with polyamidoamine (PAMAM) dendrimers bearing 64-aminofunctional groups with the purpose of maximizing potential reactive sites on the surface. The amino groups of the PAMAM-modified slide were then converted to terminal carboxylic acid groups by reacting with glutaric anhydride. The carboxylic acid of the dendrimer film was finally converted to a phosphane group by reacting with the 2-(diphenylphosphinyl)phenol. The phosphane group has a high reactive efficiency to azide-derivatized molecules. The azide-derivatized carbohydrate moieties were prepared by solid-phase synthesis using a safety-catch linker strategy. The azide-derivatized carbohydrates were found to be efficiently immobilized onto the phosphane-functionalized slide surface. A spot volume of a 0.25 nl sample arrayed on the phosphane-functionalized slide surface produced a spot size of 400 μM in diameter. A mannose-containing carbohydrate microarray was fabricated on this substrate. Carbohydrate–protein interactions were evaluated by incubating with Alexa647-labeled Con A. This shows that the immobilization of azide-derivatized carbohydrates via the Staudinger reaction is highly efficient and can be employed to detect biomolecular interactions.

Bovin's group reported a method principally different from all described above. Saccharides were immobilized inside droplets of a porous polymeric gel (26). Immobilization of amino-derivatized oligosaccharides was achieved by the formation of a covalent bond between the amino group and the growing polymer chain during photo-initiated polymerization in the presence of a cross-linking agent. The authors have demonstrated that a hydrogel carbohydrate microarray contains three different classes of glycomolecules, which are as follows: (i) oligosaccharide derivatives bearing a primary amino group, (ii) oligosaccharide–polyacrylamide conjugates bearing allyl groups, and (iii) oligosaccharide derivatives bearing 2-aminopyridine groups. All of the three types of oligosaccharide derivatives are readily subjected to covalent attachment in the same conditions during the radical process of gel formation. For hydrogel microarray manufacturing, the gel-forming monomers, that is, methacrylamide, methylenebisacrylamide, and oligosaccharide derivatives are printed onto hydrophobized glass and irradiated with UV light. The double bond of methylenebisacrylamide readily reacts with the amino group of the oligosaccharide derivatives at pH 10.5 giving rise to a Michael addition product. After polymerization, an array of individual 3D approximately 1-nl gel drops, 150 μm in diameter, and 25 μm in height was formed. The authors have shown that the 3D hydrogel provides high sensitivity in probing proteins due to the large amount of carbohydrates immobilized in the 3D hydrogel spots.

3. Affinity Immobilization of Derivatized Carbohydrates

Biotin-derivatized carbohydrates can be immobilized on a streptavidin-coated substrate through the affinity interaction of the streptavidin–biotin pair to create carbohydrate microarrays. Biotin-derivatized carbohydrates include carbohydrate ligands that are biotinylated via a short aliphatic spacer or at the peptide part of glycopeptides. Several commercially available streptavidin-coated microwell plates can be applied when biotin-derivatized carbohydrates are available, such as the streptavidin-coated 384-well plate with a well volume 25 μl (4,18) and a streptavidin-coated 192-spot slide format (27). The first was designed to be in maximal proximity to the traditional immunochemical assay using commercial streptavidin-coated black 384-well plates.

IV. Concluding Remarks

A number of carbohydrate microarray platforms have reached or are very close to the stage of the current nucleic acid-based microarrays that are readily available for practical uses. Technical issues that require immediate attention may include but are not limited to optimization of existing technologies for array construction, quality control, and technical standardization in both microarray production and application, establishment of specialized bioinformatic tools to handle the massive amount of carbohydrate microarray data, and to effectively extract diagnostic or research information from each microarray assay. Taking care of these issues would facilitate biological and medical applications of carbohydrate microarrays.

Exploring the repertoires of glycoepitopes and their receptors represents a long-term goal of carbohydrate research. How big is the repertoire of glycoepitopes? Addressing this question is one of the most important topics in the postgenomics era. It was estimated that there are about 500 endogenous glycoepitopes in mammals (17). However, this estimation did not consider the repertoires of the "hybrid" structures that are generated by protein posttranslational modification, including both *N*- and *O*-glycosylation. Furthermore, the conformational diversity of carbohydrates and microheterogeneity of carbohydrate chains substantially increases the repertoire of carbohydrate-based antigenic determinants or glycoepitopes (39,63,66). Considering carbohydrate structures of the microbial world, which are directly relevant to medicine, the sizes and diversity of the repertoires of glycoepitopes are unpredictable. Establishment of high-throughput platforms of carbohydrate microarrays provides powerful means to facilitate the identification and characterization of carbohydrate-based pathogen signatures and other biomarkers.

Joint effort by academic and industrial sectors is highly recommended to direct the establishment of libraries of monoclonal antibodies, lectins, and other carbohydrate-binding proteins. These biomolecules are critical for defining glycoepitopes and are useful for detection of glycoepitopes in living organisms. Thus, using specific immunological probes to characterize glycoepitopes is equally important to the structural determination of glycoepitopes. Similar effort has been successfully made for protein-based biomarkers. A notable example is the establishment of a large collection of monoclonal antibodies for cell differentiation antigens (CD antigens). Availability of specific probes for CD antigens, in combination with the state-of-the-art technologies of flow cytometry (Hi-D FACS) (59), has revolutionized research in cellular biology and immunology and medical applications of CD antigens, especially in the clinical diagnosis of leukemia and other human diseases. Exploring the repertoires of glycoepitopes and their cellular receptors, with the aid of carbohydrate microarray technologies and specific immunological probes, represents one of the highly active areas of postgenomics research that may last for a few decades and likely accompanied with a fruitful outcome.

Acknowledgments

This publication was made possible by Grant Number 1 R43 RR023763–01 to XZ from the National Center for Research Resources Services (NCRR), a component of the National Institutes of Health (NIH). G.T. Carroll acknowledges an NSF-IGERT fellowship and D. Wang acknowledges NIH grants U01CA128416,

HL084318–01A1 and AI064104 and funding support from the Phil N. Allen Trust at Stanford University. Its contents are solely the responsibility of the authors and do not necessarily represent the official view of NCRR or NIH.

References

1. Adams EW, Ratner DM, Bokesch HR, McMahon JB, O'Keefe BR, Seeberger PH. Oligosaccharide and glycoprotein microarrays as tools in HIV glycobiology; glycan-dependent gp120/protein interactions. Chem Biol 2004; 11:875–881.
2. Angeloni S, Ridet JL, Kusy N, Gao H, Crevoisier F, Guinchard S, Kochhar S, Sigrist H, Sprenger N. Glycoprofiling with micro-arrays of glycoconjugates and lectins. Glycobiology 2005; 15:31–41.
3. Blixt O, Head S, Mondala T, Scanlan C, Huflejt ME, Alvarez R, Bryan MC, Fazio F, Calarese D, Stevens J, Razi N, Stevens DJ, Skehel JJ, van Die I, Burton DR, Wilson IA, Cummings R, Bovin N, Wong C-H, Paulson JC. Printed covalent glycan array for ligand profiling of diverse glycan binding proteins. Proc Natl Acad Sci USA 2004; 101:17033–17038.
4. Bochner BS, Alvarez RA, Mehta P, Bovin NV, Blixt O, White JR, Schnaar RL. Glycan array screening reveals a candidate ligand for Siglec-8. J Biol Chem 2005; 280:4307–4312.
5. Brown PO, Botstein D. Exploring the new world of the genome with DNA microarrays. Nat Genet 1999; 21:33–37.
6. Bryan MC, Plettenburg O, Sears P, Rabuka D, Wacowich-Sgarbi S, Wong C-H. Saccharide display on microtiter plates. Chem Biol 2002; 9:713–720.
7. Carroll GT, Wang D, Turro NJ, Koberstein JT. Photochemical micropatterning of carbohydrates on a surface. Langmuir 2006; 22:2899–2905.
8. Chan SM, Ermann J, Su L, Fathman CG, Utz PJ. Protein microarrays for multiplex analysis of signal transduction pathways. Nat Med 2004; 10:1390–1396.
9. Chapman R, Ostuni E, Liang M, Meluleni G, Kim E, Yan L, Pier G, Warren H, Whitesides G. Polymeric thin films that resist the adsorption of proteins and the adhesion of bacteria. Langmuir 2001; 17:1225–1233.
10. Cisar J, Kabat EA, Dörner MM, Liao J. Binding properties of immunoglobulin combining sites specific for terminal or nonterminal antigenic determinants in dextran. J Exp Med 1975; 142:435–459.
11. Cohen S, Mendelson I, Altboum Z, Kobiler D, Elhanany E, Bino T, Leitner M, Inbar I, Rosenberg H, Gozes Y, Barak R, Fisher M, Kronman C, Velan B, Shafferman A. Attenuated nontoxinogenic and nonencapsulated recombinant Bacillus anthracis spore vaccines protect against anthrax. Infect Immun 2000; 68:4549–4558.
12. Crocker PR, Feizi T. Carbohydrate recognition systems: functional triads in cell-cell interactions. Curr Opin Struct Biol 1996; 6:679–691.
13. Cummings RD, Kornfeld S. The distribution of repeating [Gal beta 1,4GlcNAc beta 1,3] sequences in asparagine-linked oligosaccharides of the mouse lymphoma cell lines BW5147 and PHAR 2.1. J Biol Chem 1984; 259:6253–6260.
14. Daubenspeck JM, Zeng H, Chen P, Dong S, Steichen CT, Krishna NR, Pritchard DG, Turnbough CL, Jr. Novel oligosaccharide side chains of the

collagen-like region of BclA, the major glycoprotein of the Bacillus anthracis exosporium. J Biol Chem 2004; 279:30945–30953.
15. DeRisi J, Penland L, Brown PO, Bittner ML, Meltzer PS, Ray M, Chen Y, Su YA, Trent JM. Use of a cDNA microarray to analyse gene expression patterns in human cancer. Nat Genet 1996; 14:457–460.
16. Dochez AR, Avery OT. The elaboration of specific soluble substance by pneumococcus during growth. J Exp Med 1917; 26:477–493.
17. Drickamer K, Taylor ME. Glycan arrays for functional glycomics. Genome-Biology 2002; 3: reviews 1034.1–1034.4.
18. Dyukova V, Shilova N, Galanina O, Rubina A, Bovin N. Design of carbohydrate multiarrays. Biochim Biophys Acta 2006; 1760:603–609.
19. Ekins R, Chu F, Biggart E. Multispot, multianalyte, immunoassay. Ann Biol Clin 1990; 48:655–666.
20. Ezzell JW, Jr., Abshire TG, Little SF, Lidgerding BC, Brown C. Identification of Bacillus anthracis by using monoclonal antibody to cell wall galactose-*N*-acetylglucosamine polysaccharide. J Clin Microbiol 1990; 28:223–231.
21. Fazio F, Bryan MC, Blixt O, Paulson JC, Wong CH. Synthesis of sugar arrays in microtiter plate. J Am Chem Soc 2002; 124:14397–14402.
22. Feizi T. The antigens Ii, SSEA-1 and ABH are in interrelated system of carbohydrate differentiation antigens expressed on glycosphingolipids and glycoproteins. Adv Exp Med Biol 1982; 152:167–177.
23. Focarelli R, La Sala GB, Balasini M, Rosati F. Carbohydrate-mediated sperm-egg interaction and species specificity: a clue from the Unio elongatulus model. Cells Tissues Organs 2001; 168:76–81.
24. Fouchier RA, Kuiken T, Schutten M, Van Amerongen G, Van Doornum GJ, Van Den Hoogen BG, Peiris M, Lim W, Stohr K, Osterhaus AD. Aetiology: Koch's postulates fulfilled for SARS virus. Nature 2003; 423:240.
25. Fukui S, Feizi T, Galustian C, Lawson AM, Chai W. Oligosaccharide microarrays for high-throughput detection and specificity assignments of carbohydrate-protein interactions. Nat Biotechnol 2002; 20:1011–1017.
26. Galanina OE, Mecklenburg M, Nifantiev NE, Pazynina GV, Bovin NV. GlycoChip: multiarray for the study of carbohydrate-binding proteins. Lab Chip 2003; 3:260–265.
27. Guo Y, Feinberg H, Conroy E, Mitchell DA, Alvarez R, Blixt O, Taylor ME, Weis WI, Drickamer K. Structural basis for distinct ligand-binding and targeting properties of the receptors DC-SIGN and DC-SIGNR. Nat Struct Mol Biol 2004; 11:591–598.
28. Hakomori S. Aberrant glycosylation in cancer cell membranes as focused on glycolipids: overview and perspectives. Cancer Res 1985; 45:2405–2414.
29. Heidelberger M, Avery OT. The soluble specific substance of pneumococcus. J Exp Med 1923; 38:73–80.
30. Herrwerth S, Rosendahl T, Feng C, Fick J, Eck W, Himmelhaus M, Dahint R, Grunze M. Covalent coupling of antibodies to self-assembled monolayers of carboxy-functionalized poly(ethylene glycol): protein resistance and specific binding of biomolecules. Langmuir 2003; 19:1880–1887.
31. Houseman B, Gawalt E, Mrksich M. Maleimide-functionalized self-assembled monolayers for the preparation of peptide and carbohydrate biochips. Langmuir 2003; 19:1522–1531.

32. Houseman BT, Mrksich M. Carbohydrate arrays for the evaluation of protein binding and enzymatic modification. Chem Biol 2002; 9:443–454.
33. Kabat EA, Turino GM, Tarrow AB, Maurer PH. Studies on the immunochemical basis of allergic reactions to dextran in man. J Clin Invest 1957; 37:1160–1170.
34. Kohn M, Wacker R, Peters C, Schroder H, Soulere L, Breinbauer R, Niemeyer C, Waldmann H. Staudinger ligation: a new immobilization strategy for the preparation of small-molecule arrays. Angew Chem Int Ed 2003; 42:5830–5834.
35. Ksiazek TG, Erdman D, Goldsmith CS, Zaki SR, Peret T, Emery S, Tong S, Urbani C, Comer JA, Lim W, Rollin PE, Dowell SF, Ling AE, Humphrey CD, Shieh WJ, Guarner J, Paddock CD, Rota P, Fields B, DeRisi J, Yang JY, Cox N, Hughes JM, LeDuc JW, Bellini WJ, Anderson LJ. A novel coronavirus associated with severe acute respiratory syndrome. N Engl J Med 2003; 348:1953–1966.
36. Lee M-R, Shin I. Facile preparation of carbohydrate microarrays by site-specific, covalent immobilization of unmodified carbohydrates on hydrazide-coated glass slides. Org Lett 2005; 7:4269–4272.
37. Lueking A, Horn M, Eickhoff H, Bussow K, Lehrach H, Walter G. Protein microarrays for gene expression and antibody screening. Anal. Biochem 1999; 270:103–111.
38. MacBeath G, Schreiber SL. Printing proteins as microarrays for high-throughput function determination [see comments]. Science 2000; 289:1760–1763.
39. Mammen M, Chio S-K, Whitesides GM. Polyvalent interactions in biological systems: implications for design and use of multivalent ligands and inhibitors. Angew Chem Int Ed 1998; 37:2755–2794.
40. Mond JJ, Lees A, Snapper CM. T cell-independent antigens type 2. Annu Rev Immunol 1995; 655–692.
41. Newcombe DA, Schuerger AC, Benardini JN, Dickinson D, Tanner R, Venkateswaran K. Survival of spacecraft-associated microorganisms under simulated martian UV irradiation. Appl Environ Microbiol 2005; 71:8147–8156.
42. Oehler S, Alex R, Barker A. Is nitrocellulose filter binding really a universal assay for protein–DNA interactions? Anal Biochem 1999; 268:330–336.
43. Pacifico F, Montuori N, Mellone S, Liguoro D, Ulianich L, Caleo A, Troncone G, Kohn LD, Di Jeso B, Consiglio E. The RHL-1 subunit of the asialoglycoprotein receptor of thyroid cells: cellular localization and its role in thyroglobulin endocytosis. Mol Cell Endocrinol 2003; 208:51–59.
44. Palma AS, Feizi T, Zhang Y, Stoll MS, Lawson AM, Diaz-Rodriguez E, Campanero-Rhodes MA, Costa J, Gordon S, Brown GD, Chai W. Ligands for the beta-glucan receptor, Dectin-1, assigned using "designer" microarrays of oligosaccharide probes (neoglycolipids) generated from glucan polysaccharides. J Biol Chem 2006; 281:5771–5779.
45. Park S, Lee MR, Pyo SJ, Shin I. Carbohydrate chips for studying high-throughput carbohydrate-protein interactions. J Am Chem Soc 2004; 126:4812–4819.
46. Park S, Shin I. Fabrication of carbohydrate chips for studying protein-carbohydrate interactions. Angew Chem Int Ed 2002; 41:3180–3182.

47. Ratner DM, Adams EW, Su J, O'Keefe BR, Mrksich M, Seeberger PH. Probing protein-carbohydrate interactions with microarrays of synthetic oligosaccharides. ChemBioChem 2004; 5:379–382.
48. Robbins JB, Schneerson R. Polysaccharide-protein conjugates: a new generation of vaccines. J Infect Dis 1990; 161:821–832.
49. Schena M, Shalon D, Heller R, Chai A, Brown PO, Davis RW. Parallel human genome analysis: microarray-based expression monitoring of 1000 genes. Proc Natl Acad Sci USA 1996; 93:10614–10619.
50. Schwarz M, Spector L, Gargir A, Shtevi A, Gortler M, Altstock RT, Dukler AA, Dotan N. A new kind of carbohydrate array, its use for profiling antiglycan antibodies, and the discovery of a novel human cellulose-binding antibody. Glycobiology 2003; 13:749–754.
51. Stevens J, Blixt O, Glaser L, Taubenberger JK, Palese P, Paulson JC, Wilson IA. Glycan microarray analysis of the hemagglutinins from modern and pandemic influenza viruses reveals different receptor specificities. J Mol Biol 2006; 355:1143–1155.
52. Stevens J, Blixt O, Paulson JC, Wilson IA. Glycan microarray technologies: tools to survey host specificity of influenza viruses. Nat Rev Microbiol 2006; 4:857–864.
53. Stevens J, Blixt O, Tumpey TM, Taubenberger JK, Paulson JC, Wilson IA. Structure and receptor specificity of the hemagglutinin from an H5N1 influenza virus. Science 2006; 312:404–410.
54. Stoll D, Templin MF, Schrenk M, Traub PC, Vohringer CF, Joos TO. Protein microarray technology. Front Biosci 2002; 7:C13–C32.
55. Sylvestre P, Couture-Tosi E, Mock M. A collagen-like surface glycoprotein is a structural component of the Bacillus anthracis exosporium. Mol Microbiol 2002; 45:169–178.
56. Takahashi N. Three-dimensional mapping of N-linked oligosaccharides using anion-exchange, hydrophobic and hydrophilic interaction modes of high-performance liquid chromatography. J Chromatogr A 1996; 720:217–225.
57. Takahashi N, Wada Y, Awaya J, Kurono M, Tomiya N. Two-dimensional elution map of GalNAc-containing *N*-linked oligosaccharides. Anal Biochem 1993; 208:96–109.
58. Tomiya N, Lee YC, Yoshida T, Wada Y, Awaya J, Kurono M, Takahashi N. Calculated two-dimensional sugar map of pyridylaminated oligosaccharides: elucidation of the jack bean alpha-mannosidase digestion pathway of Man9GlcNAc2. Anal Biochem 1991; 193:90–100.
59. Tung JW, Parks DR, Moore WA, Herzenberg LA, Herzenberg LA. Identification of B-cell subsets: an exposition of 11-color (Hi-D) FACS methods. Methods Mol Biol 2004; 271:37–58.
60. Turnbull PCB. Current status of immunization against anthrax: old vaccines may be here to stay for a while. Curr Opin Infect Dis 2000; 13:113–120.
61. Van Oss CJ, Good RJ, Chaudhury MK. Mechanism of DNA (Southern) and protein (Western) blotting on cellulose nitrate and other membranes. J Chromatogr 1987; 391:53–65.
62. Wang D. Carbohydrate microarrays. Proteomics 2003; 3:2167–2175.
63. Wang D. Carbohydrate antigens. In: Meyers RA ed. Encyclopedia of Molecular Cell Biology and Molecular Medicine, Verlag GmbH & CO. KGaA Wiley-VCH, 2004; Vol. II: 277–301.

64. Wang D, Carroll GT, Turro NJ, Koberstein JT, Kovac P, Saksena R, Adamo R, Herzenberg LA, Herzenberg LA, Steinman L. Photogenerated glycan arrays identify immunogenic sugar moieties of Bacillus anthracis exosporium. Proteomics 2007; 7:180–184.
65. Wang D, Kabat EA. Carbohydrate antigens (polysaccharides). In: Regenmortal MHVV ed. Structure of Antigens. Boca Raton, New York, London, Tokyo: CRC Press, 1996; vol. 3:247–276.
66. Wang D, Kabat EA. Antibodies, specificity. In: Delves Roitt ed. Encyclopedia of Immunology. London: Academic Press, 1998; 148–154.
67. Wang D, Liu S, Trummer BJ, Deng C, Wang A. Carbohydrate microarrays for the recognition of cross-reactive molecular markers of microbes and host cells. Nat Biotechnol 2002; 20:275–281.
68. Wang D, Lu J. Glycan arrays lead to the discovery of autoimmunogenic activity of SARS-CoV. Physiol Genomics 2004; 18:245–248.
69. Webb GF. A silent bomb: the risk of anthrax as a weapon of mass destruction. Proc Natl Acad Sci USA 2003; 100:4355–4356.
70. Willats WG, Rasmussen SE, Kristensen T, Mikkelsen JD, Knox JP. Sugar-coated microarrays: a novel slide surface for the high-throughput analysis of glycans. Proteomics 2002; 2:1666–1671.
71. Williams DD, Benedek O, Turnbough CL, Jr. Species-specific peptide ligands for the detection of Bacillus anthracis spores. Appl Environ Microbiol 2003; 69:6288–6293.
72. Wu AM, Song SC, Tsai MS, Herp A. A guide to the carbohydrate specificities of applied lectins-2 (updated in 2000). Adv Exp Med Biol 2001; 491:551–585.
73. Zhou X, Zhou J. Oligosaccharide microarrays fabricated on aminooxyacetyl functionalized glass surface for characterization of carbohydrate-protein interaction. Biosens Bioelectron 2006; 21:1451–1458.

Carbohydrate Chemistry, Biology and Medical Applications
Hari G. Garg, Mary K. Cowman and Charles A. Hales

DOI: 10.1016/B978-0-08-054816-6.00017-3

Chapter 17

Carbohydrate Arrays for Basic Science and as Diagnostic Tools

TIM HORLACHER, JOSE L. DE PAZ AND PETER H. SEEBERGER

Laboratory for Organic Chemistry, Swiss Federal Institute of Technology (ETH) Zürich, HCI F315, 8093 Zürich, Switzerland

I. Introduction

Carbohydrates are one of the four classes of molecules that give rise to life (1). Carbohydrates occur in organisms primarily as proteoglycans (large, heterogeneous conjugates of sugars and proteins) or glycoconjugates (proteins or lipids attached to an oligosaccharide) (2). Proteins and lipids do not only display the sugar chains, but their functions, localization, and interactions are influenced by the carbohydrates (2,3). Carbohydrate moieties themselves act in cell homeostasis, attachment, signaling, and regulation (3,4). Thus, sugars are involved in many important processes including cell adhesion (4), development (5), fertilization (6), and inflammation (7). Not surprisingly, sugars are also involved in the mediation of various diseases; most prominent is the aberrant expression of specific carbohydrate structures in many types of cancer (8).

Viruses and bacteria take advantage of carbohydrate moieties on the human cell surface as attachment sites (9). The binding facilitates the adhesion or entry of pathogens into the host cell and is an essential part of the pathogenic life cycle. Vice versa, pathogens bear important, specific sugars on their cell surface that interact with the host environment. Several of these carbohydrate structures have been exploited for the development of vaccines (10).

Consequently, carbohydrates have gained increased attention in recent years, as sugars serve important, but still largely unknown functions in organisms and are promising drug targets and important antigens.

II. Carbohydrate Microarrays

In recent years, new techniques have been developed to address the specific challenges in carbohydrate research to unravel the functions of sugars in organisms (11,12). Novel approaches in organic chemistry have made synthetic sugars more available (13). The introduction of microarrays has been of particular aid for the investigation of carbohydrate function, as they address many of the needs and overcome many of the problems in glycomics (14–16).

Carbohydrate microarrays consist of surfaces to which sugars are coupled in a nanomolar scale. To fabricate the microarray slides, carbohydrate compounds bearing a functionalized linker are, usually, chemically synthesized. Subsequently, the sugar compounds are printed onto glass slides coated with reactive surfaces using automated arraying robots.

The microarray format bears many advantages: the carbohydrates are attached mimicking their natural presentation on the cell surface; multivalent interactions of binding molecules with the sugars are enabled, often a prerequisite of sugar interactions; and only small amounts of sugars are needed for the fabrication of the microarray slide, overcoming a major problem, as the production of sugar compounds is still one of the main limitations for glycomics research.

Carbohydrate microarrays enable sugar binding tests in a high-throughput manner. This way, high affinity interactions of sugars are identified and biological functions of these interactions can be deduced. Competition and inhibition studies allow for more detailed investigation of binding events; this enables to specify structure–function relationships and to screen for drugs. Using microarrays, a putative physiological ligand for siglec-8, a sialic acid binding lectin, was detected (17). Previously, no ligand and function were known for this protein that is important for cell function. Having identified a putative ligand, the cellular function and molecular action can be investigated easily. Using microarrays, the binding of five proteins (CD4, cyanovirin N, scytovirin, DC-SIGN, and the 2G12 antibody) to the glycoprotein gp120 of HIV was investigated (18). The molecular basis of binding was determined and specific binding patterns were identified. The findings will be of great aid to understand and block the viral entry into the host cell. Viral uptake is mediated by the binding of gp120 to CD4 on the human host cell and can be prohibited by proteins binding competitively to gp120, like cyanovirin or scytovirin.

To date, the binding of RNA and proteins and the action of enzymes with sugars on microarrays have been successfully investigated yielding results of great value. RNA and protein binding to aminoglycosides coupled to microarrays was examined (19,20). Aminoglycosides, a class of broad-spectrum antibiotics, bind to bacterial RNA and inhibit bacterial protein expression. Using the microarrays, a high-throughput platform was established to screen for and develop better aminoglycoside antibiotics. The action of enzymes modifying carbohydrates was also investigated with carbohydrate chips. Microarrays bearing *N*-acetyllactosamine were incubated with a fucosyltransferase followed by a fucose-binding lectin to monitor successful action (21). Highly effective inhibitors were detected by adding various compounds in the incubation steps.

Binding of whole cells and viruses to sugars can be examined using microarrays (22–24). Thereby, the molecular basis of cell attachment to carbohydrates can be studied on the whole cell level and without the need to purify proteins. This finding can be exploited to detect bacteria and viruses by their binding specificities and will be of great aid in identifying pathogens through their binding patterns.

Carbohydrate microarrays are not only a powerful tool for basic research but also a promising technique for medical diagnostics.

In this chapter, we will focus first on the use of glycosaminoglycan (GAG) microarrays to understand the role of this important family of polysaccharides in biological processes, and second on the employment of carbohydrate microarrays as tools for the detection of pathogens and viruses.

III. Microarrays of GAG Oligosaccharides for High-Throughput Screening of GAG–Protein Interactions

Heparin and heparan sulfate are the most complex GAGs, a family of polysaccharides that also includes chondroitin sulfate, keratan sulfate, and dermatan sulfate. GAGs have important biological functions by binding to different growth factors, cell adhesion molecules, and cytokines (25,26). To understand the way GAG sequences interact with particular proteins, it is helpful to consider the use of microarrays of GAG oligosaccharides that allow for the screening of thousands of binding events on a single slide with minimal analyte consumption.

Heparin, an anticoagulant drug, is widely recognized to be a biologically important and chemically unique polysaccharide (27,28). It is a highly sulfated, linear polymer that participates in a plethora of biological processes by interaction with many proteins (29). The heparin–antithrombin III (AT-III) interaction, for example, is responsible for heparin's anticoagulant activity (30). The interaction of heparin with fibroblast growth factors (FGFs) is crucial for regulating the activity of these signaling polypeptides that are involved in angiogenesis, cell growth, and differentiation (31). Heparan sulfate is also a potential ligand for P- and L-selectins and chemokines, key molecules involved in the adhesion of leukocytes to inflamed endothelium as well as their entry into the target tissue that is a crucial feature of inflammation processes (32,33).

Heparin consists predominantly of disaccharide repeating units of D-glucosamine (GlcN) and L-iduronic acid (IdoA), linked by α1–4 glycosidic linkages. Typically, sulfate groups at positions 2 and 6 of the GlcN unit and position 2 of

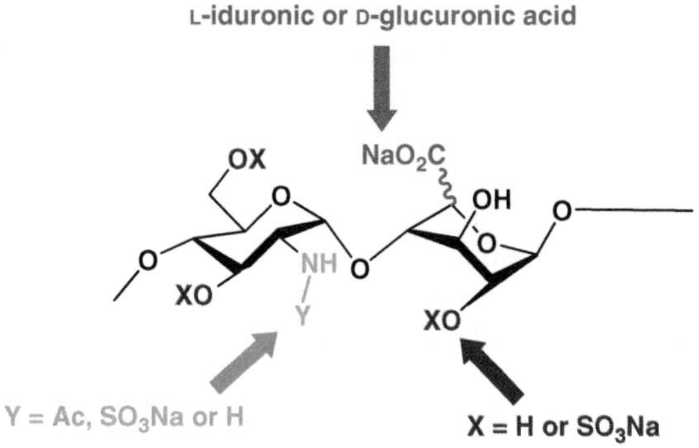

Figure 1 General structure of the disaccharide repeating unit of heparin illustrating the structural variability of this biopolymer.

the IdoA unit are present (Fig. 1). However, a number of structural variations of this trisulfated disaccharide exist and contribute to the microheterogeneity of heparin. The amino group of the glucosamine residue may be acetylated or unsubstituted. The uronic acid unit can also be D-glucuronic acid (GlcA), 2-*O*-sulfated or unsubstituted (Fig. 1). This structural variability renders heparin an extremely challenging molecule to characterize and can be responsible for the interaction of heparin with a wide variety of proteins. The chemical complexity and heterogeneity of this polysaccharide can also explain the fact that, despite its widespread medical use, the structure–activity relationship of heparin is still poorly understood.

We recently reported (34,35) the creation and use of microarrays containing synthetic heparin oligosaccharides (36). For this purpose, we developed a novel linker strategy that is compatible with the protecting-group manipulations (37) required for the synthesis of the highly sulfated oligosaccharides. A series of fully protected oligosaccharides, such as disaccharide **1**, was first synthesized as pentenyl glycosides (Fig. 2). 2-(Benzyloxycarbonylamino)-1-ethanethiol was selected for the radical elongation of the pentenyl moiety by using a catalytic amount of 2,2′-azobis(2-methylpropionitrile) (AIBN) at 75 °C. Treatment with lithium hydroperoxide and then KOH hydrolyzed the ester groups with simultaneous oxidation of sulfide into sulfone. Subsequently, the azide groups were transformed into the corresponding amines via Staudinger reduction. Next, the introduction of the *O*- and *N*-sulfate groups was achieved by treatment with SO_3–Py complex. Finally, global hydrogenolysis afforded disaccharide **6** ready for covalent immobilization onto commercially available *N*-hydroxysuccinimide-activated glass slides. When longer oligosaccharides were synthesized, such as tetra- or hexasaccharides, the deprotection–sulfation sequence was altered because the reduction of the azide protecting groups followed by simultaneous *O*- and *N*-sulfation afforded a mixture of partially sulfated sugars. *O*-sulfation with SO_3–Et_3N complex in DMF at 55 °C, followed by Staudinger reduction, and then, *N*-sulfation using SO_3–Py complex in a mixture triethylamine–pyridine gave the best results and afforded pure sulfated sequences. Following a similar approach (34,35), a small library of amine-

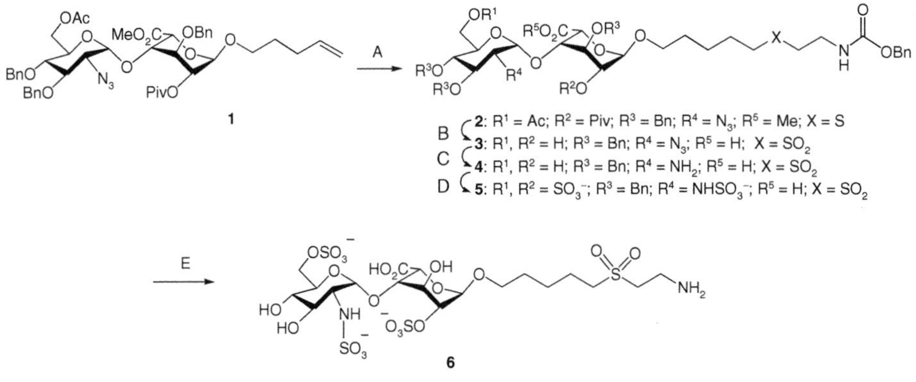

Figure 2 Synthesis of disaccharide **6**. Reagents and conditions: (A) $HS(CH_2)_2NHZ$, AIBN, THF, 75 °C, 78%; (B) LiOH, H_2O_2, KOH, MeOH, 73%; (C) PMe_3, THF, NaOH, 89%; (D) SO_3·Py, Py, 87%; and (E) H_2, Pd/C, quant.

functionalized oligosaccharides with different sequence and sulfate patterns was prepared (Fig. 3).

This linker strategy can also be employed to functionalize oligosaccharides obtained by our automated solid phase synthesis approach using the octenediol linker because the sugar is released from the solid support by olefin cross methatesis to give the corresponding pentenyl glycoside (13). This point is particularly important in order to increase the number of synthetic structures and expand the complexity and utility of the arrays. In addition, the combination of amine-functionalized glycans with *N*-hydroxysuccinimide-activated glass surfaces results in robust and reproducible covalent attachment of carbohydrates without modification of standard DNA printing protocols. The best result, that is to say, the highest signal-noise ratio was obtained by using CodeLink slides that are coated with a hydrophilic polymer containing the activated esters (Fig. 4). The three-dimensional nature of the polymer favors the accessibility of the sugar probes to be detected by proteins and minimizes nonspecific binding. The utility of these heparin chips was demonstrated by probing the carbohydrate affinity of several heparin-binding growth factors, such as acidic and basic FGFs (FGF-1 and FGF-2, respectively), that are implicated in development and differentiation of several tumors (31). The microarray experiments allowed for the evaluation of FGF binding to defined sequences, determining the influence of length and sulfation patterns on carbohydrate recognition. The results were in agreement with previously reported data (25,38). The binding assay involved initial hybridization with the FGF protein (Fig. 4). After washing away any unbound protein, the heparin array was incubated with rabbit anti-FGF polyclonal antibodies. Finally, incubation with fluorescently labeled anti-rabbit secondary antibody detected any bound protein. Scanning the slide for fluorescence produced images, where FGF binding to printed oligosaccharide spots was directly observed (Fig. 4). No binding of the antibodies without FGFs to spots was observed.

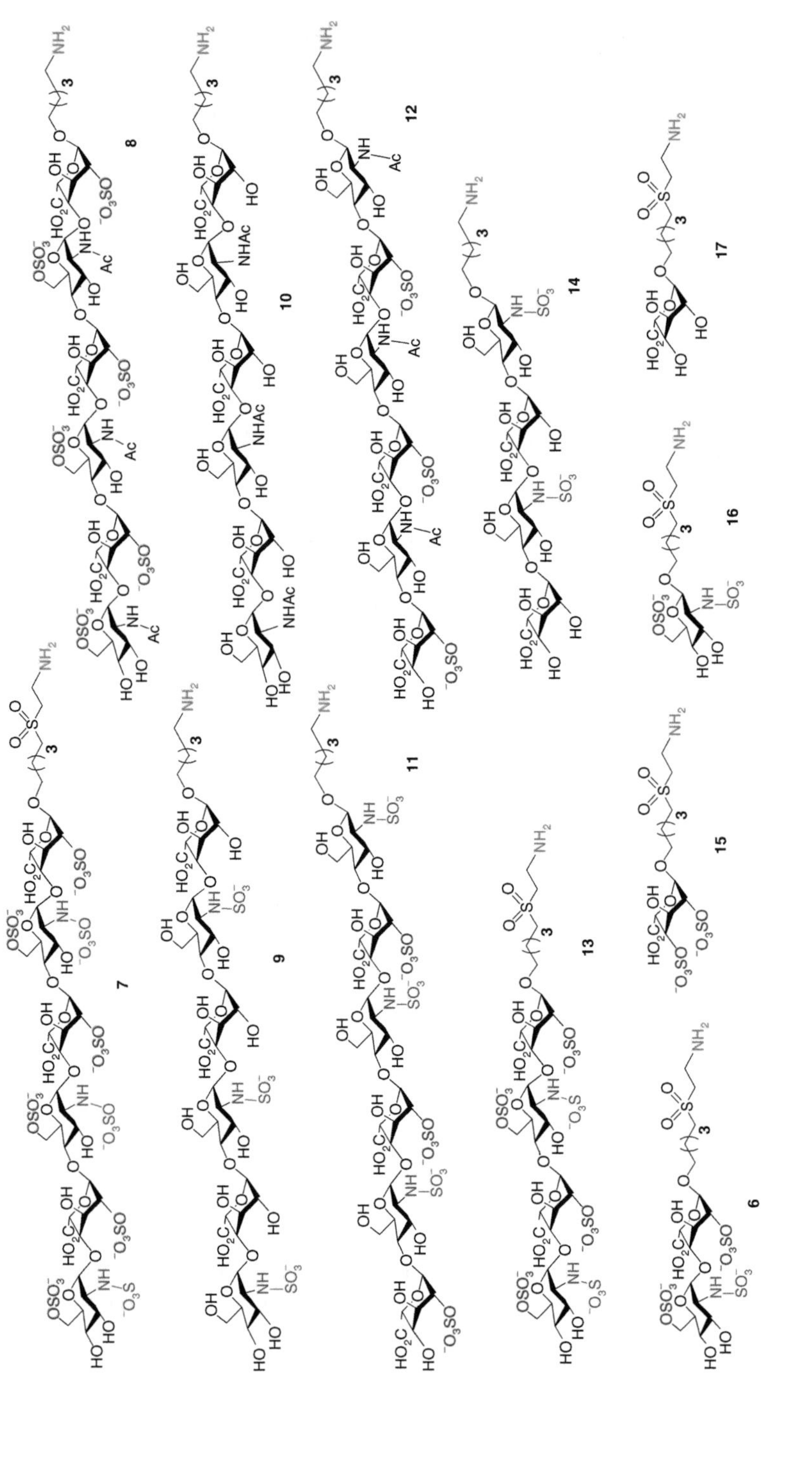

Figure 3 A small library of amine-functionalized heparin oligosaccharides ready for immobilization on a chip surface.

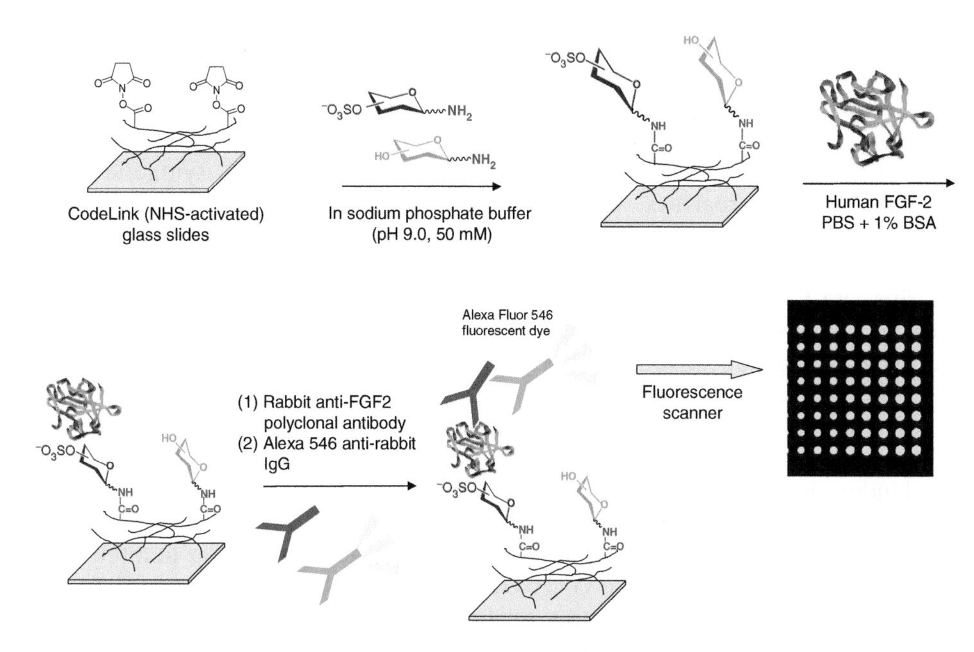

Figure 4 Optimized approach for the preparation and use of heparin microarrays.

Chondroitin sulfate is a ubiquitous component of the extracellular matrix of all connective tissues, and is also found on mammalian cell surfaces and in neural tissues (39). Chondroitin sulfate is a linear copolymer built from dimeric units composed of GlcA and 2-acetamido-2-deoxy-D-galactose (GalNAc) (Fig. 5). In the major variants, chondroitin sulfate chains contain monosulfated disaccharide units, at position 4 or 6 of the GalNAc residue. Oversulfated chains are characterized by the presence of disulfated disaccharides such as GlcA-GalNAc (4,6-di-OSO_3) (40). The sulfate group distribution within this GAG varies with the source of the polymer, and gives rise to biologically important functions

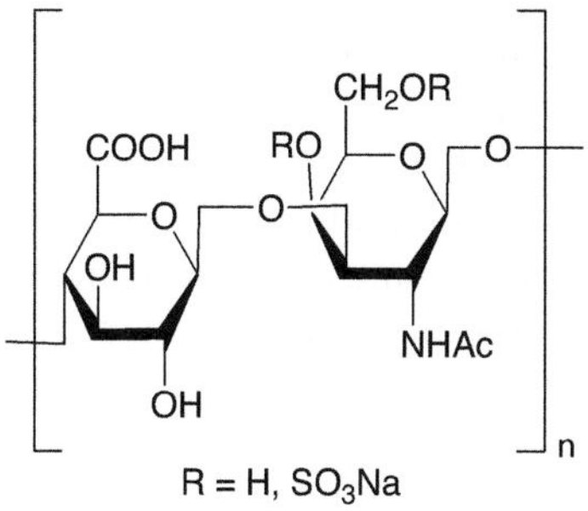

R = H, SO_3Na

Figure 5 General structure of the disaccharide repeating unit of chondroitin sulfate glycosaminoglycans.

intimately related to the position and the number of sulfate groups. These biological roles range from simple mechanical support functions to cell–cell recognition, brain development and regeneration, or binding to selectins, and many other proteins that are not completely deciphered at the molecular level (41,42).

The group of Hsieh-Wilson reported (43) the preparation of synthetic chondroitin sulfate microarrays and their use to identify a previously unknown interaction between chondroitin sulfate and tumor necrosis factor-α (TNF-α), a proinflammatory cytokine involved in numerous diseases, including rheumatoid arthritis and Crohn's disease. Chondroitin sulfate molecules displaying different sulfation sequences were synthesized with an allyl group at the reducing end of the sugar. This group was functionalized for surface immobilization by a two-step procedure: ozonolysis followed by treatment with 1,2-bis(aminooxy)ethane Next, the sugar probes were covalently attached to aldehyde-coated glass slides. The utility of these chips was validated using antibodies raised against specific sulfation motifs. The binding assay involved initial hybridization with TNF-α, followed by incubation with an anti-TNF-α antibody. Finally, incubation with fluorescently labeled secondary antibody detected any bound protein. Selective binding of TNF-α to a tetrasaccharide containing the sequence GlcA-GalNAc(4,6-di-OSO_3) was observed on the microarray. Interestingly, both tetrasaccharide and naturally occurring polysaccharide enriched in this sulfation pattern were able to inhibit the binding of TNF-α to the cell surface receptor TNFR1 and TNF-α-induced cell death.

Because the chemical synthesis of pure and defined heparin oligosaccharides is a complex and time-consuming procedure, including multiple derivatization steps, an alternative approach for microarray fabrication involving the use of naturally derived oligosaccharides is highly attractive. There are several enzymatic and chemical methods for the isolation of heparin fragments from mammalian organs (29). The most successful chemical method is depolymerization with nitrous acid (44) that cleaves heparin chains at either *N*-unsubstituted or *N*-sulfated glucosamine residues, to produce oligosaccharides containing a 2,5-anhydromannose unit at the reducing end. The aldehyde group at position 1 of the 2,5-anhydromannose unit is more reactive than aldehyde groups of reducing sugars because it is not in equilibrium with unreactive closed ring forms (45). We recently reported the preparation of microarrays containing heparin sequences derived by nitrous acid fractionation (46). The increased reactivity of the reducing end allows for the attachment of these isolated fragments to amine-coated glass surfaces either by formation of a Schiff base (Fig. 6) or via reductive amination. Fluorescently labeled heparin was employed as initial probe to test the immobilization reaction. After extensive washing, the retained fluorescence signal demonstrated that attached heparin by imine bond formation was stable enough under hybridization conditions to allow for heparin–protein binding studies. Next, several heparin fragments (octa- and decasaccharides) were obtained by depolymerization of bovine intestinal heparin and classified according to their affinity to AT-III. AT-III is a serine protease inhibitor of the blood coagulation cascade that requires heparin for full activation (30). A characteristic heparin pentasaccharide sequence, containing the crucial $GlcNSO_3(3\text{-}OSO_3)$ unit, is responsible for the binding to

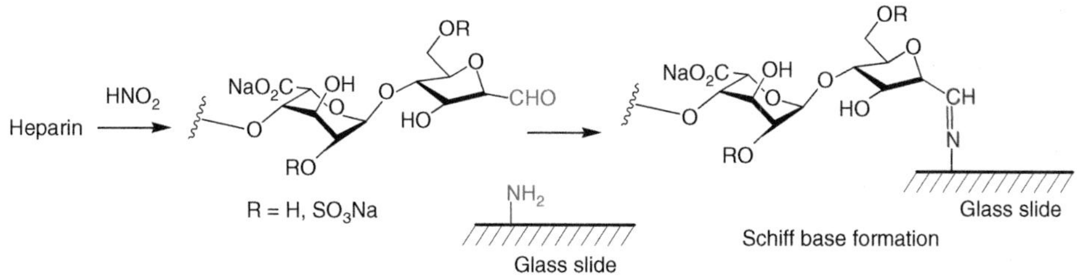

Figure 6 Immobilization of heparin oligosaccharides derived by nitrous acid depolymerization of heparin on amine-coated glass slides via formation of a Schiff base.

AT-III. Therefore, the different affinities of isolated fragments to AT-III respond to different degrees of pentasaccharide abundance. The heparin sequences were printed on amine-coated slides to prepare the corresponding heparin chips whose utility was demonstrated by incubation with proteins such as AT-III, FGF-1, and FGF-2. Interestingly, the results obtained after AT-III incubation were in agreement with those obtained by using affinity chromatography on AT-III that served to classify the sequences.

Turnbull et al. (47) described a new and simple approach for the preparation of heparin microarrays on gold surfaces. The strategy is based on the creation of a hydrazide-derivatized self-assembled monolayer on a gold surface for efficient immobilization of oligosaccharide probes via their reducing end (48). This platform was used to assess binding of specific heparin-binding proteins, such as the growth factor receptor FGFR-2, at very high sensitivity. Isolated decasaccharides from porcine mucosal heparin as well as mannose oligomers were employed in this study, demonstrating that the approach can be used with sulfated and nonsulfated sugars. For the preparation of the naturally derived heparin sequences, both nitrous acid digestion and bacterial lyase enzyme depolymerization were employed. The sugars derived from chemical digestion gave generally higher fluorescent signals, due to the presence of a more reactive aldehyde group. For negatively charged molecules such as sulfated oligosaccharides, it is likely that the probes could be adsorbed nonspecifically onto the positively charged hydrazide surface. To distinguish whether the sugars were attached by covalent bonds or by nonspecific physical adsorption, the immobilization of two heparin oligosaccharides, one of them containing a blocked reducing end by reduction with $NaBH_4$, was compared. The experiment revealed that the sugars were attached predominantly by covalent bonds.

One of the biggest problems to use gold surfaces for fluorescence-based detection experiments is the fluorescence quenching due to the metal surface. This issue was partially solved by using a long spacer between the gold surface and the immobilized sugar. Additionally, detection involving three successive protein layers (the target molecule and two subsequent antibodies) contributes to minimize fluorescence quenching. This glycochip approach provides a great degree of versatility because the gold surface is compatible with detection techniques, other than fluorescence, such as surface plasmon resonance (49) or matrix-assisted laser desorption ionization mass spectroscopy (50).

IV. Carbohydrate Microarrrays to Detect Pathogens and Viruses

Virtually all cells are surrounded by a layer of proteins and carbohydrates covering the plasma membrane to protect the cell from the surrounding and to regulate cell interactions with the environment (1). Many proteins interact with the sugars on the cell surface mediating cell signaling and cell adhesion. However, various viruses and bacteria also bear proteins that bind to the target host cell, often initializing entry into the cell (51,52).

Many bacteria adhere to human cells by binding to sugar moieties on the cell surfaces. In several cases (e.g., *Escherichia coli* and *Helicobacter pylori*), the ability to bind can define the virulence of a subspecies of microorganisms, as the property to adhere to humans cells can turn relatively harmless bacteria into hazardous pathogens causing a broad variety of diseases (9).

The binding of *E. coli* cells to sugars on microarrays was tested to assess the carbohydrate-mediated adhesion of bacteria (22). Generally, *E. coli* species are relatively harmless, commensal inhabitants of the human gut; however, several pathogenic subspecies are known causing infections and diarrhea (51). For *E. coli*, two main types of virulence factors are known: one is the expression of enterotoxins and the second factor is the ability to adhere to human cells, often based on carbohydrate binding. An example are the urovirulent strains of *E. coli* that are able to bind to monomannose residues on uroepithelical cells in the human urinary tract and cause urinary tract infections in humans (53).

To test the adhesion of *E. coli* cells to sugars on microarrays (22), a series of monosaccharides (Fig. 7) bearing an ethanolamine linker was synthesized and printed onto CodeLink amine-reactive slides in various concentrations. *E. coli* cells were fluorescently labeled using permeable, fluorescent cell dyes and incubated on the microarrays. Unbound cells were washed from the array and the slides were scanned using a fluorescent microarray scanner. It was shown that *E. coli* cells of the strain ORN178 bind specifically to monomannose on the microarray; no binding was observed to other monosaccharides. *E. coli* mutants (strain ORN209) defective in the gene for the adhesion protein (fimbrin H adhesin) did bind poorly in comparison to the strain ORN178. A decreased signal (sevenfold) was received for binding to spots of high mannose concentration; no binding could be detected at spots of low mannose concentration. Therefore, this method allows distinguishing between various *E. coli* strains and, thereby, to identify virulent strains that can bind to monomannose on epithelial cell in the urinary tract causing urinary tract infections.

Using dilution series of *E. coli* suspensions, the detection limit was determined. *E. coli* in concentrations as little as 10^5–10^6 cells can be detected using this technique. The sensitivity of this assay is still lower than the one of other methods for the identification of bacteria; however, this chip-based technique is much faster

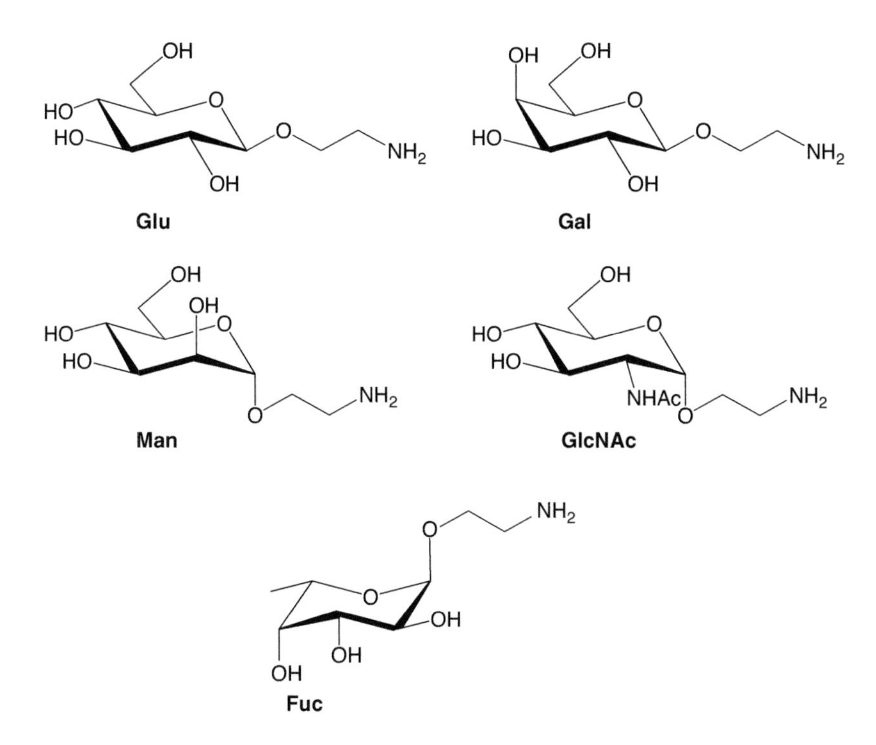

Figure 7 Series of monosaccharides used for testing of bacterial adhesion (Glu=Glucose, Gal=Galactose, Man=Mannose, GlcNAc=*N*-Acetylglucosamine, Fuc=Fucose).

and the microarray technology is rapidly improving. Inhibition studies showed that binding can be prohibited using various compounds and analogues. In these experiments, each 10^8 *E. coli* cells were incubated on the slides in presence of varied concentrations of the inhibitor. Thereby, the IC_{90} was assayed for the compounds and the best inhibitor was found to be a mannose polymer. This observation clearly demonstrates the necessity and importance of multivalency in carbohydrate-mediated cell adhesion.

Further, binding of *E. coli* present in more complex samples was examined. *E. coli* cells were mixed with serum or erythrocytes. The resulting solutions are mimicking bacteria present in human blood samples. The cells in the mixtures were harvested by centrifugation, stained, and incubated on the slides. Only binding of *E. coli* cells to mannose could be detected. The detection limit for the mixtures was again determined using dilution series and the limit was only twofold reduced in comparison to pure samples.

Bacterial binding to the sugars on the slides could be removed specifically, so that the bound microbes can be applied for further testing. Therefore, microarrays were incubated with *E. coli* cells and unbound cells were washed away. An inoculation loop was scraped over the sugar spots and streaked onto LB plates. After growth of the plates at 37 °C, *E. coli* cells could be detected on plates streaked with inoculation loops transferred from mannose spots. This technique has proven to be a powerful tool to detect pathogens in diagnostic samples (54). It enables to identify pathogenic strains and allows recovering the detected bacteria for detailed characterization.

Whole influenza viruses have also been shown to bind to carbohydrate microarrays in a preliminary study (24). The observed binding pattern matched with the previous determined binding preferences. Influenza virus is one of the most dangerous infectious pathogens leading to hundred thousands of deaths per year worldwide (55). In addition, occasional pandemic outbreaks are a fatal threat: About 50 million people died in the pandemic outbreak of 1918 (56), further pandemics occurred in 1957 and 1968. High mutation rates of the influenza virus and the occurrence of new influenza strains are major obstacles in fighting influenza. Therefore, a serious threat is caused by avian influenza strains (57). Avian influenza is now common in Southeast Asia and is spreading fast. Infected birds have been found in Europe and Africa. Avian influenza barely infects humans until now and it only transmits from birds to humans, if humans have very close contact to infected birds. However, in contrast to human influenza, the death rate for humans infected with avian influenza is exceptionally high. A new influenza strain might arise from recombination of human and avian or mutation of avian influenza strains alone that is as lethal as avian and as infectious as human influenza (58–60).

Binding of influenza viruses to the human cell surface is mediated by the hemagglutinin (HA) proteins. The attachment initiates the uptake of the virus particle into the host cell. Influenza HAs bind to carbohydrates bearing sialic acids (*N*-acetylneuraminic acid). The human influenza virus HA proteins preferentially bind to α2–6-linked sialic acids (Fig. 8A) that are found in the upper respiratory tract of humans. In contrast, avian HA proteins preferentially bind to α2–3-linked sialic acid (Fig. 8B) (61–63). Glycans of this type occur in the respiratory tract and intestine of birds. This way, species specificity of the influenza strains is generated and a barrier exists for avian strains infecting humans, as the sialic acid with an α2–3 linkage is rare on the cell surface of human epithelial cells in the readily accessible upper respiratory tract (61,64).

To examine the sequence specificity of influenza HAs, Stevens et al. tested binding preferences of eight recombinant-expressed HA proteins of human and

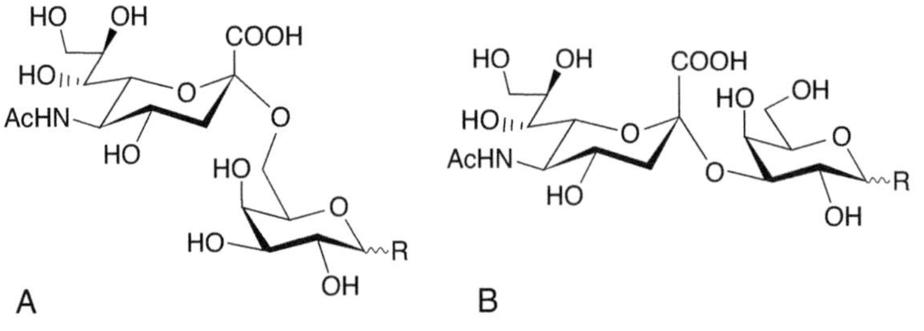

Figure 8 Linkages of sialic acid: α2–6-linked sialic acid (A) and α2–3-linked sialic acid, each linked to galactose.

avian origin using carbohydrate microarrays (65). HAs of human origin prefer α2–6-linked sialic acid moieties and avian HA proteins show specificity for α2–3-linked sialic acid. In addition, this study unraveled the fine details for the binding specificities of the HA proteins regarding linkages, length of oligosaccharides, branching patterns, and sugar moieties in the oligosaccharide chains. It became clear that each HA bears an own, specific binding pattern for oligosaccharides containing sialic acid. It was found that single mutations in the HA protein sequence can dramatically shift the binding specificity, including changes in the specificity from α2–3 linkages to α2–6 linkages. These results lead to a better understanding how binding preferences are generated and how mutations affect the evolution of new influenza strains, especially with regard to shifts in species specificities.

The finding that each HA protein and each virus has its own glycan-binding pattern may enable the identification of virus strains through their binding pattern on carbohydrate microarrays (65,66). Influenza strains are classified based on the subtypes of HA proteins and neuramidase proteins they bear using identification techniques that currently require several days. Carbohydrate microarrays are a promising tool to detect and identify influenza strains by their binding pattern to carbohydrates and enable to establish a rapid screening platform. This way, influenza strains could be classified easily, reducing time and costs, while new strains with new binding patterns can be identified. This is of particular aid to detect avian influenza strains changing their binding specificities from avian carbohydrates toward human sugar moieties. Such a technique will be of valuable aid monitoring the propagation and evolution of influenza strains and controlling epidemic outbreaks.

V. Conclusions

Carbohydrate microarrays have proven to be powerful tools for basic science and diagnostics. The microarray format enables high-throughput screening and minimizes the amount of sugar compounds as well as the quantity of binding samples, lowering cost for experiments and leading to high efficiency. Carbohydrate microarrays bear a broad application range. Various binding partners can be tested, including proteins, RNA, enzymes, viruses, and whole cells. Many aspects of binding can be addressed: binding preferences can be determined, competition studies enable drug screening, and whole cell binding is of high diagnostic value.

Therefore, carbohydrate microarrays are expected to be of great value increasing our current knowledge about carbohydrates and sugar functions in organisms. In addition, carbohydrate microarrays are effective tools for the rapid detection of microbes in pathological samples and will be of great aid in the rapid identification and characterization of infectious agents.

Acknowledgments

We thank all present and past members of the Seeberger group and our collaborators, who contributed to the results reported in this chapter. We thank the ETH

Zürich, the Swiss National Science Foundation, and the European Commission (Marie Curie Fellowship for J. L. P.) for financial support.

References

1. Roseman S. Reflections on glycobiology. J Biol Chem 2001; 276:41527–41542.
2. Varki A, Cummings R, Esko JD, Freeze H, Hart G, Marth J. Essentials in Glycobiology. Plainview, New York: Cold Spring Harbor Laboratory Press, 2002.
3. Hakomori SI. Inaugural article: the glycosynapse. Proc Natl Acad Sci USA 2002; 99:225–232.
4. Collins BE, Paulson JC. Cell surface biology mediated by low affinity multivalent protein-glycan interactions. Curr Opin Chem Biol 2004; 8:617–625.
5. Lennarz WJ. Glycoprotein synthesis and embryonic development. CRC Crit Rev Biochem 1983; 14:257–272.
6. Benoff S. Carbohydrates and fertilization: an overview. Mol Hum Reprod 1997; 3:599–637.
7. Ley K, Kansas GS. Selectins in T-cell recruitment to non-lymphoid tissues and sites of inflammation. Nat Rev Immunol 2004; 4:325–335.
8. Dube DH, Bertozzi CR. Glycans in cancer and inflammation—potential for therapeutics and diagnostics. Nat Rev Drug Discov 2005; 4:477–488.
9. Hooper LV, Gordon JI. Glycans as legislators of host-microbial interactions: spanning the spectrum from symbiosis to pathogenicity. Glycobiology 2001; 11:1R–10R.
10. Jones C. Vaccines based on the cell surface carbohydrates of pathogenic bacteria. An Acad Bras Cienc 2005; 77:293–324.
11. Ratner DM, Adams EW, Disney MD, Seeberger PH. Tools for glycomics: mapping interactions of carbohydrates in biological systems. Chembiochem 2004; 5:1375–1383.
12. Doores KJ, Gamblin DP, Davis BG. Exploring and exploiting the therapeutic potential of glycoconjugates. Chemistry 2006; 12:656–665.
13. Plante OJ, Palmacci ER, Seeberger PH. Automated solid-phase synthesis of oligosaccharides. Science 2001; 291:1523–1527.
14. Feizi T, Chai W. Oligosaccharide microarrays to decipher the glyco code. Nat Rev Mol Cell Biol 2004; 5:582–588.
15. Paulson JC, Blixt O, Collins BE. Sweet spots in functional glycomics. Nat Chem Biol 2006; 2:238–248.
16. Love KR, Seeberger PH. Carbohydrate arrays as tools for glycomics. Angew-Chem-Int-Ed-Engl 2002; 41:3583–3586.
17. Bochner BS, Alvarez RA, Mehta P, Bovin NV, Blixt O, White JR, Schnaar RL. Glycan array screening reveals a candidate ligand for Siglec-8. J Biol Chem 2005; 280:4307–4312.
18. Adams EW, Ratner DM, Bokesch HR, McMahon JB, O'Keefe BR, Seeberger PH. Oligosaccharide and glycoprotein microarrays as tools in HIV glycobiology: glycan-dependent gp120/protein interactions. Chem Biol 2004; 11:875–881.
19. Disney MD, Seeberger PH. Aminoglycoside microarrays to explore interactions of antibiotics with RNAs and proteins. Chem Eur J 2004; 10:3308–3314.

20. Disney MD, Magnet S, Blanchard JS, Seeberger PH. Aminoglycoside microarrays to study antibiotic resistance. Angew-Chem-Int-Ed-Engl 2004; 43:1591–1594.
21. Bryan MC, Lee LV, Wong CH. High-throughput identification of fucosyltransferase inhibitors using carbohydrate microarrays. Bioorg Med Chem Lett 2004; 14:3185–3188.
22. Disney MD, Seeberger PH. The use of carbohydrate microarrays to study carbohydrate–cell interactions and to detect pathogens. Chem Biol 2004; 11:1701–1707.
23. Nimrichter L, Gargir A, Gortler M, Altstock RT, Shtevi A, Weisshaus O, Fire E, Dotan N, Schnaar RL. Intact cell adhesion to glycan microarrays. Glycobiology 2004; 14:197–203.
24. Blixt O, Head S, Mondala T, Scanlan C, Huflejt ME, Alvarez R, Bryan MC, Fazio F, Calarese D, Stevens J, Razi N, Stevens DJ, Skehel JJ, van Die I, Burton DR, Wilson IA, Cummings R, Bovin N, Wong CH, Paulson JC. Printed covalent glycan array for ligand profiling of diverse glycan binding proteins. Proc Natl Acad Sci USA 2004; 101:17033–17038.
25. Raman R, Sasisekharan V, Sasisekharan R. Structural insights into biological roles of protein–glycosaminoglycan interactions. Chem Biol 2005; 12:267–277.
26. Esko JD, Selleck SB. Order out of chaos: assembly of ligand binding sites in heparan sulfate. Annu Rev Biochem 2002; 71:435–471.
27. Capila I, Linhardt RJ. Heparin–protein interactions. Angew-Chem-Int-Ed-Engl 2002; 41:391–412.
28. Noti C, Seeberger PH. Chemical approaches to define the structure–activity relationship of heparin-like glycosaminoglycans. Chem Biol 2005; 12:731–756.
29. Casu B, Lindahl U. Structure and biological interactions of heparin and heparan sulphate. In: Advances in Carbohydrate Chemistry and Biochemistry. San Diego, CA, USA: Academic Press Inc., 2001; Vol. 57:159–206.
30. Petitou M, van Boeckel CAA. A synthetic antithrombin III binding pentasaccharide is now a drug! What comes next? Angew-Chem-Int-Ed-Engl 2004; 43:3118–3133.
31. Pellegrini L. Role of heparan sulfate in fibroblast growth factor signalling: a structural view. Curr Opin Struct Biol 2001; 11:629–634.
32. Wang LC, Fuster M, Sriramarao P, Esko JD. Endothelial heparan sulfate deficiency impairs L-selectin- and chemokine-mediated neutrophil trafficking during inflammatory responses. Nat Immunol 2005; 6:902–910.
33. Parish CR. Heparan sulfate and inflammation. Nat Immunol 2005; 6:861–862.
34. Noti C, de Paz JL, Polito L, Seeberger PH. Preparation and use of microarrays containing synthetic heparin oligosaccharides for the rapid analysis of heparin-protein interactions. Chemistry 2006; 12:8664–8686.
35. de Paz JL, Noti C, Seeberger PH. Microarrays of synthetic heparin oligosaccharides. J Am Chem Soc 2006; 128:2766–2767.
36. Orgueira HA, Bartolozzi A, Schell P, Litjens R, Palmacci ER, Seeberger PH. Modular synthesis of heparin oligosaccharides. Chem Eur J 2003; 9:140–169.
37. de Paz JL, Ojeda R, Reichardt N, Martin-Lomas M. Some key experimental features of a modular synthesis of heparin-like oligosaccharides. Eur J Org Chem 2003; 3308–3324.

38. de Paz JL, Angulo J, Lassaletta JM, Nieto PM, Redondo-Horcajo M, Lozano RM, Gimenez-Gallego G, Martin-Lomas M. The activation of fibroblast growth factors by heparin: synthesis, structure, and biological activity of heparin-like oligosaccharides. Chembiochem 2001; 2:673–685.
39. Kjellen L, Lindahl U. Proteoglycans—structures and interactions. Annu Rev Biochem 1991; 60:443–475.
40. Karst N, Jacquinet JC. Stereo-controlled total synthesis of shark cartilage chondroitin sulfate D-related tetra- and hexasaccharide methyl glycosides. Eur J Org Chem 2002; 815–825.
41. Sugahara K, Mikami T, Uyama T, Mizuguchi S, Nomura K, Kitagawa H. Recent advances in the structural biology of chondroitin sulfate and dermatan sulfate. Curr Opin Struct Biol 2003; 13:612–620.
42. Gama CI, Hsieh-Wilson LC. Chemical approaches to deciphering the glycosaminoglycan code. Curr Opin Chem Biol 2005; 9:609–619.
43. Tully SE, Rawat M, Hsieh-Wilson LC. Discovery of a TNF-alpha antagonist using chondroitin sulfate microarrays. J Am Chem Soc 2006; 128:7740–7741.
44. Horton D, Philips KD. Nitrous-acid deamination of glycosides and acetates of 2-amino-2-deoxy-D-glucose. Carbohydr Res 1973; 30:367–374.
45. Powell AK, Yates EA, Fernig DG, Turnbull JE. Interactions of heparin/heparan sulfate with proteins: appraisal of structural factors and experimental approaches. Glycobiology 2004; 14:17R–30R.
46. de Paz JL, Spillmann D, Seeberger PH. Microarrays of heparin oligosaccharides obtained by nitrous acid depolymerization of isolated heparin. Chem Commun (Camb) 2006; 3116–3118.
47. Zhi ZL, Powell AK, Turnbull JE. Fabrication of carbohydrate microarrays on gold surfaces: direct attachment of nonderivatized oligosaccharides to hydrazide-modified self-assembled monolayers. Anal Chem 2006; 78:4786–4793.
48. Lee M, Shin I. Facile preparation of carbohydrate microarrays by site-specific, covalent immobilization of unmodified carbohydrates on hydrazide-coated glass slides. Org Lett 2005; 7:4269–4272.
49. Cochran S, Li CP, Fairweather JK, Kett WC, Coombe DR, Ferro V. Probing the interactions of phosphosulfomannans with angiogenic growth factors by surface plasmon resonance. J Med Chem 2003; 46:4601–4608.
50. Su J, Mrksich M. Using mass spectrometry to characterize self-assembled monolayers presenting peptides, proteins, and carbohydrates. Angew-Chem-Int-Ed-Engl 2002; 41:4715–4718.
51. Finlay BB, Falkow S. Common themes in microbial pathogenicity revisited. Microbiol Mol Biol Rev 1997; 61:136–169.
52. Forrest JC, Dermody TS. Reovirus receptors and pathogenesis. J Virol 2003; 77:9109–9115.
53. Johnson JR. Virulence factors in *Escherichia coli* urinary tract infection. Clin Microbiol Rev 1991; 4:80–128.
54. Mahal LK. Catching bacteria with sugar. Chem Biol 2004; 11:1602–1604.
55. WHO. Influenza—Fact Sheet N° 211. 2003.
56. Johnson NP, Mueller J. Updating the accounts: global mortality of the 1918–1920 "Spanish" influenza pandemic. Bull Hist Med 2002; 76:105–115.
57. WHO. Avian influenza ("bird flu")—Fact sheet; 2006.

58. Taubenberger JK, Reid AH, Lourens RM, Wang R, Jin G, Fanning TG. Characterization of the 1918 influenza virus polymerase genes. Nature 2005; 437:889–893.
59. Antonovics J, Hood ME, Baker CH. Molecular virology: was the 1918 flu avian in origin? Nature 2006; 440:E9; discussion E9–10.
60. Gibbs MJ, Gibbs AJ. Molecular virology: was the 1918 pandemic caused by a bird flu? Nature 2006; 440:E8; discussion E9–10.
61. Shinya K, Ebina M, Yamada S, Ono M, Kasai N, Kawaoka Y. Avian flu: influenza virus receptors in the human airway. Nature 2006; 440:435–436.
62. Rogers GN, D'Souza BL. Receptor binding properties of human and animal H1 influenza virus isolates. Virology 1989; 173:317–322.
63. Connor RJ, Kawaoka Y, Webster RG, Paulson JC. Receptor specificity in human, avian, and equine H2 and H3 influenza virus isolates. Virology 1994; 205:17–23.
64. Neumann G, Kawaoka Y. Host range restriction and pathogenicity in the context of influenza pandemic. Emerg Infect Dis 2006; 12:881–886.
65. Stevens J, Blixt O, Glaser L, Taubenberger JK, Palese P, Paulson JC, Wilson IA. Glycan microarray analysis of the hemagglutinins from modern and pandemic influenza viruses reveals different receptor specificities. J Mol Biol 2006; 355:1143–1155.
66. Stevens J, Blixt O, Paulson JC, Wilson IA. Glycan microarray technologies: tools to survey host specificity of influenza viruses. Nat Rev Microbiol 2006; 4:857–864.

Index

H

I

Printed and bound by CPI Group (UK) Ltd, Croydon, CR0 4YY

08/05/2025

01864785-0002